Statistical Methods for Quality

Statistical Methods for Quality

With Applications to Engineering and Management

IRWIN MILLER
Partner, Miller and Miller
Director, Opinion Research Corporation

MARYLEES MILLER
Partner, Miller and Miller

PRENTICE HALL, Englewood Cliffs, New Jersey 07632

Library of Congress Cataloging-in-Publication Data

MILLER, IRWIN, [date]
  Statistical methods for quality : with applications to engineering
  and management / IRWIN MILLER, MARYLEES MILLER.
      p.   cm.
  Includes bibliographical references and index.
  ISBN 0-13-013749-9
  1. Quality control—Statistical methods.   I. Miller, Marylees.
II. Title.
TS156.M517   1995
658.5'62'015195—dc20   94-4684

Acquisitions editor: *Jerome Grant*
Project manager: *Edie Riker*
Cover design: *Rosemarie Votta*
Production coordinator: *Trudy Pisciotti*
Editorial assistant: *Joanne Wendelken*

© 1995 by Prentice-Hall, Inc.
A Simon & Schuster Company
Englewood Cliffs, NJ 07632

Printed in the United States of America

10   9   8   7   6   5   4   3   2   1

ISBN 0-13-013749-9

Prentice-Hall International (UK) Limited, *London*
Prentice-Hall of Australia Pty. Limited, *Sydney*
Prentice-Hall Canada Inc., *Toronto*
Prentice-Hall Hispanoamericana, S.A., *Mexico*
Prentice-Hall of India Private Limited, *New Delhi*
Prentice-Hall of Japan, Inc., *Tokyo*
Simon & Schuster Asia Pte. Ltd., *Singapore*
Editora Prentice-Hall do Brasil, Ltda., *Rio de Janeiro*

CONTENTS

ANSWERS TO ODD-NUMBERED EXERCISES

INDEX

PREFACE

This book has been written for an introductory course in statistical methods for quality improvement for students of engineering and of business management. The text has been designed so that its essentials can be covered in a one-semester course that can be fitted into today's crowded undergraduate engineering or management curriculum. A full treatment of all its topics can be readily covered in two semesters.

The experience of the authors in teaching engineering and management students in university and in-service industrial settings, as well as our direct industrial experience, convinces us that more attention needs to be given to teaching the tools of modern quality practice at the undergraduate level. Most of these tools are statistical, but some are not. We have seen how quickly in-service engineers forget the statistics they studied as undergraduates because they have not been shown how to apply the theory taught them to their work. We have, therefore, restricted the statistical methods included in this text to those that our experience suggests are most frequently used in engineering for quality. We have added non-statistical material relating to quality, not normally included in traditional statistics texts, not only to provide additional context for statistical applications but also in recognition of the increasing importance of quality in today's international economy.

The interest of engineering students and in-service engineers is more likely to be retained if the tools of quality are taught methodologically, rather than theoretically, and in the context of practical applications. We believe that maintaining the theme of quality will contribute to maintaining student interest and motivation. In this spirit, we have attempted to prepare a text that can be taught by the professor of engineering concerned with modern quality practice, who may not also be a statistician. At the same time, we have tried to remain true enough to theory that a mathematical statistician will feel comfortable using this text.

Chapter 1 introduces the purpose, coverage, and methods of the text, and Chapter 2 gives a brief overview of the modern field of quality, giving a context of quality to the statistical methods to be covered later. Chapter 3 provides the basic material of descriptive statistics, with applications to quality. Chapters 4 and 5 introduce probability and some of the more widely used probability distributions in statistical applications. Chapters 6 and 7 cover sampling, sampling distributions, and the fundamentals of statistical inference. Chapters 8 to 11 introduce the statistical tools most widely applied to quality—Statistical Process Control (*SPC*), Regression Analysis, Design of Experiments, and Reliability.

A one-semester course that emphasizes engineering applications can include Chapters 1 to 5, and Chapters 8 and 9. Relevant material from Chapters 6 and 7 can be woven into Chapters 8 and 9 as needed. A one-semester course that gives more emphasis to statistical inference can be based on Chapters 1 to 7. The entire text can be covered in a more leisurely fashion in two semesters.

The mathematical background required for this text includes some familiarity with calculus. However, the text can be understood by readers not well-versed in calculus by omitting the few formulas in Chapters 5, 9, and 11 that make use of calculus notation, substituting the verbal descriptions also given.

The authors would like to express their appreciation to the D. Van Nostrand Company for its permission to reproduce the material in Table 2; to the MacMillan Publishing Company, Inc., for permission to reprint Table 4 from *Statistical Methods for Research Workers* by R. A. Fisher; to Prof. E. S. Pearson and the *Biometrika* Trustees for permission to reproduce the material in Tables 5 and 6; to B. Duncan, H. L. Harter, and *Biometrics* for permission to reproduce Table 9; and Minitab (State College, Penn.) for permission to include output from its MINITAB software package.

We wish to acknowledge the many useful suggestions made by the reviewers, William R. Astle, Colorado School of Mines; Stephen J. Biello, III, Director of Quality, Raytheon Service Co.; Larry G. David, University of Missouri–Columbia; Ahmed K. Elshennawy, University of Central Florida; Zakkula Govindarajulu, University of Kentucky; John Lawson, Brigham Young University; Thomas W. Reiland, North Carolina State University; Robert L. Schaefer, Miami University; and Robert C. Williams, Alfred University, whose careful reviews of the manuscript led to many improvements. The authors also wish to thank John E. Freund for his valuable advice and kind encouragement throughout.

IRWIN MILLER and MARYLEES MILLER

Statistical Methods for Quality

With Applications to Engineering and Management

INTRODUCTION

Where are we going?   In this chapter, we introduce the purpose and methods of this text.

In Section 1.1, *Statistics in Engineering and Management,* we define the term *statistics* and comment on its applications in engineering practice and in management.

In Section 1.2, *What Is Quality?* we define the word *quality* and contrast its meaning with *performance, reliability,* and *maintainability.*

In Section 1.3, *Quality and Statistics,* we relate some of the statistical methods used by both engineers and management to the pursuit of quality.

1.1 Statistics in Engineering and Management

The term *statistics* is one of the more confusing technical words that has come into common usage. Perhaps it is because *statistics* really has two rather distinct meanings; it is plural in its first meaning and singular in its second. *Statistics* in its *plural* meaning refers to a collection of numerical information. Closely associated with this meaning is the art of summarization and presentation of numerical data. Some aspects of this art, of importance to engineers and managers, will be introduced in Chapter 3.

Statistics in its *singular* meaning does not refer to collections of data, but denotes the art and science of collecting, interpreting, and drawing inferences from certain special sets of data, called *samples.* Most of this text deals with this more modern and scientific definition of statistics, applying statistical science to problems commonly encountered by engineers and managers in design, engineering, production, testing, and marketing.

Engineers constantly are confronted with the need to assemble data, to summarize them in ways that will convey useful information, and to draw inferences from data that will enable them to take appropriate action. In designing products or processes, engineers need to interpret specifications in terms of performance requirements. Engineering changes and trouble-shooting require effective collection and interpretation of test data. Information fed back from marketing, warranty results, and directly from users also requires correct and efficient data collection and analysis. Production of quality products involves integration of product design with production methods and machinery as well as statistical control of the production process to assure that specifications are met. Determining the causes of production problems and setting up efficient test and inspection schedules requires knowledge of statistical design of experiments.

Management is involved in all phases of quality, from product concept, through design, to marketing and customer acceptance. Effective management for quality requires an understanding of data summarizations and how to draw useful conclusions from data. Statistical interpretation of marketing and warranty results is crucial to effective product and process management.

1.2 What Is Quality?

The word **quality** has been defined by some to be "a degree of excellence," and by others as "a degree of performance to a standard." (Both definitions are given in *Webster's Third New International Dictionary.*) Thus, a high-quality product is said to be one that is equaled by few or none in its performance or appearance. High-quality furniture, for example, uses the finest of solid woods and upholstery materials; it has a deep, lustrous finish; it has parts which fit together perfectly; and it is built solidly to last a long time. Unfortunately, this definition of quality often entails high initial cost.

The modern definition of quality does not require the highest degree of excellence. Instead, it requires that a standard be determined and that the product meet the standard in every respect. Ideally, this standard should be set by the customer, and quality can be defined as "a product or service that meets the customer's expectations."

For example, a low-priced automobile may have fewer luxury features than a high-priced car, and it may be smaller. But, if it meets or exceeds the reasonable expectations of customers in the low-priced automobile market, and if it holds up well in its intended use, it can be said to be a quality automobile in spite of its size and lack of some luxury features.

To take the concept of quality a step further, let us examine quality in relation to each of the following three related concepts:

- Performance
- Reliability
- Maintainability

The **performance** of a product relates to its specific design features and capabilities. For example, an engine, designed to be capable of high horsepower and torque at relatively low RPM, often is said to be a high-performance engine. But, it is not a high-quality engine unless it delivers this performance consistently and with a minimum of downtime and scheduled maintenance. An engine designed to have low horsepower but high fuel economy and durability, can be a high-quality engine if it performs consistently to these criteria. Quality requires that a product or service deliver its designed performance consistently.

For a product or service to have high quality, it is not enough that it delivers its designed performance. This level of performance must be met consistently, over a sufficiently long period of use. Thus, a newspaper delivery service that promises delivery by 6:00 A.M. is not of high quality if the delivery time is sporadic, or if it becomes later and later as time goes on. A product or service is said to have **reliability** if it delivers its promised performance consistently, with a minimum of breakdowns, *over a sufficiently long period of use.* Quality also requires that a product or service be reliable.

If frequent or difficult maintenance is required to keep a product performing reliably, the product has poor quality. A product is said to be **maintainable** if it can be kept in an operative state with minimal repairs or adjustments. A high-quality product requires a minimum of routine maintenance and repairs that can be made with relative ease and at modest cost.

For a product or service to be of high quality, it need not be designed to have every possible feature, but it must reliably perform to its design criteria, and it must be maintainable. A high-quality product or service doesn't just happen.

Product quality begins with conception and design. A product or service that is designed to meet standards that cannot be met consistently does not have high quality. It often happens that, in their zeal to produce products with every possible feature, designers plan products that simply cannot be manufactured with quality. For example, design tolerances need to be so fine that they cannot be kept in mass production. The involvement of production engineers in the design process to insure finding production methods that are suitable to each design concept, often will improve the quality of the final product. Thus, production methods must constantly be improved so that a product conforms to the product design and that standards and specifications are continuously met or exceeded in every respect.

Product quality is proved and improved by constant measurement and testing. But quality does not end with design, manufacture, and testing; it also requires proper packaging, distribution, and marketing. A product designed and produced with high quality can lose its quality if it is improperly packaged or if its distribution system allows it to "sit out in the rain" or spend too much time "on the shelf." Finally, marketing and advertising the product with untrue or unrealistic descriptions of its features or with failure to recognize and make known safety hazards in its normal or potential use can negatively affect the quality of a product that might have been good otherwise.

Service quality requires appropriate equipment as well as an environment for consistent delivery of the promised service. For example, a towing service cannot deliver high quality if its tow trucks are not properly maintained or its drivers are inadequately trained. Poor service quality often results from a lack of service standards, shortages of personnel required to deliver to the required standards, a lack of adequate training, or improper or faulty equipment required to deliver the service.

1.3   Quality and Statistics

A study of modern quality concepts and methods can be made on several levels. It has been found that changes in traditional management methods are needed to improve quality. New attitudes of production-line workers and their increased involvement are needed for quality improvement. A greatly increased use of statistical tools has been found useful in improving and measuring quality. In this text, we shall emphasize the statistical tools necessary for quality improvement. But we shall not study statistical methods for their own sake; at every turn, we shall attempt to put statistics into the general context of quality. For this reason, we begin by giving a brief background of modern quality thinking and an introduction to some of the more important modern management methods in Chapter 2.

The role played by statistics in engineering and management was briefly summarized in Section 1.1. The role of statistics in quality is not unlike its role in engineering and management, but several statistical methods are especially useful in the pursuit of quality. We emphasize these methods in this text.

Statistical ideas are needed in engineering measurement, essential for verifying that specifications are met both in production and in testing. Control of manufacturing processes requires knowledge of the concept of variability and its statistical measures as well as an understanding of statistical inference. Research and development often is aided by applications of statistical design of experiments. Experimental design also can be helpful in finding the cause or causes of problems on the production line, as well as in product improvement.

In this text, the student will be confronted with frequently met problems in engineering and management for quality—problems for which the solutions are greatly aided by the application of statistical methods. In solving these and allied problems, the needed statistical methods will be introduced with only enough mathematical theory to be understood and not misapplied. An understanding of the basis for each statistical method and of their interrelationships with each other is critical. Most of the errors made in the application of statistics stem not from mistakes in arithmetic (computers give protection from that) but from the use of methods that are inconsistent with the source of the data or the underlying assumptions, or the use of a weak method when stronger ones are available.

Practice will be given in the interpretation of the results of applying each of the methods described. Examples will be given of each method, and computer programs will be used in applying those methods requiring heavy computation. Thus, computation will be minimized while interpretation is emphasized. Stress will be given to the assumptions underlying each method to avoid misuse of these powerful methods in practice. The extent to which the assumptions need to be met only approximately also will be pointed out.

GLOSSARY

	page		page
Maintainability	3	Quality	2
Performance	3	Reliability	3

2

QUALITY

Where have we been? In Chapter 1, we gave a brief definition of quality, and discussed its components of performance, reliability, and maintainability.

Where are we going? In this chapter, we expand on the modern philosophy and practice of quality. We introduce the philosophies of the leaders in the field, we discuss the major new management methods for quality improvement, and we summarize the cost aspects of quality, first from a traditional point of view, and then from the broader point of view of Taguchi.

Section 2.1, *Modern Philosophers of Quality,* discusses the philosophies of six of the most prominent contributors to quality.

Section 2.2, *Management Methods for Quality Improvement,* introduces some motivational programs, new organizational methods, and quality tools.

Section 2.3, *Total Quality Management,* describes a relatively new quality initiative, now mandatory for certain U.S. government procurement and becoming more accepted with time.

Section 2.4, *Group Problem Solving and Identification of Causes,* introduces the ideas of the quality-improvement team, brainstorming, and the Ishikawa fishbone diagram.

Section 2.5, *The Cost of Quality,* breaks down quality costs into those of quality improvement, non-conformance, and lost opportunity, and gives simple examples of the calculation of the cost of quality.

Section 2.6, *The Taguchi Loss Function,* gives a theory for calculating the losses associated with failure of manufactured products to meet specifications and for calculating manufacturing tolerances.

2.1 Modern Philosophers of Quality

Probably the first major ideas advancing the idea of quality came from Dr. Walter A. Shewhart in 1924 and subsequent years. Working as a physicist in the Bell Telephone Laboratories, Shewhart developed the control chart, which recognizes the important concept of variability in mass production. In 1931, Dr. Shewhart described the theory, philosophy, applications, and economic consequences of control charts in his book, *Economic Control of Quality of Manufactured Product*. The field of **quality control** got its name from this book, and it remains a major part of the modern field of quality. Shewhart control charts will be introduced in Chapter 8.

One of Shewhart's colleagues at this time was statistician Dr. W. Edwards Deming. Deming studied under Shewhart and based much of his work on Shewhart's theories. During World War II, Dr. Deming applied what he had learned by teaching quality control to over 30,000 workers, many of whom were involved in government procurement. After the war, Deming went to Japan to help guide that country's industrial redevelopment. Prior to that time, Japanese products exported to the United States and elsewhere were thought to be of inferior quality. Japanese industrialists enlisted Deming to help them turn their poor reputation around. The large number of Japanese products imported into the United States every year and their reputation for high quality attest to the success of this endeavor. The Japanese Union of Scientists and Engineers, in recognition of his important contributions, established the Deming Prize in 1950. This prize recognizes exceptional achievement in company-wide quality, strategy, management, and execution.

U.S. companies were not eligible for the Deming prize until 1984; since then, very few U.S. companies have qualified for the award. In 1987, the U.S. Congress established the **Malcolm Baldrige National Quality Award** to recognize quality achievement and excellence of U.S. companies. The European foundation for Quality Management also established a quality award in 1992.

Dr. Deming has developed a number of concepts related to quality. His fundamental philosophy is based on defect *prevention* rather than *detection*. He believes that quality improvement is an ongoing effort, based on the application of statistical methods. He argues that quality and productivity are not conflicting goals; quality means achieving a consistent and predictable standard which meets the needs of the marketplace.

As a result of his years of consulting, Deming has recognized the importance of management in the improvement of quality, and he has developed a set of guiding principles. Deming's obligations for top management are shown in the box labeled "Deming's Fourteen Points."

Dr. Joseph Juran followed Deming to Japan in 1954. Juran has become an international consultant and teacher of quality, spending much of his time teaching quality-improvement methods in the United States. Juran offers four principles for quality improvement. First, Juran cautions management not to wait until problems become insurmountable. It is better to act in advance than to wait until the feedback signal of lost customers gets to the top. He urges that quality be made as concrete and as important a target as cost and sales. Juran suggests that management study other firms that are getting good results and build on what they are doing. He wants management to set up more and more

---

DEMING'S FOURTEEN POINTS

1. Innovate. Plan products and services for the years ahead.
2. Learn and use statistical quality control.
3. Discard dependence on mass inspection for incoming and outgoing materials.
4. Most companies must drastically reduce the number of their suppliers.
5. About 85% of waste and defects are caused by management.
6. Institute modern on-the-job training.
7. Improve supervision.
8. Drive fear out of the organization.
9. Break down barriers between departments.
10. Eliminate numerical goals, slogans, and posters.
11. Look carefully at work standards; they often cause as much loss as poor materials and errors, especially if they do not take quality into consideration.
12. Instruct all employees in simple statistical methods.
13. Retrain employees in new skills to keep up with change.
14. Make maximum use of statistical knowledge and talent.

---

quality-improvement projects. Juran's four points are summarized in a box labeled "Juran's Four Points."

Another American leader in quality thought is Philip B. Crosby, author of *Quality Is Free* (see Bibliography, Appendix B). He argues that the cost savings inherent in a high-quality operation more than make up for the direct costs of pursuing quality. Such savings include reduced inspection, less rework and scrap, lower warranty costs, and higher sales resulting from increased customer satisfaction. Crosby encourages measuring the costs of *not* meeting quality standards. He insists on the exclusive use of facts, not judgment or guesswork, in making decisions regarding quality.

During the decades that Deming and Juran consulted in Japan, most Japanese scientists, engineers, and industrialists became convinced of the wisdom of their teachings, and some of them developed their own philosophies of quality.

JURAN'S FOUR POINTS

1. Do not wait to take action until the problems become insurmountable.
2. Quality should become as important a target as cost and sales.
3. Management should study other firms that are getting good results and build on what they are doing.
4. Management should set up and support quality-improvement projects to the maximum extent possible.

Among the noted Japanese philosophers of quality are Kaoru Ishikawa and Genichi Taguchi. Ishikawa has lent his name to the "fishbone" diagram, also called the "cause-and-effect diagram," for quantifying contributing factors to quality (Section 2.4). He also is a leading proponent of total quality control, a precursor to the more comprehensive idea of "total quality management" (Section 2.3). Taguchi is known for his ideas on loss functions (Section 2.6) and for his novel approaches to the design of experiments as applied to manufacturing quality.

2.2   Management Methods for Quality Improvement

The pursuit of improved quality has given rise to several organizational and methodological approaches in recent years. These approaches can be categorized into three types, as follows:

- Motivational programs
- New organizational disciplines
- Quality tools

Motivational programs often are used in the beginning of a quality-improvement program to increase awareness and generate support and commitment from employees at all levels of top management. They often take the form of meetings and conferences during which senior managers state their goals and outline the program's details. These meetings often are followed up by posting signs and banners to remind everyone in the organization of the goals and principles. Such signs often carry messages like "Quality Is Everyone's Business" or "Our Goal Is 0.001 Defects."

One such motivational program is called **Zero Defects**. This program is based on the belief that all defects are caused by error and that through extraordinary care and attention to detail the final product should contain no defects at all. Although Zero Defects was originated for use in the space program, where a single defect could cost the lives of several astronauts, it has since been applied to the production of less exotic products and even to services such as insurance.

Motivational programs alone usually result in measurable quality improvement only for limited periods of time. It is too easy for workers and managers to return to the "old ways" after the initial stimulation of the program wears off. Attitudes and methods that lead to quality improvement need to become habitual in order to have a good chance of effectiveness in the long run. Perhaps this is the reason that Deming advocates the elimination of signs and posters in his tenth point.

The **new organizational disciplines** for quality management can be summarized by the single word *teamwork*. Poor quality often results from lack of communication among the different departments within an organization. The design engineer doesn't always interact with the production engineer to assure that the product can be manufactured reliably according to specifications. The testing department may not have been made thoroughly aware of all the performance requirements of the product. The marketing department may present the product without paying sufficient attention to its capabilities or its limitations, sometimes selling it for the wrong application.

In an effort to improve communication as well as to create a sense of common purpose throughout the organization, Japanese industrialists developed the concept of **quality circles**. The quality circle is a group of workers and managers involved in the production of a given product. It is meant to function collegially, without regard to rank or status in the organization, and to provide all members with equal opportunity to make suggestions or criticisms relating to the design, production, or other aspects of the product. The quality circle is an attempt to break down the compartmentalization that has been pervasive in many companies, thus improving communications and increasing the motivation of workers at all levels to produce a quality product.

The quality circle, which has now evolved into the "quality-improvement team," is a team formed to pursue a particular project or to solve a specific problem. Membership on a team is determined solely by the disciplines required to deal with the project or problem, not by rank or department. While the quality circle is not always the final answer in the pursuit of quality improvement, its use often is an effective step leading to what best can be described as the "quality environment."

The quality environment is one in which people at all levels are dedicated to common goals. In such an environment, there are fewer levels of management, departmental barriers are lower or ideally nonexistent, and professional elitism has been removed. Also, there is less disparity in pay scales and increased emphasis on training and education.

An important part of training for quality improvement involves imparting knowledge of modern **quality tools**. Some of these tools relate to problem solving and identification of causes, and many of them involve statistical methods. Some of the problem-solving and cause-identification methods in current usage are introduced in Section 2.4. Statistical methods for quality now are widely used in engineering and management. Many of these methods are studied in this text. Chapters 3 to 7 give the basic ideas of statistics needed to understand and apply these methods, as well as direct applications to quality of many of these ideas. Chapters 8 to 11 introduce more advanced topics in statistical application of special importance to engineers and managers in the pursuit of quality.

2.3   Total Quality Management

The management philosophy and many of the methods of Section 2.2 have been consolidated and augmented by the U.S. Department of Defense into a program called **Total Quality Management (TQM)**. Most defense suppliers have adopted the principles of TQM, and its elements are gaining widespread acceptance among other manufacturing and service industries. TQM is an initiative for continuously improving quality at every level, by focusing on the following goals:

- Build and sustain a culture committed to continuous improvement.
- Satisfy customer needs and expectations.
- Require dedication, commitment, and participation from top management.

- Involve every individual in improving his or her own work processes.
- Create teamwork and constructive working relationships.
- Recognize people as the most important resource.
- Employ the best available management practices, techniques, and tools.

TQM aims to achieve *continuous improvement* by focusing on the processes that create products. The quality of the products themselves are indicators of the adequacy of their processes.

Total quality management relies on four main elements for continuous quality improvement, as shown in the inverted pyramid of Figure 2.1. The element **vision** in this figure refers to providing long-term focus and continuity to decisions and actions. It recognizes the challenges of change and competition, and it is the starting point for guiding an enterprise into the future. Whether the vision of a company is to provide the best mousetrap in the world or merely to give every customer consistent value for the money spent, it is essential for management to have a long-term, consistent vision that guides its every decision and activity.

The **principles** of TQM define the fundamental concepts that shape and guide the program. They serve as basic rules for management decisions and actions, and they provide a framework to form expectations and to judge behavior. The basic principles of TQM include a commitment to *continuous process improvement*, with emphasis on preventing defects, rather than attempting to remove them by inspection. Defect prevention requires a thorough *knowledge of the process* and a constant *focus on the user*, whether the ultimate user (the customer) or an intermediate user (the next stage of the process). All phases of quality improvement are managed *from the top down*, with a *consistency of purpose* that depends on *total involvement* by everyone involved. To help assure involvement, *teamwork* is essential to align goals and approaches, and management must make an *investment in*

FIGURE 2.1   Elements of TQM

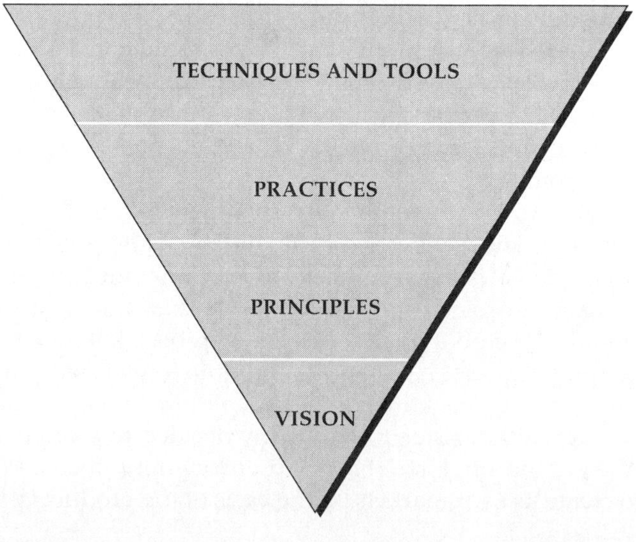

TECHNIQUES AND TOOLS

PRACTICES

PRINCIPLES

VISION

people through training, team-building, and providing an environment in which people can grow.

Among the **practices** of TQM are *planning and goal-setting* to assure consistency of purpose, sending out *consistent signals* by relating performance assessments to contributions that improve quality, and effective *communication.* Further practices include *skill building* in group dynamics, quantitative techniques, and other process improvement methods; *efficient use of resources*; and encouragement of *supplier improvement.*

Some of the **techniques and tools** of TQM are *design review,* involving participation by many disciplines, including the production and marketing departments; and *award programs* to communicate and reinforce goals. Central among TQM techniques are *statistical methods,* including data analysis, statistical process control, and design of experiments. Other techniques are *group dynamics skills,* and *"just-in-time" inventory control.*

In summary, TQM is a comprehensive program for quality improvement which requires new managerial methods and work environments, which emphasize communications and personal relationships and encourage the use of a wide variety of techniques and skills. It can be said that the elements of TQM, outlined in this section, comprise the criteria used by the U.S. government to award contracts. TQM, an initiative of the U.S. Department of Defense, is the basis used to establish the criteria for the Malcolm Baldrige National Quality Award. Competitors for this award are judged on their success in following TQM guidelines. Some of the methodology encompassed by total quality management is discussed in greater detail in the remaining sections of this chapter.

The following brief case study illustrates how the application of TQM concepts can turn a company around. An American company once was a leader in its field. But foreign competition had eroded its customer base to where it was in danger of going bankrupt. The American company's products, had been known for high quality and high performance, although they were somewhat "pricey." However, consumers were willing to pay a little extra for the quality they were getting, the reputation of the product, and the prestige that came with owning.

Then, foreign manufacturers began to introduce competitive products with superior features and a far smaller price. Although their quality had yet to be demonstrated, other products previously introduced by the same companies were regarded as being of high quality. The result was a rapid fall-off of sales for the American company.

The American company responded initially by cheapening its products in an effort to meet the price competition. Although they were able to narrow the price gap considerably, they never quite succeeded in closing it entirely. Consumers purchasing its product on the basis of its previous reputation were embittered by the poor quality of the new product. It no longer looked and felt like the premium product it once was, and it no longer had the reliability it once had. Thus, this strategy backfired, and sales plummeted even more rapidly.

To avoid disaster, the company decided to change the way it did business from the ground up. First, it restated its long-term focus. Where previously it was able to create its own markets on the basis of the products it sold, now it found that it was necessary to provide the marketplace with what it wanted, rather than what the company was willing to give it. The company wanted to maintain its vision of providing the premier product in its field, but now its vision included a long-term focus on the evolving needs of the consumer and the way the competition was addressing them. The company made a commitment to do whatever was necessary to stay ahead of the competition in design, features, and quality. To do so, it greatly improved its market research, and it set in place methods for rapidly adapting its designs to meet the needs of the marketplace as they were revealed by this research.

The company decided to spend whatever was necessary to improve its manufacturing processes and maintain an ongoing program of continuous quality improvement. Information from market research, from warranty claims, and from customer complaints was fed quickly into production to eliminate the causes of quality problems without delay. Workers were retrained, and when they saw that top management really began to care about quality, their own attitudes changed as well. Teamwork at all levels produced better ideas and, since many of the workers also were users of the product, their ideas for product and process improvement now could be rapidly integrated into the process.

The decision of top management to make the heavy initial investment necessary to achieve higher quality was courageous, especially in view of the rapidly deteriorating fortunes of the company. Their decision, and the realization that the consequence of failure was disaster, made consistency of purpose imperative. Not only were the principles and practices of TQM employed to the fullest, but the necessary techniques and tools were learned and employed. By so doing, not only was the quality of the product greatly improved, but the cost of producing it was reduced enough to make the company's products competitive with the imports.

Is this a true story? Yes, except that many details have been left out. How did it come out? Much to everyone's surprise, it worked! It worked better than anyone expected. Application of TQM turned the company around, and it even restored the company's once preeminent place in its product world. It's fun, again, to own a Harley.

This is not an isolated case example. The emphasis on quality at Ford Motor Company has helped it restore its historical market share. Other companies, both obscure and well-known have been restored to health by applying principles of TQM. Wherever management has had the courage to change its vision, to make the initial investment, and to maintain a steady course embracing the principles and practices of TQM, similar success stories have been reported.

2.4 Group Problem Solving and Identification of Causes

Quality improvement often requires the identification and solution of problems. For example, an automobile manufacturer, in testing a new model, finds that the braking distance is greater than the design requirement. The design, manufacturing, and test engineers each have their own ideas as to how to "fix" the problem, but no definitive solution immediately suggests itself.

The product manager decides to assemble a **quality-improvement team (QIT)** to "brainstorm" the problem and to propose a solution. The membership of the team is carefully selected to represent all aspects of the problem; the team probably includes, as a minimum, a design engineer, a production engineer, a representative of marketing, a cost analyst, and several production personnel. (There is a tendency in "elitist" organizations to omit production personnel from problem definition and problem solving. But, anyone who has spent time at production facilities soon discovers that it is the production-floor people who are most familiar with factory practices and everyday problems.) The team will select a leader who has demonstrated abilities as a facilitator. The leader will keep the team focused on the issue at hand and will have the ability to interact effectively with other team members. The team also should include a member who understands the problem in the larger context of other issues. A secretary is appointed from among the members of the team to record the proceedings of each meeting, to maintain all data summaries, and to disseminate action items and agendas.

The team will decide what is needed to solve the problem, collect relevant data, determine influential factors, generate potential solutions, select the "best" solution, recommend it to management, and, if approved, see that it is effectively implemented. In determining influential factors, or potential causes, it is often useful to conduct one or more **brainstorming** sessions. It is important to set a time limit for such sessions—their efficiency diminishes rapidly after too much time elapses. The round-robin approach is used. Each member has an opportunity to propose an influential factor or to pass. All ideas are recorded without evaluation or criticism. The purpose of the session is to generate "raw" ideas; discussion and evaluation take place later. The session ends when no new ideas are proposed or when the time limit has been reached.

Note that the process so far described avoids the common "rushing to a solution" behavior that often results in poor or temporary solutions. A systematic approach is less likely to overlook more effective potential solutions, and it requires less time in the long run.

After potential causes have been listed, the QIT ranks them and does a cause-and-effects analysis. This analysis often consists of constructing an Ishikawa "fishbone" diagram. In such a diagram it is customary to divide the potential causes into four main categories. For manufacturing applications, these categories are *Methods, Machinery, Manpower,* and *Material.* For service applications, they are *Equipment, Policies, Procedures,* and *People.* Major potential causes are listed under each of these categories, subsidiary causes are listed under some or all of the major causes, and so forth. It is important to note that causes are listed initially without regard to their probability of being responsible for the effect—excess braking distance in this example. The list is intended deliberately to be exhaustive. Most likely causes are identified later by a process of collecting data and expert opinion to pare down the more exhaustive list into a short list for further investigation.

Suppose the QIT lists the following possible major causes under *Methods*: Design, Production control, and Test conditions. Under *Machinery* the following are listed: Disk/drum turning and Tolerance-measuring equipment. Under *Manpower*: Training, Supervision, and Inspection; and under *Material*: Brake pads and Brake fluid.

These causes are diagrammed in a "tree" diagram, listing each cause as a branch eventually leading back to the effect, the main branch which bears a statement of the problem. Figure 2.2 shows the resulting **cause-and-effect diagram**,

CAUSE   **EFFECT**

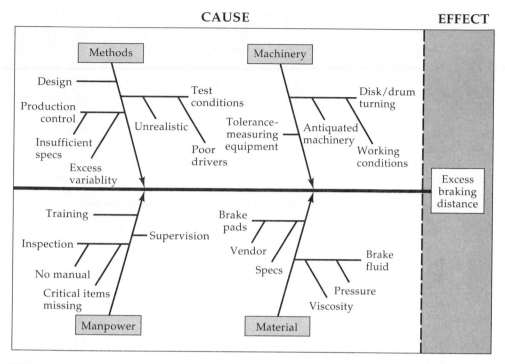

FIGURE 2.2   Fishbone Diagram for Manufacturing

resembling a fishbone. Note from Figure 2.2 that several subsidiary causes were added under some of the main causes. For example, under Production control, Insufficient specs and Excess variability were added.

To illustrate a cause-and-effect diagram for a service application, suppose there have been life-threatening delays in getting the results of hospital laboratory analyses to attending physicians. A quality-improvement team, consisting of laboratory, nursing, administrative, and medical personnel, might have produced the fishbone diagram shown in Figure 2.3.

After listing the potential causes and their relationships in a fishbone diagram, the quality-improvement team discusses each one, eliminating those not supported by data. Sometimes the team will eliminate a cause solely on the basis of an opinion rendered by a member of the team or an expert whom the team has consulted. Such action, in the absence of confirming data, should be taken with great caution and only after the team has considered whether the expert is advising objectively or is merely protecting a department or an enshrined point of view.

Returning to the braking-distance example, let us suppose that all causes listed in Figure 2.2 have been eliminated, except Excess variability under Production control and Vendor under Brake pads. Now, the team gathers data on various brake-pad measurements that relate to stopping distance, using direct measurements of pads supplied by the current vendor and several competitive vendors. Suppose the analysis of a statistically designed experiment (Chapter 10), supervised by the team, shows that the coefficient of friction of the current vendor's pads has excessive variability when compared to that of other vendors. It is further

CAUSE   EFFECT

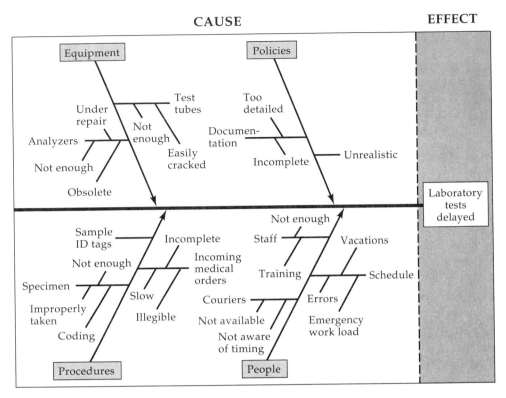

FIGURE 2.3   Fishbone Diagram for a Service

determined that this variability is sufficient to account for the random but excessively frequent failure of the brakes to pass the stopping-distance test.

The QIT recommends to management that the brake-pad vendor be
changed. The team's responsibilities do not end here, however. The QIT will use
argument and analysis to convince management to accept the recommendation.
Once adopted, the change will be monitored by the team to assure that the problem has been solved. Only then can the efforts of the QIT be considered successful,
and the team can disband.

EXERCISES

2.1   Which of the following studied under Shewhart: Deming, Juran, Crosby,
Ishikawa, Taguchi?

2.2   Who is the author of the "quality is free" concept?

2.3   Who can be said to be "the father of quality control"?

2.4   Which of the philosophers of quality regards statistics as most
important?

2.5   a.   Name three modern management methods for quality.

b.   Which method appears to have the least lasting effect?

c.   Which involves statistics?

d. Which involves teamwork?

2.6 Name the four main elements of total quality management.

2.7 Under what auspices did TQM originate?

2.8 Name at least two practices of TQM that are used to support each of the four elements.

2.9 Consider the problem of poor television reception. "Brainstorm" the problem (preferably with some of your classmates) and list as many possible causes as you can think of.

2.10 Suppose a bank branch has received too many complaints about long lines at the teller windows. Use brainstorming to list as many potential causes as possible.

2.11 Referring to Exercise 2.9, construct a "fishbone" diagram, displaying the relationships of the possible causes.

2.12 Construct a fishbone diagram for the causes found in Exercise 2.10.

2.13 Consider the problem of an employee's chronic lateness in coming to work. Brainstorm to list as many possible causes as possible, and construct a fishbone diagram.

2.14 The light bulbs in your home seem to have too short a life. List the possible causes and construct a fishbone diagram.

2.15 Practice "teamwork" by getting together with a group of your classmates to discuss a quality problem with which you have some familiarity. List possible causes, rank them in order of importance, and construct a fishbone diagram. To the extent of your knowledge and with the aid of whatever data are available, select one or two most likely causes, and make a specific recommendation for further investigation or immediate action.

2.5   The Cost of Quality

The total expenditure on quality of a business unit is not simply the direct cost of conforming to specifications. This cost also should include money spent to correct mistakes as well as the direct and indirect costs incurred by not having good quality. The objective of a quality-improvement program is to *reduce* the total cost of quality by reducing the incidence of error and improving customer relations, even though the cost of conforming to specifications may increase.

The three main components of the cost of quality can be summarized as follows:

COMPONENTS OF THE COST OF QUALITY

1. The cost of quality improvement
2. The cost of non-conformance
3. The cost of lost opportunity

The **cost of quality improvement** includes all expenditures incurred to assure that customer requirements are met. These costs are incurred to prevent the waste that accompanies scrap and rework and the loss of business that occurs when customer needs and expectations are not met. Thus, the cost of quality improvement can be thought of as an investment—an expenditure made with the expectation of monetary returns.

Quality-improvement costs include the cost of plant and equipment required to produce products or services according to specifications. Some examples are improved lighting, noise and pollution control, more modern machinery, and better tools. These costs also include money spent on training and education to assure a more knowledgeable and efficient work force. They include the costs of communication and supervision, such as blueprints, specification sheets, newsletters, meetings, and quality-improvement teams. In short, the costs of quality improvement encompass the costs of "doing it right the first time."

The **cost of non-conformance** includes the cost of scrap and rework made necessary by errors in production. These costs can be considerable. A scrapped part represents a loss in material or purchased parts. Both scrap and rework involve losses in labor time, resulting in the need for a larger labor force to produce a given amount of product, or expensive overtime (or both) together with additional plant, machinery, and indirect costs of labor. Even in modestly sized production facilities, these costs can add up to millions of dollars each year.

Non-conformance costs include excess inspection to verify that parts, assemblies, and finished product are in conformance with specifications. In industries producing high-technology products, especially electronics, the number of inspectors can approach the number of production-line workers. Inspection is not only expensive but also often fails to catch defects. While sampling inspection is most frequently used (Section 4.6), even 100% inspection has been shown to miss as much as 10% of the defects. Some electronics products, such as computers, have many hundreds of circuit elements embedded in integrated circuits. Imagine trying to find a defective element or soldered connection under a microscope when inspecting them all. It has been found that "doing it right the first time" can greatly reduce the cost of inspection (even eliminate inspection entirely in some cases), while decreasing the number of defective products shipped.

Non-conformance costs also include warranty costs. In some products, such as automobiles and major appliances, warranty costs can amount to a considerable fraction of the original production cost. Studies have shown that the cost of maintenance of certain military equipment greatly exceeds the purchase price; some of these costs are now borne by the manufacturer under warranty.

The **cost of lost opportunity** may be the largest cost of all, albeit the most difficult one to measure. Lost opportunities result when a customer cancels an order as a result of dissatisfaction with the product or service, or when a rental or leasing contract is canceled. Lost opportunity also occurs when a customer no longer will purchase a product or service from a given supplier because of the poor quality of a recently purchased product. Entire corporations, some large and famous, have gone out of business when the quality of their product or service was allowed to deteriorate, or when the competition offered a better quality product at a competitive price.

The following example shows how some of the hard-to-quantify costs of quality can influence a decision to make a quality-improvement investment.

> *Example.* An appliance manufacturing company uses an old machine to cut sheet metal for its washers and dryers. The cut edges frequently have burrs that must be removed. As a result, each cut piece of sheet metal must be inspected for burrs, and those found to have burrs must be sent to a separate department for de-burring. A machine-tool salesman demonstrates a new sheet-metal cutter that will eliminate burrs and costs $750,000.
>
> The accounting department determines that the current annual cost of non-conformance is $15,000 for inspection time, $24,000 for de-burring labor, and $1,500 for equipment. It figures that it can amortize the new cutting machine over a 10-year period, at an annual cost of $75,000 (ignoring interest). Thus, the annual quality-improvement cost is $75,000 and the annual reduction in the cost of noncompliance is $15,000 + $24,000 + $1,500 = $40,500. Inasmuch as the directly measurable cost of quality would increase by $34,500 per year, this department recommends against purchasing the new machine.
>
> The production manager, however, knows that the de-burring process is not perfect—some small burrs are not found by inspection, and the de-burring tool (a hand-held grinder) sometimes deforms the metal slightly or does not remove the entire burr. An examination of final inspection records shows that the annual cost of retouching washers and dryers on final inspection is $20,000. While this potential saving still leaves a cost-of-quality increase of $14,500, the production manager believes that some warranty costs are caused by the de-burring problem (warranty records are insufficient to make an exact determination), and process slowdowns sometimes result from the discovery of burrs (again, records are not kept which would make it possible to quantify this loss). He recommends in favor of purchasing the cutting machine.
>
> The sales manager has been concerned about loss of market share. In an effort to understand the loss, she has hired a survey research company to interview consumers, comparing her company's machines with those of the competition. High on the list of negative comparisons was "general appearance." In consultation with the production manager, she has concluded that the difficulty of applying a superior finish to de-burred sheet metal plays a significant role in the "appearance" problem. Although she cannot quantify the business loss associated with the de-burring problem, her instinct leads her to recommend strongly in favor of the new machine.
>
> The chief executive officer ordered the machine purchased.

The cost of doing business can be reduced by improving the quality of a service, as shown in the following example.

Example. Many states or localities levy sales taxes on purchased products, and some states tax services. Customers such as educational or religious institutions, not-for-profit companies, governmental agencies, and so forth often are exempt from paying sales taxes. Normally, the tax is collected from the customer by the seller of the product or service and periodically paid by the seller to the taxing authority. If a customer incorrectly claims an exemption, by the time a tax audit reveals this information it is often impossible or impractical for the seller to collect the tax from the customer. The failure to recognize that the taxing authority is a customer, and to satisfy its requirements, can cost a large company millions of dollars annually. Suppose the failure of a given company to satisfy this requirement costs it $2.5 million per year. The company decides to hire two auditors to check the certification of each customer claiming sales-tax exemption. The total annual cost of these auditors is $130,000, including their salaries and benefits and the cost of their office space. Subsequent experience shows that the cost of incorrect tax exemptions has been reduced from $2.5 million to $900,000 per year.

(a) What is the cost of quality improvement?

(b) What is the reduction in the cost of quality?

Solution. (a) The cost of quality improvement is $130,000.

(b) Since the cost of noncompliance has been reduced from $2.5 million to $900,000, the cost of quality is reduced to

$$\$900,000 + \$130,000 = \$1,030,000$$

which means an annual saving of $1,470,000.

EXERCISES

2.16 a. List the components of cost of quality.

   b. For each of the following, state to which component it belongs:

i.	scrap	vi.	customer complaints
ii.	on-the-job training	vii.	high warranty costs
iii.	a TQM program	viii.	"Zero Defects" program
iv.	decreased sales	xi.	low productivity
v.	excessive inspection	x.	decisions based on data

2.17 After the assembly of electronic components, 12% of the 20,000 produced per year are scrapped on final test. Poor lighting has been determined by a QIT to be the most likely cause of this high rejection rate. It would cost $50,000 to install adequate lighting in the assembly department; this cost can be amortized over 10 years. It costs $15 for each scrapped component. It is estimated on the basis of experience in a sister plant that improved lighting can reduce the scrap rate by 25%.

   a. What is the annual cost of the improved lighting?

b. What is the current cost of scrapped final assemblies?

c. What would be the estimated cost of scrapped final assemblies if new lighting were installed?

d. What is the quality cost of installing the lighting?

2.18 A consulting company finds that about 30% of their assignments come from repeat clients, and the average value of such assignments is $50,000. A new department is instituted to review and edit all outgoing final reports, at an annual cost of $200,000. The rate of repeat assignments then improves to 45% and the value of such engagements increases to $60,000 on average. What is the annual cost of quality in this instance, if the company has an average of 500 assignments per year?

2.19 An automobile manufacturer spends $50 million to improve the quality of a new model. Its warranty costs drop from $250 to $150 per unit as a result. In a model year when 400,000 units are sold, what is the cost of quality related to this improvement?

2.20 Referring to Exercise 2.19, sales of the model increase to 500,000 units, believed to be a result of the quality improvement. If there is a profit of $1,000 per unit, was the $50 million expenditure justified?

2.6   The Taguchi Loss Function

Loss Function

A keystone of the quality philosophy of Genichi Taguchi involves losses resulting from poor quality in manufacturing. Under the Taguchi principle, quality can be measured by the losses a product or service imparts to society. Thus, quality should be measured by a loss function of some kind. The loss function not only facilitates the calculation of losses resulting from poor quality but also can be used to determine manufacturing tolerances.

Whenever a "target" specification is not met, even if the given attribute of the product is within specification limits, there is a measurable loss. Taguchi assumes that the loss associated with a given product is proportional to the *square* of the deviation of the measured attribute of that product from its target value. This assumption gives rise to a quadratic loss function of the form

$$\textbf{\textit{Loss Function:}} \quad L(y) = K(y - M)^2$$

In this formula, K is a proportionality constant, called the **loss parameter**, y is the measured value of the **product attribute**, and M is its **target value**.

To illustrate, consider the simple problem of manufacturing a shaft that must fit inside a sleeve. To satisfy the customer, the diameter of the shaft must not be too small, so it will not fit too loosely into the sleeve, or too large, so it can be inserted without forcing. Thus, the product attribute in this example is the diameter of the shaft. Suppose that the minimum acceptable diameter is 1.25 cm and the maximum is 1.30 cm. Then the target value of the diameter is $M = (1.25 + 1.30)/2 = 1.275$ cm. The **customer's tolerance** is the amount of variation allowable for a product attribute to remain within the customer's specification limits, or ±0.025 cm in this example.

Manufacturer's Specifications

If the manufacturer decides to produce the shaft to a diameter of 1.275 cm with a tolerance of ±0.025 cm, there could be a risk of making too many shafts with diameters outside of the customer's specifications of 1.25–1.30 cm. Thus, a manufacturer seeking high quality should set up the shaft-making process to somewhat tighter limits. Taguchi's loss function can be used to determine what the **manufacturer's tolerance** should be.

Suppose the customer's tolerance is $\pm T_1$ and it is desired to find what the manufacturer's tolerance, $\pm T_2$, should be. These tolerances are shown in the graph of the loss function in Figure 2.4. (In our example, $M = 1.275$ cm and T_1 is the distance between M and either specification limit, or $T_1 = 0.025$ cm.) To determine the manufacturer's tolerance, it is necessary to introduce a notion of cost. Suppose C_1 represents the customer's cost of having been shipped a defective shaft and C_2 represents the manufacturer's cost of producing a defective shaft (including the cost of scrap and/or rework). In all our examples, C_1 will be greater than C_2. Under Taguchi's philosophy, the cost of a discrepant product is assumed by "society," and he includes the full cost of producing the product in the customer's cost, as well as other costs, such as lost time, inspection, and excess inventory.

FIGURE 2.4   Taguchi Loss Function

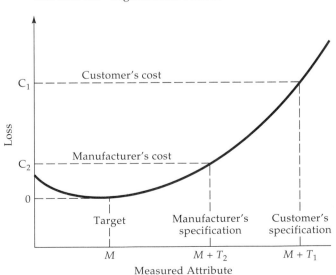

The customer realizes the cost C_1 whenever the shaft diameter lies at or outside the specification limits $M \pm T_1$; in particular, when the diameter equals $M + T_1$, the customer's loss is

$$C_1 = L(M + T_1) = K(M + T_1 - M)^2$$

or

$$C_1 = KT_1^2$$

Similarly, it can be shown that

$$C_2 = KT_2^2$$

The value of the loss parameter, K, can be found by solving either of these last two equations for K, obtaining

---

Loss Parameter: $K = \dfrac{C_1}{T_1^2} = \dfrac{C_2}{T_2^2}$

In our example, suppose the customer's cost of a discrepant shaft is $C_1 = \$40$, and the manufacturer's cost of scrapping an out-of specification shaft is $C_2 = \$10$. The value of the loss parameter is given by

$$K = \frac{40}{.025^2} = \$64,000$$

The specific loss function for this example is obtained by substituting $K = 64,000$ and $M = 1.275$ in the general formula for the loss function, obtaining

$$L(y) = 64,000 \, (y - 1.275)^2$$

expressed in dollars.

The quality loss for a shaft of any given diameter now can be calculated. For example, when a shaft has a diameter of 1.5 cm, the loss is $64,000 \, (1.5 - 1.275)^2 = \$3,240$. Since the diameter of this shaft is much greater than the upper specification limit, the loss is very large. If another shaft has a diameter of 1.272 cm, the associated loss is only $64,000 \, (1.272 - 1.275)^2 = \$0.58$. Note that there is a loss, even for a product within specification limits, unless the attribute measured is exactly at the target value.

The required manufacturing tolerance now can be found. From the equation for the loss parameter,

$$\frac{C_1}{T_1^2} = \frac{C_2}{T_2^2}$$

The manufacturer's tolerance is obtained by solving for T_2, with the following result:

$$\textit{Manufacturer's Tolerance:} \qquad T_2 = T_1 \sqrt{\frac{C_2}{C_1}}$$

(Note that the manufacturer's tolerance can be determined from the customer's tolerance only if the *ratio* of the two costs is known.)

To illustrate the calculation of a manufacturer's tolerance when the customer's tolerance and the costs are known, suppose the customer's tolerance for the diameter of the shaft is ±0.025 cm, that is, $T_1 = 0.025$. We shall assume that the costs are as before, that is $C_1 = \$40$ and $C_2 = \$10$. Thus, the manufacturer's tolerance should be

$$T_2 = .025 \sqrt{\frac{10}{40}} = 0.0125 \text{ cm}$$

That is, the manufacturer should set up the process to the specification limits 1.2625 cm–1.2875 cm, or half the tolerance that the customer requires.

> *Example.* A drawer for a cabinet must have a nominal width of 14 inches. The manufactured width must be no less than 13.94 inches and no greater than 14.06 inches so it can fit into the drawer guides without binding. The loss to the furniture dealer (the customer) is \$75 if a cabinet is delivered with a drawer that does not fit, and the manufacturer's loss is \$15. Find the drawer-width specifications that the manufacturer should use to produce a quality product, and calculate the loss if a drawer is delivered with a width of 14.1 inches.
>
> *Solution.* First, note that $C_1 = \$75, C_2 = \$15$, and $T_1 = 0.06$ inch. Substituting into the equation for the manufacturer's tolerance,
>
> $$T_2 = .06 \sqrt{\frac{15}{75}} = 0.027 \text{ inch}$$
>
> Thus, the manufacturer should make the drawer width to within about 0.027 in. The loss, if a drawer is delivered with a width of 14.1 inches, is calculated from the loss function. First, the equation for the loss parameter is used to find K, with the result $K = 75/(0.06)^2 = \$20,833$. The loss is given by
>
> $$L(14.1) = 20,833 (14.1 - 14)^2 = \$208.33$$

One-sided Tolerances

So far, only the case of two-sided tolerances has been considered; that is, the case where there are both upper and lower specification limits. In manufacturing a structural beam to a specified strength in psi, for example, interest is focused on whether a particular beam has a strength equal to or *greater* than the specification. Or, in making a sheet of steel, the steelmaker is concerned about whether the amount of impurities is *less* than a specified percentage. Taguchi refers to this case as "smaller is better," while the case of the structural beam is an example of "larger is better."

The loss functions for these two cases are assumed by Taguchi to be as follows:

$$\textbf{\textit{Smaller Is Better:}} \quad L(y) = Ky^2 \quad \text{where} \quad K = \frac{C_1}{T_1^2}$$

$$\textbf{\textit{Larger Is Better:}} \quad L(y) = \frac{K}{y^2} \quad \text{where} \quad K = C_1 T_1^2$$

In the smaller is better case, the manufacturer's tolerance is

$$T_2 = T_1 \sqrt{\frac{C_2}{C_1}}$$

and in the larger is better case, it is

$$T_2 = T_1 \sqrt{\frac{C_1}{C_2}}$$

Example. For a heat of steel, suppose the amount of impurities is to be less than 0.5%. The cost to the consumer of steel from a "bad heat" is $1,000, and the producer's cost is $500. Find (a) the value of the loss parameter, (b) the quality loss for a heat of steel having 0.6% impurities, and (c) the manufacturer's tolerance.

Solution. (a) The loss parameter is $K = 1000/(.5)^2 = \$4,000$, and

(b) the quality loss of a heat of steel with 0.6% impurities is $L(.6) = 4,000(.6)^2 = \$1,440$.

(c) The steel mill should make heats to a tolerance of less than $.5(500/1000)^{1/2} = 0.35\%$ impurities.

EXERCISES

2.21 a. Graph the loss function implicit in the philosophy that a product is acceptable if it is "within specs" and worthless if it is not.

b. Compare this function with the Taguchi loss function.

2.22 The Taguchi loss function places a severe penalty on product that is very far outside of specifications.

a. Why?

b. Can you think of other loss functions (sketch some) that might be preferable?

2.23 Find the value of the loss parameter, K, in each case:

   a. C_1 = $30, and T_1 = 0.5 pound.

   b. C_2 = $8, T_1 = 0.02 inch, and T_2 = 0.01 inch.

   c. The customer's cost of a product "out of spec" is $30, and the customer's specification is 16 to 17 ounces.

   d. The producer's cost is $17 and the producer's tolerance is ±4%.

   e. The customer's cost is $75 and the customer's specification is < 1%.

   f. The customer's cost is $20 and the customer's specification is > 2,000 psi.

2.24 a. What is the target in the smaller is better case?

   b. What is the target in the larger is better case?

2.25 What should the manufacturer's tolerance be in the following?

   a. in Exercise 2.23(a) if C_2 = $20 and there are both upper and lower specification limits

   b. in Exercise 2.23(c) if the manufacturer's cost is $12

   c. in Exercise 2.23(e) if the manufacturer's cost is $50

   d. in Exercise 2.23(f) if the producer's cost is $17

2.26 A food product should have no more than 15% animal fat. If the food manufacturer produces to a tolerance of 3%, and the manufacturer's cost of a rejected lot is $250, what is the customer's cost of a lot having more than 15% fat?

2.27 An over-the-counter pharmaceutical should have at least 50% active ingredient. To what tolerance should the drug manufacturer produce if his cost of a "bad" lot is $55 and the consumer's cost is $75?

2.28 If it is known that the manufacturer's specifications are 2.49 to 2.51 inches, and the manufacturer's and customer's costs are $30 and $45, respectively, what are the customer's specifications?

2.29 An integrated circuit should have a speed of at most 12 nanoseconds. Find the manufacturer's tolerance if the manufacturer's and customer's costs are $1.50 and $3.00, respectively.

2.30 The customer's specifications for the diameter of a copper wire are 9 to 11 mils, and the customer's and manufacturer's costs for an out-of-spec reel of wire are $250 and $150, respectively.

   a. Find the value of the loss parameter.

   b. Find the manufacturer's tolerance.

   c. Graph the loss function, showing the two costs and both tolerances on the graph.

2.31 Suppose the loss function were changed to $L(y) = K(y - M)^{3/2}$.

    a. Derive a formula for the loss parameter, K.

    b. To what tolerance should the manufacturer produce if the consumer's and producer's costs of an out-of-spec product are C_1 and C_2, respectively?

REVIEW EXERCISES

Review exercises will appear after each chapter. They can be used for informal review of the chapter, or as two practice examinations, designed to be taken over a time period of about one hour. The odd-numbered exercises comprise one examination; the even-numbered exercises comprise the other.

2.32 a. Name four American philosophers of quality.

    b. Name two Japanese philosophers of quality.

2.33 Match each name with the appropriate contribution to quality:

Shewhart	four points for quality improvement
Deming	fishbone diagrams
Juran	"quality is free"
Crosby	14 obligations for management
Taguchi	quality control
Ishikawa	loss function

2.34 Summarize each of the seven goals of TQM, using one word if possible, but no more than three words.

2.35 Name the four main elements of TQM and give an example of each.

2.36 What are the four main branches of a fishbone diagram?

2.37 What is the purpose of a fishbone diagram? (Confine your answer to one or two sentences.)

2.38 Give four examples of quality-improvement costs.

2.39 Give three examples of non-conformance costs.

2.40 A plant manufacturing defense-electronic equipment employs 500 inspectors. The annual cost of this department averages $300,000 for management, and $40,000 per inspector. Suppose the introduction of statistical process control (SPC) and a change in culture, which makes each production operator responsible for his or her own work, results in a reduction of the need for inspection so that the plant now requires only 200 inspectors. The annual costs of SPC introduction (training and equipment) and the motivational and training programs needed to change the production-line culture are $2.5 million. What is the cost of quality in this instance?

2.41 A home heating-oil company decides to institute automatic delivery to those customers wishing it. The customer-tracking required by this program requires hiring an additional clerk, at an annual cost of $35,000. After one year of operation, the company found that the number of lost customers diminished by 21, and the number of new customers exceeded the average of the past five years by 12. The average annual gross profit per customer is $200. Find the cost of quality.

2.42 Find the loss parameter when the customer's tolerance is 1.5 and customer cost is $50.

2.43 Find the loss parameter when the manufacturer's tolerance is 0.9 and manufacturer's cost is $14.50.

2.44 What should be the manufacturer's tolerance when the customer's specifications and cost are 11.9 to 12.1 ounces and $12.50, respectively, and the manufacturer's cost is $8.00?

2.45 What should be the manufacturer's tolerance if the customer's tolerance is ±.05 cm and the customer's and manufacturer's costs are $17.50 and $28.00, respectively?

GLOSSARY

	page		page
Brainstorming	14	New Organizational Disciplines	9
Cause-and-Effect Diagram	14	Product Attribute	21
Cost of Lost Opportunity	18	Quality Circle	10
Cost of Non-Conformance	18	Quality Control	7
Cost of Quality Improvement	18	Quality Tools	10
Customer's Tolerance	22	Quality-Improvement Team	13
Deming's Fourteen Points	8	Smaller Is Better	25
Larger Is Better	25	Target Value	21
Loss Function	21	Total Quality Management	10
Loss Parameter	21	Practices	12
Malcolm Baldrige National		Principles	11
Quality Award	7	Techniques and Tools	12
Manufacturer's Tolerance	22	Vision	11
Motivational Programs	9	Zero Defects	9

3

DATA ANALYSIS

Where have we been?   Statistics has been defined in Chapter 1 as ". . . the art and science of collecting, interpreting, and drawing inferences from certain special sets of data, called *samples*." The term *sample* will be defined in Chapter 6, and methods for drawing inferences from samples will be introduced in Chapter 7. But, first, we must set the stage.

Where are we going?   In this chapter, we introduce the most frequently used methods for summarizing and interpreting raw data, whether or not they arise from samples. We will find that we can apply these methods directly to problems involving quality, even before we are further armed with the underpinnings of statistical inference. (In the following two chapters, we will introduce the notions of probability and probability distributions necessary to build a foundation for the science of statistical inference.)

Section 3.1, *Measurement and Variability,* discusses errors of measurement, accuracy and precision, and introduces the idea of variability.

Section 3.2, *Exploration of Raw Data*, discusses the need to summarize data and introduces stem-and-leaf displays, dot diagrams, and Pareto analysis.

Section 3.3, *Frequency Distributions,* introduces distributions of data by the frequency of occurrence of their values.

Section 3.4, *Graphs of Frequency Distributions*, constructs pictorial representations of frequency distributions, including histograms and bar charts.

Section 3.5, *Descriptive Measures,* defines and measures the location of a data collection and its variation for both raw data and data given in the form of frequency distributions.

Section 3.6, *Application to Quality,* applies some of the concepts of this chapter to the identification of frequently encountered quality problems and their probable causes.

3.1 Measurement and Variability

Every measurement, whether it be of the speed of light, the mass of a neutron, or simply the length of a table, is subject to error. The word *error* used in this context, does not necessarily mean "mistake." Whenever repeated measurements are made of the same entity, experience shows that the results will differ if the measurements are fine enough. If the diameters of "identical" steel rods are measured with a ruler calibrated only to the nearest sixteenth of an inch, there may be no apparent difference in successive measurements, but if a caliper that measures to within ¼₄ inch is used, differences probably will become apparent.

Error in measurement can arise from several sources. First, the measurement apparatus itself can be imperfect. For example, a photometric method may be used to measure the thickness of an oxide film deposited on a silicon disk. This method requires careful calibration and it may be sensitive to various ambient conditions as well as the exact position on the disk. Second, the entity measured might change from measurement to measurement. For example, thermal expansion and contraction can cause the true length of a steel beam to vary from measurement to measurement. Finally, there is the possibility of a mistake. A measuring instrument can be misread or it can be incorrectly calibrated.

Often it is difficult to determine the source or sources of error. The imperfection of all human endeavor suggests that error can never be eliminated entirely. However, a better understanding of the components of error can aid in its control. The major components of error can be illustrated by examining errors in target shooting. The rifle is fired at a target with the objective of "hitting the bull's-eye," but some (if not all) shots miss the target. There is a "spread" of results, as shown in Figure 3.1. This figure shows two distinct patterns. In the first pattern, (Figure 3.1 [a]), the shots seem to be centered on the bull's-eye, but most of them miss, some by a considerable distance. The second pattern (Figure 3.1 [b]) shows shots that are closer together but not centered on the bull's-eye.

In Figure 3.1(a) it appears that the pattern, although centered on the bull's-eye, exhibits a great deal of **variability**. In contrast, the pattern of Figure 3.1(b) shows less variability, but its location is off center. The distinction between these two patterns illustrates the difference between accuracy and precision. Measurements are said to be **accurate** if their tendency is to center around the actual value of the entity being measured. Measurements are **precise** if they differ from one another by a small amount.

FIGURE 3.1 Accuracy and Precision

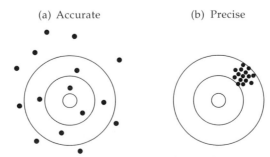

(a) Accurate   (b) Precise

Without *both* accuracy and precision, there is error. But errors resulting from inaccuracy usually have different causes and remedies than do errors resulting from imprecision, or excess variability. For this reason, it is important to know the nature of an error when attempting to take remedial action. For example, at a production station where holes are punched in sheet metal for later insertion of screws, the distance from the edge of the sheet to hole 1 is specified to be 0.5 ± 0.02 cm. When this distance was measured on 20 sheets of punched metal selected from a day's production, it was found that the 20 measurements were nearly identical, but they averaged 0.58 cm, well outside acceptable tolerances. Usually, a simple adjustment in the location of the punch will restore accuracy to the process.

Suppose, instead, it was found that the average distance was 0.498 cm, but that the 20 measurements were considerably different from one another, ranging from 0.43 to 0.58 cm. Identifying the source of this variability can be difficult. Perhaps the punch "wobbles" because it is not properly seated in its holder; perhaps it is dull and "skates" as it makes contact with the metal; perhaps the sheet metal is not held securely in the punch press. There well may be other causes, or a combination of causes, that could explain the observed lack of precision.

This example points out the importance of classifying error into the categories of accuracy and precision. However, the occurrence of one category does not preclude the other. Inaccuracy and imprecision often occur together, and their contributions to error must be separately measured if they are to be controlled efficiently. Measures of accuracy and precision are developed in Section 3.5. The roles of these measures in understanding how data distribute themselves are discussed in Chapter 4, *Probability Distributions.* This knowledge will be applied to the control of industrial processes in Chapter 8, *Statistical Process Control.* Excess variability often is more difficult to correct than lack of accuracy. Methods for separating and identifying sources of variability are useful in correcting excess variability. Such methods are described in Chapter 10, dealing with the design of experiments.

3.2 Exploration of Raw Data

Stem-and-Leaf Displays

When confronted with raw data, often consisting of a long list of measurements, it is difficult to understand what the data are informing us about the process, product, or service which gave rise to them. The following data, giving the response times of 30 integrated circuits (in picoseconds), illustrate this point:

Integrated Circuit Response Times (ps)

4.6	4.0	3.7	4.1	4.1	5.6	4.5	6.0	6.0	3.4
3.4	4.6	3.7	4.2	4.6	4.7	4.1	3.7	3.4	3.3
3.7	4.1	4.5	4.6	4.4	4.8	4.3	4.4	5.1	3.9

Examination of this long list of numbers seems to tell us little other than, perhaps, the response times are greater than 3 ps or less than 7 ps. (If the list contained several hundred numbers, even this information would be difficult to elicit.)

A start at exploring data can be made by constructing a **stem-and-leaf display**. To construct such a display, the first digit of each response time is listed in a column at the left, and the associated second digits are listed to the right of each first digit. For the response-time data, we obtain the following stem-and-leaf display:

3	7 4 4 7 7 4 3 7 9
4	6 0 1 1 5 6 2 6 7 1 1 5 6 4 8 3 4
5	6 1
6	0 0

In this display, each row is a **stem** and the numbers in the column to the left of the vertical line are called **stem labels**. Each number on a stem to the right of the vertical line is called a **leaf**.

The stem-and-leaf display allows examination of the data in a way that would be difficult, if not impossible, from the original listing. For example, it can quickly be seen that there are more response times in the range 4.0 to 4.9 ps than any other, and that the great majority of circuits had response times of less than 5 ps. This method of exploratory data analysis yields another advantage; namely, there is no loss of information in a stem-and-leaf display.

The first two stems of this stem-and-leaf display contain the great majority of the observations, and more detail might be desirable. To obtain a finer subdivision of the data in each stem, a **double-stem display** can be constructed by dividing each stem in half so that the leaves in the first half of each stem are 0, 1, 2, 3, and 4, and those in the second half are 5, 6, 7, 8, and 9. The resulting double-stem display looks like this:

3*f*	4 4 4 3
3*s*	7 7 7 7 9
4*f*	0 1 1 2 1 1 4 3 4
4*s*	6 5 6 6 7 5 6 8
5*	6 1
6*	0 0

The stem labels include the letter *f* (for *first*), to denote that the leaves of this stem are 0–4, and *s* (for *second*) to denote that the leaves are 5–9. The asterisk is used with stem labels 5 and 6 to show that all 10 digits are included in these stems.

Dot Diagrams

The **dot diagram** is another approach to initial data exploration that also preserves all information. To construct a dot diagram of the response-time data, a horizontal axis is drawn with a scale sufficient to cover the full range of the data. Then, a dot

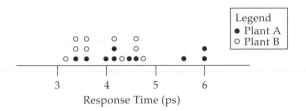

FIGURE 3.2 Dot Diagram

is placed above the line, locating it appropriately on the scale. If two or more observations have the same value, the corresponding dots are lined up vertically.

The dot diagram is especially useful when comparing two sets of data. To illustrate, suppose the first two rows of the response-time data on page 31 represent observations from two different manufacturing plants; the first row (10 observations) comes from plant *A*, represented by the black circles in Figure 3.2, and the second row comes from plant *B*, represented by the white circles in this figure. It appears from Figure 3.2 that plant *B* produced circuits with lower response times than did plant *A*.

Pareto Analysis

Another method of data exploration is useful in the initial stages of quality-improvement programs to prioritize problems. To illustrate, suppose a list of the results in the final inspection of 100 refrigerators shows the following defects:

Defects in the Final Inspection of Refrigerators	
Light inoperative	4
Compressor inoperative	1
Poor door fit	3
Missing insulation	1
Missing/bent shelves	2
Interior finish marred	5
Metal finish—bubbles	11
Metal finish—scratches	6
Metal finish—dents	2

The numbers appearing to the right are the frequencies of occurrence of the different types of defect. To obtain a **Pareto diagram**, the defects are listed in the order of their frequency of occurrence, and the results are plotted, as illustrated in Figure 3.3. The appearance of this figure is not unusual; typically, over half of the defects in a product are accounted for by only two or three elements.

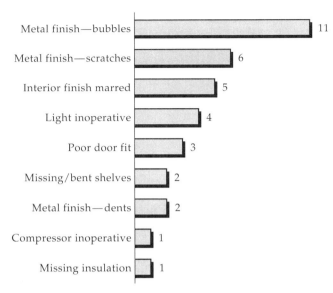

FIGURE 3.3 Pareto Diagram

This method has come to be known as "Pareto analysis" after the Italian economist who found that personal fortunes were very unevenly distributed. His discovery that about 80% of the wealth was controlled by 20% of the population has come to be known as the "80–20 rule," or "Pareto's law." Other social phenomena, such as wealth, crime, and drug use, also are very unevenly distributed in most societies.

It has been found by experience that most potential causes of poor quality are similarly distributed. In the refrigerator example, almost half of the defects discovered on final inspection were of only two kinds—scratches and bubbles in the metal finish. In the pursuit of quality improvement, it would seem reasonable to concentrate on correcting these two problems first; their elimination will reduce the total defect rate by almost 50%.

In quality improvement, there is always the temptation to pursue the easiest problem or the one that might have the most political impact, regardless of its effect on overall quality. This temptation can be avoided by adherence to the tenet of quality: "Pursue the vital few, rather than the trivial many." Thus, a quality-improvement program should begin by making a list of all quality problems, gathering data on the frequency of occurrence of these problems, and performing a Pareto analysis to determine which are the "vital few" whose solution will yield the greatest improvement.

Computer Applications

Examples will be given, using computers, whenever a method is introduced in the text requiring extensive computation. Frequently used statistical software for microcomputers includes MINITAB, SASS, RS-1, MICROSTAT, STATPACK, and others. In this book, MINITAB software will be used to illustrate computer methods.

Several columns of data may be entered with *MINITAB* software. To enter data into the first column, the command SET C1 is given at the prompt MTB >, and then the data are entered, separated by spaces or by commas. Thus, the following sequence of commands and data entries places the response times given on page 31 into column 1:

SET C1

4.6 4.0 3.7 4.1 4.1 5.6 4.5 6.0 6.0 3.4 3.4 4.6 3.7 4.2 4.6 4.7
4.1 3.7 3.4 3.3 3.7 4.1 4.5 4.6 4.4 4.8 4.3 4.4 5.1 3.9

END

MINITAB software can be used to obtain stem-and-leaf displays. The command STEM-AND-LEAF C1 will produce the following display:

STEM-AND-LEAF DISPLAY OF C1
LEAF DIGIT UNIT = .1000

1	3T	3
4	3F	444
8	3S	7777
9	3.	9
14	4*	01111
(2)	4T	23
14	4F	4455
10	4S	6667
5	4.	8
4	5*	1
3	5T	
3	5F	
3	5S	6

In this display, the first column cumulates the numbers of observations from the top and from the bottom. *MINITAB* constructs a five-stem display, dividing each stem unit into five parts, labeled *, T, F, S, and . .

EXERCISES

3.1   How are accuracy and precision measured?

3.2   Measurements made on the thickness of the tin plate at various locations on a sheet of plated steel show excess variation, even though their average is within specifications. Would an adjustment of the time spent in the plating bath solve this problem? Explain.

3.3   The following are the percentages of tin in measurements made on 24 solder joints:

61	63	59	54	65	60	62	61	67	60	55	68
57	64	65	62	59	59	60	62	61	63	58	61

   a. Construct a stem-and-leaf diagram using 5 and 6 as the stem labels.

   b. Construct a double-stem display.

   c. Which is more informative?

3.4 Suppose the first row of 12 observations in Exercise 3.3 came from solder connections made at station 105 and the second row came from station 107. Use a dot diagram to determine whether you should suspect a difference in the soldering process at the two stations.

3.5 Two different lathes turn shafts to be used in electric motors. Measurements made of their diameters (in cm) are:

Lathe A:	1.42	1.38	1.40	1.41	1.39	1.44	1.36	1.42	1.40
Lathe B:	1.47	1.31	1.56	1.33	1.29	1.46	1.28	1.51	

Construct a dot diagram to see if you should suspect that the two lathes are turning out shafts of different diameters.

3.6 If you found a difference between the lathes in Exercise 3.5, which lathe is producing higher-quality product? Why?

3.7 Inspection of 1,000 integrated circuits for 18 circuit parameters shows the following number of instances in which each parameter had a value outside of specification limits:

Parameter:	RG	RI	GDS	CGS	T	LG	LD	PGI	CGI	ZDI
No. Failed:	18	15	4	21	58	20	7	131	24	212

   a. Use Pareto analysis to determine which parameter should be investigated first.

   b. What percentage of all failures are represented by this parameter?

3.8 Suppose the ZDI and PGI problems (Exercise 3.7) have been solved so that the occurrence of failures of these parameters has become negligible. Do a new Pareto analysis and determine which parameter(s) should be investigated next.

3.9 Use *MINITAB* or some other computer software to construct a stem-and-leaf display for the combined data of Exercise 3.5.

3.10 Use *MINITAB* or some other computer software to construct a stem-and-leaf display for the following data representing the time to make coke (in hours) in successive runs of a coking oven.

7.8	9.2	6.4	8.2	7.6	5.9	7.4	7.1	6.7	8.5
10.1	8.6	7.7	5.9	9.3	6.4	6.8	7.9	7.2	10.2
6.9	7.4	7.8	6.6	8.1	9.5	6.4	7.6	8.4	9.2

The following exercises introduce new material, expanding on some of the topics covered in the text.

3.11 *Stem labels with two or more digits.* The following are temperatures (in °F) in successive runs of an oven used for drying enameled metal:

124	135	115	129	141	119	137	145	128	
140	136	128	117	133	146	127	130	124	132

Construct a stem-and-leaf display, using 11, 12, 13, and 14 as the stem labels.

3.12 *Continuation.* Would a double-stem display have produced a more desirable stem-and-leaf display in Exercise 3.11? Why?

3.3 Frequency Distributions

On page 35 the five-stem display shows how the leaves are distributed among the stems of the stem-and-leaf display. In discussing Pareto analysis, the word *distributed* was used to describe how data are arranged with respect to different kinds of defects. In both cases, the data are distributed by some attribute or category. In the second case, the data are categorized by a non-numerical attribute, and the resulting distribution is an example of a **categorical distribution**. In the first case, the categories (stems) are numerical, and the stem-and-leaf diagram is an example of a **numerical distribution**.

Example. Fifty-six accidents at a steel mill were grouped according to severity. The following accident summary is a further example of a categorical distribution:

Minor—no time lost	23
Minor—time lost	16
Major—no permanent disability	7
Major—permanent disability	9
Fatal	1

Numerical data can be grouped according to their values in several other ways in addition to stem-and-leaf displays or dot diagrams. A **frequency distribution** groups numerical data into classes having definite lower and upper limits. The construction of a frequency distribution is easily facilitated with a computer program such as *MINITAB* (see page 35). The following discussion may be omitted if a computer program is used to construct frequency distributions.

To construct a frequency distribution, first a decision is made about the number of classes to use in grouping the data. The number of classes can be chosen to make the specification of upper and lower class limits convenient. Generally, the number of classes should increase as the number of observations becomes larger, but it is rarely helpful to use fewer than 5 or more than 15 classes.

The smallest and largest observations that can be put into each class are called the **class limits**. In choosing class limits, it is important that the classes do not overlap, so there is no ambiguity about which class contains any given observation. Also, enough classes should be included to accommodate all observations. Finally, the observations are tallied to determine the **class frequencies**, the number of observations falling into each class. The steps in creating a numerical distribution can be summarized as follows:

1. Decide on the number of classes to be used.
2. Choose non-overlapping, equally spaced class limits that include all the observations.
3. Find the class frequencies by tallying the observations.

Example. Construct a frequency distribution of the following compressive strengths (in psi) of concrete samples, given to the nearest 10 psi:

4890	4830	5490	4820	5230	4860	5040	5060	4500	5260
4610	5100	4730	5250	5540	4910	4430	4850	5040	5000
4600	4630	5330	5160	4950	4480	5310	4730	4700	4390
4710	5160	4970	4710	4430	4260	4890	5110	5030	4850
4820	4550	4970	4740	4840	4910	5200	4880	5150	4890
4900	4990	4570	4790	4480	5060	4340	4830	4670	4750

Solution. Since the smallest observation is 4260 and the largest is 5540, it will be convenient to choose seven classes, having the class limits 4200–4390, 4400–4590, . . . , 5400–5990. (Note that class limits of 4200–4400, 4400–4600, etc., are not used because they would overlap and assignment of 4400, for example, would be ambiguous; it could fit into either of the first two classes.) The following table exhibits the results of tallying the observations; that is, counting the number that fall in each class:

Class Limits		Tally	Frequency
4200	– 4390	///	3
4400	– 4590	⫫⫫ //	7
4600	– 4790	⫫⫫ ⫫⫫ //	12
4800	– 4990	⫫⫫ ⫫⫫ ⫫⫫ ////	19
5000	– 5190	⫫⫫ ⫫⫫ /	11
5200	– 5390	⫫⫫ /	6
5400	– 5590	//	2
	Total		60

Although the tally was shown in the foregoing example, it is customary to omit the tally when presenting frequency distributions in their final form. The tally will not be shown in the remainder of the text. Note that the class limits in this example were rounded, as were the original data, to the nearest 10 psi. In general, class limits are always given to the same number of decimal places or significant digits as the original data. If the original data had been given to the nearest pound per square inch, the class limits 4200–4399, 4400–4599, and so forth, would have been used.

The midpoint between the upper class limit of a class and the lower class limit of the next class is called a **class boundary**. Class boundaries, rather than class marks, are used in constructing cumulative distributions (Exercise 3.21, page 41). The interval between successive class boundaries is called the **class interval**; it can also be defined as the difference between successive lower class limits or successive upper class limits. (Note that the class interval is *not* obtained by subtracting the lower class limit of a class from its upper class limit.) A class can be represented by a single number, called the **class mark**. This number is calculated for any class by averaging its upper and lower class limits.

Once data have been grouped into a frequency distribution, each observation in a given class is treated as if its value is the class mark of that class. In so doing, its actual value is lost; it is known only that its value lies somewhere between the class limits of its class. Such an approximation is the price paid for the convenience of working with a frequency distribution. It will be observed in Section 3.5 that the approximation usually causes acceptable errors in calculating descriptive measures.

> *Example.* For the frequency distribution of compressive strengths of concrete given in the example on page 38, find (a) the class boundaries, (b) the class interval, and (c) the class mark of each class.

> *Solution.*   a. The class boundaries of the first class are 4195 4395. The class boundaries of the second through the sixth classes are 4395–4595, 4595–4795, 4795–4995, 4995–5195, and 5195 –5395, respectively. The class boundaries of the last class are 5395–5595. Note that the lower class boundary of the first class is calculated as if there were a class below the first class, and the upper class boundary of the last class is calculated as if there were a class above it. Also note that, unlike class limits, the class boundaries overlap.

> b. The class interval is 200, the difference between the upper and lower class boundaries of any class. It also can be found by subtracting successive lower class limits, for example $4400 - 4200 = 200$ psi, or by subtracting successive upper class limits, for example, $4590 - 4390 = 200$.

> c. The class mark of the first class is $(4200 + 4390)/2 = 4295$; it is $(4400 + 4590)/2 = 4495$ for the second class; and the class marks are 4695, 4895, 5095, 5295, and 5495 for the remaining five classes. Note that the class in-

terval, 200, also is given by the difference between any two successive class marks.

EXERCISES

3.13 The following are the drying times (minutes) of 100 sheets coated with polyurethane under various ambient conditions:

45.6	50.3	55.1	63.0	58.2	65.5	51.1	57.4	60.4	54.9
56.1	62.1	43.5	63.8	64.9	59.9	63.0	67.7	53.8	57.9
61.8	52.2	61.2	51.6	58.6	73.8	53.9	64.1	57.2	75.4
55.9	70.1	46.2	63.6	56.0	48.1	62.2	58.8	50.8	68.1
51.4	73.9	66.7	42.9	71.0	56.1	60.8	58.6	70.6	62.2
59.9	47.5	72.5	62.0	56.8	54.3	61.0	66.3	52.6	63.5
64.3	63.6	53.5	55.1	62.8	63.3	64.7	54.9	54.4	69.6
64.2	59.3	60.6	57.1	68.3	46.7	73.7	56.8	62.9	58.4
68.5	68.9	62.1	62.8	74.4	43.8	40.0	64.4	50.8	49.9
55.8	66.8	67.0	64.8	57.6	68.3	42.5	64.4	48.3	56.5

Construct a frequency distribution of these data, using eight classes.

3.14 Eighty pilots were tested in a flight simulator and the time for each to take corrective action for a given emergency was measured in seconds, with the following results:

11.1	5.2	3.6	7.6	12.4	6.8	3.8	5.7	9.0	6.0	4.9	12.6
7.4	5.3	14.2	8.0	12.6	13.7	3.8	10.6	6.8	5.4	9.7	6.7
14.1	5.3	11.1	13.4	7.0	8.9	6.2	8.3	7.7	4.5	7.6	5.0
9.4	3.5	7.9	11.0	8.6	10.5	5.7	7.0	5.6	9.1	5.1	4.5
6.2	6.8	4.3	8.5	3.6	6.1	5.8	10.0	6.4	4.0	5.4	7.0
4.1	8.1	5.8	11.8	6.1	9.1	3.3	12.5	8.5	10.8	6.5	7.9
6.8	10.1	4.9	5.4	9.6	8.2	4.2	3.4				

Construct a frequency distribution of these data.

3.15 Find the class boundaries, the class interval, and the class marks of the frequency distribution constructed in Exercise 3.13.

3.16 Find the class boundaries, the class interval, and the class marks of the frequency distribution constructed in Exercise 3.14.

3.17 The following are the number of highway accidents reported on 30 successive days in a certain county:

6	4	0	3	5	6	2	0	0	12	3	7	2	1	1
0	4	0	0	0	1	8	0	2	4	7	3	6	2	0

Construct a frequency distribution of these data. Identify the class boundaries, the class marks, and the class interval.

The following exercises introduce new material, expanding on some of the topics covered in the text.

3.18 *Percentage distributions.* A **percentage distribution** is obtained from a frequency distribution by replacing each frequency by 100 times the ratio of that frequency to the total frequency. Construct a percentage distribution using the reaction-time data of Exercise 3.14.

3.19 *Continuation.* Construct a percentage distribution using the drying-time data of Exercise 3.13.

3.20 *Continuation.* Percentage distributions are useful in comparing two frequency distributions having different total frequencies. Construct percentage distributions from the following two frequency distributions and determine whether the distributions of daily absences in the two departments follow similar patterns.

	FREQUENCIES	
Class Limits	*Shipping Department*	*Security Department*
0 – 1	26	18
2 – 3	18	11
4 – 5	10	7
6 – 7	4	3
8 – 9	2	1
Totals	60	40

3.21 *Cumulative distributions.* A **cumulative distribution** is constructed from a frequency distribution by replacing each frequency with the sum of the frequency of the given class and the frequencies of all classes above it, and representing each class by its upper class boundary. Construct a cumulative distribution using the data of Exercise 3.13.

3.22 *Continuation.* Construct a cumulative distribution using the data of Exercise 3.14.

3.23 *Continuation.* Construct cumulative percentage distributions from the frequency distributions of absences given in Exercise 3.20.

3.24 *Unequal class intervals.* The small number of observations greater than 7 in Exercise 3.17 may cause some difficulty in constructing a frequency distribution. To keep class intervals equal, one is faced with the dilemma of creating either too many classes for only 30 observations, or using a small number of classes with excessive loss of information in the first few classes. In such cases, one is tempted to drop the rule of equal-size classes, using a larger interval for the last class.

a. If that were done, what would the resulting frequency distribution become?

b. Is there a unique class interval?

3.25 *Continuation.* The following are the times to failure of 38 light bulbs, given in hours of operation.

150	389	345	310	20	310	175	376	334	340
332	331	327	344	328	341	325	2	311	320
256	315	55	345	111	349	245	367	81	327
355	309	375	316	336	278	396	287		

a. Dropping the rule that class intervals must be equal, construct a frequency distribution from these data.

b. Can you find the class mark of every class?

3.4 Graphs of Frequency Distributions

The "shape" of a frequency distribution can play an important role in characterizing data. (This point will be developed in greater detail in Section 3.6). Graphs representing frequency distributions are capable of revealing much about the structure of data, diagnosing quality problems, and verifying assumptions needed to perform statistical tests.

Histograms

Among the most widely used graphs of frequency distributions are **histograms**. A histogram is used to represent **continuous data**, that is, data that theoretically can take on any value in an interval. Histograms represent each class by means of a bar whose area is proportional to the class frequency and whose width equals the class interval. As histograms are easily constructed by computer programs such as *MINITAB* (see page 35), the reader may wish to omit the following discussion involving the compressive-strength data given in the example on page 38.

To construct a histogram representing these data, first a line segment is drawn; then points representing the class boundaries of each class are marked off on this line, namely 4195, 4395, 4595, . . . , 5595. Between each successive pair of

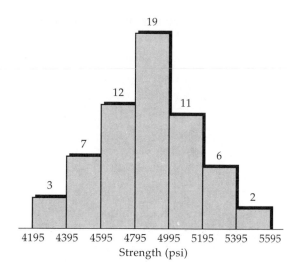

FIGURE 3.4   Histogram of Compressive Strengths

boundaries, a bar is drawn whose *area* is proportional to the frequency of the corresponding class. Since the class intervals are equal in this case, it is sufficient to draw bars whose *heights* are proportional to the class frequencies. The resulting histogram is shown in Figure 3.4.

This histogram shows that the data are concentrated near the middle of the interval from 4195 to 5595. The class frequencies decline more-or-less symmetrically on either side of the center. The shape of the histogram also resembles that of the cross-section of a bell. **Bell-shaped distributions** of data are special kinds of **symmetric distributions**, and they occur frequently in many applications. An important special class of bell-shaped theoretical distributions will be studied in Chapter 5.

Not all histograms are bell-shaped. To illustrate, suppose a wire is soldered to a board and pulled with continuously increasing force until the bond breaks. The forces (in grams) required to break the solder bonds are as follows:

Force Required To Break Solder Bonds (grams)									
19.8	13.9	30.4	16.4	11.6	6.9	14.8	21.1	13.5	5.8
10.0	17.1	8.5	16.6	23.3	12.1	18.8	15.4	14.4	23.8
14.2	26.7	11.8	22.9	12.6	19.8	7.0	10.7	12.2	17.7
19.0	14.9	24.0	12.0	9.1	13.8	18.6	26.0	17.4	13.3

As the first step in drawing a histogram, the following frequency distribution is constructed:

Breaking Strength (grams)	Frequency
5.0 – 9.9	5
10.0 – 14.9	16
15.0 – 19.9	11
20.0 – 24.9	5
25.0 – 29.9	2
30.0 – 34.9	1
Total	40

The class boundaries are 4.95, 9.95, . . . , 34.95, and the corresponding histogram is shown in Figure 3.5. Notice that the histogram shown in this figure is not bell-shaped or even symmetric. This histogram exhibits a right-hand "tail," suggesting that while most of the solder bonds had moderate or low breaking strengths, a few had breaking strengths that were much greater than the rest. Histograms with a long tail on the right or a long tail on the left are said to be **skewed**. A histogram exhibiting a long right-hand tail is said to have **positive skewness** and a histogram with a long left-hand tail has **negative skewness**. Positive skewness is not rare in engineering or quality data. Distributions of product lifetimes and the results of many kinds of stress tests often exhibit asymmetric distributions. It is also common for distributions of incomes to have positive skewness; this

FIGURE 3.5 Histogram of Solder-Bond Strengths

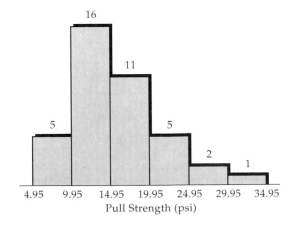

observation is consistent with the original observation that gave rise to Pareto's law (page 34).

It is tempting to combine the last two classes in the frequency distribution of solder-bond strengths. The combined class then will have a class interval of 10 grams, *twice* that of the other classes. In order that all classes continue to be represented by bars whose *areas* are proportional to the class frequencies, the height of the corresponding bar in the histogram of the new distribution should be proportional to 1.5, half the frequency of the combined class. When two or more classes are combined, class limits, class boundaries, and class marks change, but they remain uniquely defined. It remains possible to draw a histogram and the methods for grouped data, to be described in Section 3.5, remain valid.

Some data collections include a few observations that are very much larger or smaller than the main body of data. To accommodate such data in a frequency distribution, either the data must be grouped coarsely, or several classes having 0 frequencies are required. Open class intervals, such as "25.0 or greater," or "less than 10" sometimes are employed in such cases. When open class intervals are used, it is not possible to draw a histogram, and the methods of Section 3.5 for grouped data are inapplicable. For these reasons, open class intervals should be avoided.

Bar Charts

As was pointed out on page 42, histograms are used to represent *continuous* numerical data. Some numerical data can take on only certain values but not all values on a segment of the real line. Such numerical data are called **discrete data**. For example, the number of absences per day, shown in the frequency distributions of Exercise 3.20 on page 41, can take on only the integer values 0, 1, 2, . . . For discrete data, and also for categorical data, graphical representations very much like histograms are used. Such a graph is called a **bar chart**.

The procedure for constructing a bar chart is similar to that of constructing a histogram, with the exception that the bars are drawn between the class limits rather than the class boundaries. Thus, the bars are separated by gaps. For numerical data, the class limits could be labeled, or one could label each bar with a description of the corresponding class. For example, for the data on absences, the first bar could have been labeled "0 or 1," and so forth. A bar chart for the shipping-department absence data, given in Exercise 3.20 on page 41, is shown in Figure 3.6. Where there are no numerical class limits, such as for the refrigerator data on page 33, verbal descriptions must be used. For example, the first bar might be labeled "light inoperative," and successive bars would be labeled with the appropriate defect.

Computer Applications

An illustration of the construction of histograms with *MINITAB* software makes use of the response-time data given on page 31. It is assumed that these data have

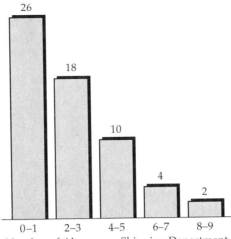

Number of Absences – Shipping Department

FIGURE 3.6 Bar Chart

been entered into column 1 (page 35). The command HISTOGRAM C1 will yield the following:

MIDDLE OF INTERVAL	NUMBER OF OBSERVATIONS	
3.2	1	*
3.6	7	* * * * * * *
4.0	6	* * * * * *
4.4	6	* * * * * *
4.8	6	* * * * * *
5.2	1	*
5.6	1	*
6.0	2	* *

EXERCISES

3.26 a. Construct a histogram of the reaction times of pilots from the data in Exercise 3.14 on page 40.

b. What can be said about the shape of this histogram?

3.27 a. Construct a histogram of the drying times of polyurethane from the data in Exercise 3.13 on page 40.

b. What can be said about the shape of this histogram?

3.28 Use the data of Exercise 3.26 to illustrate that class marks are given by the midpoint between successive class boundaries as well as the midpoint between successive class limits.

3.29 Using the data of Exercise 3.27, show that the class marks also are given by the midpoint between successive class boundaries.

3.30 Construct a histogram using the solder-joint data in Exercise 3.3 on page 35.

3.31 Construct a bar chart for the refrigerator-inspection data given on page 33.

3.32 Construct a bar chart representing the data on integrated circuits given in Exercise 3.7 on page 36.

3.33 a. Construct a bar chart representing the accident data of the example on page 37.

     b. Why should a bar chart be used rather than a histogram?

3.34 a. Construct a bar chart representing the frequency distribution of shipping department absences given in Exercise 3.20 on page 41.

     b. Would it have made sense to use a histogram instead? Why?

3.35 a. Combining the first two rows of the data for the response times given on page 31, construct a histogram.

     b. How would you describe the shape of the histogram?

     c. Referring to the dot diagram of Figure 3.2, what can you say about the relationship between the shape of the histogram and what you observe in the dot diagram?

3.36 a. Combining the data for both lathes in Exercise 3.5 on page 36, construct a histogram.

     b. How would you describe the shape of the histogram?

     c. If you have not already done so, make a dot diagram representing the combined data. How are the dot diagram and the shape of the histogram related?

3.37 Use *MINITAB* or some other computer software to construct a histogram of the oven temperatures given in Exercise 3.11 on page 36.

3.38 Use *MINITAB* or some other computer software to construct a histogram of the coking-time data given in Exercise 3.10 on page 36.

3.39 Use *MINITAB* or some other computer software to construct a histogram of the drying-time data in Exercise 3.13 on page 40.

The following exercises introduce new material, expanding on some of the topics covered in the text.

3.40 *Frequency polygons.* A plot of the points (x, f), where x represents the class mark of a given class in a frequency distribution and f represents its frequency, is called a **frequency polygon**. Construct a frequency polygon using the data in Exercise 3.14 on page 40.

3.41 *Continuation.* Construct a frequency polygon from the data in Exercise 3.13 on page 40.

3.42 *Ogives.* A plot of the cumulative frequency (see Exercise 3.21 on page

41) on the y axis and the corresponding upper class boundary on the x axis is called an **ogive.**

    a. Construct an ogive for the drying times given in Exercise 3.14 on page 40.

    b. Construct an ogive representing the cumulative percentage distribution.

3.43 *Continuation.*

    a. Construct an ogive for the data of Exercise 3.13 on page 40.

    b. Using the same set of axes, re-label the y axis so that the same graph also shows the ogive of the percentage distribution of reaction times.

3.44 *Unequal class intervals.*

    a. Construct a frequency distribution and the corresponding histogram for the data of Exercise 3.25 on page 42, using the class limits 0–49, 50–99, . . . , 350–399. Note the concentration of data in only one class.

    b. Now, change the class limits to 0–99, 100–199, 200–299, 300–324, 325–349, and 350–399. Re-draw the histogram using bar heights proportional to the class frequencies.

    c. Does this new histogram correctly represent the distribution? Why?

3.45 *Continuation.*

    a. Re-draw the second histogram in Exercise 3.44 so the *areas* of the bars are proportional to the class frequencies.

    b. Does this histogram more fairly represent the distribution?

3.5   Descriptive Measures

In the previous section, it has been shown that a frequency distribution can be characterized somewhat by its shape, and that this shape becomes more evident from a histogram or a bar chart. Numerical measures, as well as diagrams, are useful in describing a data set. In this section, several frequently used numerical measures will be introduced to describe the location and variation of data.

The Mean

To introduce the most common measure of location, imagine that a histogram has been cut out of a piece of cardboard, as illustrated in Figure 3.7. The balance point of the histogram can be found by calculating the x coordinate of the centroid of the cardboard shape. This coordinate is defined to be

$$\bar{x} = \frac{\sum\limits_{i=0}^{K} x_i \cdot m_i}{\sum\limits_{i=0}^{K} m_i}$$

where K is the number of classes, x_i is the center of bar i, and m_i, is its mass. The centers of the bars are represented by the class marks and, if the cardboard has uniform mass, the masses of the bars, m_i, are proportional to the class frequencies, f_i. By analogy, then, the centroid of data which have been grouped into a frequency distribution is given by

$$\text{Mean (grouped data): } \quad \bar{x} = \frac{\sum\limits_{i=1}^{K} x_i f_i}{\sum\limits_{i=1}^{K} f_i}$$

The **mean** of a frequency distribution seems like a reasonable measure of the location or "center" of the data underlying that distribution. Note that the mean is measured in the same units as those of the original data.

FIGURE 3.7 "Balance Point" of Histogram

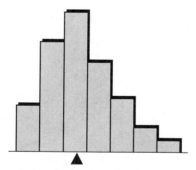

Example. Find the mean of the coking-time data in Exercise 3.10 on page 36.

Solution. First, the data are grouped into a frequency distribution. To compute the mean, three columns are formed, as follows:

Class Marks, x_i	Class Frequencies, f_i	$x_i f_i$
5.45	2	10.90
6.45	7	45.15
7.45	10	74.50
8.45	5	42.25
9.45	4	37.80
10.45	2	20.90
Totals	30	231.50

Division of 231.50 by 30 produces the result $\bar{x} = 7.72$ hours, rounded to two decimal places.

The mean was rounded to two decimal places in this example because the original data were given to the nearest 0.1 hour. For large sets of data, it is reasonable to round the mean to one more significant digit than the original data.

On page 39 it was noted that every observation in a given class is treated as if its value were equal to the corresponding class mark. This approximation is inherent whenever data are grouped into a frequency distribution. To calculate the mean of a set of data *exactly*—that is, without the approximation intrinsic in grouping data—one has only to imagine that each individual observation constitutes its own class, having the frequency 1. Then $K = n$, where n is the total number of observations, and each $f_i = 1$. Thus, the equation for the centroid of ungrouped data becomes

$$\text{Mean (ungrouped data):} \quad \bar{x} = \frac{\sum_{i=1}^{n} x_i}{n}$$

Applying this formula to the raw (ungrouped) coking-time data given in Exercise 3.10 on page 36, we obtain

$$\bar{x} = (7.8 + 9.2 + 6.4 + \ldots + 9.2)/30 = 7.76.$$

Notice the closeness of the mean 7.72, computed from grouped data, to the exact value, 7.76. (The error introduced into the mean by approximating the exact data with a frequency distribution is only about 0.5 percent in this case.)

The Median

While the mean is generally preferred as a *measure* of location, for some distributions the mean gives a poor *description* of the center of a data set. For example, suppose a department in a plant includes ten persons, eight production-line workers, a foreman, and an engineer. Assume further that the eight workers have annual gross incomes (to the nearest thousand dollars) of 16, 15, 18, 13, 17, 14, 20, and 15, the foreman earns $25,000, and the engineer earns $42,000. (Exercise 3.46 on page 57 asks to show that the mean salary of the department is $19,500.) Eight of the ten members of the department earn less than the mean, while only two earn more! The mean always has the desirable property that it is the centroid of the data, but in this example, it is a poor *descriptive* measure.

A better measure for the purpose of description would be the value in the "middle," such that half the observations exceed this value, and half are less. The middle value is called the **median**, and, like the mean, it is measured in the same units as the original observations.

To calculate the median, the observations are ranked from smallest to largest, and the observation is found that occupies the middle rank. If the number of observations is odd, there will be a unique middle one; but if n is even, there are two middlemost observations, and the median is given by their mean.

Example. Find the median of the salary data on page 50.

Solution. First, the data are ranked as follows:

Rank:	1	2	3	4	5	6	7	8	9	10
Observation:	13	14	15	15	16	17	18	20	25	42

Since $n = 10$ is an even number, the mean of the two middlemost observtions, the fifth and sixth largest, is calculated, obtaining the value $(16 + 17)/2 = 16.5$, or \$16,500, for the median.

A word of caution—it is easy to slip into the mistake of calculating the median to be the middle *rank* (or the mean of the two middle ranks) instead of using the *observation* that occupies that rank. In this example, the median (in thousands of dollars) is 16.5, not 5.5! Also, it should be pointed out that the median and the mean are identical for perfectly symmetric distributions; the median is less than the mean for most distributions that are skewed to the right, and greater than the mean for most left-hand skewed distributions. Although some exceptions can be found, a comparison of the mean and the median of a set of data usually gives a relatively quick indicator of its skewness.

Variance and Standard Deviation—Ungrouped Data

Knowledge of the location of a data set tells us only where its center is located on the scale of numbers, and nothing about its variability. To illustrate, the following data are presented, giving the thickness of the chrome plating (in mm) at twelve different locations on two different automobile bumpers:

						Thickness of Chrome Plating (mm)					
A:	1.5	2.1	1.8	1.4	1.0	1.7	1.9	2.2	1.9	1.7	2.3	1.5
B:	1.8	1.7	1.6	1.7	1.8	1.6	1.9	1.8	1.7	1.8	1.9	1.7

The reader is asked to verify that the mean of each data set is 1.75 mm. However, one set of data is more variable than the other, as shown by the difference in the "spreads" of the two data sets in the dot diagram in Figure 3.8.

The idea of concentration of data about their mean can be employed to construct a useful measure of variation, basing this measure on the "distance" of the

FIGURE 3.8   Difference in Variability

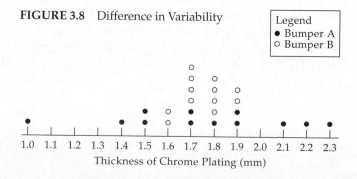

Thickness of Chrome Plating (mm)

observations from their mean. To be specific, the quantity $(x_i - \bar{x})$ is defined to be the **deviation** of observation i from the mean. Variation could be measured by finding the mean value of these deviations, except for the fact that their sum is zero. The proof of this statement is straight-forward. First, each term can be summed separately. Then, since $\bar{x}$ equals $1/n$ times the sum of all values of x_i, the first sum can be written as $n\bar{x}$. In summing the second term, $\bar{x}$ is added n times, also obtaining $n\bar{x}$. The difference of these terms is zero. Thus, it has been proved that the positive and negative deviations always cancel each other to produce the sum of zero.

This cancellation can be avoided by squaring each deviation and measuring variation by finding the mean of the *squared* deviations. But, now another issue presents itself. How many *different* deviations really are averaged to find the mean of the squared deviations? If any $n - 1$ deviations are given and all n deviations must sum to zero, the remaining deviation is determined. Thus, there are actually only $n - 1$ *independent* deviations. For this reason (and other, theoretical reasons) the mean of the squared deviations is obtained by dividing by $n - 1$ instead of by n. To measure variation, then, the following formula is used.

$$\text{Defining Formula for the Variance: } \quad s^2 = \frac{\sum_{i=1}^{n}(x_i - \bar{x})^2}{n - 1}$$

This measure of variation, denoted by s^2, is called the **variance**.

If the observations are measured in, say, inches, their variance will be measured in square inches. To return to the original metric, it is necessary to take the square root of the variance. The resulting *root-mean-squared deviation* is called the **standard deviation,** and this measure of variability is given by

$$\text{Standard Deviation: } \quad s = \sqrt{\frac{\sum_{i=1}^{n}(x_i - \bar{x})^2}{n - 1}}$$

Computation of the variance or the standard deviation can be tedious, especially for large data sets. First, it is necessary to find the mean. Then, the deviation from the mean must be calculated for each observation. Finally, these deviations are squared and summed, the resulting sum is divided by n -1, and the square root is taken. Modern computers take most of the tedium out of this task. However, when only a calculator is available, the calculation of s^2 and s can be made somewhat less daunting with the use of the following algebraically equivalent formula for the variance:

$$\text{Computing Formula for the Variance: } \quad s^2 = \frac{n\sum_{i=1}^{n}x_i^2 - \left(\sum_{i=1}^{n}x_i\right)^2}{n(n - 1)}$$

Example.   Use the computing formula for the variance to find the standard deviation of the coking-time data given in Exercise 3.10 on page 36.

Solution.   To use this formula, first find the sum of the observations, and the sum of their squares, obtaining

$$\sum_{i=1}^{30} x_i = 232.8, \qquad \sum_{i=1}^{30} x_i^2 = 1847.16$$

Substituting these values into the computing formula, the result

$$s^2 = \frac{(30)\,(1847.16) - (232.8)^2}{(30)\,(29)} = 1.40$$

is obtained. The standard deviation is the square root of the variance, or $s = 1.18$.

Variance and Standard Deviation—Grouped Data

The standard deviation also can be found from grouped data, data that have been grouped into a frequency distribution. The procedure is similar to that shown on page 49 for calculating the mean of grouped data, except that a fourth column is added for the purpose of cumulating the sum of *squares*.

Example.   Find the standard deviation of the data whose mean has been computed from grouped data in the example on page 49.

Solution.   Adding the column $x^2 f$, the table on page 49 becomes:

Class Marks, x_i	Class Frequencies, f_i	$x_i f_i$	$x_i^2 f_i$
5.45	2	10.90	59.4050
6.45	7	45.15	291.2175
7.45	10	74.50	555.0250
8.45	5	42.25	357.0125
9.45	4	37.80	357.2100
10.45	2	20.90	218.4050
Totals	30	231.50	1,838.2750

Substituting these totals into the computing formula for the variance, $s^2 = [(30)(1,838.275) - (231.5)^2]/(30)(29) = 1.789$. The standard deviation is found by taking the square root of the variance, obtaining $s = 1.34$.

Note that as many significant digits as possible were carried in the intermediate calculations; rounding the result to two decimal places occurs only at the end. Even the computing formula for the variance, which has less error propagation than the defining formula, is prone to the propagation of round-off errors.

Also note that the error associated with grouping in this example is somewhat larger for the standard deviation than it was for the mean. In the previous example, $s = 1.18$ was obtained for the ungrouped data, whereas $s = 1.34$ was obtained for the grouped data; the error in the standard deviation associated with grouping is almost 14% in this case. Errors of this magnitude in estimating standard deviations are rarely of concern in practical applications.

The Range

One might ask: "Why go through all this effort to calculate the standard deviation when a simple and straightforward measure of variability is given by the difference between the largest and the smallest observations?" This measure, sometimes used to describe variability, is called the **range** of the data. The value of the range, however, is entirely dependent on only two of the observations, whereas the standard deviation makes use of them all. Thus, if daily measurements are made of solids suspended in the water of a reservoir, the range might be 5 parts per million for the 30 observations taken in one month, and 25 ppm the next month. One erroneous assay or one unusual measurement (perhaps taken on the day of a bad storm) could have a much greater effect on the range than on the standard deviation.

For very small data sets, up to about five observations, the range will do almost as good a job of measuring variation as the standard deviation. In fact, for two observations, one is simply a constant multiple of the other (see Exercise 3.60 on page 57). When statistical process control is studied in Chapter 8, it will be found that decisions often are made on the basis of many data groups, each of small size, and it is traditional to use the range to measure variability. In such cases, and *only* in such cases, the ease of calculation of the range sometimes outweighs the minor loss in its precision relative to that of the standard deviation. In the age of the ubiquitous computer, even this argument has lost much of its force.

Box-and-Whisker Plots

The distribution of a set of data can be summarized graphically using descriptive measures. Such a data summarization is called a **box-and-whisker plot** and, like the dot diagram, it starts with a horizontal number line. This plot places a box, centered at the "center" of the data and including the central portion of the observations within the box confines. Then, the plot indicates the full range of the data by placing horizontal lines, or "whiskers" on each end of the box.

The box-and-whisker plot is designed to be insensitive to unusually large or small observations. Such observations, called **outliers**, are sufficiently far from the main body of data that they are believed not to be part of the process from which the main data originated. They may have resulted from measurement errors or transcription errors, or some other process unlike that which gave rise to the remaining data.

The first step in constructing a box-and-whisker plot is to identify the *upper* and *lower fourths* of the data. If n, the number of observations is even, the upper fourth is the median of the $n/2$ largest observations and the lower fourth is the median of the $n/2$ smallest observations. If n is odd, these are the medians of the largest and smallest $(n + 1)/2$ observations, respectively. Then, the *fourth spread*, given by

$$f_s = \text{upper fourth} - \text{lower fourth}$$

is calculated. Using a horizontal measurement axis, a box is drawn extending from the lower fourth to the upper fourth. The "center" of the box is indicated by a vertical line segment placed at the median of all the data, as shown in Figure 3.9. Then, a line is drawn from the smallest observation to the left-hand end of the box, *as long as the smallest observation is within a distance of 1.5f_s from the left-hand end of the box*. Similarly, a line is drawn to the right of the box to the largest observation that is within a distance of $1.5f_s$ from the right-hand end of the box.

Any observation that falls outside a distance of $1.5f_s$ from the closest edge of the box is regarded as an outlier. If an outlier lies within $3f_s$ of the nearest edge of the box, it is regarded as a *mild outlier* and is indicated by a white circle; otherwise it is called an *extreme outlier*, indicated by a black circle. Figure 3.9 shows a box-and-whisker plot with one "mild" outlier to the left of the box and an "extreme" outlier to its right.

Box-and-whisker plots give a visual indication of the location and dispersion of data that is relatively insensitive to outliers, and they are relatively easy to construct. Such plots are especially useful in making visual comparisons of two or more sets of data.

Example. Construct a box-and-whisker plot representing the following 20 measurements made of the thickness of cold-rolled steel plates (in cm):

0.94	0.88	0.91	1.02	0.89	0.98	0.95	0.94	0.80	0.99
1.06	0.87	0.89	0.93	0.83	0.94	1.26	0.99	1.01	0.93

FIGURE 3.9   Typical Box-and-Whisker Plot

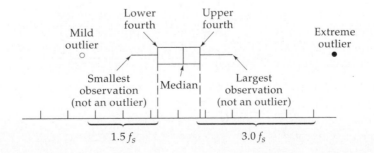

Solution.   There are 20 observations; the upper fourth of the data is the median of the ten largest observations, or 0.99. The lower fourth of the data is the median of the ten smallest observations, or 0.89. The median of all 20 observations is the mean of the 10th and 11th largest observations, or $(0.94 + 0.94)/2 = 0.94$. Thus, the box extends from 0.89 to 0.99, and the "center" is placed at 0.94, as shown in Figure 3.10. Next, we calculate $f_s = 0.99 - 0.89 = 0.10$, $1.5f_s = 0.15$, and $3f_s = 0.30$. No observation has a value less than $0.89 - 0.15 = 0.74$. Thus, the left-hand "whisker" extends from the smallest observation (0.80) to the left-hand end of the box. However, the observation 1.26 exceeds $0.99 \pm 0.15 = 1.14$. Thus, the right-hand whisker extends from the right-hand end of the box to 1.06, the next largest observation. A white circle is placed at 1.26 to indicate a "mild" outlier at that point.

FIGURE 3.10   Box-and-Whisker Plot for Steel-Thickness Data

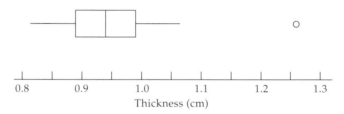

Computer Applications

The response-time data have already been set into column 1, using *MINITAB* software (page 35). If the command DESCRIBE C1 is given, the resulting computer output will include the following:

N	30
MEAN	4.317
MEDIAN	4.250
STDEV	0.701
MAX	6.000
MIN	3.300
Q3	4.600
Q1	3.700

The values Q_1 and Q_3 are the first and third quartiles of the data, which will be defined in Exercise 3.69 on page 58.

A box-and-whisker plot also can be obtained with *MINITAB*. First, we enter the data with, for example, the command SET C1. Then, we give the command BOXPLOT C1. Using *MINITAB* to obtain a boxplot of the data given in the example on page 55, we obtain the following BOXPLOT:

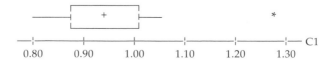

EXERCISES

3.46 Calculate the mean of the salaries given on page 50.

3.47 a. Find the mean of the ungrouped compressive-strength data given in the example on page 38.

   b. Using the frequency distribution of these data given on page 38, find the mean of the grouped data. How do these means compare?

3.48 a. Find the mean of the integrated-circuit response times given on page 31.

   b. Group the data into a frequency distribution, find the mean from the grouped data, and compare the two means.

3.49 Find the mean and the standard deviation of the number of highway accidents given in Exercise 3.17 on page 40.

3.50 Find the mean and the standard deviation of the failure times of light bulbs given in Exercise 3.25 on page 42.

3.51 Find the standard deviations of the thickness of chrome plating given on page 51 for bumper *A* and for bumper *B*, and compare these standard deviations.

3.52 Find the mean and the standard deviation of the solder-bond strengths given on page 43.

3.53 Find the mean and the standard deviation of the data on shipping-department absences from the frequency distribution given in Exercise 3.20 on page 41.

3.54 Redo Exercise 3.50 using grouped data and compare results.

3.55 Redo Exercise 3.51 using grouped data and compare results.

3.56 Redo Exercise 3.52 using the frequency distribution given on page 44 and compare results.

3.57 Find the median and the range of the data of Exercise 3.17 on page 40.

3.58 Find the median and the range of the data of Exercise 3.25 on page 42.

3.59 Find the ranges for each of the bumpers mentioned on page 51 and compare these results with the standard deviations calculated in Exercise 3.51.

3.60 Show that the range is a constant multiple of the standard deviation when $n = 2$ and find this multiple.

3.61 a. Construct a box-and-whisker plot of the solder-bond breaking strengths given on page 43.

   b. How does this plot reveal the skewness of these data?

3.62 Construct a box-and-whisker plot of the data on highway accidents given in Exercise 3.17 on page 40.

3.63 Use *MINITAB* or some other statistical program to calculate the mean and the standard deviation of the data given in Exercise 3.13 on page 40.

3.64 Use *MINITAB* or some other statistical program to calculate the mean and the standard deviation of the data given in Exercise 3.14 on page 40.

3.65 Use *MINITAB* or some other statistical program to make a box-and-whisker plot of the data of Exercise 3.13 on page 40.

3.66 Use *MINITAB* or some other statistical program to make a box-and-whisker plot of the data of Exercise 3.14 on page 40.

The following exercises introduce new material, expanding on some of the topics covered in the text.

3.67 *Coefficient of variation.* The quantity

$$CV = 100 \cdot \frac{s}{\bar{x}}$$

is called the **coefficient of variation**, expressing the standard deviation as a percentage of the mean. Find the coefficient of variation of the data of Exercise 3.49.

3.68 *Continuation.* Find the coefficient of variation of the data of Exercise 3.50.

3.69 *Quartiles.* The three **quartiles** Q_1, Q_2, and Q_3 divide a set of data into four equal parts; that is, one quarter of the observations are less than the first quartile, Q_1, one quarter lie between the first and second quartiles, and so forth. Therefore, the second quartile, Q_2, is the median of the data. If the number of observations is divisible by four, two observations will straddle any given quartile. It is customary in such cases to use the mean of the two observations to approximate the quartile. Find the quartiles of the highway-accident data given in Exercise 3.17 on page 40.

3.70 *Continuation.* Find the quartiles of the solder-bond strengths given on page 43.

3.71 *Continuation.* Use *MINITAB* or some other statistical program to find the quartiles of the data of Exercise 3.25 on page 42.

3.72 *Percentiles.* **Percentiles** are similar to quartiles (Exercise 3.69) except that they divide the data set into 100 equal parts.

   a. Find the 10th and 90th percentiles of the highway- accident data of Exercise 3.17 on page 40.

   b. What proportion of the data lie between these percentiles?

3.73 *Continuation.*

   a. Find the 5th and 95th percentiles of the failure times given in Exercise 3.25 on page 42.

   b. What proportion of the light bulbs lasted for the number of hours given between these percentiles?

3.74 *Continuation.* Use *MINITAB* or some other statistical program to help in calculating the 17th and 65th percentiles of the data of Exercise 3.25 on page 42.

3.75 *Weighted mean.* The **weighted mean** is given by

$$\overline{x}_w = \frac{\sum_{i=1}^{k} w_i x_i}{\sum_{i=1}^{k} w_i}$$

where w_i is a weight given to observation i. Five shipments of bolts contained 110, 150, 95, 80, and 125 bolts, respectively. The mean weights of bolts in these shipments were 2.6, 2.8, 2.4, 3.1, and 2.5 ounces, respectively. What is the mean weight of the 560 bolts in all five shipments?

3.6 Application to Quality

The shape of a histogram is a valuable guide in the search for causes in the early stages of an investigation into a quality problem. In addition, comparison of a frequency distribution with specifications makes it apparent whether a problem is associated with poor accuracy or with lack of precision.

It has been suggested that error can never be eliminated entirely. For this reason, it is customary to separate errors into two kinds, **assignable cause** and **residual error**. Some examples of assignable cause are differences in material, ambient conditions, and vendors. Most assignable causes result from human error, whether it be the result of lack of training, fatigue, poor working conditions, or poor instructions. Residual error is that error which remains after all assignable causes have been identified and removed. Residual error is associated with limitations of measuring equipment and with general "background variability" that cannot be assigned to any specific cause. Improvements in technology can reduce residual error by improving measuring methods and by identifying and removing additional sources of background variation.

Histogram Shapes

Experience has taught that the distribution of residual errors is normally bell-shaped. (A specific formula has been found to describe it; this "normal" distribution will be described in Chapter 5.) Thus, a histogram that is not bell-shaped sometimes indicates the presence of assignable cause. To illustrate, suppose a histogram, constructed from the depths of borings made by several automatic boring machines, resembles that of Figure 3.11. This histogram is characterized by having two peaks. Its shape suggests that, perhaps, it is the result of combining two bell-shaped distributions having different means. Possibly, the depth of one or more of the boring machines has been set incorrectly. The clue given by the histogram should lead to taking boring-depth data separately from each machine. Perhaps it is then discovered that one of the boring machines had been set incorrectly. After eliminating this assignable cause, the histogram of the boring depths probably will become bell-shaped, and, of course, the variability of the process will be greatly decreased.

A histogram is said to have a **mode** if the frequency of any class exceeds that of its two surrounding classes. The histogram in Figure 3.11 is bimodal; that is, it has two modes. A histogram exhibiting more than two modes is said to be multi-modal. Such histograms often result from the operation of several assignable causes. Each cause generates its own distribution, each with its own mean, and the histogram that results from combining the underlying data sometimes shows as many modes as there are assignable causes. In especially severe problems, where many assignable causes are at work, there will be so many different modes that the resulting histogram may appear to be "flat," having many bars of nearly equal heights.

A skewed histogram, having only one tail, may or may not be related to assignable cause. Some examples of naturally skewed data include the duration of telephone calls, the time intervals between emissions of radioactive particles from a material, and the strength of some materials. Sometimes skewed distributions are the result of a particular choice of metric. Transformation of the data, perhaps by taking logarithms, may produce bell-shaped data in such cases. It is important for engineers working with a process to become familiar with types of measurements that naturally produce skewed distributions and types of measurements for which skewed distributions are a cause for concern.

Truncation of data often gives rise to a distribution that is abruptly "cut off" at some value. Data truncation occurs when observations greater than or less than a given magnitude are removed from a data set. There are several common causes of truncation. Measuring instruments which are incapable of making measurements above or below a given value can cause the resulting measurements to "pile up" at or near that value, producing a truncated distribution. A distribution of measurements made on products after they have been inspected to a given specification limit usually is truncated at the specification limit. Inspection of incoming material from vendors sometimes produces truncated distributions. When this happens, there is evidence that the vendor is not shipping the normal run of parts, but rather is selecting parts that just meet specifications, perhaps reserving the

FIGURE 3.11   Bimodal Histogram

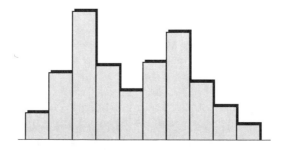

better parts for another customer whose specifications are tighter. Finally, data truncation can be caused by human failing, such as neglect in transmitting all the data. Such neglect can be the result of oversight, or it can be a deliberate effort to hide mistakes.

Specification Limits

Specification limits, added to the horizontal axis of a histogram, help to determine whether quality problems exist or are imminent. In examining a histogram to which specification limits have been added, it is useful to take into account not only its shape but also its location and variability relative to these limits.

Figure 3.12 shows four histograms to which two specification limits have been added, a lower specification limit, denoted by *LSL*, and an upper limit, *USL*. The histogram in Figure 3.12(a) is approximately bell-shaped and centered well within the specification limits. Thus, it is unlikely that there are assignable causes of variation operating. The process is producing measurements having a mean

FIGURE 3.12   Histograms and Specification Limits

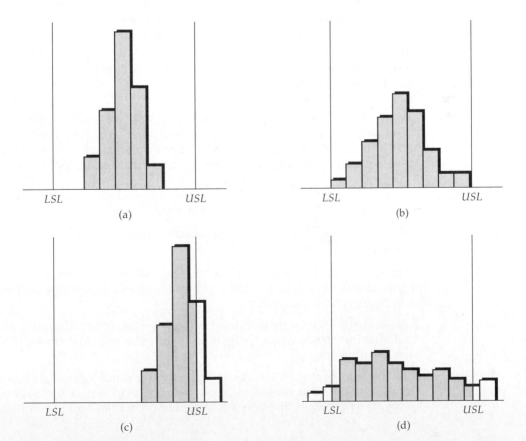

value midway between the specification limits, and the variation of the measurements is small enough that all observed values are well within specifications.

The histogram in Figure 3.12(b) has all the properties of that in Figure 3.12(a), except for increased variability. Note from this histogram that, while all the measurements are within specification limits, several are in the two classes adjacent to these limits. In such cases, it is likely that some product is, or soon will be produced outside of specification limits. If the tails of the histogram actually had been outside of specifications, a more clear-cut signal of quality problems would have been given, but a histogram like that of Figure 3.12(b) should be considered as an early warning that such problems are imminent.

A histogram like that in Figure 3.12(c) arises from a process that is off-center, illustrating an error of accuracy rather than one of precision. This error causes a portion of the products to be defective, as indicated by the unshaded area in the figure. Re-centering the process, so that the resulting measurements have a mean centered between the specification limits, will eliminate the defective products.

The histogram of Figure 3.12(d) illustrates a problem of multiple assignable causes. Although the process appears to be centered, and the measurements follow a more-or-less symmetrical distribution, there will be defective product (indicated by the unshaded area). Since there do not appear to be a limited number of assignable causes for this variability (the histogram does not exhibit a small number of clear-cut modes), the source of the unacceptable variability may be a large number of causes, or perhaps the process simply is incapable of producing to the design limits.

Numerical measures relating to comparisons of histograms with specification limits are introduced in Section 8.5.

> *Example.* Measurements were made of transaction times at the teller windows of a bank, producing the histogram shown in Figure 3.13. The bank's goal is to have no transaction last more than five minutes. (Special windows are used for lengthy transactions, such as business deposits.)
>
> a. How does this histogram differ from the desired one?
>
> b. What causes may account for this histogram?
>
> c. Will elimination of these causes necessarily produce a bell- shaped distribution?
>
> *Solution.* a.   The histogram is bimodal. Elimination of the cause underlying the distribution that gives rise to the second mode probably will result in an acceptable histogram.
>
> b. The second mode could be caused by one or more tellers requiring additional instruction or accepting transactions that should go to the special window.
>
> c. A skewed distribution of transaction times may remain after the causes of bimodality have been eliminated, because most transactions are of very short duration, and only a few require more than a minute or two.

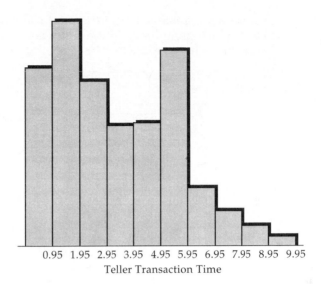

0.95 1.95 2.95 3.95 4.95 5.95 6.95 7.95 8.95 9.95
Teller Transaction Time

FIGURE 3.13 Teller Transaction Times

EXERCISES

3.76 What shape would normally be expected in a histogram arising from a process that

    a.   is centered and no assignable causes are present

    b.   is not centered and no other assignable causes are present

    c.   is not capable of meeting specifications but is properly centered and no assignable causes are present

    d.   has a single assignable cause

    e.   has multiple assignable causes

3.77 Sketch a histogram for two specification limits illustrating

    a.   a process not centered

    b.   a process having excess variability

    c.   a process that is both off-centered and too variable

    d.   bimodality, with excess variability

    e.   no defects, but not acceptable quality

3.78 A vendor supplies electrical connectors to length specifications of 1.25 ± .10 cm. Another customer has tighter specifications, namely ± 0.05 cm. Sketch a histogram that might result from measuring incoming connectors.

3.79 A coating, used on certain military equipment to protect it from adverse field conditions, is specified to be at least 0.10 inch thick. There is a high-scrap rate in production; thus, material with coatings found to be too thin are being removed from the production line before final assembly. Sketch a histogram that might result when coating thicknesses are measured on the final product.

3.80 The following are measurements of the Brinnel hardness of 100 samples of metal:

Brinnel Hardness	Frequency
5000 – 5499	9
5500 – 5999	16
6000 – 6499	24
6500 – 6999	20
7000 – 7449	18
7500 – 7999	13

a. Draw a histrogram and include the specifications 5000 – 8000 on the drawing.

b. What problem is apparent from this histogram?

3.81 The following is a frequency distribution of the number of complaints received by a television-cable company each week over a one-year period.

No. Complaints	Frequency
0 – 1	16
2 – 3	11
4 – 5	3
6 – 7	5
8 – 9	12
10 – 11	4
12 – 13	1

a. What clue is given by this distribution concerning the nature of the complaint problem?

b. What should be the goal of an investigation into this problem?

REVIEW EXERCISES

These review exercises can be used for informal review of the chapter, or as two practice examinations, designed to be taken over a time period of about one hour. The odd-

numbered exercises comprise one examination; the even-numbered exercises comprise the other.

3.82 The following are mileages (to the nearest thousand miles) recorded on eight truck tires between recaps:

| 24 | 31 | 18 | 27 | 22 | 36 | 41 | 29 | 35 |

a. Make a stem-and-leaf display.

b. Find the mean and the standard deviation.

c. Find the median and the range.

3.83 The following is the number of defects found in each of 10 lots of manufactured product:

| 3 | 7 | 4 | 0 | 5 | 8 | 3 | 2 | 1 | 5 |

a. Draw a dot diagram.

b. Find the mean and the median.

c. Find the standard deviation and the range.

3.84 The grades of 47 students taking a statistics course are:

81	74	68	71	87	92	56	76	65	70	77	79
47	61	80	95	71	78	99	79	64	73	72	84
93	42	88	59	80	84	97	52	67	77	75	83
66	95	71	86	62	76	50	85	64	75	60	

a. Group these grades into a frequency distribution.

b. Find the class boundaries, the class interval, and the class marks of the frequency distribution.

c. Find the mean and the standard deviation of the grouped data.

d. Sketch a histogram.

3.85 The following are 40 determinations of the hourly emission (in tons) of sulfur oxides from a factory:

1.1	0.5	0.9	0.7	2.2	1.5	1.0	0.3	1.2	1.8
0.9	1.8	2.4	1.6	1.2	0.6	1.0	1.4	1.2	1.7
1.0	0.8	1.6	1.4	0.7	2.0	0.5	0.7	2.1	1.5
1.9	1.6	1.4	0.4	1.3	1.8	1.5	1.2	1.7	2.0

a. Group these data into a frequency distribution, and sketch a histogram.

b. Find the mean and the standard deviation of the grouped data.

c.   * Make a cumulative distribution and a percentage distribution.

3.86 * Find the third quartile of the data given in Exercise 3.82.

3.87 *What is the second quartile of the data given in Exercise 3.83?

3.88 Make a box-and-whisker plot of the data given in Exercise 3.84.

3.89 Make a box-and-whisker plot of the data given in Exercise 3.85.

3.90 a.   Sketch a histogram that might result from a process that is centered between its specification limits, but has excess variability.

b.   Shade the area(s) corresponding to defective products.

3.91 a.   Sketch a histogram that might result from a process that has acceptable variability but is not centered between its specification limits.

b.   Shade the area(s) corresponding to defective products.

GLOSSARY

	page		page
Accuracy	30	Mean	49
Assignable Cause	59	Median	50
Bar Chart	45	Mode	60
Bell-Shaped Distribution	43	Numerical Distribution	37
Box-and-Whisker Plot	54	Ogive	48
Categorical Distribution	37	Outliers	54
Class Boundaries	39	Pareto Diagram	33
Class Frequencies	38	Percentage Distribution	41
Class Interval	39	Percentiles	58
Class Limits	37	Precision	30
Class Mark	39	Quartiles	58
Coefficienct of Variation	58	Range	54
Continuous Data	42	Residual Error	59
Cumulative Distribution	41	Skewness	44
Deviation	52	Standard Deviation	52
Discrete Data	45	Stem	32
Dot Diagram	32	Stem Labels	32
Double-Stem Display	32	Stem-and-Leaf Display	32
Frequency Distribution	37	Symmetric Distribution	43
Frequency Polygon	47	Truncation	60
Histogram	42	Variability	30
Leaf	32	Variance	52
		Weighted Mean	59

*This material is covered only in exercises.

4

PROBABILITY DISTRIBUTIONS

Where have we been? Chapter 3 gave us methods for the analysis and interpretation of raw data, and we applied some of these methods to problems involving quality. Thus, we have already begun to see how even the simplest statistical methods can be used for quality improvement.

Where are we going? Before we can apply more advanced statistical methods to problems of quality improvement, we need to extend the idea of distributions, introduced in Section 3.3. By relating distributions to probability, we set the stage for the development and application of statistical methods based on random samples. Such methods form the basis of modern statistics and its applications to quality.

Section 4.1, *Sample Spaces and Random Variables*, contrasts data distributions with probability distributions, defining the sample space of outcomes of an experiment and the relation of random variables to probability.

Section 4.2, *Probability*, reviews the basic ideas of probability, giving the axioms of probability, some elementary theorems, and some of their applications.

Section 4.3, *Probability Distributions*, unifies the ideas of probability and random variables, defining a probability distribution, its mean, and its standard deviation.

Section 4.4, *The Binomial Distribution*, solves a frequently encountered problem leading to the binomial distribution, describes its properties, including its mean and standard deviation, and illustrates its applications.

Section 4.5, *The Poisson Distribution*, introduces the Poisson distribution as an approximation of the binomial distribution, giving its properties and several of its applications.

Section 4.6, *Application to Quality*, discusses the pros and cons of acceptance sampling, and shows how the binomial distribution is used to obtain the operating characteristic curve of a sampling plan.

4.1 Sample Spaces and Random Variables

The notion of data distributions was introduced in Chapter 3, where it was shown that their characteristics such as shape, location, and variability can be helpful in diagnosing quality problems and suggesting approaches to their solution. These ideas will be explored further in this chapter by introducing models of data distributions, examining their properties, and beginning to discover the breadth of application of these models, called "theoretical distributions." Perhaps of even greater importance, this chapter will show how theoretical distributions can be used to find probabilities, making it possible to reduce the guesswork of drawing conclusions from data to a more exact, scientific approach based on probabilities.

Whenever data are gathered, whether they be data on a process, a product, or a service, it can be said that an **experiment** has been performed. The word *experiment* is used here in its most general sense, meaning any activity that can lead to one or more specific outcomes, or results. For example, the simple observation of the number of defective parts in a lot and the more complex scientific procedure for measuring the specific gravity of a fluid, both can be considered experiments. Some experiments involve only casual, unplanned observation, while others are carefully designed for the purpose of drawing inferences from their outcome. Methods for drawing inferences are given in Chapter 7. The design and analysis of such experiments, vital to quality improvement, will be presented in Chapter 10. This chapter gives some of the necessary background for drawing such inferences and designing experiments to support them.

Sample Spaces

The collection or set of all outcomes that can result from an experiment is called a **sample space**. A sample space is like any other set, except that its elements are outcomes of an experiment, such as a head or a tail when a coin is tossed, the number of defectives in a lot containing manufactured products, or the value obtained when an electrical parameter of a transistor is measured. An **event** is a subset of the sample space; thus, an event is a collection of outcomes. The event "2 or more heads will occur" when a coin is tossed five times includes 4 of the 6 total outcomes comprising the sample space for this experiment. If a lot contains 50 parts, the event "the lot contains no more than 3 defectives" contains the outcomes 0, 1, 2, and 3 defectives in a sample space containing the 51 outcomes, 0, 1, 2 , . . . , 50 defectives.

The usual set notation also applies to sample spaces. The **union** of events A and B, denoted by $A \cup B$, means that event which occurs when *either A or B* occur. The **intersection** of events A and B, denoted by $A \cap B$, means the event which occurs when *both A and B* occur. The **complement** of event A, denoted by A', is the event which occurs when *A does not occur*. Two events are said to be **mutually exclusive** when their intersection is the empty set; that is, when they cannot both occur as an outcome of a given experimental observation.

Example.   a.   List all the outcomes in the sample space for rolling a single die.

b.   Which of these outcomes belong to the event "the die comes up an even number"?

    c.   What is the intersection of this event with the event "the die comes up greater than 4"?

Solution. a.   A die has six sides, having 1, 2, . . . , 6 spots, respectively. The sample space can be represented by the outcomes $S = \{1,2,3,4,5,6\}$.

    b.   The event "the die comes up an even number" is the subset $A = \{2,4,6\}$ of the sample space.

    c.   The event "the die comes up greater than 4" is the subset $B = \{5,6\}$. The intersection of A and B is the subset $A \cap B = \{6\}$.

Random Variables

A *variable*, in mathematics, is a place holder, a symbol that can take on any value selected from a specified set. Functions of variables allow general statements to be made about how the elements of two or more sets are related to one another under specified conditions. To make general statements about data, it is helpful to introduce the notion of a **random variable**; it facilitates the definition of the underlying or theoretical distributions that give rise to data.

    A random variable associates outcomes with probabilities. Thus, the number of heads resulting from 10 consecutive tosses of a balanced coin can be regarded as a random variable because it is possible to calculate the probability that the number of heads will take on any given value from 0 to 10. If the outcome of a given sequence of 10 coin tosses is "5 heads," the number five is said to be a *value* of this random variable.

    Example.   Find the values and the associated probabilities for the random variable defined as the number of failures in two flights of a test rocket, if the probability of failure on any flight is .6.

    Solution.   The sample space has the following four outcomes, where F denotes failure and S denotes success: FF FS SF SS. Using results which will be explained in Section 4.2, the probability of the outcome FF is $(.6)(.6) = .36$, for the outcome FS it is $(.6)(.4) = .24$, and for the remaining two outcomes the probabilities are $(.4)(.6) = .24$ and $(.4)(.4) = .16$, respectively. The variable "the number of successes in two flights" can take on three values, namely 0, 1, or 2. The value 0 corresponds to the outcome FF, with probability .36; the value 1 corresponds to the union (FS or SF), with probability $.24 + .24 = .48$; and the value 2 corresponds to the outcome SS, with probability .16. Since each value of the variable has been associated with a probability, it is a *random* variable.

As illustrated in this example, it is customary to define random variables so that their values are numbers, rather than categories.

    Sample spaces and events in sample spaces are needed to define probabilities. Random variables are needed to define theoretical distributions, also called "probability distributions," which are necessary to understand how reliable inferences can be drawn from data. Thus, probability is the thread that runs through most statistical methods. It is assumed that the reader is familiar with some of the basic concepts of probability, which will be reviewed briefly in the next section.

4.2 Probability

Axioms of Probability

The word *probability* is a part of everyday language, but it is difficult to define this word without using the word *probable* or its synonym, *likely* in the definition. Webster's *Third New International Dictionary* defines *probability* as follows: "the quality or state of being probable." If the concept of probability is to be used in science and engineering, it becomes necessary to have a more exact, less circular definition.

The following definition of probability allows calculations to be made with probabilities, so that if the probability of a given event is known, it is possible to calculate the probabilities of other, related events. A **probability** is a number associated with each outcome in a sample space, denoted by the letter S. Probabilities are denoted as $P(A)$, or $P(B)$, and so on, where A and B are specific outcomes in the sample space under consideration. Such numbers must obey the following rules:

AXIOMS OF PROBABILITY

1. For each event A in S, $0 \leq P(A) \leq 1$.

2. $P(S) = 1$.

3. If A and B are any mutually exclusive events in S, then $P(A \cup B) = P(A) + P(B)$.

The first axiom states that a probability is a real number that can take on values in the interval from 0 to 1. The second axiom requires that the probability associated with the entire sample space must be 1. It can be thought of as a statement that all possible events have been included in S. The third axiom states that the probability of occurrence of either of two mutually exclusive events is the sum of the probabilities assigned to each event individually. Axiom 3 is an axiom of *additivity*, as it states that probabilities of mutually exclusive events simply can be added to produce the probability of their union.

Calculation of Probabilities

It is remarkable that the entire structure of probability theory can be built on these three simple axioms. Nevertheless, these axioms are somewhat disappointing because they do not actually give the probabilities of the various outcomes. However, three methods of determining these probabilities are in wide use.

The **classical method** of determining probabilities assumes that all outcomes in the sample space can be assigned equal probabilities. In a sample space consisting of n outcomes, s of which are favorable, or "successes," this method defines the probability of a success to be s/n. This definition was developed to solve problems involving gambling, for which the assumption of equal probabilities, or "equal likelihood," usually is true, at least to a good approximation.

To illustrate the determination of probabilities by the classical method, let us calculate the probability that the top card will be an ace in a well-shuffled deck of 52 playing cards. There are 52 cards in the deck, or 52 outcomes, of which 4 are aces, or 4 are "successes." Thus, the required probability is $4/52 = 1/13$. As a further illustration, the probability of throwing a 7 with a pair of dice is $6/36 = 1/6$, because there are 36 ways that two dice can be thrown and the events (1,6), (2,5), (3,4), (6,1), (5,2), and (4,3) are the 6 results that add to 7.

An empirical and often useful way to calculate probabilities is given by the **relative-frequency method.** This method interprets the probability of an event to be the proportion of times the event will occur in a long run of repeated experiments. Application of the relative-frequency method of determining probabilities requires a well-documented history of the outcomes of an event over a known number of experimental trials. In the absence of such a history, a series of experiments can be planned and their results observed. For example, the probability that a lot of manufactured product will contain at most three defectives is estimated to be .90 if, in 90% of many previous lots, the number of defectives was three or less. In the absence of such a history, observations can be made on the number of defectives observed in the inspection of future lots.

A more recently employed method of calculating probabilities is called the **subjective method.** With this method, a personal or subjective assessment is made of the probability of an event which is difficult to estimate in any other way. Managers often use this method, whether consciously or not, to assign rough probabilities to the success of a business enterprise. Juries use this method when determining guilt "beyond a reasonable doubt." Subjective probabilities are best used when other methods fail and with the application of a high level of expertise.

Some Elementary Theorems

The following "laws of probability" facilitate calculation of the probabilities of certain events when the probabilities of other events are known. These laws can be proved by applying the axioms; although the proofs are straightforward, they shall be omitted here, and only the statements of these theorems will be given, together with some brief examples of their application.

The first theorem generalizes axiom 3 to include more than two events. It simply states that probabilities of mutually exclusive events are additive for any *finite* number of events.

The second theorem, sometimes called the **special law of multiplication**, states that the probabilities of two independent events can be multiplied to find the

---

THEOREMS OF PROBABILITY

1. If $A_1, A_2, \ldots, A_n$ are mutually exclusive events in S, then
$$P(A_1 \cup A_2 \cup \ldots \cup A_n) = P(A_1) + P(A_2) + \ldots + P(A_n)$$

2. If A and B are *independent* events in S, then
$$P(A \cap B) = P(A) \cdot P(B)$$

3. If A and B are any events in S, then
$$P(A \cup B) = P(A) + P(B) - P(A \cap B)$$

4. If A is an event in S, then
$$P(A') = 1 - P(A)$$

---

probability of their intersection. Two events are said to be **independent events** if the probability of one of them is not changed when it is known that the other has occurred. For example, if a part is drawn from a bin containing 50 parts of which 5 are defective, the probability of choosing a defective part is .10. If two parts are drawn in succession, replacing the first part before drawing the second, the probability that they are *both* defective is $(.10)(.10) = .01$, according to the second theorem. But if the first part is not replaced, the bin will contain 49 parts before the second part is drawn, and either 4 or 5 of them will be defective, depending upon the outcome of the first drawing. The probability of drawing a defective part the second time is either 4/49 or 5/49, *depending on the first result*. Thus, the two events "a defective on the first drawing" and "a defective on the second drawing" are not independent when drawing without replacement, and the second theorem does not apply.

The third theorem, sometimes called the **general law of addition** of probabilities, shows how to find the probability of the union of two events when they are not mutually exclusive. To illustrate this theorem, suppose warranty-card data shows that a personal computer is used for recreation in 75% of the households purchasing the product, for business in 35% of the households, and for both recreation and business in 18% of the households. Applying the third theorem, the probability that, in a given household, the computer is used for recreation or business is $.75 + .35 - .18 = .92$.

To illustrate the fourth theorem, the probability will be found that a household having a personal computer does not use it for business. If A represents the event that such a household uses the computer for business, then the event that it does *not* use it for business is the complementary event, A'. According to the fourth theorem, $P(A') = 1 - P(A) = 1 - .35 = .65$. An important corollary of this theorem states that the probability of the empty set is zero. This corollary follows immediately from the fact that the empty set is the complement of the sample space, and, by the second axiom, $P(S) = 1$. Applying the fourth theorem, P(empty set) $= 1 - P(S) = 1 - 1 = 0$.

> *Example.* If A, B, and C are mutually exclusive events, $P(A) = .2$, $P(B) = .3$, and $P(C) = .4$, find

a. $P(A \cup B \cup C)$ b. $P(A')$ c. $P(A \cap B)$

Solution. a.   By the first theorem,

$$P(A \cup B \cup C) = .2 + .3 + .4 = .9$$

b.   By the fourth theorem, $P(A') = 1 - P(A) = 1 - .8 = .2$.

c.   Since A and B are mutually exclusive, they have no outcomes in common, and $A \cap B$ is the empty set. Thus, $P(A \cap B) = 0$.

EXERCISES

4.1 a.   List all possible outcomes in the sample space for tossing three coins.

   b.   Which outcomes comprise the event "two or more heads"?

4.2 a.   How many outcomes are there in the sample space for drawing a card from a standard deck of playing cards?

   b.   How many of these outcomes comprise the event "a spade is drawn"?

4.3 A manufactured part can have two kinds of defect, critical and noncritical. Two parts are chosen, and it is determined which have either kind of defect, or both, or none.

   a.   What is the sample space?

   b.   Which of the outcomes in this sample space are contained in the event "neither part has a critical defect"?

4.4 Suppose a television set must have at least one of the following attributes: color, stereo sound, and remote control.

   a.   List the sample space of outcomes for a given television set.

   b.   How many of these outcomes belong to the event "the television set has stereo sound"?

   c.   Are the three events—color, stereo sound, and remote control—mutually exclusive?

4.5 A card is drawn from a deck of 52 playing cards.

   a.   Are the events "the card is a diamond" and "the card is an ace" independent events?

   b.   Are they mutually exclusive? Why?

4.6 a.   Using the sample space of Exercise 4.1, find the intersection of the event B = "three or more heads" with A = "2 or more heads."

   b.   Are these events mutually exclusive? Why?

4.7 Assume that a screw, selected from a bin containing 100 screws, is chosen so that the probability of selection of any screw is equal to that of any other. If two of the screws have defective threads, what is the probability of selection of a defective screw on the first selection?

4.8 What is the probability that, in two tosses of a balanced coin, both will be heads?

4.9 Experience shows that it snows on 85% of the days that snow is forecast. Estimate the probability it will not snow on a day when snow is forecast.

4.10 If A, B, and C are mutually exclusive events, and $P(A) = .25$, $P(B) = .4$, and $P(C) = .35$ find

    a. $P(A \cap B)$ b. $P(A \cup C)$ c. $P(A \cup B')$

4.11 If $P(A) = .6$, $P(B) = .2$, and $P(C) = .1$, and A, B, and C are mutually exclusive, find

    a. $P(A \cup B \cup C)$ b. $P(A \cap B)$ c. $P(A' \cup C)$

4.12 If $P(A) = .25$, $P(B) = .45$, and $P(A \cap B) = .1$, find

    a. $P(A \cup B)$ b. $P[(A \cap B)']$

4.13 If $P(A) = .42$, $P(A \cup B) = .65$, $P(A \cap B) = .14$, find

    a. $P(B)$ b. $P(A')$

4.14 An experiment has four mutually exclusive outcomes, A, B, C, and D. Is each of the following assignments of probabilities permissible? Why?

    a. $P(A) = .2$, $P(B) = .3$, $P(C) = .4$, $P(D) = .5$.

    b. $P(A) = .1$, $P(B) = .2$, $P(C) = .3$, $P(D) = .4$.

    c. $P(A) = P(B) = P(C) = .15$, $P(B \cup D) = .70$.

    d. $P(A \cup B) = .95$, $P(C \cup D) = .10$.

4.15 An experiment has three mutually exclusive outcomes, A, B, and C. Is each of the following assignments of probabilities permissible? Why?

    a. $P(A) = .32$, $P(B) = .18$, $P(C) = .50$.

    b. $P(A \cup C) = .67$, $P(B) = .18$.

    c. $P(A \cup B) = .38$, $P(B \cup C) = .88$.

    d. $P(A \cup B \cup C) = 1$.

4.16 If $P(A) = .45$, $P(B) = .35$, and A and B are independent events, find $P(A \cap B)$.

4.17 a. Find the probability of drawing 2 consecutive red balls from an urn containing 10 balls, of which 6 are red and 4 are white, if the first ball is replaced before the second is drawn.

    b. Are the two events independent? Why?

4.18 a. Referring to Exercise 4.17, find the probability of drawing 2 red balls when the first ball is not replaced before the second ball is drawn.

    b. Are the two events independent? Why?

The following exercises introduce new material, expanding on some of the topics covered in the text.

4.19 *Conditional probability.* The **conditional probability** of event A, given event B, denoted by $P(A \mid B)$, is the probability that A occurs, given that B has occurred. Find the conditional probability of drawing 1 red ball on the second draw from an urn containing 10 red balls and 5 green balls, given that the first draw resulted in a green ball and the ball was not replaced.

4.20 *General law of multiplication.* The probability that A and B will occur is given by $P(A \cap B) = P(A)P(B \mid A)$. [Equivalently, for a conditional probability, $P(B \mid A) = P(A \cap B) \, / \, P(A)$.] If A and B are independent events and $P(A) \neq 0$, show that the special law of multiplication implies that $P(A \mid B) = P(A)$.

4.21 *Continuation.* A bin contains 100 parts to be inspected. If the percentage defective is assumed to be 1.0%, find the probability that the first two parts are not defective when drawing the parts from the bin *without* replacement.

4.3 Probability Distributions

If a random variable can take on numerical values on a continuum, such as the real line, or an interval on the real line, it is said to be a **continuous random variable**. If a random variable is not continuous, it is said to be a **discrete random variable**. In practice, discrete random variables occur when the occurrences of certain phenomena can be counted, while continuous random variables arise when something is measured, such as length, mass, or temperature.

A continuous random variable takes on infinitely many values; for instance, all the real numbers from 1 to 2. A discrete random variable can take on a finite number of values or a countable infinity of values. (By "countable infinity" we mean as many values as there are whole numbers.) The discrete random variable representing the number of heads when a coin is tossed 10 consecutive times takes on one of the 11 values $0, 1, \ldots, 10$. But the random variable that represents the number of emissions of a beta particle from a radioactive source can take on any of the values $0, 1, 2, \ldots$. Since no practical upper limit can be given for the number of beta particles emitted, it is convenient to think of this discrete random variable as having infinitely many values.

Two probability distributions of discrete random variables will be introduced in this chapter, the binomial distribution in Section 4.4, and the Poisson distribution in Section 4.5. An important distribution of a continuous random variable, the normal distribution, will be introduced in Chapter 5.

Probability distributions of discrete random variables, or discrete probability distributions, are similar to percentage distributions of data (Exercise 3.18 on page 41), except that the percentages are replaced by probabilities. Interpreting the observations as values of a random variable, and using the relative-frequency method for estimating the probability that an observation will fall into a given class, a percentage distribution of a set of data can be thought of as an estimate of the probability distribution of the underlying random variable.

In general, a **probability distribution** is a function whose arguments are values of a random variable, and whose functional values are probabilities. Thus, the table

x	1	2	3	4	5	6
$f(x)$	.15	.25	.30	.15	.10	.05

gives a probability distribution whose arguments are values of a random variable x, that can equal any integer value from 1 to 6, and whose values are given by $f(x)$. Note that $f(x)$ in this table represents a probability. Thus, for example, the statement: "$f(2) = .25$" means "the probability that the random variable will take on the value 2 is .25." Note also that each value of $f(x)$ lies between 0 and 1, and that all values of $f(x)$ sum to 1, as required by the first two axioms for probabilities, respectively.

The conditions for *any* function to represent a probability distribution are summarized as follows:

Probability Distributions:

1. Each value of $f(x)$ must lie between 0 and 1, inclusive.
2. The sum of all values of $f(x)$ must equal 1.

Probability distributions can be given in the form of tables, or they can be expressed by means of equations. The equation

$$f(x) = \frac{x^2}{30}, \quad for \ x = 1, 2, 3, 4$$

represents a probability distribution. The reader will be asked in Exercise 4.22 on page 78 to show that its values are non-negative real numbers that sum to 1. Note that the range of values that can be assumed by the random variable is given together with the equation for the probability distribution. Without this range of values, there is no way of assuring that the values of $f(x)$ add to 1. Thus, the range of the random variable is an integral part of the specification of a probability distribution.

Probability distributions are represented pictorially by a chart similar to a histogram, called a **probability histogram**. In a probability histogram, the areas of the bars represent probabilities instead of frequencies or percentages. A probability histogram representing the distribution whose equation is given above is shown in Figure 4.1.

Mean and Standard Deviation of a Probability Distribution

In Chapter 3 it was seen that the mean of a set of data can be regarded as the centroid of the data. The same idea is used to define the mean of a probability distribution. Given the probability distribution $f(x)$, its **mean** is defined to be

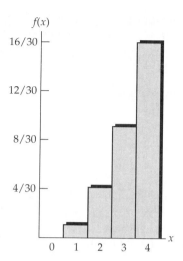

FIGURE 4.1 Probability Histogram

$$\textit{Mean of a Probability Distribution:} \quad \mu = \sum_{all\ x} xf(x)$$

Just as the mean of a set of data measures the "center" of the data, the mean of a distribution measures the center of the distribution. The units of measurement for the mean are, as for data, the same as those for the values of the original random variable. In this formula, the Greek letter μ is used instead of $\bar{x}$. It is customary to use Greek letters to denote the mean and the standard deviation of probability distributions and Latin letters to denote the mean and standard deviation of data.

The definition of the standard deviation of a probability distribution also borrows an idea from physics. If one imagines a mass spun about its centroid, its moment of inertia is the sum of the squared distance from the centroid of each element of mass times the mass of the element. Thinking of a probability histogram, the elements of mass are the bars, their masses are proportional to their areas, and the centroid of the chart is μ. A probability distribution having greater variability will have a probability histogram whose bars have greater mass farther away from the centroid than that of a distribution with lesser variability; thus, it will have a larger moment of inertia. This argument leads to the following definition of the **variance** of a probability distribution:

$$\textit{Variance of Probability Distribution} \quad \sigma^2 = \sum_{all\ x} (x - \mu)^2 f(x)$$

Note the use of the Greek σ^2 for the variance of a probability distribution in place of s^2, used for the variance of data. The **standard deviation** is defined for probability distributions, as it is for data, to be the square root of the variance, and it is denoted by the Greek letter σ.

Example. Find the mean and the standard deviation of the probability distribution whose equation is given on page 76.

Solution. The mean is given by

$$\mu = \sum_{x=0}^{4} x \cdot \frac{x^2}{30} = \frac{0 + 1 + 8 + 27 + 64}{30} = \frac{10}{3}$$

and the variance is

$$\sigma^2 = \sum_{x=0}^{4} \left(x - \frac{10}{3}\right)^2 \cdot \frac{x^2}{30} = \frac{0 + \left(-\frac{7}{3}\right)^2 + \left(-\frac{4}{3}\right)^2 \cdot 4 + \left(-\frac{1}{3}\right)^2 \cdot 9 + \left(\frac{2}{3}\right)^2 \cdot 16}{30} = \frac{31}{45}$$

Thus, the standard deviation is

$$\sigma = \sqrt{\frac{31}{45}} = 0.83$$

EXERCISES

4.22 Show that the values of the probability distribution whose equation is given on page 76 are non-negative and that they sum to 1.

4.23 Does the following table represent a probability distribution? Why?

x	0	1	2	3	4	5	6
$f(x)$	.1	.2	.3	.4	.3	.2	.1

4.24 Does the following table represent a probability distribution? Why?

x	2	3	4	5	6	7	8
$f(x)$	0	.2	−.3	.4	.5	.1	.1

4.25 Draw a probability histogram representing the probability distribution

x	0	1	2	3	4	5
$f(x)$	.05	.10	.25	.30	.20	.10

4.26 Draw a probability histogram representing the following probability distribution:

x	1	2	3	4	5	6	7
$f(x)$	.10	.15	.20	.30	.15	.08	.02

4.27 Find the mean and the standard deviation of the probability distribution given in Exercise 4.25.

4.28 Find the mean and the standard deviation of the probability distribution given in Exercise 4.26.

4.29 Find the mean and the standard deviation of the probability distribution having the equation

$$f(x) = x/15, \qquad \text{for } x = 0, 1, 2, 3, 4, 5$$

4.30 Find the mean and the standard deviation of the probability distribution having the equation

$$f(x) = (x-2)^2/10, \text{ for } x = 0, 1, 2, 3, 4$$

4.31 Suppose

$$f(x) = kx, \qquad \text{for } x = 0, 1, 2, \ldots, 10$$

Find the value of k if $f(x)$ represents a probability distribution.

4.32 Represent the probability distribution of the number of heads in four tosses of a balanced coin by means of a table or an equation.

4.33 Find the mean and the standard deviation of the probability distribution defined in Exercise 4.31.

4.34 Find the mean and the standard deviation of the probability distribution defined in Exercise 4.32.

The following exercises introduce new material, expanding on some of the topics covered in the text.

4.35 *Random numbers.* A **random digit** can be defined as a value of the discrete uniform distribution

$$f(x) = \frac{1}{10} \quad \text{for } x = 0, 1, 2, \ldots, 9$$

What is the probability that a random variable having this distribution will assume the value k, where k is an integer with $0 \le k \le 9$?

4.36 *Continuation.* Multi-digit random numbers can be formed by performing arithmetic operations on random digits. The *MINITAB* command RANDOM K OBS., PUT INTO C1; simulates a random sample of k observations, putting them in C1. The semicolon at the end of this command asks for a subcommand. The subcommand UNIFORM 0,1. assures that the sample will be uniformly distributed on the interval (0,1). Use *MINITAB* or some other statistical software to generate 7 random digits having values less than or equal to 10. *Hint*: Multiply C1 by 10 and round to the nearest integer.

4.4 The Binomial Distribution

An example of a discrete probability distribution that can be derived directly from probability theory is the **binomial distribution**. This distribution arises in a variety of practical problems, and it is the basis for control charts for attributes, to be introduced in Chapter 8.

The binomial distribution is the solution to problems of the following kind. A coin is tossed 20 times and the probability of 10 heads is desired. Or 50 purchasing managers are interviewed and it is desired to know the probability that more than half of them will project an increase in their purchases in the next quarter. Or, perhaps 20 finished products are selected from a large lot and subjected to final inspection, and the probability is required that no more than two of them will fail.

These problems have the following conditions in common:

```
                CONDITIONS FOR BINOMIAL DISTRIBUTION:

 1. The experiment consists of n repetitions, or trials.
 2. Each trial can have only one of two possible outcomes. (The two
    outcomes must be mutually exclusive.)
 3. The probability of a given outcome is the same for each trial.
 4. The trials are independent; that is, the probability of obtaining a
    given result on any trial does not depend upon the results of the
    previous trials.
```

Sometimes, it is convenient to refer to one of the outcomes as a "success" and the other as a "failure" without implying that either outcome is desirable or undesirable.

To illustrate, in the aforementioned coin-tossing problem, there are $n = 20$ trials, and the probability of a success (a head) is .50 on each trial if the coin is balanced. The trials are independent, as the probability of a head on any toss does not depend on the results of the previous tosses. With respect to the purchasing managers, $n = 50$ and it is assumed that the managers' projections are given independently of one another. It is further assumed that the pre-interview probabilities of the managers projecting increased purchases are equal. In the inspection problem, $n = 20$. The probability of a defective product is not constant if the products are drawn from the lot without replacement, but these probabilities differ only slightly from one another if the lot is very large compared to the number of products selected. Thus, the assumption of equal probabilities is made as an approximation. If products in the bin containing the lot are thoroughly mixed before products are selected for inspection, the assumption of independence of successive selections is at least approximately correct.

The general statement of problems leading to the binomial distribution is as follows: *Find the probability of obtaining* x *successes in* n *independent trials for which the probability of a success on each trial is* p. To solve the problem, it is first noted that the probability of the complementary outcome, a failure, is $1 - p$. Since the trials are independent, the probability of two consecutive successes is p^2; for a string of three consecutive successes this probability is p^3, and for x successes in a row it is p^x. Similarly, the probability that the remaining $(n - x)$ trials will be failures is $(1 - p)^{n-x}$. Independence of the trials allows the multiplication of these two probabilities to obtain the probability $p^x(1 - p)^{n-x}$ of x successes, followed by $(n - x)$ failures.

The last probability given does not solve the problem, however, since there are many ways to obtain x successes and $n - x$ failures other than x successes followed by $n - x$ failures. For example, there could be 5 successes, followed by 4 failures, followed by $(x - 5)$ successes, followed by $(n - x - 4)$ failures. It is evident that each of these other ways involves the same number of successes, x, and the same number of failures, $n - x$. Thus, *the probability of each ordering of* x *successes and* n − x *failures is* $p^x(1 - p)^{n-x}$. To obtain the probability of x successes and $n - x$ failures in *any* order, it is necessary to count the number of possible orders and to add the probability $p^x(1 - p)^{n-x}$ as many times as this count.

The number of combinations of n things, taken x at a time is given by the *combinatorial coefficient*

$$\binom{n}{x} = \frac{n!}{x! \, (n - x)!}$$

where, for any positive integer m, $m!$, in words "m factorial," is defined as

$$m! = m(m - 1) \cdots (2)(1) \qquad \text{and } 0! = 1$$

[For example, $4! = (4)(3)(2)(1) = 24$.] The combinatorial coefficient counts the number of orderings of x successes and $n - x$ failures; thus, the required probability is:

Binominal Distribution:

$$b\,(x; n, p) = \frac{n!}{x! \, (n - x)!} \, p^x \, (1 - p)^{n-x} \qquad \text{for } x = 0, 1, 2, \ldots, n$$

The binomial distribution is the probability distribution of the number of successes in n independent trials, where p is the probability of a success on any trial. The number of successes can take on the values $0, 1, 2, \ldots, n$. The notation $b(x;n,p)$ indicates that x is a value of the random variable having the binomial distribution, and n and p are the **parameters** of this distribution. Thus, the binomial distribution actually is a *two-parameter family of distributions*; a particular binomial distribution in this family is determined by the values of n and p.

Example. Find the probability of obtaining 4 heads in 10 tosses of a balanced coin.

Solution. There are 10 independent trials of an event (heads) having the probability of .5 on each trial. The required probability is given by $b(4;10,.5) = \dfrac{10!}{4! \; 6!} \, (.5)^4(.5)^6 = .2051$.

Probability histograms for the binomial distributions $b(x;6,.5)$ and $b(x;6,.25)$ are shown in Figure 4.2. Notice that the probability histogram is symmetric for $p = .5$, but it is positively skewed for $p = .25$. In general, binomial distributions will be symmetric whenever $p = .5$; and they will be skewed with long right-hand tails whenever $p < .5$, and with long left-hand tails whenever $p > .5$.

FIGURE 4.2 Binomial Distributions

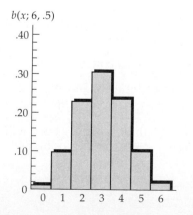 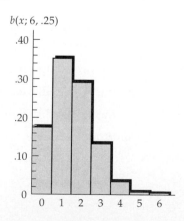

Cumulative Binomial Distribution

The binomial distribution allows us to calculate the probability that the number of successes, x, will take on a specific value. In most applications, there is a need to compute the probability that x will exceed a given value or that x will be less than a certain value. For example, suppose it is desired to know the probability that there will be fewer than 3 defectives in a lot of 20 parts, when past experience has shown that the probability of a defective part is .05. Since the outcomes $0, 1, 2, \ldots,$ 20 defectives are mutually exclusive, the required probability is given by the sum of the probabilities for 0, 1, and 2 defectives:

$$b(0;20,.05) + b(1;20,.05) + b(2;20,.05)$$

This sum can be written as

$$B\,(2;\,20,\,.05) = \sum_{x=0}^{2} b(x;\,20,\,.05)$$

In general, the **cumulative binomial distribution** is given by

---

Cumulative Bionomial Distribution: $\quad B\,(x;\,n,\,p) = \sum_{k=0}^{x} b\,(k;\,n,\,p)$

Computations involving the binomial distribution can be tedious, especially when n is large. To facilitate such computations, the cumulative binomial distribution has been tabulated in Table 1 (Appendix A). This table gives values of $B(x;n,p)$ for values of n from 2 through 20, and values of p from .05 to .95, in increments of .05. This table can be used to find the probability of exactly x successes, at least x successes, and x or more successes. To find the probability of *exactly* x *successes in* n *trials,* use can be made of the formula

$$b(x;n,p) = B(x;n,p) - B(x - 1;n,p)$$

The probability of *at least* x *successes in* n *trials* is given directly by $B(x;n,p)$, and the probability of x *or more successes in* n *trials* is given by $1 - B(x - 1;n,p)$.

Example. If the probability is .25 that a torsion bar will fail when the torque exceeds 500 foot-pounds, find

a. the probability that exactly 4 of 15 bars tested at this level will fail,

b. the probability that 4 or more bars will fail, and

c. the probability that at most 6 bars will fail.

Solution. a. The probability that exactly 4 bars will fail is given by $B(4;15,.25) - B(3;15,.25) = .6865 - .4613 = .2252$, using Table 1.

b. The probability that 4 or more bars will fail is given by $1 - B(3;15,.25)$ $= 1 - .4613 = .5387$.

c. The probability that at most 6 bars will fail is given by $B(6;15,.25) = .9434$.

Mean and Standard Deviation

The mean of the binomial distribution is found by substituting $b(x;n,p)$ for $f(x)$ in the equation defining the mean of the probability distribution given on page 77. The resulting mean is given by

$$\mu = \sum_{x=0}^{n} x \cdot \frac{n!}{x! \,(n-x)!} \, p^x \, (1-p)^{n-x}$$

It can be shown that the summation becomes simply

> **Mean of the Binomial Distribution:** $\quad \mu = np$

The standard deviation is given by

$$\sigma = \sqrt{\sum_{x=0}^{n} (x - np)^2 \, \frac{n!}{x! \,(n-x)!} \, p^x \, (n-p)^{n-x}}$$

It can be shown that this complicated formula is algebraically equivalent to

> **Standard Deviation of the Bionomial Distribution:** $\quad \sigma = \sqrt{np\,(1-p)}$

Example. Find the mean and the standard deviation of the distribution of the number of "shorted" transistors in a lot of 100 transistors, if the percentage of shorted transistors produced is 15%.

Solution. This can be regarded as 100 independent trials, with the probability .15 of obtaining a shorted transistor on any given trial. Thus, $n = 100$ and $p = .15$. The mean number of shorted transistors is $(100)(.15) = 15$, and the standard deviation is $\sqrt{100\,(.15)\,(.85)} = 3.57$.

In this example, it has been shown that the mean number of shorted transistors in a lot of 100 transistors is 15. Since lot sizes may vary, it may be of greater interest to know the mean *proportion* of transistors in a lot of any size. If the random variable x has the binomial distribution, it can be shown that the random variable having the values x/n, giving the *proportion* of successes in n trials, has the mean

Mean of a Proportion: $\mu_{\frac{x}{n}} = p$

and the standard deviation

Standard Deviation of a Proportion: $\sigma_{\frac{x}{n}} = \sqrt{\dfrac{p\,(1-p)}{n}}$

Thus, the mean proportion of shorted transistors in the preceding example is $p = 0.15$ and its standard deviation is $\sqrt{(.15)\,(.85)\,/\,100} = 0.036$.

Computer Applications

Binomial probabilities and cumulative binomial probabilities are calculated by *MINITAB* software. The command sequence

MTB > PDF;

SUBC > BINOMIAL N = 10, p = .27.

produces a table of binomial probabilities. To obtain cumulative probabilities, the initial command PDF is replaced by CDF.

BINOMIAL PROBABILITIES FOR N=10 AND P=.2700

K	P(X = K)	P(X LESS OR = K)
0	.0430	.0430
1	.1590	.2019
2	.2646	.4665
3	.2609	.7274
4	.1689	.8963
5	.0750	.9713
6	.0231	.9944
7	.0049	.9993
8	.0007	.9999
9	.0001	1.0000
10	.0000	

EXERCISES

4.37 If a random variable has the binomial distribution with $p = .22$, find the probability of 2 successes in 3 independent trials.

4.38 Find the probability of 1 success in 4 independent trials if the probability of a success on any given trial is .43.

4.39 The probability that a sales call produces a sale is .5. Find the probability that no sales will result from 5 consecutive sales calls.

4.40 The probability that a light bulb will survive at least 1,000 hours of use before burning out is found by experience to be .75. Find the probability that one of 4 light bulbs will burn out before 1,000 hours.

4.41 A container is filled with 50 packages to be shipped to two different destinations. Can the binomial distribution be used to find the probability that 10 of the first 20 packages removed from the container, without replacement, will go to a specific one of the destinations? Why?

4.42 a. A bin contains 100 parts; 30 of them are to be used in assembling product A, and 70 are to be used for product B. Can the binomial distribution be used to find the probability that 15 of the first 50 parts removed from the bin, without replacement, will be used for product A? Why?

   b. Would your answer be different if each part were replaced before the next one were drawn? Explain.

4.43 A lot containing 30 manufactured products is examined for *major* defects. Whenever a product is found to contain such defects, it is set aside. Ten products are removed from the remainder of the lot and examined for *minor* defects. Each product is replaced before the next is removed.

   a. Can the binomial distribution be used to find the probability that fewer than two of the products removed from the remainder lot will contain *minor* defects?

   b. If so, will this probability necessarily apply to removal of products from the original lot of 30? Why?

4.44 Sketch a probability histogram of the binomial distribution having $n = 5$ and $p = .8$.

4.45 Sketch a probability histogram of the binomial distribution having $n = 5$ and $p = .2$.

4.46 Use Table 1 (Appendix A) to find the following probabilities:

   a. $B(6;15,.4)$ c. $b(2;9,.15)$

   b. $B(12;18,.75)$ d. $b(8;17,.45)$

4.47 Use Table 1 to find the following probabilities:

   a. $B(10;11,.65)$ c. $b(8;16,.5)$

   b. $B(12;14,.8)$ d. $b(11;20,.65)$

4.48 If a random variable has the binomial distribution with $n = 15$ and $p = .45$, use Table 1 to find the probability that its value will

   a. be less than or equal to 8

   b. be greater than or equal to 6

  c. exceed 4

  d. be less than 10

  e. be exactly 5

  f. lie between 4 and 10, inclusive

4.49 If a random variable has the binomial distribution with $n = 18$ and $p = .7$, use Table 1 to find the probability that its value will

  a. be less than or equal to 9

  b. equal or exceed 11

  c. be greater than 15

  d. be less than 13

  e. equal 12

  f. lie between 10 and 14, inclusive

4.50 Find the mean and the standard deviation of the binomial distribution in Exercise 4.38.

4.51 Find the mean and the standard deviation of the binomial distribution in Exercise 4.37.

4.52 Find the mean and the standard deviation of the number of light bulbs burning out in Exercise 4.40.

4.53 Find the mean and the standard deviation of the number of sales calls resulting in a sale in Exercise 4.39.

4.54 In Exercise 4.40, find the mean and the standard deviation of the proportion of light bulbs that will burn out.

4.55 a. Find the mean proportion of sales calls resulting in a sale in Exercise 4.39.

  b. Find the standard deviation of this proportion if there are 50 sales calls.

4.56 a. If the number of trials is doubled, what happens to the standard deviation of the binomial distribution?

  b. In general, if n is multiplied by the factor k in a binomial distribution having the parameters n and p, what statement can be made about the standard deviation of the resulting probability distribution?

4.57 To reduce the standard deviation of the binomial distribution by half, what change must be made in the number of trials?

4.58 If experience shows that the proportion of convertibles to total automobiles on a used car lot is .07 on any given day, use Table 1 to find the probability that there will be more than 3 convertibles on a day that the lot contains 20 automobiles.

4.59 What is the probability that 15 of 16 lawn mowers will start on the first pull if the manufacturer's claim is true that first-pull starting can be expected 90% of the time?

4.60 A fire chief claims that 60% of the calls are false alarms.

    a. Find the probability that, in 15 consecutive calls, there are fewer than 6 false alarms.

    b. Would this result tend to substantiate or to cast doubt on the chief's claim? Why?

4.61 A manufacturer claims that, at most 10% of the time, a given product will sustain fewer than 500 hours of use before requiring service. Twelve products were tested, and it was found that 2 of them required service before 500 hours of use. Use Table 1 to find the probability that 2 or more of 12 products would require service before 500 hours, and comment on the validity of the manufacturer's claim.

4.62 a. Use a computer program to calculate the probability that between 40 and 60 heads will be obtained when a balanced coin is tossed 100 times.

    b. Would you be surprised if fewer than 25 heads appeared? Why?

4.63 a. Use a computer program to calculate the probability that there will be more than 5 integrated circuits not meeting electrical specifications in a lot of 150 circuits, if the percentage of defectives is assumed to be 1%.

    b. Would such a result cast doubt on this assumption? Why?

4.64 a. Use a computer program to calculate the probability that more than 15% of 80 business telephone calls will last longer than five minutes if is assumed that 10% of such calls last this long.

    b. Can this result be regarded as evidence that more than 10% of the calls last longer than 5 minutes?

4.65 a. Use a computer program to calculate the probability of rolling between 10 and 20 sevens in 90 rolls of a pair of balanced dice. (The probability of rolling a seven on any given roll is 1/6.)

    b. Would it surprise you if more than 22 sevens were rolled? Why?

The following exercises introduce new material, expanding on some of the topics covered in the text.

4.66 *Geometric distribution.* Suppose, in a sequence of trials satisfying the assumptions underlying the binomial distribution, we wish to know the number of trials required for the first success to occur. Show that the probability that the first success occurs on trial x is given by the **geometric distribution**

$$g(x; p) = p(1 - p)^{x-1} \qquad \text{for } x = 1,2,3, \ldots$$

Hint: The first $x - 1$ trials must result in failures.

4.67 *Continuation.* After an initial engine test, the rotor blades of a jet engine are removed and examined for flaws. If experience has shown that the probability of a flawed rotor blade is .005, what is the probability that it will require 100 inspections to find the first flawed blade?

4.68 *Multinomial distribution.* Suppose, in a sequence of n independent trials, there are k possible outcomes on any given trial, having the probabilities $p_1, p_2, \ldots, p_k$, with $p_1 + p_2 + \ldots + p_k = 1$. Using methods similar to those used in deriving the binomial distribution, it can be shown that the probability of obtaining x_1 outcomes of type 1, x_2 outcomes of type 2, $\ldots$, x_k outcomes of type k is given by the **multinomial distribution**

$$f(x_1, x_2, \ldots, x_k) = \frac{n!}{x_1! \, x_2! \cdots x_k!} \, p_1^{x_1} p_2^{x_2} \cdots p_k^{x_k} \quad \textit{with} \sum_{i=1}^{k} x_i = n$$

Suppose that manufactured units are classified into three categories: defective, standard quality, and high quality. If the probabilities of the three kinds of quality are .05, .85, and .10, respectively, find the probability of obtaining 1 defective unit, 12 standard quality units, and 2 high-quality units in a production run of 15 units.

4.69 *Hypergeometric distribution.* The binomial distribution can be used for taking samples from a finite population with replacement. Suppose a sample of size n is taken without replacement from a population having N elements. If there are a "successes" in the population, it can be shown that the probability that the sample contains x successes is given by

$$h(x; n, a, N) = \frac{\binom{a}{x}\binom{N-a}{n-x}}{\binom{N}{n}} \quad \textit{for } x = 0, 1, \ldots, n$$

(Refer to page 81 for the definition of the *combinatorial coefficient.*) This equation defines the **hypergeometric distribution** whose parameters are the sample size n, the population size N, and the number of "successes" in the lot a. Defining the draw of a red ball as a "success," find the probability that there will be 3 red balls in a sample of size 5 chosen without replacement from an urn containing 5 red balls and 4 black balls.

4.70 *Continuation.* Find the exact probability of obtaining no defectives in drawing a sample of size 5 without replacement from a lot containing 8 parts, of which 1 is defective. Approximate this probability with the binomial distribution (assume the sample was chosen with replacement) and compare results.

4.5 The Poisson Distribution

In many important applications, the value of p in the binomial distribution is small, and the number of trials, n, is large. When inspecting product, for example, large numbers of product often are inspected, and the proportion of defectives is expected to be very small. In controlling the number of defects, such as the number of bubbles in a long sheet of enameled steel, the number of defects per unit (perhaps one foot of the sheet) is expected to be small, and many units are inspected. Calculations involving the binomial distribution can become extremely difficult when n is large; thus, it is useful to have a relatively simple approximation for the binomial distribution in such cases.

To find a reasonable approximation, as the parameter p of the binomial distribution becomes close to zero while the parameter n becomes arbitrarily large, it is useful to keep their product, np, fixed at some value, λ. This product is the mean of the binomial distribution; keeping np fixed while simultaneously letting p approach zero and n approach infinity leads to the limit approached by the binomial distributions having the mean $\lambda = np$.

The limiting form of the binomial distribution having parameters n and p, as $n \to \infty$ and $p \to 0$, while their product np remains fixed at the value λ, is called the **Poisson distribution**.

The equation for this distribution is

Poisson Distribution: $f(x; \lambda) = \dfrac{\lambda^x e^{-\lambda}}{x!}$ *for $x = 0, 1, 2, \ldots$*

This is a single-parameter family of distributions, having the parameter λ. In this equation, e is the base of the system of natural logarithms, $e \approx 2.718282$.

Since the mean of the binomial distribution, np, was kept fixed at the value λ throughout the limiting process, the **mean of the Poisson distribution** is λ. The variance of the binomial distribution approximated by this Poisson distribution is $np(1 - p)$. Throughout the approach to the limit, np is kept fixed at the value λ, and p approaches 0. Thus, the variance approaches λ. It follows that the **variance of the Poisson distribution** is λ and its standard deviation is $\sqrt{\lambda}$.

A probability histogram of the Poisson distribution having the parameter $\lambda = 1$ is shown in Figure 4.3. The histogram in this figure typifies all Poisson distributions by exhibiting skewness, with a long right-hand tail. Poisson distributions typically assign relatively high probabilities to a few small values of the random variable, and small probabilities to many large values.

> *Example.* A random variable has the binomial distribution with parameters 100 and .02. Find the exact probability that this random variable will have a value of 3. Compare this result with the approximate probability obtained from the corresponding Poisson distribution.
>
> *Solution.* Substituting $n = 100, p = .02$, and $x = 3$ into the equation for the binomial distribution, the required probability is
>
> $$b(3; 100, .02) = \frac{100!}{3! \, 97!} (.02)^3 (.98)^{97} = .182$$

Substituting $\lambda = (100)(.02) = 2$ and $x = 3$ into the equation for the Poisson distribution, the approximate probability is

$$f(3;2) = \frac{2^3 \cdot e^{-2}}{3!} = .180$$

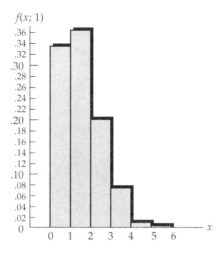

FIGURE 4.3   Poisson Distribution with $\lambda = 1$

The approximation in this example has an error of only .002, or about 1%. The approximation of binomial probabilities by the Poisson distribution usually is good whenever $n \geq 20$ and $p \leq .05$. If $n \geq 100$, the approximation is good as long as $np \leq 10$.

The Poisson distribution is tabulated in Table 2 (Appendix A) for values of λ from 0.02 to 25 in various increments. This table gives values of the **cumulative Poisson distribution.**

Cumulative Poisson Distribution: $F(x;\lambda) = e^{-\lambda} \sum_{k=0}^{x} \frac{\lambda^k}{k!}$, for $x = 0, 1, 2, \ldots$

Table 2 is used in much the same way as Table 1 is. The probability that a random variable having the Poisson distribution with parameter λ assumes the value x is given by $F(x;\lambda) - F(x - 1;\lambda)$. The probability that such a random variable assumes a value greater than x is given by $1 - F(x;\lambda)$.

    Example.   If the average number of bubbles per 100 feet in a coil of enameled steel is 0.1, find the probability that a 100-foot section will contain (a) at most 1 bubble, (b) no bubbles.

    Solution. a.   Approximating this probability with the Poisson distribution having the mean $\lambda = 0.1$, the probability of at most 1 bubble is given by $F(1;0.1)$ which is found in Table 2 to equal .995.

b. The probability of no bubbles is given by $f(0;0.1) = F(0;0.1) = .905$.

Note in this example it was not necessary to know the values of n and p. To calculate probabilities with the Poisson distribution, it is enough to know the value of its mean, the parameter λ. In the next example, the Poisson distribution is used to approximate a binomial probability.

> *Example.* If the proportion defective on inspection of stampings for an automotive quarter panel is 0.4%,
>
> a. What is the probability of obtaining 3 or more defective quarter panels on a day when 350 panels are stamped?
>
> b. What are the mean and the standard deviation of the number of defective panels?
>
> *Solution.* a. Using the Poisson approximation to the binomial, with $\lambda = (350)(.004) = 1.40$, the required probability is given by $1 - F(2;1.40) = 1 - .833 = .167$.
>
> b. The mean number of panels is $\lambda = 1.4$ panels, and the standard deviation is $\sqrt{1.4} = 1.18$.

It would be remiss to leave this brief introduction to the Poisson distribution without mentioning this distribution's important applications in addition to serving as an approximation of the binomial distribution. The Poisson distribution can be thought of as the distribution of "rare events," and it is often an appropriate model for such probabilities. This distribution also arises in queuing theory, where probabilities involving the lengths of waiting lines and arrival times are calculated. A random process, a physical process wholly or partly controlled by a chance mechanism, which satisfies assumptions analogous to those underlying the Poisson distribution, is called a **Poisson process**.

Computer Applications

To obtain values of the Poisson distribution with *MINITAB* software, the command sequence

```
MTB > PDF ;
SUBC > POISSON, MU = 1.38.
```

will produce the first and second columns of the table given at the top of page 92. Changing the initial command from PDF to CDF will produce the corresponding cumulative distribution shown in the first and third columns.

POISSON PROBABILITIES FOR MEAN = 1.380

K	P(X = K)	P(X LESS OR = K)
0	.2516	.2516
1	.3472	.5988
2	.2396	.8283
3	.1102	.9485
4	.0380	.9865
5	.0105	.9970
6	.0024	.9994
7	.0005	.9999
8	.0001	1.0000

EXERCISES

4.71 Use Table 2 (Appendix A) to find the following probabilities:

    a. $F(4;3.0)$ c. $f(4;2.4)$

    b. $F(18;12.5)$ d. $f(5;8.0)$

4.72 Use Table 2 to find the following probabilities:

    a. $F(11;10.0)$ c. $f(5;8.5)$

    b. $F(30;25.0)$ d. $f(4;1.1)$

4.73 If a random variable has the Poisson distribution with a mean of 5.2, use Table 2 to find the probability that the random variable will assume a value greater than 7.

4.74 Use Table 2 to find the probability that a random variable having the Poisson distribution with a mean of 13.5 will assume a value greater than or equal to 10.

4.75 Use Table 2 to find the probability that a random variable having the Poisson distribution with parameter 7.6 will assume a value between 4 and 9, inclusive.

4.76 Use Table 2 to find the probability that a random variable having the Poisson distribution with $\lambda = 17$ will assume a value greater than 10 but less than 20.

4.77 A container is filled with 1,000 bolts, of which 10 have stripped threads. Use the Poisson distribution to approximate the probability of finding no more than one bolt with stripped threads in 100 bolts selected from the container.

4.78 a. The probability that an insured will have a claim in any year is .005. If an insurance company has 2,000 policyholders, use the Poisson distribution to approximate the probability that there will be more than 15 claims in a given year.

    b. What is the mean number of yearly claims?

4.79 The mean waiting time between customers at a gasoline station is 5 minutes. If the number of customers in a unit of time can be assumed to have the Poisson distribution, find the probability that there will be at least 10 customers in a given hour.

4.80 A conveyor belt containing peanuts is continuously scanned for the presence of discolored peanuts, which may contain aflatoxin. If such a peanut is found every ten minutes on average, what is the probability of finding no more than 5 discolored peanuts in any 60-minute period?

4.81 Sketch a probability histogram of the Poisson distribution having the mean $\lambda = 2.0$.

4.82 Sketch a probability histogram of the Poisson distribution having the mean 3.4.

4.83 Find the standard deviation of the Poisson distribution $f(x;16)$.

4.84 Find the standard deviation of the Poisson distribution whose mean is 12.5.

4.85 a. Use a computer program to calculate the exact probability of obtaining 5 or more defectives in inspection of 1,000 products when the probability that a given product will be defective is .006.

   b. Using the same program, find the Poisson approximation for this probability, and compare the two results.

4.86 a. Use a computer program to approximate the probability calculated in Exercise 4.64 on page 87.

   b. Compare the two results.

4.87 a. Use a computer program to approximate the probability found in Exercise 4.63 on page 87.

   b. Would your conclusion about the assumption of 1% defectives change as a result of this approximation?

4.88 a. Use a computer program to approximate the probability of rolling between 10 and 20 sevens in 90 rolls of a pair of balanced dice. (The probability of rolling a seven on any given roll is 1/6.)

   b. Compare this probability with that obtained in Exercise 4.65 on page 87.

The following exercises introduce new material, expanding on some of the topics covered in the text.

4.89 *Poisson process.* The Poisson distribution can be applied to a process occurring in continuous time, called the **Poisson process.** If the assumptions underlying the binomial distribution are satisfied, and the probability of success in a small interval of time Δt is proportional to the length of the interval, that is, $p = \alpha \Delta t$, then the probability of x successes in time T is given by $b(x;n,p)$, with $n = \dfrac{T}{\Delta t}$. As $n \to \infty$, this probability approaches the corresponding Poisson probability with parameter $\lambda = np = \dfrac{T}{\Delta t}\alpha \Delta t = \alpha T$. Note that α is the mean number of successes per unit of time. Find the probability that a Poisson process with parameter $\alpha = 0.1$ will have at most 1 success in 10 units of time.

4.90 *Continuation.* In a rolling mill, an average of 0.06 imperfections are spotted per foot of sheet metal produced. What is the probability of obtaining more than 10 imperfections in 100 feet of sheet metal?

4.6   Application to Quality

Modern quality practice is based on the principle: "Make it right the first time." If this principle were followed perfectly, finished products would contain no defects, and there would be no need for final inspection. Indeed, some production facilities in Japan have been able to eliminate the need for final inspection entirely. Careful design for producibility, teamwork starting with the design and going through production and distribution, good use of statistical process control, and the assumption of responsibility for defect prevention by each line worker instead of by inspectors, can reduce defects in finished products to nil or nearly nil.

Until a high degree of confidence has been achieved that a product has been produced without defects, however, at least some final inspection will remain necessary. Unfortunately, final inspection to a given level of defective product makes it too easy to fall into the habit of accepting that level of defectives as unavoidable, with the resulting loss of interest in quality improvement. Thus, final inspection should be regarded only as necessary to *verify* that quality-improvement methods are working, and to measure the extent that quality has continuously been improved. Inspection *never* should be relied upon as a means of improving the quality of a product by weeding out the defectives. It is a dangerous delusion to believe that quality can be inspected into a product!

To illustrate this important point, let us examine the fallacy of 100% inspection. Defects often are missed, even in 100% inspection, as a result of inspector fatigue, inadequate instrumentation, differences in the interpretation of what constitutes a defect, and the desire to "pass" borderline product. In fact, 100% inspection has been found to miss as many as 40% of the defects, and it has been reported that 50% or more of the defects have been missed for some complex systems. Military contracts, recognizing this fact, sometimes call for a second 100% inspection of lots containing excess defectives. But even this seemingly unerring approach can be inadequate.

Pursuing this argument with specific numbers, suppose that 100% inspection shows that 10% of the products contain defects, and experience shows that 40% of the defectives are missed on the first round. Thus, the original level of defectives actually was $10/.6 = 17\%$, not 10%. Thus, if 100 products are inspected and 10%, or 10 products are weeded out in the first 100% inspection, there will be 7 defective products left to be found in the second 100% inspection of the remaining 90 products. But only 60%, or about 4 of them, will be found in the second 100% inspection, leaving 3 defective products. The defect level in the product shipped to the customer, then, is 3 defectives in $100 - 10 - 4 = 86$ products shipped, or 3.5%. Anyone interested in producing a quality product will not accept a level of defectives in shipped product as high as 3.5% (or even 1%).

Since inspection never should be used to improve quality by weeding out defectives, the role of inspection should be confined to estimating the defect level. It has been shown in many instances that **sampling inspection** can do at least as good a job of estimating the defect level as does 100% inspection. A well-designed

sampling plan reduces the cost of inspection; sampling reduces fatigue and elimi-
nates some of the other causes of missed defectives by making the process of
inspection easier to manage.

In sampling inspection, a specified sample of a lot of manufactured product is inspected under controlled, supervised conditions. If the number of defectives in the sample exceeds a given **acceptance number**, the lot is rejected. (A
rejected lot may be subjected to closer inspection, but it is rarely scrapped.) A **sampling plan** consists of a specification of the number of items to be included in the
sample taken from each lot, and a statement about the maximum number of defectives allowed before rejection takes place.

The probability that a lot will be accepted by a given sampling plan, of
course, will depend upon p, the actual proportion of defectives in the lot. Since the
value of p is unknown, we shall calculate the probability of accepting a lot for several different values of p. Suppose a sampling plan requires samples of size n from
each lot, and that the lot size is large with respect to n. Suppose, further, that the
acceptance number is c; that is, the lot will be accepted if c defectives or fewer are
found in the sample. The probability of acceptance, the probability of finding c or
fewer defectives in a sample of size n, is given by the binomial distribution to a
close approximation. (Since sampling inspection is done without replacement, the
assumption of equal probabilities from trial to trial, underlying the binomial distribution, is violated. But if the sample size is small relative to the lot size, this
assumption is nearly satisfied.) Thus, for large lots, the probability of accepting a
lot having the proportion of defectives p is closely approximated by

$$\textbf{\textit{Probability of Acceptance:}} \quad L(p) = \sum_{k=0}^{c} b(k;n,p) = B(c;n,p)$$

This formula simply states that the probability of c or fewer defectives in the
sample is given by the probability of 0 defectives, plus the probability of 1 defective, . . . , up to the probability of c defectives, with each probability being approximated by the binomial distribution having the parameters n and p.

It can be seen from the formula for $L(p)$ that, for a given sampling plan
(sample size, n, and acceptance number, c) the probability of acceptance, depends
upon p, the actual (unknown) proportion of defectives in the lot. Thus a curve can
be drawn that gives the probability of accepting a lot as a function of the lot
proportion defective, p. This curve, called the **operating characteristic curve**, or
OC curve, defines the characteristics of the sampling plan.

To illustrate the construction of an OC curve, let us consider the sampling
plan having $n = 20$ and $c = 3$. That is, samples of size 20 are drawn from each lot,
and a lot is accepted if the sample contains 3 or fewer defectives. Referring to the
line in Table 1 (Appendix A) corresponding to $n = 20$ and $x = 3$, the probabilities
that a random variable having the binomial distribution $b(x;20,p)$ will assume a
value less than or equal to 3 for various values of p are as follows:

p	.05	.10	.15	.20	.25	.30	.35	.40	.45
$L(p)$	.9841	.8670	.6477	.4114	.2252	.1071	.0444	.0160	.0049

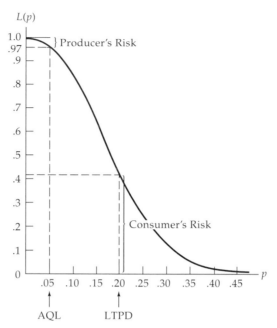

FIGURE 4.4   OC Curve

A graph of $L(p)$ versus p is shown in Figure 4.4.

      Inspection of the OC curve given in Figure 4.4 shows that the probability of acceptance is quite high (greater than .9) for small values of p, say values less than about .10. Also, the probability of acceptance is low (less than .10) for values of p greater than about .30. If the actual proportion of defectives in the lot lies between .10 and .30, however, it is somewhat of a tossup whether the lot will be accepted or rejected. An "ideal" OC curve would be like the one shown in Figure 4.5. In this figure, there is no "gray area"; that is, it is certain that a lot with a given small value of p or less will be accepted, and it is certain that a lot with a value of p greater than the given value will be rejected. By comparison, the OC curve of Figure 4.4 seems to do a poor job of discriminating between "good" and "bad" lots. In such cases, a better OC curve can be obtained by increasing the sample size, n.

FIGURE 4.5   "Ideal" OC Curve

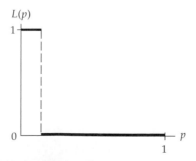

The OC curve of a sampling plan never can be like the ideal curve of Figure 4.5 with finite sample sizes, as there always will be some error associated with sampling. However, sampling plans can be evaluated by choosing two values of p considered to be important and calculating the probabilities of lot acceptance at these values. First, a number, p_0, is chosen so that a lot containing a proportion of defectives less than or equal to p_0 is desired to be accepted. This value of p is called the **acceptable quality level**, or **AQL**. Then, a second value of p, p_1, is chosen so that we wish to reject a lot containing a proportion of defectives greater than p_1. This value of p is called the **lot tolerance percentage defective**, or **LTPD**. We shall evaluate a sampling plan by finding the probability that a "good" lot (a lot with $p \le p_0$) will be rejected and the probability that a "bad" lot (one with $p \ge p_1$) will be accepted.

The probability that a "good" lot will be rejected is called the **producer's risk**, and the probability that a "bad" lot will be accepted is called the **consumer's risk.** The producer's risk expresses the probability that a "good" lot (one with $p < p_0$) will erroneously be rejected by the sampling plan. It is the risk that the producer takes as a consequence of sampling variability. The consumer's risk is the probability that the consumer erroneously will receive a "bad" lot (one with $p > p_1$).

Suppose an AQL of .05 is chosen ($p_0 = .05$). Then, it can be seen from Figure 4.4 that the given sampling plan has a producer's risk of about .03, since the probability of *acceptance* of a lot with an actual proportion defective of .05 is approximately .97. Similarly, if an LTPD of .20 is chosen, the consumer's risk is about .41. This plan obviously has an unacceptably high consumer's risk—over 40% of the lots received by the consumer will have 20% defectives or greater. To produce a plan with better characteristics, it will be necessary to increase the sample size, n, to decrease the acceptance number, c, or both. The following example shows what happens to these characteristics when c is decreased to 1, while n remains fixed at 20.

Example. Find the producer's and consumer's risks corresponding to an AQL of .05 and an LTPD of .20 for the sampling plan defined by $n = 20$ and $c = 1$.

Solution. First, we calculate $L(p)$ for various values of p. Referring to Table 1 (Appendix A) with $n = 20$ and $x = 1$, we obtain the following table:

p	.05	.10	.15	.20	.25	.30	.35	.40	.45
$L(p)$	.7358	.3917	.1756	.0692	.0243	.0076	.0021	.0005	.0001

A graph of this OC curve is shown in Figure 4.6. From this graph, we observe that the producer's risk is $1 - .7358 = .2642$, and the consumer's risk is .0692.

Reduction of the acceptance number from 3 to 1 obviously has improved the consumer's risk, but now the producer's risk seems unacceptably high. Evidently, a larger sample size is needed.

The preceding example has been somewhat artificial owing to the limit of $n = 20$ in Table 1. It would be quite unusual to specify an LTPD as high as .20 (20% defectives), and much higher sample sizes than 20 usually are used for acceptance

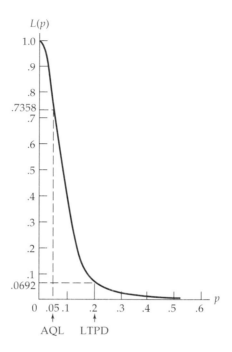

FIGURE 4.6   OC Curve for Example on Page 97

sampling. In practice, OC curves have been calculated for sampling plans having many different combinations of n and c. Choice then is made of the sampling plan whose OC curve has as nearly as possible the desired characteristics, AQL, LTPD, consumer's risk, and producer's risk.

EXERCISES

4.91 a.   If 30% of the defectives in a lot of 1,000 products are missed on 100% inspection, how many defective products are found when inspecting a lot having 10% defectives?

   b.   How many are missed?

4.92 A lot containing 15% defectives is inspected 100%, eliminating the defectives found. The lot is re-inspected 100% and any additional defectives that are found are removed from the lot. If 25% of the defectives are missed in each inspection, what is the percentage of defectives remaining in the reduced lot after the second inspection?

4.93 A sampling inspection program has a .10 probability of rejecting a lot when the true proportion of defectives is .01, and a .95 probability of rejecting the lot when the true proportion of defectives is .03. If .01 is the AQL and .03 is the LTPD, what are the producer's and consumer's risks?

4.94 The producer's risk in a sampling program is .05 and the consumer's risk is .10, the AQL is .03 and the LTPD is .07.

   a.   What is the probability of accepting a lot whose true proportion of defectives is .03?

b. What is the probability of accepting a lot whose true proportion of defectives is .07?

4.95 Suppose the acceptance number in the example on page 95 is changed from 3 to 2. Keeping the producer's risk at .05 and the consumer's risk at .10, what are the new values of the AQL and the LTPD?

4.96 From Figure 4.4 on page 96,

a. Find the producer's risk if the AQL is .10.

b. Find the LTPD corresponding to a consumer's risk of .05.

4.97 Sketch the OC curve for a sampling plan having a sample size of 15 and an acceptance number of 2.

4.98 Sketch the OC curve for a sampling plan having a sample size of 10 and an acceptance number of 1.

4.99 Sketch the OC curve for a sampling plan having a sample size of 8 and an acceptance number of 0.

4.100 If the AQL is .1 and the LTPD is .25 in the sampling plan given in Exercise 4.98, find the producer's and consumer's risks.

4.101 Find the AQL and the LTPD of the sampling plan in Exercise 4.97 if both the producer's and consumer's risks are .10.

4.102 a. In Exercise 4.98, change the acceptance number from 1 to 0 and sketch the OC curve.

b. How do the producer's and consumer's risks change if the AQL is .05 and the LTPD is .3 in both sampling plans?

REVIEW EXERCISES

These review exercises can be used for informal review of the chapter, or as two practice examinations, designed to be taken over a time period of about one hour. The odd-numbered exercises comprise one examination; the even-numbered exercises comprise the other.

4.103 If A, B, and C are mutually exclusive events in the sample space S, express the following probabilities in terms of $P(A)$, $P(B)$, and $P(C)$.

a. $P(A \cup B)$ d. $P(S)$

b. $P(A \cap B)$ e. $P(C')$

c. $P(B \cup C)$ f. $P(A \cup B')$

4.104 If A and B are independent events, with $P(A) = .4$ and $P(B) = .3$, find

a. $P(A \cap B)$ c. $P(A' \cap B)$

b. $P[(A \cap B)']$ d. $P(A \cup B)$

4.105 Three playing cards are chosen from a well-shuffled deck of 52 cards, replacing each card and re-shuffling before the next card is drawn.

a. Make a table showing the probability distribution of the number of spades drawn.

b. Find the mean and the variance of this distribution.

4.106 Given the probability distribution having the equation

$$f(x) = k/x \qquad \text{for } x = 1, 2, 3, 4$$

a. Find the value of k.

b. Find the mean and the standard deviation of this distribution.

c. Find the probability that x assumes a value less than 3.

4.107 If x is a value of a random variable having the binomial distribution with $n = 12$ and $p = .25$, use Table 1 to find the probabilities of the following events:

a. $x \le 5$ b. $x = 9$ c. $x > 7$ d. $0 \le x \le 6$

4.108 Given that x is a value of a random variable having a binomial distribution, find the following:

a. x, if $b(x;6,.75) = .2967$

b. p, if $B(4;8,p) = .7396$

c. n, if $B(19;n,.85) = .9612$

d. the probability that $2 < x < 6$, if $n = 15$ and $p = .25$.

4.109 In a two-candidate election, 55% of the voters voted for the winning candidate. If 12 voters are interviewed in a post-election poll, what is the probability of interviewing between 5 and 7 people who voted for the winning candidate?

4.110 a. If the probability that a fisherman will catch a salmon weighing over 500 pounds during a day of deep-sea fishing is .15, what is the mean number of such salmon caught by a boat containing 18 fishermen?

b. What is the probability that they will catch more than 4 such fish? Assume that once a fisherman catches such a salmon, he or she quits fishing.

4.111 A study of 5,000 patients shows that 750 of them have certain abnormalities in their blood chemistry.

a. In hospitals containing 100 patients, what is the mean proportion of patients with these abnormalities?

b. What is the standard deviation of this proportion?

4.112 It has been estimated that 50% of fatal highway accidents are caused by drunk drivers. What is the standard deviation of the proportion of accidents caused by drunk drivers in a study of 400 accidents?

4.113 If x is a value of a random variable having the Poisson distribution with a mean of 5.2, use Table 2 (Appendix A) to find the probabilities of the following events:

a. $x \le 3$ b. $x = 1$ c. $x > 2$ d. $1 \le x \le 3$

4.114 Given that x is a value of a random variable having a Poisson distribution, find the following:

a. λ, if $F(30;\lambda) = .936$

b. x, if $f(x;2) = .180$

c. the probability that $2 < x < 6$, if the mean is 6.0

4.115 Find an approximation for the probability that between 8 and 12 aces will be drawn in 130 draws from a deck of 52 playing cards, where each card drawn is replaced and the deck is re-shuffled before each draw.

4.116 If the probability of finding a "rare" coin in circulation is .005, what is the probability of finding at least one such coin in a jar containing 50 coins?

4.117 Sketch the OC curve for a sampling plan with $n = 5$ and $c = 0$. What is the AQL if the producer's risk is .10?

4.118 What is the LTPD of a sampling plan with $n = 11$ and $c = 2$ if the consumer's risk is .2?

GLOSSARY

	page		page
Acceptable Quality Level	97	Mutually Exclusive Events	68
Acceptance Number	95	Operating Characteristic	
Binomial Distribution	79	Curve	95
Classical Method	71	Parameter of Distribution	81
Complementary Event	68	Poisson Distribution	89
Conditional Probability	75	Poisson Process	93
Consumer's Risk	97	Probability	70
Continuous Random Variable	75	Probability Distribution	76
Cumulative Binomial		Probability Histogram	76
Distribution	82	Producer's Risk	97
Cumulative Poisson		Random Digit	79
Distribution	90	Random Variable	69
Discrete Random Variable	75	Relative-Frequency	
Event	68	Method	71
Experiment	68	Sample Space	68
General Law of Addition	72	Sampling Inspection	94
Geometric Distribution	87	Sampling Plan	95
Hypergeometric Distribution	88	Special Law of	
Independent Events	72	Multiplication	71
Intersection of Events	68	Standard Deviation of:	
Lot Tolerance Percentage		Binomial Distribution	83
Defective	97	Probability Distribution	77
Mean Of:		A Proportion	84
Binomial Distribution	83	Subjective Method	71
Poisson Distribution	89	Variance of:	
Probability Distribution	76	Poisson Distribution	89
A Proportion	84	Probability Distribution	77
Multinomial Distribution	88	Union of Events	68

5

CONTINUOUS PROBABILITY DISTRIBUTIONS

Where have we been? We have laid the foundation for further exploration of problems involving quality by reviewing probability theory and introducing some important probability distributions for discrete measurements. But our task is incomplete.

Where are we going? Most data taken in connection with problems of quality are continuous. Thus, we must extend the idea of probability distributions to include continuous data. We then can discover the important applications of the normal distribution, and introduce several other continuous distributions that are met in the solution of problems in quality and reliability.

Section 5.1, *Distributions of Continuous Random Variables,* contrasts continuous probability distributions with those involving discrete random variables, and characterizes probabilities as areas under curves in the continuous case.

Section 5.2, *The Normal Curve of Error,* introduces the normal distribution and some of its properties.

Section 5.3, *Probabilities and the Normal Distribution,* defines a standardized random variable and the standard normal distribution, showing how probabilities involving normally distributed random variables can be calculated using a table of the standard normal distribution.

Section 5.4, *Checking Data for Normality,* shows how normally distributed data appear on a probability plot or on a normal-scores plot and how to transform non-normal data to data that are approximately normally distributed.

Section 5.5, *Other Continuous Distributions,* introduces three important distributions: the continuous uniform distribution, the lognormal distribution, and the exponential distribution, identifying their properties and some of their most useful applications.

Section 5.6, *Application to Quality,* applies the normal distribution to the identification of discrepant observations, to reducing the proportion of defective product, and to testing the reasonableness of a claim made about the quality of a product or service.

5.1 Distributions of Continuous Random Variables

The distinction between a continuous and a discrete random variable was described on page 75. This distinction creates profound differences in the interpretation of probabilities with continuous and discrete random variables.

Probabilities are given directly for discrete random variables. Thus, if $f(x)$ is a value of a discrete probability distribution function, this value can be interpreted directly as the probability that the associated random variable takes on the value x. But probabilities cannot be given in this direct fashion for distributions of continuous random variables.

To see why, let us contrast the graphs of discrete and continuous probability functions. The graph of a discrete probability distribution was called a probability histogram on page 76, and it is a sequence of bars as shown in Figure 4.1. Each bar represents a distinct value of the associated random variable, and its area represents the probability that the random variable will take on that value. For a continuous probability distribution, however, the graph of $f(x)$ versus x is a continuous curve. Probabilities remain defined as areas, but now they are areas under a curve rather than areas of distinct bars. For example, the probability that the continuous random variable having the distribution given by $f(x)$ will take on a value from 1 to 2 is given by the area under the curve representing $f(x)$ from $x = 1$ to $x = 2$.

The probability that the continuous random variable having the distribution given by $f(x)$ will assume a value between x_0 and $x_0 + \Delta x$ is given by an area under $f(x)$ having a base of width Δx, as shown in Figure 5.1. As $\Delta x \to 0$, we obtain the probability that the random variable assumes the value x_0 *exactly*. But, the corresponding area also approaches zero; therefore the associated probability is zero. From this we can see that probabilities for continuous random variables can be non-zero only for intervals having non-zero width. For this reason, whenever x is a value of a continuous random variable, $f(x)$ is referred to as a **probability density function,** rather than a probability distribution. Its values give probability *densities*, rather than probabilities.

As a physical analogy to this idea, consider a rod of uniform material and circular cross-section with continuously varying diameter. Any slice of the rod, no matter how thin, will have a mass given by the mean density from the left to the

FIGURE 5.1 Probabilities Associated with Continuous
Random Variables

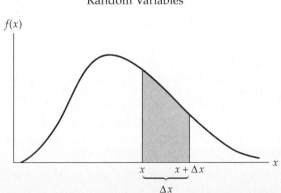

right endpoint of the slice, times the length of the slice. A slice of width zero will have zero mass. Thus, the density of the rod is not its mass, but a number which helps us calculate the mass of a slice of given thickness cut from the rod.

If $f(x)$ is a value of a discrete function, the conditions it must satisfy in order to represent a probability distribution were summarized on page 76. Some changes in these conditions are necessary for a probability density function. First, no longer is it necessary for $f(x)$ to be less than 1, since it is a probability density function and not a probability. Second, the *area* under $f(x)$ over its entire range of definition must equal 1; thus, the word *sum* in the second condition on page 76 is replaced by the word *integral*. Summarizing, the conditions for $f(x)$ to represent a probability density on the interval $a < x < b$, where a and b are any real numbers, are as follows:

Probability Density Function:

(1) For every value of x in the interval $a < x < b$, $f(x) \geq 0$.

(2) The integral of $f(x)$ from a to b must equal 1.

    Example.   a.   Show that

$$f(x) = \frac{3}{37}(x-4)^2 \quad \text{for } 0 < x < 1$$

represents a probability density function, and (b) find the probability that $0.1 < x < 0.2$.

    Solution.   a.   Since x appears only in the term $(x - 4)^2$, $f(x) \geq 0$ for all values of x. Also,

$$\int_0^1 \frac{3}{37}(x-4)^2\,dx = \frac{3}{37}\left[\frac{(1)^3}{3} - 4(1)^2 + 16(1)\right] = 1$$

     b.   The required probability is given by

$$\int_{0.1}^{0.2} \frac{3}{37}(x-4)^2\,dx = 0.12$$

    EXERCISES

5.1   a.   Show that

$$f(x) = 2x \qquad \text{for } 0 < x < 1$$

represents a probability density function.

     b.   Sketch a graph of this function, and indicate the area associated with the probability that $0 < x < 0.5$.

c. Calculate the probability that $0 < x < 0.5$.

5.2 a. Show that

$$f(x) = 3x^2 \qquad \text{for } 0 < x < 1$$

represents a probability density function.

b. Sketch a graph of this function, and indicate the area associated with the probability that $0.5 < x < 1$.

c. Calculate the probability that $0.5 < x < 1$.

5.3 a. Show that

$$f(x) = e^{-x} \qquad \text{for } 0 < x < \infty$$

represents a probability density function.

b. Sketch a graph of this function, and indicate the area associated with the probability that $1 < x < 2$.

c. Calculate the probability that $1 < x < 2$.

5.4 If a random variable has the density function given in Exercise 5.3, find the probability that its value equals 1.

5.2 The Normal Curve of Error

On page 59 it is stated that the distribution of residual errors is normally bell-shaped. Empirical studies of error over a period of several centuries led to a description of what was called "the normal curve of error," a symmetric, bell-shaped distribution of differences between observed and "accepted" values, usually of physical measurements. If no assignable causes of variability were present, this distribution was centered on the mean value 0, but its variance depended upon the quantity being measured.

The normal curve of error originally was studied by the German mathematician Karl Gauss (1777–1855), who found an equation for a probability distribution that was closely followed by empirical distributions of error. Also, it was found that this equation is a limiting form, as $n \to \infty$, of the distribution of the random variable having the values

$$z = \frac{x - np}{\sqrt{np - (1 - p)}}$$

where x is a value of a random variable having the binomial distribution with parameters n and p. Thus, the distribution found by Gauss not only provides a model for error but also serves as an approximation of the binomial distribution for large values of n.

This distribution, sometimes called the Gaussian distribution, is more frequently referred to as the **normal distribution**, and its equation is

$$\textbf{\textit{Normal Distribution:}} \quad n(x;\mu,\sigma^2) = \frac{1}{\sqrt{2\pi}\,\sigma}\, e^{-(x-\mu)^2/2\sigma^2}, \quad \text{for} \ -\infty < x < \infty$$

It can be seen from this equation that normal distributions are a two-parameter family of distributions whose parameters are μ and σ.

Although the random variable having the binomial distribution is discrete, that of the normal distribution is continuous. The distribution of a continuous random variable cannot be represented by a probability histogram. To show every value the associated random variable can take on, such a histogram would require infinitely many bars, each having the width zero. However, pictorial representations of continuous distributions are possible. For example, the normal distribution can be represented pictorially by drawing a smooth curve having the coordinates $[x, n(x;\mu,\sigma)]$, where $n(x;\mu,\sigma)$ is given by the equation for the normal distribution and x is a value of this normally distributed random variable. Such a curve is called a **normal curve**, and the normal curve representing the normal distribution having the mean 5 and the standard deviation 2, is shown in Figure 5.2.

The mean and the standard deviation of a continuous distribution can be defined by formulas like those given on page 77 for discrete distributions, except that the sums are replaced by integrals. Using these definitions, it can be shown that the parameter μ of the normal distribution is its mean, and the parameter σ is its standard deviation.

5.3 Probabilities and the Normal Distribution

In Section 5.1 we saw that probabilities are found for any continuous random variable by finding areas under the curve representing its probability density function. For example, for a random variable having the normal distribution with the mean 2 and the standard deviation 5, the probability that it takes on a value between 3 and 4 is the area between the limits 3 and 4 under the normal curve having mean 2 and standard deviation 5. (Since the probability of obtaining exactly 3 or exactly 4 is zero, it no longer matters whether or not the interval includes either of its endpoints.) Thus, to compute probabilities involving the normal distribution, it is

FIGURE 5.2 Normal Curve

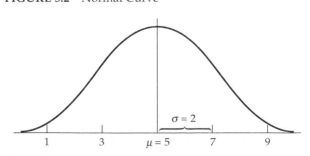

necessary to find areas under the normal curve. Such areas are given by integrals; the integral

$$P(a \leq x \leq b) = \frac{1}{\sqrt{2\pi}\sigma} \int_a^b e^{-(t-\mu)^2 / 2\sigma^2} \, dt$$

gives the probability that a normally distributed random variable will assume a value between a and b. The integral

$$N(x; \mu, \sigma) = \frac{1}{\sqrt{2\pi}\sigma} \int_{-\infty}^x e^{-(t-\mu)^2 / 2\sigma^2} \, dt$$

gives the probability that a normally distributed random variable having the mean μ and the standard deviation σ will assume a value less than x. The function $N(x;\mu,\sigma)$, defined by this integral, is called the **cumulative normal distribution** and, unlike the normal density function, it does give probabilities directly.

The Standard Normal Distribution

It is not possible to perform the integrations required to find probabilities using the normal distribution. To facilitate the calculation of normal probabilities, it is necessary to approximate the corresponding integrals using methods of numerical integration, and then tabulating the results. In constructing such tables, it would appear that tabulation of areas under many normal curves would be required, each table having a different mean and standard deviation. The resulting tables, to be of any practical use, would fill volumes. It is possible to avoid this problem, however, by tabulating only the distribution of the **standardized random variable** whose values are given by the following

---

Value of a Standardized Random Variable: $\quad z = \dfrac{x - \mu}{\sigma}$

If x is a value of a normally distributed random variable having the mean μ and the standard deviation σ, it can be shown that the corresponding standardized random variable also is normally distributed; however, its mean is 0, and its standard deviation is 1. This normal distribution, having $\mu = 0$ and $\sigma = 1$, is called the **standard normal distribution.** The equation for the standard normal distribution is

---

Standard Normal Distribution: $\quad n(z; 0, 1) = \dfrac{1}{\sqrt{2\pi}} e^{-x^2/2}, \quad \text{for} \ -\infty < z < \infty$

---

and the **standard normal curve** is shown in Figure 5.3. It is necessary only to tabulate areas under the standard normal curve to find probabilities involving *any* normal distribution.

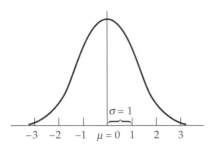 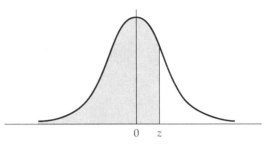

FIGURE 5.3   Standard Normal Curve   **FIGURE 5.4**   Areas under the Standard Normal
                                            Curve Given by Table 3

      Table 3 (Appendix A) gives values of the cumulative standard normal dis-
tribution; that is, it gives the probability that a standard normal random variable
assumes a value less than z. Thus, Table 3 gives left-hand tail areas under the standard normal curve from $-\infty$ to z. The values of z given in this table range from 0
to 3.50 in increments of .01 and the table also includes the values 4.0, 5.0, and 6.0.
Table 3 is entered by looking up z to one decimal place in the left-hand column, and
then finding the second decimal place in the top row. The body of the table gives
values of the cumulative standard normal distribution, $N(x;0,1)$, corresponding to
values of z. Areas given by Table 3 are illustrated by the shaded area in Figure 5.4.

      Example.   Find the area under the standard normal curve to the left of
$z = 1.26$.

      Solution.   This area is given by Table 3 by looking up 1.2 in the left-hand
column and 0.06 in the top row, to obtain .8962.

      In the preceding example, a value of z was given, and we were asked to
find the corresponding area under the standard normal curve. Now, we shall find
a value of z corresponding to a given area under the curve. The value of z that cuts
off a right-hand tail of area p is denoted by z_p, as shown in Figure 5.5. Such values
can be found by entering the body of Table 3 to find the closest entry to $1 - p$, and
then by identifying the corresponding value of z.

FIGURE 5.5   Definition of Z_p

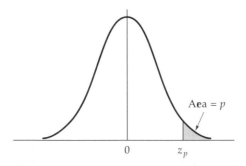

Example. Find $z_{.05}$.

Solution. The body of Table 3 shows the entries .9495, corresponding to $z = 1.64$, and .9505, corresponding to $z = 1.65$. By interpolation, $z_{.05} = 1.645$.

To find the probability that a normally distributed random variable having mean μ and standard deviation σ assumes a value less than x, first the value of the variable is standardized by computing

$$z = \frac{x - \mu}{\sigma}$$

and then the corresponding area is found from Table 3. To illustrate, suppose it is desired to find the probability that a random variable having the mean 5 and the standard deviation 3 assumes a value less than 7. First the standardized value $z = \dfrac{3 - 2}{5} = 0.67$ is calculated, then Table 3 is entered at $z = 0.67$, finding the entry .7486 in the body of the table. Thus, the probability that a normally distributed random variable having mean 5 and standard deviation 3 will assume a value less than 7 is .7486.

Calculation of Probabilities

The probability that a normally distributed random variable having a given mean and standard deviation lies in any interval (a,b) can be calculated with the aid of Table 3, as illustrated in the following example.

Example. Find the probability that a random variable having the normal distribution with mean 2 and standard deviation 5 will assume a value between 3 and 4.

Solution. The required probability is the area under the normal curve from 3 to 4. Equivalently, it is the value of the cumulative normal distribution at $x = 4$, minus the value at $x = 3$. To find these values with the aid of Table 3, first it is necessary to standardize these two values of x. The standardized lower endpoint of the interval $(3,4)$ is given by $z = \dfrac{3 - 2}{5} = .02$ and the standardized upper endpoint is $z = \dfrac{4 - 2}{5} = 0.4$. The required probability is $N(0.4;0,1) - N(0.2;0,1) = .6554 - .5793 = .0761$.

Table 3 contains only positive values of z. To find probabilities involving negative values of z, use is made of the symmetry of the normal distribution. By symmetry, the area under the standard normal curve from z to ∞ is equal to the area under the curve from $-\infty$ to $-z$. That is, the area of the left-hand tail of the curve equals the area of the corresponding right-hand tail. But the area under the entire curve must be 1; thus, the area under the right-hand tail is $1 - N(z;0,1)$. This result can be expressed by the following equation:

> *Areas for Negative Values of z:* $N(-z;0,1) = 1 - N(z;0,1)$

Example. Find the probability that a random variable having the normal distribution with mean 1 and standard deviation 4 will assume a value between 0 and 1.

Solution. Standardizing the endpoints of the interval (0,1), the new interval from $\dfrac{0-1}{4} = -0.25$ *to* $\dfrac{1-1}{4} = 0$ is obtained. The required probability is given by $N(0) - N(-.25) = .5000 - [1 - N(.25)] = .5000 - (1 - .5987) = .0987$.

Calculation of probabilities involving the normal distribution can be simplified by sketching the area corresponding to the required probability, breaking it up into areas that can be found in Table 3, and entering Table 3 to find these areas. Thus, the process of finding the probability that a random variable having the normal distribution with mean μ and standard deviation σ can be summarized by the following steps:

FINDING AREAS UNDER THE NORMAL CURVE

Step 1. To find the probability that a random variable having the normal distribution with mean μ and standard deviation σ assumes a value in the interval (a,b), standardize the endpoints of the required interval by calculating

$$z_1 = \frac{a - \mu}{\sigma}; \qquad z_2 = \frac{b - \mu}{\sigma}$$

Step 2. Draw a sketch of the standard normal distribution, shading the area under the curve from z_1 to z_2.

Step 3. Break the shaded area into parts whose areas can be found directly from Table 3.

Step 4. Add or subtract the areas found, as needed to find the required area.

This method of finding normal probabilities is illustrated in the following example.

Example. Find the probability that a random variable having the normal distribution with mean −2.5 and standard deviation 1.6 will assume a value between −3 and 1.

Solution. First, we calculate $z_1 = \dfrac{-3 - (-2.5)}{1.6} = -0.32$ and $z_2 = \dfrac{1 - (-2.5)}{1.6} = 2.19$. Then, a sketch is drawn like that in Figure 5.6. It is clear from this figure that the required area can be found by subtracting the area in the left-hand tail up to $z = -0.32$ from the area in the left-hand tail up to $z = 2.19$. By symmetry, the first area is the same as the area of the right-hand tail, from $+0.32$ to ∞. This area is 1 minus the left-hand tail area up to 0.32, or $1 - .6255 = .3745$. The second area is given directly by Table 3 to be .9857. Thus, the required probability is $.9857 - .3745 = .6112$.

FIGURE 5.6   Area Required for Example on Page 110

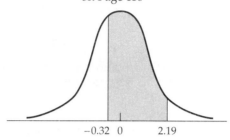

Computer Applications

Probabilities involving the normal distribution can be obtained directly with *MINITAB* software. To illustrate, suppose we want to obtain the probability that a normally distributed random variable having the mean 3.5 and the standard deviation 1.8 will assume a value between 2.4 and 6.5. First, we find values of the cumulative normal distribution $N(x;3.5,1.8)$. Using *MINITAB* Release 8.2, first we enter the values 2.4 and 6.5 in column C1. Then we give the command CDF FOR VALUES IN C1, STORE RESULTS IN C2;.

The semicolon at the end of the command tells *MINITAB* that a subcommand is required, and the prompt SUBC > then appears. At this prompt, we give the subcommand NORMAL, mu = 3.5, sigma = 1.8. · (Note the period placed at the end of the subcommand.)

The command PRINT C1, C2 now will produce the output:

```
     C1        C2
     2.4    .289257
     6.5    .9 5
```

The numbers in C2 are values of the cumulative normal distribution having the mean 3.5 and the standard deviation 1.8 for $x = 2.4$ and 6.5, respectively. To obtain the required probability, we subtract .289257 from .957485 to obtain the result .668228.

EXERCISES

5.5   If z is a value of the random variable having the standard normal distribution, find the probabilities of the following events:

a. $z \leq 1.0$ c. $z > 1.0$ e. $z < -1.0$

b. $z < 1.0$ d. $1.0 < z < 2.0$ f. $-1.0 < z < 1.0$

5.6   Find the probability that a random variable having the standard normal distribution will assume a value

a.   less than 2.0   d.   between 0 and 2.0

b.   less than or equal to 2.0   e.   less than −2.0

c.   greater than 2.0   f.   between −2.0 and 2.0

5.7   If a random variable has the normal distribution with $\mu = 7.5$ and $\sigma = 2.5$, find the probability that it will take on a value

a.   less than 4.5   c.   between 8.0 and 10.0

b.   greater than 8.5   d.   between 4.0 and 9.0

5.8   If a random variable has the normal distribution with mean −2 and standard deviation 0.75, find the probability that it will take on a value

a.   less than or equal to −1   c.   between −1.5 and 1.5

b.   greater than 0   d.   between −3 and −1

5.9   Find the probability that a random variable having the normal distribution with mean 100 and standard deviation 15 will take on a value less than 80 or greater than 120.

5.10 If a random variable is normally distributed with mean 325 and standard deviation 27, find the probability that it will assume a value *outside* the interval (300, 350).

5.11 Find

a. $z_{.01}$ b. $z_{.025}$

5.12 Find

a. $z_{.10}$ b. $z_{.001}$

5.13 Find the quartiles of the standard normal distribution. (See Exercise 3.69 on page 58 for the definition of quartiles.)

5.14 Find the probability that a random variable having the normal distribution will take on a value no further from its mean than one standard deviation.

5.15 a.   Use a computer program to find the probability that a random variable having the normal distribution with mean 10.8 and standard deviation 4.2 will assume a value between 5.1 and 16.9.

b.   Interpolate in Table 3 to find this probability and compare your result with the more exact value found in part a.

5.16 a.   Use a computer program to find the probability that a random variable having the normal distribution with mean 6.84 and standard deviation 1.25 will assume a value between 5.11 and 16.93.

b.   Interpolate in Table 3 to find this probability and compare your result with the more exact value found in part a.

5.4 Checking Data for Normality

In many of the applications of statistics to problems in quality, engineering, and management it is assumed that the data are approximately normally distributed. This assumption will be made in Section 5.6 and in many of the methods to be discussed in later chapters of this book. Thus, it is important to make sure that the assumption of normality can, at least reasonably, be supported by the data. Since the normal distribution is symmetric and bell-shaped, examination of the histogram picturing the frequency distribution of the data is useful in checking the assumption of normality. If the histogram is not symmetric, or if it is symmetric but not bell-shaped, the assumption that the data set comes from a normal distribution cannot be supported. Of course, this method is subjective; data that appear to have symmetric, bell-shaped histograms may not be normally distributed.

Another, somewhat less subjective method for checking data, one that is more effective in detecting departures from normality, is the **normal-probability plot**, or its close relative, the **normal-scores plot**. In each case, the observations are ordered by size and plotted against an appropriate scale. When constructing a probability plot, the "normal-probability" scale is provided by special graph paper, called "normal-probability paper." The scale for a normal-scores plot is constructed by transforming each observation to a "normal score" so that ordinary graph paper can be used.

Normal-Probability Plot

Construction of a normal-probability plot will be illustrated with the following data, giving the diameters (in inches) of 10 shafts turned on a lathe:

    1.42   1.38   1.40   1.41   1.39   1.44   1.34   1.42   1.40   1.47

To test whether it is reasonable to regard these data as coming from a normal distribution, the first step is to order the data, from smallest to largest. The next step involves finding the locations of the ordered data points on the normal-probability scale, called the "plotting points." In this example there are $n = 10$ observations. Thus, the probability interval from 0 to 1 must be divided into 10 equal intervals, namely the intervals from 0 to .1, .1 to .2, . . . , .9 to 1. The plotting points are the centers of these intervals, or .05, .15, . . . , .95 In general, if there are n observations, the i-th largest observation is plotted with the following ordinate:

Plotting Points for a Normal-Probability Plot:
$$y_i = \frac{2i - 1}{2n} \quad \text{for} \quad i = 1, 2, \ldots, n$$

If a normal-probability plot closely approximates a straight line, it is reasonable to assume that the underlying data are at least approximately normally distributed. To test the shaft-diameter data for normality, the data are arranged by size, and the coordinates for making a probability plot of the 10 shaft diameters are tabulated, as follows:

x:	1.34	1.38	1.39	1.40	1.40	1.41	1.42	1.42	1.44	1.47
y:	5	15	25	35	45	55	65	75	85	95

In this table, the plotting points are expressed as percentages to correspond to the probability scale of most commercially available normal-probability graph papers. A graph of these values of x, plotted against the corresponding values of y on normal-probability graph paper, is shown in Figure 5.7. The 10 points shown on the

FIGURE 5.7   Normal-Probability Plot

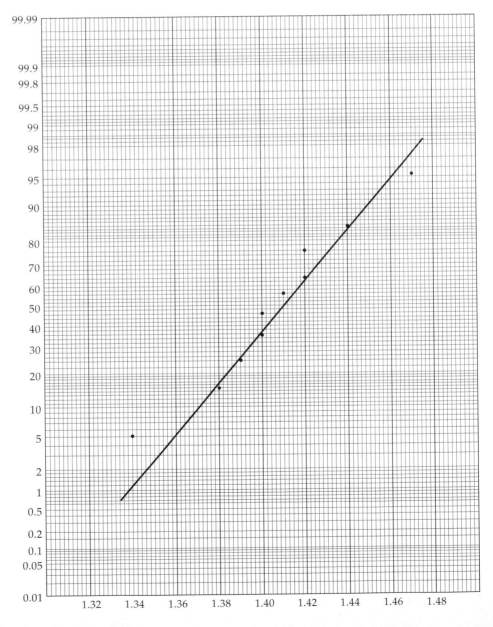

graph closely follow a straight line, and it seems reasonable to assume that the shaft diameters are, at least approximately, normally distributed.

A quick approximation of the mean of the data can be found from a normal-probability plot by first drawing a straight line through the plotted points. Then, since the normal curve is symmetric, the mean equals the median; thus, the mean is the value of x corresponding to the y value of 50% on the line. To find the standard deviation, use is made of the fact that the area under the normal curve to the left of one standard deviation above the mean is approximately .84. Thus, the standard deviation of the data can be approximated by finding the x value corresponding to 84% on the line and subtracting it from the mean value previously found. The mean and standard deviation of the 10 shaft diameters have been estimated from the graph to be 1.41 and 0.03, respectively. Computation of their actual values gives $\bar{x} = 1.409$ and $s = 0.031$. Of course, estimates of the mean and the standard deviation obtained from a probability plot are of little value if the data do not closely follow a straight line.

Normal-Scores Plot

The normal-scores plot makes use of ordinary graph paper rather than normal-probability paper. But instead of calculating plotting points for the normal-probability scale, it now becomes necessary to find **normal scores**, z_p. If n observations are ordered from smallest to largest, they divide the area under the normal curve into $n + 1$ equal parts, each having the area $1/(n + 1)$. The normal score for the first of these areas is the value of z such that the area under the standard normal curve to the left of z is $1/(n + 1)$, or $-z_{1/(n+1)}$. Thus, the normal scores for $n = 4$ observations are $-z_{.20} = -0.84$, $-z_{.40} = -0.25$, $z_{.40} = 0.25$, and $z_{.20} = 0.84$. The ordered observations then are plotted against the corresponding normal scores on ordinary graph paper.

> *Example.* Find the normal scores and the coordinates for making a normal-scores plot of the following six observations:
>
> 3,   2,   7,   4,   3,   5
>
> *Solution.* Since $n = 6$, there are 6 normal scores, as follows: $-z_{.14} = -1.08$, $-z_{.29} = -0.55$, $-z_{.43} = -0.18$, $z_{.43} = 0.18$, $z_{.29} = 0.55$, and $z_{.14} = 1.08$. When the observations are ordered and tabulated together with the normal scores, the following table results:

Observation:	2	3	3	4	5	7
Normal score:	−1.08	−0.55	−0.18	0.18	0.55	1.08

Finding the coordinates for a normal-scores plot can be very tedious. The tedium can be reduced by making use of a cumulative percentage distribution of the data (see Exercises 3.21 and 3.23 on page 41), and finding normal scores only for the cumulative percentages at the class boundaries. To illustrate, the compressive strengths of concrete samples given in the example on page 38 will be used. The cumulative percentage distribution is as follows:

Class Boundary	Cumulative Percentage	Normal Score
4395	5	−1.64
4595	17	−0.95
4795	37	−0.33
4995	69	0.50
5195	87	1.13
5395	97	1.88

A graph of the class boundaries versus the normal scores is shown in Figure 5.8. It can be seen from this graph that the points lie in an almost perfect straight line, strongly suggesting that the underlying data are very close to being normally distributed.

Transforming Data to Normality

Sometimes a normal-scores plot showing a curve can be changed to a straight line by means of an appropriate transformation. The procedure involves identifying the type of transformation needed, making the transformation, and then checking the transformed data by means of a normal-scores or normal-probability plot to see if they can be assumed to have a normal distribution.

When data appear not to be normally distributed because of *too many large values*, the following transformations are good candidates to try:

logarithmic transformation $u = \log(x)$

square-root transformation $u = \sqrt{x}$

reciprocal transformation $u = \dfrac{1}{x}$

FIGURE 5.8   Normal-Scores Plot

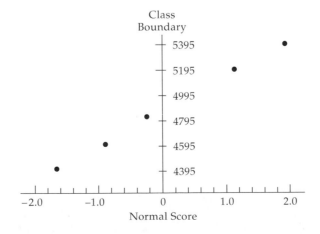

Normal Score

When data exhibit *too many small values*, the following transformations may produce approximately normal data:

power transformation	$u = x^a$, where a $>$ 1
exponential transformation	$u = a^x$, where a $>$ 1

On rare occasions, it helps to make a linear transformation of the form $u = a + bx$ first, and then to use one of the indicated transformations. This strategy becomes necessary when some of the data have negative values and logarithmic, square-root, or certain power transformations are to be tried. However, making a linear transformation alone cannot be effective. It can be shown that, if x is a value of a normally distributed random variable, then the random variable having the values $a + bx$ also has the normal distribution. Thus, a linear transformation alone cannot transform non–normally distributed data into normality.

Example. Make a normal-scores plot of the following data. If the plot does not appear to show normality, make an appropriate transformation, and check the transformed data for normality.

$$54.9 \quad 8.3 \quad 5.2 \quad 32.4 \quad 15.5$$

Solution. The normal scores are −0.95, −0.44, 0, 0.44, and 0.95. A normal-scores plot of these data (Figure 5.9[a]) shows sharp curvature. Since two of the five values are very large compared with the other three values, a logarithmic transformation (base 10) was used to transform the data to

$$1.74 \quad 0.92 \quad 0.72 \quad 1.51 \quad 1.19$$

A normal-scores plot of these transformed data (Figure 5.9[b]) shows a nearly straight line, indicating that the transformed data are approximately normally distributed.

FIGURE 5.9 Normal-Scores Plots for above Example

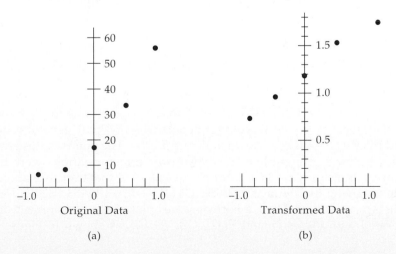

(a) Original Data

(b) Transformed Data

If lack of normality seems to result from one or more outliers, a single large observation, a single small observation, or both, it is not likely that the data can be transformed to normality. It is difficult to give a hard-and-fast rule for identifying outliers. For example, it would be inappropriate to define an outlier as an observation whose value is more than three standard deviations from the mean, since such an observation can occur with a reasonable probability in a large enough number of observations taken from a normal distribution. However, an observation that clearly does not lie on a straight line defined by the other observations in a normal-scores plot can be considered an outlier. Outliers also can be identified by making a box-and-whisker plot of the observations, using the rules given for identification of outliers on page 55. In the presence of suspected outliers, normal-scores plots of the data should be re-drawn after the outlier or outliers have been omitted to verify that the remaining data are, indeed, consistent with the assumption of a normal distribution.

Outlying observations may result from several causes, such as an error in recording data, an error of observation, or an unusual event such as a particle of dust settling on a material during thin-film deposition. There is always a great temptation to drop outliers from a data set entirely on the basis that they do not seem to belong to the main body of data. But an outlier can be as informative about the process from which the data were taken as the remainder of the data. Outliers which occur infrequently, but regularly in successive data sets, give evidence that should not be ignored. For example, a hole with an unusually large diameter might result from a drill not having been inserted properly into the chuck. Perhaps the condition was corrected after one or two holes were drilled, and the operator failed to discard the part with the "bad" hole, thus producing one or two outliers. While outliers sometimes are separated from the other data for the purpose of performing a preliminary analysis, they should be discarded only after a good reason for their existence has been found.

Computer Applications

Normal scores and normal-score plots can be obtained with *MINITAB* software. To illustrate the procedure, the following 20 numbers are entered with the command and data-entry instructions

```
SET C1:
0 215 31 7 15 80 17 41 51 3 58 158 0 11 42 11 17 32 64 100
END
```

Then the command NSCORES C1 PUT IN C2 is given to find the normal scores and place them in the second column. A normal-scores plot, generated by the command PLOT C1 VS C2, is shown in Figure 5.10(a). The points in this graph clearly do not follow a straight line. Several power transformations were tried in an attempt to transform the data to normality. The cube-root transformation $u = x^{1/3}$, made by giving the command RAISE C1 TO THE POWER .3333 PUT IN C3, seemed to work best. Then, a normal-scores plot of the transformed data can be generated with the command PLOT C3 VS C2, as shown in Figure 5.10(b). It

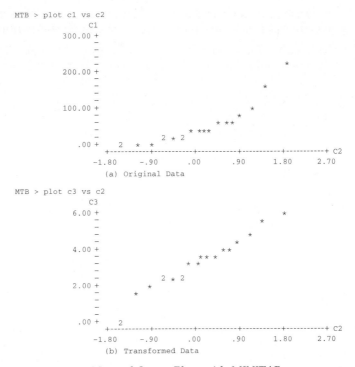

FIGURE 5.10 Normal-Scores Plots with *MINITAB*

appears from this graph that the cube-roots of the original data are approximately normally distributed.

EXERCISES

5.17 Check the following data for normality by finding normal scores and making a normal-scores plot:

3.9   4.6   4.5   1.6   4.2

5.18 Check the following data for normality by finding normal scores and making a normal-scores plot:

36   22   3   13   31   45

5.19 Find the plotting points for a normal-probability plot when there are 19 observations.

5.20 Find the plotting points for a normal-probability graph when there are 9 observations.

5.21 Make a normal-probability plot or a normal-scores plot of the 10 salaries given on page 50. Is it reasonable to assume that the data are normally distributed?

5.22 The weights (in pounds) of seven shipments of bolts are

37   45   11   51   13   48   61

Make a normal-probability plot or a normal-scores plot of these weights. Can they be regarded as having come from a normal distribution?

5.23 Make a normal-probability plot or a normal-scores plot of the grouped data on solder-bond breaking strengths given in the example on page 43. Do these data appear to be normally distributed?

5.24 Make a normal-probability or a normal-scores plot of the grouped coking-time data given in Exercise 3.10 on page 36. Is it reasonable to assume that the coking times are normally distributed?

5.25 Make an appropriate transformation of the following data and check to see if the transformed data seem to be normally distributed: 13, 18, 21, 24, 25, 30, 42.

5.26 Make an appropriate transformation of the following data and check to see if the transformed data appear to be normally distributed: 1.1, 5.0, 10.3, 15.5, 21.8, 38.2.

5.27 Use a computer program to make a normal-scores plot of the raw (ungrouped) integrated-circuit response times given on page 31. If they do not appear to be normally distributed, try a few transformations to see which one works best.

5.28 Use a computer program to make a normal-scores plot of the reaction times given in Exercise 3.14 on page 40. If they do not appear to be normally distributed, try a few transformations to see which one works best.

5.5   Other Continuous Distributions

So far, distributions of continuous random variables have been illustrated only by the normal distribution. While this distribution has great importance in applications, many other continuous distributions have been identified that are useful in describing a variety of different kinds of data. In this section, we shall illustrate some of them by describing the continuous uniform distribution, the lognormal distribution, and the exponential distribution.

The Continuous Uniform Distribution

The calipers on automotive disk brakes make sliding contact with the disk each time the brakes are applied. The initial contact point of the disk for a given caliper can be assumed to be random for each brake application; that is, there is no reason to believe that any one point on the circumference of the disk is more likely to be worn by contact with the calipers than any other.

To construct a probability model for this phenomenon, let us define x to be the value of a random variable that gives the location of initial caliper contact at each point on the circumference of the disk. Then, the probability that x takes on a given value in any small subsection of length Δx is the same no matter where this subsection is located on the disk's circumference.

Now, let us imagine "unrolling" the circumference of the disk onto a line segment, placing its left-hand end at the point α. The right-hand end will be at the point β, where $\beta - \alpha$ equals the length of the circumference. In keeping with our assumption of "equal wear," the probability density of the random variable giving the point of caliper contact on the circumference of the disk should take on the same value for every point between α and β, and it should equal zero elsewhere. Thus, the probability density representing disk wear should have the equation

> **Uniform Distribution:** $\qquad f(x) = \dfrac{1}{\beta - \alpha} \qquad$ for $\alpha < x < \beta$

where $f(x) = 0$ outside this interval. This probability density function is illustrated in Figure 5.11.

The probability that the random variable having the uniform distribution will take on a value between a and b, where $\alpha < a < b < \beta$, is given by the integral

$$\int_a^b \frac{1}{\beta - \alpha} dx = \frac{b - a}{\beta - \alpha}$$

The area under the uniform density function corresponding to this probability is shown in Figure 5.12. It can be seen that, for any segment of the circumference of the disk, the probability of a given amount of total wear is proportional to the length of the segment, and the probabilities of a given amount of wear on segments of equal length are equal.

To obtain the mean of the uniform distribution we can apply the formula for the mean given on page 77, substituting an integral for the sum. We thus obtain

$$\mu = \int_\alpha^\beta x \cdot \frac{1}{\beta - \alpha} dx = \frac{\alpha + \beta}{2}$$

FIGURE 5.11 Uniform Distribution

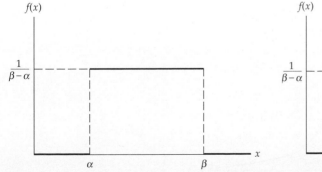

FIGURE 5.12 Probabilities with Uniform Distribution

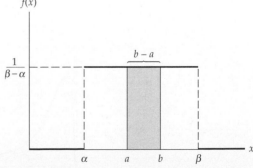

The variance is found similarly by integrating

$$\sigma^2 = \int_\alpha^\beta (x - \mu)^2 \frac{1}{\beta - \alpha} dx = \frac{(\beta - \alpha)^2}{12}$$

Example.   Assume that a length of extruded wire has a single defect, and that defects are uniformly distributed over the length of the wire. If a spool of wire is 10 feet long, find the probability that the defect is located in the first 2 feet of wire.

Solution.   If we let $\alpha = 0$, $\beta = 10$, $a = 0$, and $b = 2$, the required probability is given by

$$\frac{b - a}{\beta - \alpha} = \frac{2 - 0}{10 - 0} = .2$$

The Lognormal Distribution

The lognormal distribution occurs in practice whenever a random variable is such that the logarithms of its values are normally distributed. The density function of the lognormal distribution is given by

$$\boxed{\textit{Lognormal Distribution: } f(x) = \frac{1}{\sqrt{2\pi}\beta} x^{-1} e^{-(\ln x - \alpha)^2 / 2\beta^2} \text{ for } x \rangle 0, \ \beta \rangle 0}$$

where $\ln x$ is the natural logarithm of x. Since logarithms are taken, x cannot assume a value less than or equal to zero. A graph of a typical lognormal distribution is shown in Figure 5.13.

To find the probability that a random variable having the lognormal distribution will take on a value between a and b, where $0 < a < b$, we must evaluate the integral

FIGURE 5.13   Lognormal Distribution

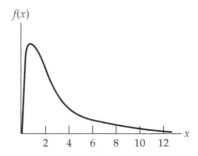

$$\int_a^b \frac{1}{\sqrt{2\pi}\beta} x^{-1} e^{-(\ln x - \alpha)^2 / 2\beta^2} dx$$

Making the change of variable $y = \ln x$, with $dy = x^{-1} dx$, we obtain

$$\int_{\ln a}^{\ln b} \frac{1}{\sqrt{2\pi}\beta} e^{-(y - \alpha)^2 / 2\beta^2} dy$$

To recognize this integral, we refer to the integral on page 107, which gives the probability that a normally distributed random variable having mean μ and standard deviation σ will assume a value between a and b. We thus recognize this integral as giving the probability that a normally distributed random variable having the mean α and the standard deviation β will assume a value between $\ln a$ and $\ln b$.

It can be shown that the mean of the lognormal distribution can be expressed directly in terms of its parameters. The mean is given by

$$\mu = e^{\alpha + \beta^2/2}$$

and its variance is

$$\sigma^2 = e^{2\alpha + \beta^2} (e^{\beta^2} - 1)$$

Example.   If experience shows that the natural logarithms of processing times for jobs submitted to a certain mainframe computer have the normal distribution with a mean of 1.5 and a standard deviation of 0.4, find the probability that such a time will lie between 3 and 6 minutes.

Solution.   The required probability is equivalent to the probability that a normally distributed random variable with mean 1.5 and standard deviation 0.4 will assume a value between $\ln 3 = 1.099$ and $\ln 6 = 1.792$. Standardizing each of these values to get $z_1 = \dfrac{1.099 - 1.5}{0.4} = -1.00$ and $z_2 = \dfrac{1.792 - 1.5}{0.4} = 0.73$, and using Table 3, we find the required probability to be $.7673 - (1 - .8413) = .6086$.

Example.   If a random variable has the lognormal distribution with parameters $\alpha = 1.8$ and $\beta = 0.4$, find (a) the probability that this random variable will assume a value between 1 and 5, and (b) the mean of this distribution.

Solution.   a. The required probability is given by the area under the normal curve having the mean 1.8 and the standard deviation 0.4 that lies between $\ln 1 = 0$ and $\ln 5 = 1.609$. To find this area, we first standardize 1.609, obtaining. $z = \dfrac{1.609 - 1.8}{.4} = -0.23$. The area is found from Table 3 to be $.1 - .5910 = .4090$. (b) The mean of this lognormal distribution is $e^{1.8 + 0.4^2/2} = 6.53$.

The Exponential Distribution

The exponential distribution plays an important role in reliability theory (Chapter 11) and in queuing theory (page 91) where it can be shown that, under the assumption that arrivals to a queue have a Poisson distribution, the waiting times between successive arrivals have an exponential distribution. The density function of this distribution is given by

---

Exponential Distribution: $f(x) = \alpha e^{-\alpha x}$ for $x > 0$, $\alpha > 0$

---

and $f(x) = 0$ elsewhere. Graphs of exponential distributions are shown in Figure 5.14 for several values of the parameter α.

FIGURE 5.14   Exponential Distributions

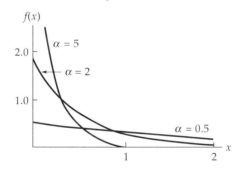

The mean of the exponential distribution can be found by evaluating the integral

$$\mu = \int_0^\infty x \cdot \alpha e^{-\alpha x} dx$$

Integrating by parts gives the result $\mu = 1/\alpha$. The variance of the exponential distribution can be shown to be $\sigma^2 = 1/\alpha^2$.

Example.   The time intervals between the emissions of beta particles in a nuclear reaction has the exponential distribution with a mean of 5 milliseconds. Find (a) the probability that there will be at most an 8 millisecond wait between the emission of a given beta particle and the succeeding particle, and (b) the standard deviation of the waiting time between beta particles.

Solution.   a. Since the mean is 5 milliseconds, the exponential parameter α has the value $1/5 = 0.2$. The required probability is given by the integral

$$\int_{c}^{\theta} .2e^{-2x}dx = 1 - e^{-1.6} = .798$$

b. The standard deviation of this distribution is given by $\sqrt{1/\alpha^2} = 1/\alpha$ = 5 milliseconds.

EXERCISES

5.29 Repetitions of a certain measurement are uniformly distributed on the interval from 5 to 20.

a. Find the probability of obtaining a measurement between 10 and 12.

b. What are the mean and the standard deviation of such measurements?

5.30 Suppose bids for a construction contract are found to be uniformly distributed from C to 2C, where C is the cost of doing the job. Find the probability that a given contractor will bid a job within 20% of the mean bid.

5.31 Show that the area under the continuous uniform density function equals 1.

5.32 If the natural logarithm of a random variable has the normal distribution with a mean of 1.8 and a standard deviation of 0.4, find the probability that the random variable will take on a value less than 2.5.

5.33 Current gains of a certain transistor have the lognormal distribution with $\alpha = 2$ and $\beta = 0.1$. Find the probability that a transistor has a current gain that lies between 6.5 and 8.0.

5.34 A certain measurement has the exponential distribution with $\alpha = 1.6$.

a. Find the mean and the standard deviation of the distribution of this measurement.

b. Find the probability that a given measurement will assume a value between 0.4 and 2.5.

5.35 The time to failure of a certain kind of transistor has the exponential distribution with $\alpha = 0.005$. Find the probability that the life of such a transistor will exceed 250 hours.

5.36 a. Show that the area under any exponential distribution equals 1.

b. Derive the formula for the mean of the exponential distribution given on page 124.

5.6 Application to Quality

The normal distribution is a time-tested model for error; thus, it is not surprising that normally distributed data, or approximately normally distributed data, occur frequently in applications. The success of a quality-improvement program often depends upon the analysis of measurements taken to detect or confirm quality problems or to verify statements involving quality. When measurements are made on products or services selected from homogeneous groups, such as a lot of manufactured product, a specified production machine or tool, or a given station dispensing a service, the resulting data often are normally distributed, at least to a good degree of approximation. Probabilities calculated from such distributions can be used to identify discrepant observations, to reduce the proportion of defective product, and to test the reasonableness of a claim made about the quality of a product or service.

Identification of Discrepant Observations

When describing data that are normally distributed, some remarkable and useful conclusions can be drawn from calculating the probability that the data will take on values within only a few standard deviations from its mean. The probability that a random variable having the normal distribution with mean μ and standard deviation σ will take on values within k standard deviations from its mean is equivalent to the probability that it will assume a value in the interval $\mu \pm k\sigma$. To find this probability, it is necessary first to find z values for the endpoints of the interval. For the lower endpoint, $z = \dfrac{\mu - k\sigma - \mu}{\sigma} = -k$ and for the upper endpoint, $z = \dfrac{\mu + k\sigma - \mu}{\sigma} = k$. Probabilities for $k = 1, 2, 3, 4, 5$, and 6 have been obtained from Table 3, and they are summarized in the following table:

PROBABILITY THAT A RANDOM NORMAL VARIABLE TAKES ON A VALUE WITHIN k STANDARD DEVIATIONS OF ITS MEAN	
k	*Probability*
1	.6826
2	.9544
3	.9974
4	.99994
5	.9999994
6	.999999998

From this table, it is evident that a normally distributed random variable will take on a value within two standard deviations of its mean with a probability greater than .95. This probability exceeds .99 for three standard deviations, and it approaches certainty for four or more standard deviations. This table also emphasizes an important role played by the standard deviation. Knowledge only of the mean and the standard deviation of a set of data reveals a great deal of information about the values that the data will take on. Even for non-normal distributions, it can be shown that the resulting observations often are clustered within only a few standard deviations from their mean (see Exercise 5.50 on page 131).

Example.   Specifications require that the thickness of asphalt on a road be at least 3 inches, with a standard deviation of no more than 0.15 inch. Believing that there is a quality problem involving the paving contractor, an engineer takes a single core sample, obtaining a thickness of 2.5 inches. Is the engineer's suspicion justified?

Solution.   It will be assumed that asphalt thicknesses in randomly distributed core samples are approximately normally distributed. If specifications are barely met, the mean of this distribution will be 3.0 inches, and its standard deviation will be 0.15 inch. A measurement of 2.5 inches is $\frac{3.0 - 2.5}{0.15}$ = 3.3 standard deviations less than the mean. Since the probability of obtaining a value within 3 standard deviations of the mean is given by the table on page 126 to be .9974, the probability of a value greater than 3 standard deviations from the mean is only .0026. The result 2.5 is in the left-hand tail of the normal distribution more than 3 standard deviations from the mean; thus, its probability is less than .0026/2 = .0013 if the contractor is adhering to specifications. Not one to believe in miracles, the engineer concluded that the contractor is not meeting the specification.

In this example, the assumption was made that the asphalt thicknesses in core samples are approximately normally distributed. Before acting on the conclusion reached in the example, the engineer should demonstrate that this assumption is reasonable. It is likely that previous experience with the same kind of road construction will provide data which can be checked for normality using one of the methods of Section 5.4.

The following example extends this concept to the detection of an outlying observation.

Example.   One hundred observations are taken of the burst strength (in pounds) of cardboard cartons. Previous experience with such tests indicates that the distribution of burst strengths is approximately normal with a mean of 135 pounds and a standard deviation of 8 pounds. Would an observation of 102 pounds in this sample be surprising?

Solution.   The observation of 102 pounds is just over four standard deviations less than the mean. The probability that a single observation would take on a value four or more standard deviations less than its mean is $(1 - .99994)/2 = .00003$. Even in a data set containing 100 observations, such an observation would be extremely rare. The probability of obtaining 1 or more observations more than four standard deviations from the mean in a data set containing 100 observations, can be approximated by the Poisson distribution with $\lambda = (100)(0.00003) = 0.003$. This probability is nearly zero. (It would require a sample size of more than 1,000 before so small an observation would be even modestly likely.)

It is always difficult to know what action to take when an outlying observation appears. As indicated on page 118, it can be dangerous to simply discard such an observation from the sample without further investigation. In the foregoing example, an intermittent quality problem could have caused the occasional carton to be unusually weak. Thus, while the observation 102 pounds clearly does not belong to the distribution representing ordinary production, it might be a symptom of an infrequent, intermittent quality problem that needs correction. Only if further investigation shows that the observation 102 pounds is the result of an error of measurement or recording should this datum be discarded and ignored.

Reducing the Proportion of Defective Product

The following example illustrates how the normal distribution can be used to estimate the proportion of defectives produced by a manufacturing process.

> *Example.* If the copper thickness of parts coming from a plating operation is approximately normally distributed with a mean of 2.68 mm and a standard deviation of 0.12 mm, what proportion of the parts will be produced outside the specification limits 2.5 – 3.0?
>
> *Solution.* The probability that a part will be produced with a copper thickness outside the interval (2.5, 3.0) is required. This probability is obtained by finding $z_1 = \dfrac{2.5 - 2.68}{0.12} = -1.50$ and $z_2 = \dfrac{3.0 - 2.68}{0.12} = 2.67$, and by using Table 3 to find the required probability. This probability is $(1 - .9332) + (1 - .9963) = .0705$. Thus, about 7% will be produced outside the specification limits, or 93% will be within specifications.

Centering the process so that its mean lies midway between the specification limits can reduce the proportion defective, as illustrated in the following example.

> *Example.* The process described in the preceding example appears to be centered improperly; its mean is 2.68, but the midpoint of the specification limits is 2.75. What improvement in quality could be expected if the process were centered at 2.75?
>
> *Solution.* If the mean of the normal distribution of plating thicknesses were changed to 2.75, z_1 would become $\dfrac{2.5 - 2.75}{0.12} = -2.08$, and z_2 would become $\dfrac{3.0 - 2.75}{0.12} = 2.08$. The probability of a part having a copper thickness outside of the specification limits 2.75 ± 0.25 would become $(1 - .9812) + (1 - .9812) = .0376$. The proportion of defectives would be reduced to about 4%.

Further reductions in the proportion of defectives would require process improvements that would reduce its variability. The following example shows how to calculate what further reduction in the percentage defective would result from a decrease in the standard deviation of the process from 0.12 to 0.10.

> *Example.* What proportion of defectives could be expected in the plating operation of the preceding example if the process mean is held at 2.75 while its standard deviation is reduced to 0.10?

Solution. Now, $z_1 = \dfrac{2.5 - 2.75}{0.10} = -2.5$, and $z_2 = \dfrac{3.0 - 2.75}{0.10} = 2.5$. The probability of producing a part outside of specifications becomes $(1 - .9938) + (1 - .9938) = .0124$. The proportion defective has been reduced to about 1.2%.

Testing Claims about Quality

The need often occurs to test the reasonableness of a statement about the distribution from which measurements are taken. For example, a salesman may claim that the gasoline additive he is promoting is capable of increasing gasoline mileage by 5 miles per gallon. Or an oceanographer claims that the temperature of the ocean at a certain depth is 10° Centigrade. Or an accountant claims that she can process 25 tax returns per day. In each case, a claim is made about a parameter of some distribution, and data can be taken to verify whether or not the claim is reasonable in the face of the data. A general method for dealing with issues of this kind, called "tests of significance," will be introduced in Chapter 7.

> *Example.* If measurements made of the ocean temperature in a given locality and at a certain depth are approximately normally distributed with a mean of 15°C and a standard deviation of 5°C, is it reasonable to expect a reading of 10°C?
>
> *Solution.* The probability of obtaining a reading of 10°C *or lower* is found by calculating $z = \dfrac{10 - 15}{5} = -1$, and using Table 3 to find the probability $1 - .8413 = .1587$. Thus, although the observation 10°C seems low, almost 16% of all observations arising from a normal distribution with $\mu = 15$ and $\sigma = 5$ will be this low or lower, and it would seem unreasonable to become overly suspicious of such an observation.

The following is a further example illustrating the testing of a claim about quality.

> *Example.* The mean fill weight of cans of tomato juice is required to be 24 ounces. Previous experience shows that the standard deviation of the fill weights are normally distributed with a standard deviation of 0.5 ounce. A consumer advocate finds a can of juice purchased in the supermarket to have a fill weight of 22 ounces, and she claims that the cans are underfilled. Does this observation provide evidence that she is correct?
>
> *Solution.* The claim can be tested by finding the probability of obtaining such a small fill weight; that is, a fill weight of 22 ounces or less in a can selected from a process having $\mu = 24$ and $\sigma = 0.5$. It is observed that 22 ounces is 4 standard deviations less than the mean of 24 ounces. Thus, it would be practically impossible to obtain a can having a weight this small or smaller if the process mean were 24 ounces. The consumer advocate has evidence from this can that she is correct.

Applications of the normal distribution are so widespread that it is easy to fall into the mistake of believing that all data are, at least approximately, normally distributed. There are many practical situations giving rise to data that are not normally distributed. Examples are the distribution of service times of equipment, which usually is exponential, the distribution of measurements on a logarithmic scale, which may be lognormal, and distributions exhibiting more than one assignable cause of variability. Some frequently encountered non-normal distributions already have been introduced in this chapter.

Any assumption about the underlying distribution of data always should be checked to assure that it is at least approximately true. Probabilities, such as those calculated in the examples given in this section, cannot be relied upon if the underlying distribution is very different from the assumed distribution.

EXERCISES

5.37 If a set of data has the normal distribution, within how many standard deviations of the mean can *exactly*

   a. 90%   b. 95%   c. 99%

of the observations be expected to lie?

5.38 Assume that a very large data set has the normal distribution. An observation is selected blindfolded from this data set.

   a. What is the probability that its value lies within 1.96 standard deviations from the mean of the data?

   b. What is the probability that two or more of 10 observations similarly selected will have a value more than 1.96 standard deviations from the mean?

5.39 Construct a table like that on page 126, giving the probabilities that a normally distributed random variable will take on a value *more than k* standard deviations away from its mean.

5.40 Construct a table like that on page 126, giving the probabilities that a normally distributed random variable will take on a value more than k standard deviations *above* its mean.

5.41 Assume the process described in the example on page 128 is properly centered. Calculate the process standard deviation required to reduce the percent defective to 0.5%.

5.42 Measurements made on the product produced by a certain production process are normally distributed and centered between their specification limits. The standard deviation of these measurements is 1/6 of the distance between the upper and lower specification limits. What percentage defective can be expected?

5.43 A boring machine bores holes in a block of metal to a specified depth. Past measurements have shown these depths to be approximately normally distributed with a standard deviation of 0.005 inch. Specification limits for the depth are 1 inch $\pm$ .02 inch.

   a. At what depth should the boring machine be set to minimize the proportion of defectives?

   b. What is the minimum proportion of defectives?

c. What would be the proportion of defectives if the machine were set at a depth of 1.005 inches?

5.44 Suppose the boring machine of Exercise 5.43 were improved to reduce the standard deviation of the hole depths to 0.002 inch.

a. What would be the minimum proportion of defectives it could be expected to produce?

b. What proportion of defectives would the machine produce if it were set at 1.015 inches?

5.45 A stamping machine produces can tops to a diameter specified to be between 2.99 and 3.01 inches. If the distribution of these diameters is approximately normal, and if the process is centered to produce mean diameters of 3.00 inches, what standard deviation must the diameters have so that 99% of the can tops will be produced within specifications?

5.46 The oven temperature at a standard setting is used to test a toaster oven; the specified temperature, in degrees F, is 200° ± 15°. If these temperatures have been found to be normally distributed with a mean value of 200°, what must be their standard deviation to assure that no more than 95% of the toaster ovens will require adjustment after testing?

5.47 A college president claims that the students at her college have a mean score of 350 on an aptitude test whose results are known to be normally distributed with a standard deviation of 35. If the record of one of the students is selected blindfolded from the files, would it be surprising to find an aptitude-test score less than 300? Why?

5.48 A new inventory policy is claimed to reduce the mean number of stored parts on any given day from 1,500 to 1,000. Past records show that the daily number of parts in inventory is approximately normally distributed with a standard deviation of 150 parts. The warehouse is inspected 30 days after the new policy has been put in place. Would it be surprising to find 1,350 parts in inventory? Why?

5.49 Use a computer to find the probability of finding one or more observations in a set of 100 normally distributed observations that are 3.5 or more standard deviations from the mean of the distribution from which the observations were selected. Would you be surprised to find such a result?

The following exercise introduces new material, expanding on some of the topics covered in the text.

5.50 *Chebyshev's theorem.* An inequality, called **Chebyshev's theorem,** shows the role played by the mean and the standard deviation of *any* distribution in determining the range of data. This theorem states that, for an observation chosen at random from any distribution having the mean μ and the standard deviation σ, the probability that its value will deviate from μ by less than $k\sigma$ is at least $1 - 1/k^2$.

a. Use this theorem to find the minimum probability that an observation will differ from the mean of its distribution by less than four standard deviations.

    b. Construct a table analogous to the table on page 126 and compare the Chebyshev minimum probabilities with the exact probabilities for the normal distribution.

REVIEW EXERCISES

These review exercises can be used for informal review of the chapter, or as two practice examinations, designed to be taken over a time period of about one hour. The odd-numbered exercises comprise one examination; the even-numbered exercises comprise the other.

5.51 a. Show that

$$f(x) = 5x^4 \qquad \text{for } 0 < x < 1$$

    represents a probability density function.

    b. Find the probability that a random variable having this density function will assume a value between 0.9 and 1.

5.52 a. What must be the value of the constant A for

$$f(x) = A\mathbf{x} \qquad \text{for } 0 < x < 5$$

    to represent a probability density function?

    b. Find the probability that a random variable having this density function will assume a value between 2 and 3.

5.53 If x is a value of a random variable having the normal distribution with a mean of 10 and a standard deviation of 2, find the probability that

    a. $x < 11.5$ b. $x \geq 12.5$ c. $8.5 < x < 10.8$

5.54 Find the probability that $0 \leq x < 1$ if x is a value of a random variable having the normal distribution with

    a. $\mu = 0$ and $\sigma = 1$

    b. $\mu = -1$ and $\sigma = 1$

    c. $\mu = 0.5$ and $\sigma = 2$

5.55 Find approximate values for

    a. $z_{.75}$ b. $z_{.15}$

5.56 Find approximate values for

    a. $z_{.80}$ b. $z_{.25}$

5.57 Find the normal scores for making a normal-scores plot of five observations.

5.58 Sketch a normal-scores plot of the observations

    13, 17, 11, 15

5.59 The thickness of O-rings manufactured by a certain process is known to be normally distributed with a mean of 0.125 inch and a standard deviation of 0.005 inch. Specifications for this attribute are 0.124 ± 0.010 inch. What is the proportion of out-of-specification O-rings produced by this process?

5.60 What will the proportion of out-of-specification products become if the mean of the process described in Exercise 5.59 is centered between the specification limits?

5.61 In a survey of 1,225 households, the mean number of hours of daily television watching was found to be 5.9 hours, with a standard deviation of 2.1 hours. The distribution of hours of television watching also was found to be approximately normal. If a friend claims that his household watches television less than 1.5 hours a day, would you doubt his word? Why?

5.62 The claim is made that a process can produce a bolt with diameters having a normal distribution with a mean of 0.55 inch and a standard deviation of 0.01 inch. A bolt is selected from stock and is found to have a diameter of 0.58 inch. Do you suspect the truth of the claim? Why?

5.63 a. Find the probability that a random variable having the exponential distribution with the parameter $\alpha = 0.1$ will take on a value greater than 5.

   b. What is the mean value of this random variable?

5.64 a. Find the probability that a random variable having the uniform distribution on the interval (2, 4) will take on a value between 3 and 3.5.

   b. What is the mean value of this random variable?

GLOSSARY

6

SAMPLING AND SAMPLING DISTRIBUTIONS

Where have we been? We have completed our study of probability and probability distributions for both discrete and continuous random variables. We have seen, in an informal way, how these ideas can be applied to making decisions based on data. Now we are ready to formalize the decision process, introducing the notion of sampling, which provides the basis for statistical science.

Where are we going? In this chapter, first we define what is meant by the term *random sample*. Then we discuss the distributions of several important sample "statistics," such as the sample mean and the sample variance. In the next chapter, we shall use these "sampling distributions" to develop and apply the statistical science of inference.

Section 6.1, *Random Samples*, defines the terms *population* and *random sample* and discusses methods for drawing such samples, as well as some of the pitfalls involved in sampling.

Section 6.2, *The Large-Sample Sampling Distribution of the Mean*, makes use of an experimental approach to develop the sampling distribution of the mean of a random sample, leading to the central limit theorem.

Section 6.3, *The Small-Sample Sampling Distribution of the Mean*, states and applies the sampling distribution of the *t* statistic, involving the mean of a small sample, where the population variance is unknown.

Section 6.4, *Sampling Distributions Involving Variances*, introduces the chi-square statistic involving the sample variance and the *F* statistic, involving the ratio of two sample variances.

6.1 Random Samples

$\mathbf{I}$t should not be surprising that observations are made for the purpose of obtaining information about one or more parameters of a distribution, such as its mean and its standard deviation. For example, we may wish to obtain information about the mean number of complaints received from customers each month to determine whether a quality-improvement program is succeeding. Or we may wish to know about the standard deviation of measurements made on a critical dimension to control the variability of this dimension.

To construct a theory about how observations yield useful information about parameters of a distribution, we need first to be assured that the observations "are taken from that distribution." More precisely, the observations are considered to be values of a set of random variables, and we need to be sure that each such random variable has the probability distribution whose parameters are of interest. In such cases, we say that the observations are "representative" of the common distribution from which they are taken.

To understand better what is meant by the word *representative*, it is helpful to introduce the idea of a **population**. The word *population* has come into statistics from studies of human populations, such as the population of people living in a certain city. Today, it means any set of objects or measurements whose properties we desire to describe. Thus, one can speak of the population of goods in a warehouse, the population of policies underwritten by an insurance company, or the population consisting of all possible measurements that can be made of the length of a beam. Some populations, such as goods in a warehouse or insurance policies, are **finite populations**; others, such as all the measurements that possibly can be made of the length of a beam, are considered to be **infinite populations**.

If a representative sample is to be taken from a finite population, it is enough to assure that all members of the population have an equal chance of being included in the sample. This criterion not only appeals to a sense of fairness but also allows the use of probabilities in making statements about the population on the basis of a sample. Such a sample is called a **random sample** and, for a sample taken from a finite population, it is defined formally as follows:

Random Sample (finite population):

The set of observations $x_1, x_2, \ldots, x_n$ is a random sample of size n from a finite population if it is chosen so that all subsets of n elements chosen from the population have the same probability of being selected.

This definition does not work for infinite populations because the probability of any given *finite* subset is zero. However, infinite populations can be represented by the common probability distribution of the random variables whose values are observed, and the definition of a random sample from an infinite population can be stated in terms of this distribution. The idea that a random sample should represent the population from which it is taken is preserved if every member of the sample is a value of a random variable that has the same distribution as the population, and if these random variables are independent. Thus, the definition of a random sample from an infinite population can be stated formally as follows:

Random Sample (infinite population):

A set of n observations constitutes a random sample of size n from the probability distribution $f(x)$, if:

1. Each observation is a value of a random variable having the distribution $f(x)$.

2. The n random variables are independent.

While it is practically impossible to take a purely random sample, there are several methods that can be employed to assure that a sample is close enough to randomness to be useful in representing the distribution from which it came. In selecting a sample from a production line, *systematic sampling* can be used to select units at evenly spaced periods of time or having evenly spaced run numbers. In selecting a random sample from products in a warehouse, a *two-stage sampling process* can be used, numbering the containers and using a random device, such as a set of random numbers generated by a computer, to choose the containers. Then, a second set of random numbers can be used to select the unit or units in each container to be included in the sample. There are many other methods, employing mechanical devices or computer-generated random numbers, that can be used to aid in selecting a random sample.

Selection of a sample that reasonably can be regarded as random sometime requires ingenuity, but it always requires care. Care should be taken to assure that only the specified distribution is represented. Thus, if a sample of product is meant to represent an entire production line, it should not be taken only from the first shift. Care should be taken to assure independence of the observations. Thus, the production-line sample should not be taken from a "chunk" of products produced at about the same time; they represent the same set of conditions and settings, and the resulting observations are closely related to each other. Human judgment in selecting samples usually includes personal bias, often unconscious, and such judgments should be avoided. Whenever possible, the use of mechanical devices or random numbers (see Exercise 4.35 on page 79) is preferable to methods involving personal choice.

Once a random sample has been chosen from the population of interest, we wish to generalize from information contained in the sample to parameters of the population. In making such generalizations, we usually calculate a sample quantity such as the sample mean, $\bar{x}$, using it to make statements about a population parameter, such as the population mean, μ. Such a quantity, calculated from a sample, is called a **statistic**. Since the results of a sample are controlled largely by chance, the values of sample statistics such as $\bar{x}$ and s will vary from sample to sample. Thus, each statistic will have a distribution reflecting the size of the sample from which it is obtained as well as the definition of the statistic itself. In the remainder of this chapter, we shall study the sampling distribution of the sample mean for both large and small samples, and the sampling distribution of the sample standard deviation and the ratio of two sample standard deviations.

6.2 The Large-Sample Sampling Distribution of the Mean

Mean and Standard Deviation of a Sample Mean

As stated in the previous section, a random sample is taken from a population for the purpose of obtaining information about the parameters of the distribution of that population. Since a fundamental description of a population is given by its mean, μ, it is useful to explore the properties of the sample mean as a descriptor of μ.

Suppose $x_1, x_2, \ldots, x_n$ are a random sample of size n taken from a population having the mean μ and the standard deviation σ. By the definition of a random sample, these observations can be considered to be values of n independent random variables, each having the mean μ and the standard deviation σ. The quantity $\bar{x} = \dfrac{x_1 + x_2 + \ldots + x_n}{n}$ is the value of a new random variable called the sample mean. A comparison of the distribution of the sample mean with that of the underlying population will be useful in discovering the relation between the sample mean and the mean of the population from which the sample was taken. We shall begin this comparison by examining the mean and the variance of the sample mean.

Without proof, $\bar{x}$ is a value of a random variable whose distribution has the mean

$$\textit{Mean of Distribution of a Sample Mean:} \quad \mu_{\bar{x}} = \mu$$

If the population from which a random sample is taken is *infinite*, the standard deviation of the distribution of the sample mean is

$$\textit{Standard Deviation of Distribution of a Sample Mean:} \quad \sigma_{\bar{x}} = \frac{\sigma}{\sqrt{n}}$$

This latter quantity also is called the **standard error of the mean**. (The standard deviation of the distribution of the mean of a sample taken from a *finite* population is given in Exercise 6.19 on page 144.)

These results show how the basic parameters of the distribution of a sample mean are related to those of the population giving rise to the sample. The two distributions have the same mean, μ, and the standard deviation of the sample mean is smaller than that of the population by the factor $\dfrac{1}{\sqrt{n}}$. These results suggest that the sample mean will approximate the population mean with perfect accuracy and with a precision that improves as the sample size increases. The improvement in precision, however, is not linear with the sample size. To halve the standard deviation of the mean, for example, it is necessary to *quadruple* the size of the sample.

Example. In a random sample of 64 observations from a distribution having a mean of μ = 6.85 and the standard deviation 1.49, find the standard deviation of the sample mean.

Solution. The standard deviation of the sample mean is $\dfrac{1.49}{\sqrt{64}} = 0.19$.

Central Limit Theorem

It has been seen that the distribution of the mean of a random sample has the same mean as the population from which the sample was taken, and its standard deviation is the population standard deviation, divided by the square root of the sample size. But much more can be said about the distribution of the mean of a random sample.

To create an example of such a distribution, random samples were taken from the population having the discrete distribution

x	0	1	2	3
$f(x)$	.40	.30	.20	.10

A probability histogram of this discrete probability distribution is shown in Figure 6.1. The mean and the standard deviation of this population can be found by applying the formulas given on page 77, as follows:

$$\mu = \sum_{i=1}^{4} x f(x) = (0)(.40) + (1)(.30) + (2)(.20) + (3)(.10) = 1$$

$$\sigma = \sqrt{\sum_{i=1}^{4} (x - \mu)^2 f(x)} =$$

$$\sqrt{(0 - 1)^2(.40) + (1 - 1)^2(.30) + (2 - 1)^2(.20) + (3 - 1)^2(.10)} = 1$$

FIGURE 6.1 Probability Histogram of Population for Sampling Experiment

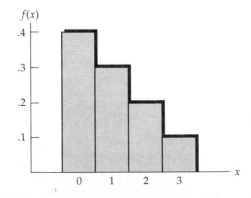

Continuing the example, random samples were generated from the given population with the aid of a computer program written for this purpose. Ten "random digits" from 0 to 9 were generated. (Random digits have virtually equal probabilities of occurring.) These random digits can be used to generate the distribution of the population from which we are sampling by defining a new random variable as follows. The new variable has the value 0 if the value of the random digit is 0, 1, 2, or 3; the value 1 if it is 4, 5, or 6; the value 2 if it is 7 or 8; and the value 3 if it is 9. Thus, the probabilities that the new random variable will take on the value 0 is .4, the value 1 is .3, and so forth, exactly duplicating the probability distribution of the population.

Thirty random samples, each of size 2, were simulated by the computer program described in the previous paragraph. Then, 30 samples of size 20 each were simulated, and finally 30 random samples of size 100 each were simulated. The means of each of these 90 random samples were calculated, with the following results:

Sample Size	Means									
2	2.0	1.5	2.0	0.5	0.0	1.5	2.5	0.5	1.0	0.0
	1.0	2.0	1.5	1.0	1.5	1.5	0.5	0.0	0.0	1.0
	1.0	1.0	2.0	0.5	1.5	1.0	0.5	1.0	2.0	0.5
20	0.70	1.00	1.10	1.25	1.20	0.90	1.20	0.80	0.75	1.00
	0.95	1.45	1.15	0.75	1.50	1.20	0.95	0.70	0.80	0.95
	0.60	1.10	0.95	0.90	0.70	0.80	1.20	1.35	1.15	1.15
100	1.11	0.96	0.88	0.98	1.10	1.04	1.05	0.94	0.83	0.95
	1.14	0.94	0.93	0.98	0.95	1.00	1.03	0.96	0.93	0.75
	0.83	0.88	1.14	1.15	0.93	1.01	1.00	0.99	1.22	1.01

As a check on the theory about the mean and the standard deviation of the mean of a random sample given in this section, the mean and the standard deviation for the 30 means in each sample-size group were calculated. They were compared with the theoretical values, with the following results:

Sample Size	Mean		Standard deviation	
n	*Actual*	*Theory*	*Actual*	*Theory* $\left(\frac{1}{\sqrt{n}}\right)$
2	1.083	1	0.696	0.707
20	1.007	1	0.234	0.224
100	0.987	1	0.104	0.100

(Since the mean and the standard deviation of the distribution from which these samples were taken both equal 1, the theoretical mean and standard deviation, given in this table for a sample of size n, are 1 and $\frac{1}{\sqrt{n}}$, respectively.) Notice the good agreement of the actual means and standard deviations with their theoretical values, especially as the sample size increases.

To pursue this experiment further, histograms of the three sets of sample means were constructed. The histograms are shown in Figure 6.2. (Figure 6.1 gave the probability histogram of the distribution from which the samples were taken.) The histograms representing the sample results were scaled to have equal areas, and normal curves, having the mean 1 and the standard deviation $\frac{1}{\sqrt{n}}$, were superimposed on the histograms. Figure 6.2 shows that the histograms of the distribution of the sample means do not mirror the shape of the probability histogram of

FIGURE 6.2   Sampling Distributions of the Mean

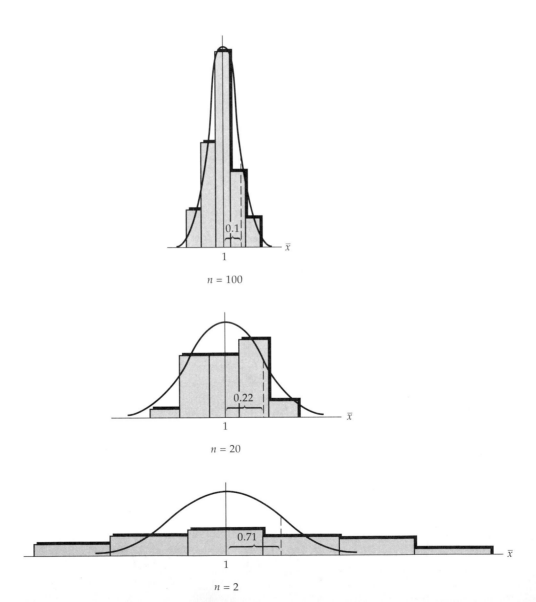

the population distribution. Instead, these histograms more and more resemble normal curves as the sample size is increased.

The empirical results just obtained can be generalized to include sampling from any distribution whose mean and standard deviation are defined. It can be proved that a random variable closely related to the mean of a random sample of size n has a distribution that approaches the standard normal distribution as n becomes large. The **standardized mean**, having the value

$$z = \frac{\bar{x} - \mu}{\sigma / \sqrt{n}}$$

is obtained by subtracting the mean of its distribution, μ, from the sample mean and dividing by the standard deviation of the sample mean, $\frac{\sigma}{\sqrt{n}}$. It is a random variable whose distribution approaches the standard normal distribution as the sample size, n, becomes large. This important result is stated more formally as follows:

Central limit theorem:

If $\bar{x}$ is the mean of a random sample of size n taken from a population whose distribution has the mean μ and the standard deviation σ, then the standardized mean

$$z = \frac{\bar{x} - \mu}{\sigma / \sqrt{n}}$$

is the value of a random variable whose distribution approaches the standard normal distribution as $n \to \infty$.

This **central limit theorem** confers the great advantage of not needing to make any assumptions about the distribution of a population when finding probabilities about values that can be taken on by the mean of a large sample. An illustration of how this theorem can be applied is given in the following example:

Example. To verify that a process of manufacturing cams is properly centered, a random sample of 100 cams was taken from the production line, and their eccentricities were measured. If the standard deviation of this sample was 0.005 inch, what is the probability of obtaining a mean value between 0.249 inch and 0.251 inch if the process center is 0.250 inch?

Solution. It is assumed that the sample has been taken from a population whose mean is $\mu = 0.250$. Since the population standard deviation is unknown, the sample standard deviation will be used in its place; that is, it will be assumed that $\sigma = 0.005$. Given a sample size of 100, it is reasonable to assume that the random variable having the values

$$z = \frac{\bar{x} - 0.250}{0.005 / \sqrt{100}} = \frac{\bar{x} - 0.250}{0.0005}$$

has the standard normal distribution. Thus, the probability that $0.249 < \bar{x}$ < 0.251 is equivalent to the probability that z lies between $\dfrac{0.249 - 0.250}{0.0005} =$ −2.0 and $\dfrac{0.251 - 0.250}{0.0005} = 2.0$. The required probability is obtained from Table 3 (Appendix A) to be $.9772 - (1 - .9772) = .9544$.

Note, in this example, that it is not necessary to know the distribution of cam eccentricities to apply the standard normal distribution in obtaining a probability involving the mean. The sample size is large enough so that the central limit theorem insures that the normal distribution provides an excellent approximation, even when the true (unknown) standard deviation of the population is approximated by the sample standard deviation. When the sample size is small, however, the central limit theorem does not provide an accurate enough approximation of such probabilities. Methods for finding probabilities involving the mean of a random sample whose size is small will be introduced in Section 6.3. As a common rule of thumb, if $n > 30$, errors are small enough that the central limit theorem can be used.

EXERCISES

6.1 Cans of food, stacked in a warehouse, are sampled to determine the proportion of damaged cans. Explain why a sample that includes only the top can in each stack would not be a random sample.

6.2 An inspector chooses a sample of parts coming from an automated lathe by visually inspecting all parts, and then including 10% of the "good" parts in the sample with the use of a table of random digits.

 a. Why does this method not produce a random sample of the production of the lathe?

 b. Of what population can this be considered to be a random sample?

6.3 Sections of aluminum sheet metal of various lengths, used for construction of airplane fuselages, are lined up on a conveyer belt that moves at a constant speed. A sample is selected by taking whatever section is passing in front of a station at five-minute intervals. Explain why this sample may not be random; that is, it is not an accurate representation of the population of all aluminum sections.

6.4 A process error may cause the oxide thicknesses on the surface of a silicon wafer to be "wavy," with a constant difference between the wave heights. What precautions are necessary in taking a random sample of oxide thicknesses at various positions on the wafer to assure that the observations are independent?

6.5 Four reinforced concrete sections are joined end to end to form a structure that is to have a length of 20 feet. Assume that the sections were chosen at random from a population of sections whose lengths have a distribution with a mean of 5 feet and a standard deviation of 1 inch.

 a. What is the mean length of the structures so formed?

 b. What is the standard deviation of the mean lengths?

6.6 How large a random sample is required from a population whose standard deviation is 3.5 so that the sample mean will have a standard deviation of 0.7?

6.7 In Exercise 6.5, what is the standard deviation of the mean lengths of random samples of 36 sections?

6.8 A random sample of size 100 has a mean of 15.8 and a standard deviation of 4.2.

   a. Can the standard deviation of the sample mean be known exactly?

   b. Using the sample standard deviation in place of the population standard deviation, find the approximate standard deviation of the sample mean.

6.9 A random sample of size 100 is taken from a population whose mean is $\mu = 25$ and whose standard deviation is $\sigma = 5$. Find the probability that the mean of this sample will exceed 26.

6.10 What is the probability that the mean of a random sample of size 50 will lie outside the limits 10 – 15 if the sample is taken from a population whose mean and standard deviation are 12 and 16, respectively?

6.11 What is the probability that the mean of a random sample of size 64, taken from a population whose standard deviation is 10, will differ from the population mean by more than 2?

6.12 A random sample of 75 observations is chosen from a population whose distribution has a standard deviation of 20. Find the probability that the sample mean will lie in an interval within 5 units from the population mean.

6.13 A random sample of 36 resistors is taken from a production line manufacturing resistors to a specification of 40 ohms, with a standard deviation of 1 ohm. If the sample mean resistance differs from the specified resistance by more than 0.5 ohm, would you question whether the line is meeting specifications? Why?

6.14 To test whether a manufacturer's claim is valid that a spring can withstand over 10,000 compressions before losing its elasticity, a sample of 40 springs is subjected to alternate compression and release. If the sample mean number of compressions before the spring fails to return to a preset height is 9,812, and the sample standard deviation is 958, has serious doubt been cast on the claim?

6.15 In a random survey of 1,225 households, the mean number of hours of daily television watching is 5.9 hours, with a standard deviation of 1.1 hours. Is this result consistent with the claim that households watch television at least 6 hours daily? Why?

6.16 The claim is made that a hinge lubricant will survive at least 5,000 cycles of opening and closing before the hinge squeaks. One hundred test samples produce a number of cycles before squeaks occur that are approximately normally distributed with a mean of 4,500 cycles and a standard deviation of 150 cycles. Does this result constitute strong evidence that the claim is false? Why?

The following exercises introduce new material, expanding on some of the topics covered in the text.

6.17 *Mean of a weighted sum.* If, for $i = 1, 2, \ldots, k$, x_i are independent random variables having the means μ_i, and the variances σ_i^2, and a_i are constants, the **weighted sum** $a_1 x_1 + \ldots + a_k x_k$ is a value of a random variable having the mean $a_1 \mu_1 + \ldots + a_k \mu_k$ and the variance $a_1^2 \sigma_1^2 + \ldots + a_k^2 \sigma_k^2$. Find the mean and the standard deviation of the length of a steel beam made up of 5 randomly chosen sections, 1 of which comes from a population of beams with a mean length of 5 feet and a standard deviation of 1 inch, 2 have mean lengths of 6 feet and standard deviations of 1.5 inches, and 3 have a mean length of 10 feet and a standard deviation of 2.0 inches.

6.18 *Continuation.* By choosing appropriate values of k and a_i, use the result given in Exercise 6.17 to

  a. find the mean of the sum of two independent random variables

  b. find the mean of the difference between two independent random variables

  c. obtain the results given on page 137 for the mean and the standard deviation of the distribution of a sample mean

6.19 *Finite populations.* The standard deviation of the sample mean of a random sample of size n, taken from a population of size N having the standard deviation σ, is given by

$$s_{\bar{x}} = \sqrt{\frac{N - n}{N - 1}} \cdot \frac{\sigma}{\sqrt{n}}$$

Find the standard deviation of the mean of a sample of size 10 taken from a population of size 40 whose standard deviation is 5.

6.20 *Continuation.* The quantity $\dfrac{N - n}{N - 1}$ is called the **finite population correction factor**.

  a. Find the value of this factor for $n = 10$ and $N = 1,000$.

  b. Show that this factor approaches 1 as $N \rightarrow \infty$.

6.3 The Small-Sample Sampling Distribution of the Mean

The central limit theorem gives a large-sample approximation for the distribution of the statistic

$$z = \frac{\bar{x} - \mu}{\sigma / \sqrt{n}}$$

Thus, application of this theorem requires that σ, the population standard deviation, be known. For large samples, the sample standard deviation, s can be substituted for σ with little error. However, for small samples (as a practical matter, samples of size 30 or less), we must rely on other theory for the sampling distribution of the mean.

For small samples, instead of the sampling distribution of z, we shall use the sampling distribution of the statistic

---

t Statistic: $\quad t = \dfrac{\bar{x} - \mu}{s / \sqrt{n}} \quad$ with $\quad \nu = n - 1 \; d.f.$

The statistic t differs from z in that s is substituted for σ in the denominator. Unfortunately, very little is known about the distribution of t unless we make the assumption that *the sample comes from a normal population*. Under this assumption, the distribution of t is called the **t distribution**. The t distribution actually is a one-parameter family of distributions having the parameter, ν, called the **degrees of freedom**, abbreviated *d.f.* The degrees of freedom for a given t distribution depends upon the sample size, with $\nu = n - 1$.

The t distribution has a mean of 0, and it is symmetrical, with a shape similar to that of the standard normal distribution. The standard deviation of the t distribution is larger than 1, that of the standard normal distribution, but it approaches 1 as the sample size (and, thus, ν) approaches infinity. A sketch of the t distribution for $\nu = 5$ is compared with the standard normal curve in Figure 6.3.

The t distribution is tabulated in Table 4 (Appendix A). This table gives values of t_α; that is, values of t that cut off right-hand tails of area $\alpha = .10, .05, .025,$ .01, and .005 for degrees of freedom from 1 to 29, as shown in Figure 6.4. The last line of Table 4 is labeled "inf.," corresponding to $\nu = \infty$. It should be noted that the values of t given in this line are identical to z_α, the corresponding values of the standard normal distribution. For example, the value of t corresponding to a right-hand tail of area .025 is 2.228 for 10 degrees of freedom and 1.96 for ∞ degrees of freedom. (Note that the value of $z_{.025}$ also is 1.96.)

The t statistic has the t distribution only when the sample is taken from a normal population. Thus, it is important to check the sample to see if this assump-

FIGURE 6.3 t Distribution

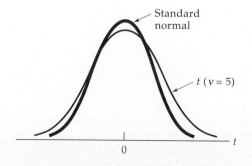

FIGURE 6.4   Values of t Given by Table 4

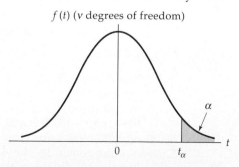

tion is at least approximately correct before applying this distribution to the solution of problems involving the means of small samples. Fortunately, in most applications, probabilities obtained with the t distribution are not very sensitive to the assumption of normality; thus, even rough agreement of the data with a normal distribution usually will be satisfactory.

The following example illustrates how the t distribution can be used to make statements about the mean of a normal population.

Example. A random sample of size $n = 9$, taken from a normal population, has a standard deviation of 1.8. What is the probability that the difference between the sample mean and the population mean exceeds 1.1?

Solution. The probability that $(\bar{x} - \mu) > 1.1$ is equivalent to the probability that

$$t = \frac{\bar{x} - \mu}{1.8 / \sqrt{9}} = \frac{\bar{x} - \mu}{0.6}$$

with $9 - 1 = 8$ degrees of freedom exceeds $1.1/0.6 = 1.83$. The closest entry in Table 4 with 8 degrees of freedom is 1.860, corresponding to $\alpha = .05$. Thus, the required probability is approximately .05.

The following example shows how the t distribution can be used to verify the validity of a statement made about the mean of a normal distribution on the basis of a small sample.

Example. The claim is made that a hinge lubricant will survive at least 5,000 cycles of opening and closing before the hinge squeaks. A test involving 15 test samples produces numbers of cycles before the hinge squeaks that are approximately normally distributed with a mean of only 4,512 cycles and a standard deviation of 145 cycles. Does this result constitute strong evidence that the claim is false?

Solution. The sample of size $n = 15$ resulted in a sample mean of $\bar{x} = 4,512$ and a sample standard deviation of $s = 145$. If the mean of the population from which the sample were taken actually was 5,000, the corresponding value of t would be

$$t = \frac{4,512 - 5,000}{145 / \sqrt{15}} = -13.03$$

Referring to Table 4 with $15 - 1 = 14$ degrees of freedom, we see that the probability that t exceeds 2.977 is only .005. From the symmetry of the t distribution, it can be seen that the probability that t will take on a value less than -13.03 is much less than .005. Thus, either the population mean is 5,000 and a very low probability sample was taken, or, as most people would conclude, the sample gives strong evidence that the population mean actually is less than 5,000 and the claim is false.

EXERCISES

6.21 Find the value of the t statistic with 20 *d.f.* which will be exceeded with probability .01.

6.22 Find the value of the t statistic with 12 *d.f.* which will be exceeded with probability .05.

6.23 A random sample of size 15, taken from a normal population, has the mean of 136. If the sample standard deviation is 16.6, what is the probability that the sample mean will exceed 145.2?

6.24 A random sample of size 25, taken from a normal population, has the mean of 38.4. If the sample standard deviation is 5.8, what is the probability that the sample mean will exceed 41.3?

6.25 The claim is made that the mean breaking strength of an alloy exceeds 150 psi. To test this claim, 9 samples are subjected to fracture tests with a resulting mean breaking strength of 138 psi and a sample variance of 28 psi.

    a. Can this sample result reasonably be obtained if the claim is true?

    b. What assumption are you making in answering part a?

6.26 A random sample of 15 dummy artillery shells are fired at a target, and the miss distances are measured. The sample mean miss distance is 24 feet and the sample standard deviation is 5.8 feet.

    a. Is the claim "The mean miss distance does not exceed 20 feet" reasonable?

    b. What assumption must be made to answer part a?

6.4 Sampling Distributions Involving Variances

The Chi-Square Distribution

So far we have discussed only sampling distributions involving means. However, if we had taken the variances of the 90 random samples discussed on page 139, we would also have obtained experimental sampling distributions of the variances. Although, as we have discovered, the theoretical sampling distribution of the mean approaches a normal curve as the sample size becomes large, this result cannot be true for the variance, since s^2 cannot be negative. Like the small-sample distribution of the mean, little is known about the distribution of sample variances unless we make the assumption that the sample came from a normal population. Under that assumption, it can be proved that

$$\textit{Chi-Square Statistic:} \quad \chi^2 = \frac{(n-1)s^2}{\sigma^2} \quad \text{with} \quad \nu = n - 1 \ d.f.$$

has the **chi-square distribution**. The parameter ν is called the **degrees of freedom** for the chi-square distribution, analogous to the degrees of freedom previously defined for the *t* distribution. A typical chi-square distribution is shown in Figure 6.5. Note that this distribution has the value 0 for values of chi-square less than or equal to zero. For positive values of chi-square, the distribution is positively skewed, with a long right-hand tail such that values of the distribution approach zero as chi-square approaches infinity.

Values of chi-square are tabulated in Table 5 (Appendix A) for various values of v and selected right-hand tail areas. The quantity χ^2_{α}, given in Table 5, is that value of chi-square which cuts off a right-hand tail under the chi-square distribution of area α, as shown in Figure 6.5. Unlike the *t* statistic, the distribution of the chi-square statistic is relatively sensitive to departures from normal populations, especially when the distribution of the underlying population has a long tail. Thus, transformation to near normality should be made before using a table of the chi-square distribution to obtain probabilities involving sample variances.

The following example illustrates how the chi-square distribution can be used to find the probability that a sample variance will exceed a given value. Note in this example that an assumed value of σ^2, the population variance, must be given, and that it must be assumed that the sample is taken from a normal population.

Example. The thickness of the oxide coating on a silicon wafer used for the manufacture of integrated circuits must be controlled to within narrow tolerances. Suppose the variance of these values throughout the wafer must not exceed 0.01 Å. What is the probability that a random sample of 10 measurements taken on a wafer whose oxide-coating thickness actually has a variance of 0.01 Å will have a sample variance of 0.02 Å or greater?

Solution. Substituting into the formula for chi-square, we get

$$\chi^2 = \frac{(10 - 1)\ (0.02)}{0.01} = -18.0$$

Examination of Table 5 with $v = 10 - 1 = 9$ degrees of freedom, we see that the probability of obtaining a value of chi-square of 18.0 or larger lies between $\alpha = .05$ and $\alpha = .025$. Such a sample result would be unlikely.

Suppose, in the preceding example, that the criterion for rejection of a wafer were set at an oxide-thickness variance of 0.02 or greater in samples of size 10 from a wafer. Then, the probability of rejecting a wafer when the variance of its

FIGURE 6.5 Values of χ^2 Given by Table 5

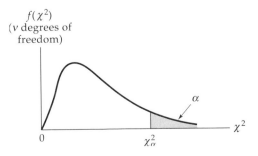

oxide thickness actually is only 0.01 would be less than .05. In other words, such a criterion would reject less than 5% of the "good" wafers.

The F Distribution

Closely related to the problem of finding the distribution of a sampling variance is that of finding the sampling distribution of the *ratio* of two variances. Knowledge of this distribution will allow us to compare two variances obtained from independent samples; the need to determine whether a given sample variance is equal to another independent variance arises when we encounter the analysis of variance tests associated with the design of experiments (Chapter 10).

If two independent random samples are taken from normal populations having the same variance, it would be expected that the ratio of the sample variances would have a value close to 1. To determine whether such a ratio is "close enough" to 1, we use the following sampling statistic.

$$\textit{F Statistic:} \quad F = \frac{s_1^2}{s_2^2}$$

This statistic has the **F distribution** provided that the samples are *independent* and they come from *normal populations with the same variance*. The F distribution is a two-parameter family of distributions having the parameters $v_1 = n_1 - 1$, called the **numerator degrees of freedom** and $v_2 = n_2 - 1$, called the **denominator degrees of freedom**. A typical F distribution is shown in Figure 6.6.

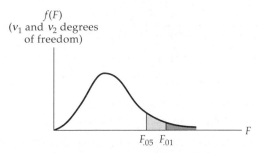

FIGURE 6.6   Values of F Given by Table 6

Unlike the t and the chi-square distributions, the F distribution has two parameters; thus, it is usually tabulated only for two right-hand tail areas, $\alpha = .05$ and $\alpha = .01$. Values of $F_{.05}$ and $F_{.01}$, which cut off right-hand tail areas of .05 and .01 for various combinations of v_1, and v_2 are given in Table 6(a) and Table 6(b), respectively. Figure 6.6 shows a typical area under the F distribution given by Table 6. The comment made on page 148 regarding transformation of the population distribution to near normality also holds for the F distribution.

Most tables of the *F* distribution, such as Table 6, give only *right-hand* tail areas under the *F* distribution. Such tables can be used only to determine the probability that a given ratio of two sample variances will be *greater* than a given value. For this reason, whenever possible, problems involving the *F* statistic are phrased so that the larger sample variance is placed in the numerator.

Example. If two independent random samples of size $n_1 = 21$ and $n_2 = 26$ are taken from a normal population, find the probability that the first sample will have a variance at least twice as large as that of the second sample.

Solution. From Table 6(a) with 20 numerator degrees of freedom and 25 denominator degrees of freedom, we find $F = 2.01$. Thus, the required probability is approximately .05.

EXERCISES

6.27 A claim that the variance of a normal population is $\sigma^2 = 30$ is to be rejected whenever the variance of a random sample of size $n = 25$ exceeds 49.2. What is the probability that the claim will be rejected?

6.28 A claim that the standard deviation of a normal population is $\sigma = 5$ is to be rejected whenever the standard deviation of a random sample of size $n = 15$ exceeds 7.2. What is the probability that the claim will be rejected?

6.29 A random sample of 10 observations is taken from a normal population having a standard deviation of 6.5. Find the approximate probability of obtaining a sample standard deviation between 3.14 and 8.94.

6.30 Find the value of *F* from Table 6 corresponding to the following:

    a. $\nu_1 = 20, \nu_2 = 11, \alpha = .01$

    b. $\nu_1 = 6, \nu_2 = 15, \alpha = .05$

    c. $n_1 = 8, n_2 = 31, \alpha = .05$

    d. $n_1 = 5, n_2 = 7, \alpha = .01$

6.31 Independent random samples, each of size 11, are taken from normal populations having the same variance.

    a. What is the approximate probability that a *given* sample variance will be at least three times that of the other?

    b. What is the approximate probability that *either* sample variance will be at least three times larger than the other?

6.32 The claim is made that normal population 1 has a larger variance than that of normal population 2. Independent random samples of sizes 13 and 16, respectively, are taken from these populations, obtaining sample standard deviations of 7.5 and 4.1, respectively. Could this result have reasonably been obtained if the claim were true? Why?

REVIEW EXERCISES

These review exercises can be used for informal review of the chapter, or as two practice examinations, designed to be taken over a time period of about one hour. The odd-numbered exercises comprise one examination; the even-numbered exercises comprise the other.

6.33 Suppose representatives to a state university students' convention were selected by choosing two students at random from each university in the state. Does every state university student have an equal chance of being chosen? Why?

6.34 a. List the different samples of size $n = 2$ that can be chosen from a population having $N = 5$ members.

   b. If a sample of size 2 is chosen at random from this population, what is the probability that a given sample will be drawn?

6.35 A random sample of size 64 is taken from a population having the mean $\mu = 164$ and the standard deviation $\sigma = 11$. Use the central limit theorem to obtain an approximation for the probability that the mean of the sample will not exceed 160.

6.36 Find the approximate probability that, in a large sample, the sample mean will be within 1.5 standard deviations of the population mean.

6.37 It is claimed that the mean of a certain population is $\mu = 5.0$ and its standard deviation is $\sigma = 2.0$. A random sample of 100 observations, taken from this population, has a mean of 4.5. Do you believe the claim? Why?

6.38 It is claimed that the mean and the standard deviation of a certain population are $\mu = 12$ and $\sigma = 6.0$. If the sample mean in a random sample of 49 observations taken from this population had a value of 14.8, would you believe the claim? Why?

6.39 If the sample size in Exercise 6.37 were changed to 18, would your conclusions remain the same? Why?

6.40 Would your conclusions in Exercise 6.38 remain unchanged if the sample size were 28? Why?

6.41 If a random sample of 18 observations is taken from a normal population having the standard deviation $\sigma = 17.5$, find the approximate probability that the sample standard deviation will exceed 24.5.

6.42 For what sample size will the probability be .01 that the variance of a random sample from a normal population exceeds 3 times the population variance?

6.43 Two independent random samples of sizes $n_1 = 9$ and $n_2 = 16$ are taken from the same normal distribution. What is the probability that the variance of the first sample will be at least four times as large as the variance of the second sample?

6.44 The claim is made that a given normal population has a smaller variance than that of a second normal population. Two independent random samples, each of size 25, are taken from these populations, obtaining sample variances of 11.1 and 16.9, respectively. Could this result have reasonably been obtained if the claim were true? Why?

DRAWING INFERENCES FROM SAMPLES

Where have we been?   Building on the ideas of probability and probability distributions, we have defined random samples. Then, we have introduced the idea of statistics and their sampling distributions.

Where are we going?   Now, we're ready to put together our knowledge of probability and sampling to formalize the scientific core of statistics—how sample data are used to draw inferences about their underlying populations.

Section 7.1, *Statistical Inference,* describes how statistical science approaches the problems encountered in making statements about a population on the basis of sample results.

Section 7.2, *Estimation of a Population Mean*, gives properties of the sample mean as an estimate of the population mean, gives methods for calculating confidence limits for a mean, and gives methods for finding the sample size required to estimate a population mean.

Section 7.3, *Estimation of a Population Variance*, shows how to estimate a population variance, both with a single number derived from the sample data and with a confidence interval.

Section 7.4, *Estimation of a Proportion*, gives methods for calculating confidence limits for a proportion.

Section 7.5, *Significance Tests Involving Means*, gives methods for testing whether a sample mean is significantly different from an assumed value of the mean of the population from which the sample was taken.

Section 7.6, *Significance Tests Involving Variances*, gives to tests for the significance of the difference between a sample variance and an assumed value, and for differences involving two variances.

Section 7.7, *Significance Tests Involving Proportions,* gives tests of the significance of the difference between a sample proportion and an assumed value, and for differences among two or more proportions.

Section 7.8, *Application to Quality*, gives perspective to the methods of this chapter in the context of quality improvement.

7.1   Statistical Inference

A random sample taken from a given population was defined on page 136. Also, it was stated on page 137 that the mean of a random sample taken from a population having the mean μ and the standard deviation σ has a distribution whose mean also equals μ, and whose standard deviation equals $\dfrac{\sigma}{\sqrt{n}}$. Thus, the sample mean has a distribution whose mean is the same as that of the population from which the sample was taken, and a standard deviation which diminishes as the sample size increases. Such results suggest that sample statistics can be effective in describing parameters of the population from which the sample was taken.

The body of theories and methods for making statements about parameters of a population on the basis of random samples drawn from that population is called **statistical inference**. This field of inquiry has been subdivided into two main parts, **estimation** and **tests of hypotheses**. One problem of estimation involves making use of a sample result to find a single acceptable value, or **point estimate** of the value of a parameter of a population, such as its mean or its standard deviation. Another problem of estimation involves finding a pair of numbers, so it can be said with a given degree of confidence that the interval contained between these numbers includes the value of the population parameter. Such an interval is called a **confidence interval**.

Tests of hypotheses use sample results to test the validity of statements made about population parameters. (Since the value of a population parameter is unknown, we must seek evidence from a random sample about the credibility of such statements.) Here, we shall introduce special kinds of tests of hypotheses, called "significance tests," discussing their applications to problems of checking the validity of statements on the basis of the results of a random sample. In this chapter, we shall consider inferences concerning means, variances, and proportions. First we shall discuss point and interval estimation of population means, standard deviations, and proportions. Then, we shall introduce methods for testing hypotheses concerning these population parameters.

7.2   Estimation of a Population Mean

When attempting to estimate the value of a population parameter on the basis of a sample, two kinds of error can be made, namely **bias** and **random error**. Bias occurs when repeated estimates have values consistently larger or smaller than the value of the parameter to be estimated. Bias in estimation is analogous to the idea of lack of accuracy, introduced on page 30, where it was shown that repeated measurements tend to cluster about a value other than the target value. A method for estimating a population parameter is said to be **unbiased** if the sampling distribution of the resulting estimate has the same mean as the parameter being estimated. For example, the sample mean, $\bar{x}$, is an unbiased point estimate of the mean of the population from which the sample was taken. This statement follows from the fact that the mean of the sampling distribution of $\bar{x}$ is μ, the same as that of the population from which the sample was taken.

Even if estimates are unbiased, however, random variability can lead to errors in estimating a parameter. Random error is like lack of precision, defined on page 30, where repeated measurements are "spread out," that is, not close to each other. The random error of a sample statistic is measured by the variance of its distribution. For example, the standard deviation of the distribution of the sample mean is $\frac{\sigma}{\sqrt{n}}$. Thus, the means of repeated samples will tend to differ less from sample to sample if σ, the population standard deviation, is smaller in value. Of greater importance, *the random error of the sample mean tends to decrease as the sample size,* n, *becomes larger*.

Statistical estimation seeks to find methods for estimating a specified parameter (or parameters) of a given population that minimize bias and random error. A common approach to point estimation is to find a statistic, based on the results of a random sample, that is both unbiased and has the smallest possible sampling variability, as measured by the variance of its sampling distribution. It can be shown that, in most practical situations, the sample mean has the smallest variance of all unbiased linear statistics for estimating a population mean.

When a random sample has been taken and the value of a sample statistic has been calculated to give a point estimate of a population parameter, it cannot be expected that the resulting point estimate will exactly equal the population parameter. Thus, it is useful to express the error in estimation by means of a confidence interval. Such an interval gives upper and lower limits for the value of the estimated parameter, along with a probability statement about these limits. Since the true value of the parameter remains unknown, the probability statement must refer to the method for obtaining the confidence limits, and not to the probability that the parameter value actually lies within the given limits. Thus, the statement "95% confidence limits for the population mean are 4 and 5", means that, in repeated random samples of the same size taken from the same population, 95% of the limits obtained will include the value of the population parameter. It does not mean that there is a 95% chance that μ lies between 4 and 5.

Large-Sample Confidence Limits for μ

If the sample size is large ($n \geq 30$), the central limit theorem can be used to state that, to a good approximation,

$$z = \frac{\bar{x} - \mu}{\sigma / \sqrt{n}}$$

is a value of a random variable whose distribution is the standard normal distribution. Thus, with probability $1 - \alpha$, this random variable will assume a value between the limits $-z_{\alpha/2}$ and $z_{\alpha/2}$, or

$$-z_{\alpha/2} < \frac{\bar{x} - \mu}{\sigma / \sqrt{n}} < z_{\alpha/2}$$

Multiplying each member of this double inequality by the positive quantity $\frac{\sigma}{\sqrt{n}}$, subtracting $\bar{x}$ from each term, and multiplying each term by -1, we can write this inequality in the form

$$\overline{x} - z_{\alpha/2} \cdot \frac{\sigma}{\sqrt{n}} < \mu < \overline{x} + z_{\alpha/2} \cdot \frac{\sigma}{\sqrt{n}}$$

The endpoints of this inequality are called "$100(1 - \alpha)$ percent confidence limits for μ." These limits are such that the probability is $1 - \alpha$ that the random interval they define will contain μ, the unknown mean of the population from which the sample was obtained.

Confidence Limits for μ (large samples): $\overline{x} \pm z_{\alpha/2} \cdot \dfrac{\sigma}{\sqrt{n}}$

As previously stated, the probability statement we are making here does not refer to the probability that μ lies between the calculated limits. Either it does or it does not! Since we do not know the value of μ, all that can be said is that the *method* for finding the limits produces intervals that will include μ $100(1 - \alpha)$% of the time. It is for this reason that these limits are called "confidence limits" rather than "probability limits."

> *Example.*   If a random sample of 100 observations from a population with the standard deviation 14.5 has a mean of 8.2, find 95% confidence limits for the population mean.
>
> *Solution.*   If $1 - \alpha = .95$, then $z_{\alpha/2} = z_{.025} = 1.96$. The 95% confidence limits are $8.2 \pm 1.96\dfrac{14.5}{\sqrt{100}}$, or 5.4 and 11.0. Thus, we can state with 95% confidence that the population mean lies between 5.4 and 11.0.

In this example, it was necessary to know the population standard deviation in order to obtain confidence limits. Since σ generally is unknown, it is customary to substitute the sample standard deviation, s, for σ in the formulas for the confidence limits. The approximation that results is excellent for large samples—those having sizes of 30 or more as a rule of thumb.

Occasionally, prior estimates of σ are substituted for σ, instead of the sample standard deviation, s. The standard deviation of large numbers of previous samples, taken from the same population, provides an estimate of σ that may be better than an estimate obtained from a single sample. The danger in making use of prior estimates of σ lies in the possibility that the current sample may have been taken under different conditions than the previous samples, and the population standard deviation might be different from that of the current sample.

Small-Sample Confidence Limits for μ

The assumption that the standardized sample mean is approximately normally distributed can be highly inaccurate when s is substituted for σ and the sample size, n is small ($n < 30$). However, for samples drawn from a normal population, the statistic

$$t = \frac{\bar{x} - \mu}{s/\sqrt{n}}$$

has the t distribution with $\nu = n - 1$ degrees of freedom. Thus, confidence intervals for μ can be found using small samples by substituting s, *the sample standard deviation*, for σ in the formula given on page 156 for large-sample confidence limits, and using t (with $\nu = n - 1$ degrees of freedom) in place of z to obtain the following confidence limits:

Confidence Limits for μ (small samples): $\quad \bar{x} \pm t_{\alpha/2} \cdot \dfrac{s}{\sqrt{n}}$

It is important to remember that these formulas will give exact $1 - \alpha$ confidence limits only if the distribution from which the sample was taken was a normal distribution. Before calculating confidence limits for small samples, it is good practice to check the sample data for normality, using one of the methods described in Section 5.4. As pointed out on page 146, however, the distribution of the random variable having the value t is reasonably well approximated by the t distribution for samples from distributions that are only approximately normal.

> **Example.**   A test of the bending strength of a turned shaft is a destructive test, requiring a measurement of the minimum force (in pounds) required to bend the shaft in a standard testing machine. Thus, a sample of only five shafts is selected at random from production to obtain a preliminary estimate of bending strength, with the following results (in pounds):
>
> $$1{,}101 \quad 1{,}315 \quad 1{,}520 \quad 1{,}216 \quad 1{,}360$$
>
> Find 90% confidence limits for the mean bending strength of shafts produced by this process.
>
> **Solution.**   First, the five measurements should be checked for normality, using a normal-scores plot. (Excellent agreement with normality was found using *MINITAB* statistical software.) The sample mean was calculated to be $\bar{x} = 1{,}302.4$, and the sample standard deviation was $s = 157.2$. The 90% confidence limits are found by substituting $\bar{x} = 1{,}302.4$, $s = 157.2$, and $t_{.05} = 2.132$ with 4 degrees of freedom into the formula for small-sample confidence limits given above. The resulting 90% confidence limits are $1{,}302.4 \pm 2.132 \dfrac{157.2}{\sqrt{5}}$, or $1{,}152.5$ and $1{,}452.3$.

To avoid confusion as to whether to use t or z in calculating confidence limits, it is a good idea *always* to use t. If the degrees of freedom are too large to be shown in Table 4 (over 29), the value ∞ can be used for ν. You can see from a comparison of Tables 3 and 4 that for $\nu > 29$ degrees of freedom, $t_{\alpha}(\nu > 29 \text{ d.f.}) \approx t_{\alpha}(\infty \text{ d.f.}) = z_{\alpha}$.

Sample Size for Estimating a Mean

The maximum error, with probability $1 - \alpha$, in estimating a population mean with the mean of a sample taken from that population is the half-width of the corresponding confidence interval, or

$$E = z_{\alpha/2} \cdot \frac{\sigma}{\sqrt{n}}$$

The size of the random sample necessary to estimate μ to within a desired maximum error of E with probability $1 - \alpha$ can be obtained by solving this equation for n, obtaining

---

Sample Size for Estimating a Mean: $n = \left[\dfrac{z_{\alpha/2} \cdot \sigma}{E} \right]^2$

In using this formula, the calculated value of n should be rounded up to the next integer value.

Application of this result is not straightforward, however, since σ usually is unknown. The value of σ often can be estimated from prior data of a similar kind. Sometimes, it is necessary to make an educated guess. Once a sample size has been determined, the sample is taken and the sample standard deviation has been computed, s can be compared with the original estimated value of σ. If s is very much larger than this value, a new calculation can be made for n, and an additional sample can be taken to build up the combined sample size to the required value.

Example. If it is desired to estimate the mean electrical resistance of a spool of wire to within a maximum error of 0.1 ohm with 95% confidence, and it can be assumed that $\sigma = 0.5$ ohm, find the required sample size.

Solution. For 95% confidence, $z_{.025} = 1.96$ is used. Since the maximum error is to be 0.1 ohm, $E = 0.1$. The standard deviation has been calculated from prior records to be 0.5 ohm. Then, substituting into the formula for the sample size of the mean, with the confidence level 95%,

$$n = \left[\frac{(1.96)(0.5)}{0.1} \right]^2 = 96.04$$

Since the sample size must be an integer, this result is rounded up to the nearest integer. Thus, to be sure with confidence level 95% that the random sample will give a mean value within 0.1 ohm of the population mean, on the assumption that the population standard deviation is 0.5 ohm, a random sample of size 97 is required.

A reasonable estimate of σ was available in the foregoing example. The following example illustrates the procedure when only a rough guess can be made.

Example. Suppose, in the last example, that no previous data were available from which to estimate σ. It is guessed on the basis of engineering experience that σ has a value of about 0.4, and a random sample of size

$$n = \left[\frac{(1.96)\ (0.4)}{0.1}\right]^2 = 61.5$$

or $n = 62$ is taken. Suppose, now, that the standard deviation of the sample of 62 observations has been calculated to be $s = 0.49$. Make a new calculation of the required sample size and determine the size of an additional random sample, if one is needed.

Solution. A new calculation for the sample size, using $\sigma = 0.49$, gives

$$n = \left[\frac{(1.96)\ (0.49)}{0.1}\right]^2 = 93$$

Thus, an additional 31 observations are required to complete the required sample of size 93. Special care is needed so that the supplementary sample is taken at random from the same population that gave rise to the initial sample of 62 observations.

Computer Applications

Confidence limits for the mean can be found with several statistical software packages. Using *MINITAB* software, the coking-time data given in Exercise 3.10 on page 36 are entered into column C1. (It shall be assumed that these data are the results of a random sample.) The command TINTERVAL WITH 99% CONFIDENCE IN COL C1 produces the following output:

	N	MEAN	STDEV	SEMEAN	99 PERCENT C.I.
C1	30	7.76	1.18	0.22	(7.16, 8.36)

Thus, 99% confidence limits for the mean of the population of coking times are 7.16 and 8.36 hours.

EXERCISES

7.1 Suppose the statistic g_1 has been proposed as a point estimate of the parameter η of a given population.

a. What must be said about the sampling distribution of g_1 for it to be an unbiased point estimate of η?

b. Suppose g_2 also is an unbiased estimate for η, but that the variance of the sampling distribution of g_2 is smaller than that of g_1. Which estimate would you prefer for η? Why?

7.2 Suppose 3.8 and 4.7 are 90% confidence limits for a given parameter of a distribution.

   a. Can the probability be found that the value of this parameter lies in the interval (3.8, 4.7)?

   b. Why?

   c. State in words to what the probability .90 refers in this context.

7.3 A random sample consisting of 49 observations has the mean 137 and the standard deviation 21. Find 95% confidence limits for the population mean.

7.4 A random sample consisting of 64 observations has the mean 11.3 and the standard deviation 3.6. Find 99% confidence limits for the population mean.

7.5 A random sample of size $n = 25$ is taken from a normal distribution. If the sample mean is 1.85 and the sample standard deviation is 0.52, find 90% confidence limits for the population mean.

7.6 A random sample of size $n = 25$ is taken from a normal population. If the sample mean and the standard deviation are 15.9 and 3.8, respectively, find 95% confidence limits for the population mean.

7.7 A random sample is to be taken from a population whose standard deviation is believed to be 0.5. How large a sample should be taken to estimate the mean of this population to within a maximum error of 0.1 with 95% confidence?

7.8 How large a sample should be taken to estimate the mean of a population having the standard deviation 3.7 to within a maximum error of 1.0 with 90% confidence?

7.9 Assume the data given in Exercise 3.11 on page 36 are the results of taking a random sample from a normal population. Use *MINITAB* or other statistical computer software to find 90% confidence limits for the population mean.

7.10 Assume the data given in Exercise 3.14 on page 40 come from a random sample. Use *MINITAB* or other statistical computer software to find 95% confidence limits for the mean of the underlying population.

7.3   Estimation of a Population Variance

In substituting s for σ in the t statistic, we were using the sample standard deviation as a point estimate of the population standard deviation. This substitution would seem reasonable if it could be shown that s has "desirable" properties as an estimate of the population standard deviation, σ. As previously discussed, such desirable properties include lack of bias and small variance in comparison to that of other unbiased estimates. In fact, it can be shown that s is slightly biased as an estimate of σ, although the bias approaches 0 as the sample size becomes large. However, it also can be shown that s^2 is an unbiased estimate of the population *variance,* σ^2. Since s^2 has a variance that is small compared to that of other unbiased estimates for σ^2, the sample variance frequently is used to estimate the variance of the population from which the sample was taken. Even though it is slightly biased, the sample standard deviation commonly is used to estimate a population standard deviation.

The sample range R, defined on page 54, also is used to estimate a population standard deviation under certain circumstances. Given a sample of size n from a *normal population*, it can be shown that the sampling distribution of the range has the mean $d_2\sigma$ and the standard deviation $d_3\sigma$, where d_2 and d_3 are constants that depend upon the sample size, as shown in the following table for $n = 2, 3, \ldots , 10$:

n	2	3	4	5	6	7	8	9	10
d_2	1.128	1.693	2.059	2.326	2.534	2.704	2.847	2.970	3.078
d_3	0.853	0.888	0.880	0.864	0.848	0.833	0.820	0.808	0.797

Since the mean of the sampling distribution of R is $d_2\sigma$, it follows that R/d_2 is a statistic that gives an unbiased estimate of σ. It can be shown that the variance of this statistic is nearly as small as that of s for very small samples ($n \leq 5$), but it becomes much greater as the sample size increases. For this reason, use of the range to estimate a population standard deviation is strongly discouraged, except for very small samples. Such use usually is confined to control charts (Chapter 8), and then only when sample sizes are small and computational ease is important.

> **Example.** Use the range of the bending-strength data given in the example on page 157 to estimate the population standard deviation, and compare your result with the sample standard deviation.

> **Solution.** The largest and smallest sample observations are 1,520 and 1,101, respectively. Thus, the range is $R = 1,520 - 1,101 = 419$. Since $d_2 = 2.326$ for $n = 5$, the range estimate of σ is $R/d_2 = 419/2.326 = 180.1$, somewhat larger than $s = 157.2$.

Confidence limits for σ can be based on the statistic

$$\chi^2 = \frac{(n-1)s^2}{\sigma^2}$$

which has the chi-square distribution with $n - 1$ degrees of freedom. Arbitrarily using tail areas of equal probability, as shown in Figure 7.1, we can assert with probability $1 - \alpha$, that the inequality

$$\chi^2_{1-\alpha/2} < \frac{(n-1)s^2}{\sigma^2} < \chi^2_{\alpha/2}$$

is true. When this inequality is solved for σ^2, we get the following inequality, whose endpoints provide $100(1 - \alpha)$ confidence limits for σ^2.

$$\textit{Confidence Limits for } \sigma^2: \qquad \frac{(n-1)s^2}{\chi^2_{\alpha/2}} < \sigma^2 < \frac{(n-1)s^2}{\chi^2_{1-\alpha/2}}$$

Confidence limits for the population standard deviation can be obtained by taking the square root of each member of this inequality.

FIGURE 7.1   Chi-Square Values for $1 - \alpha$ Confidence Limits for σ^2

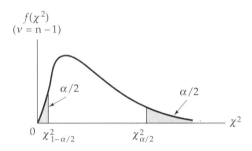

Example. Find 95% confidence limits for the standard deviation of the population of electrical resistances mentioned in the example on page 158, based on the initial sample of size 62.

Solution. In this example, a random sample of size $n = 62$ had a standard deviation of $s = 0.49$. For 95% confidence limits, $\alpha = .05$. Although the appropriate chi-square statistic has $\nu = 62 - 1 = 61$ degrees of freedom, to a close approximation, we shall use 60 degrees of freedom in Table 5, obtaining $\chi^2_{.025} = 83.298$ and $\chi^2_{.975} = 40.482$. Thus, the 95% confidence limits for σ are given by

$$\sqrt{\frac{61 \cdot (0.49)^2}{83.298}} \quad \text{and} \quad \sqrt{\frac{61 \cdot (0.49)^2}{40.482}}$$

or 0.42 and 0.60. This means that we can be 95% confident that σ, the population standard deviation, lies in the interval from 0.42 to 0.60.

EXERCISES

7.11 Four test runs of a new outboard motor design at a speed of 24 knots showed that it used one gallon of fuel in the following time periods: 12, 14, 11, and 16 minutes.

   a. Use the variance of this sample to estimate the variance of the population of all possible test runs involving one gallon of fuel.

   b. Estimate the population variance using the sample range, and compare results with the sample variance.

7.12 Three cigarettes of a certain brand were measured for tar content, with the following results: 13.6, 12.8, and 14.0 mg.

    a. Use the variance of this sample to estimate the variance of the population of all possible tar measurements on this brand of cigarette.

    b. Estimate the population variance using the sample range, and compare results with the sample variance.

7.13 A random sample of 25 diesel engines produced by a certain manufacturer had a mean thermal efficiency of 30.8% and a standard deviation of 2.1%. Find a 99% confidence interval for the variance of the thermal efficiencies of all engines produced by this manufacturer.

7.14 Random measurements made on 30 specimens of nylon yarn from a spinning machine showed a standard deviation of 1.68 denier. Find a 90% confidence interval for the standard deviation of deniers associated with this spinning machine.

7.15 A random sample from the files of a company's machinery orders shows that orders were filled in 21, 32, 39, 33, 45, 28, 19, 28, 17, and 26 days. Find a 95% confidence interval for the standard deviation of the population of all order-fill times for this company's machinery.

7.16 Find a 99% confidence interval for the variance of the population from which came the following random sample of scores attained by students in an in-service training course: 74, 85, 70, 86, 58, 91, 83, 77, 72, 84, 69, 79.

7.4 Estimation of a Proportion

The problem of estimating a proportion can be regarded as one of estimating the parameter p of a binomial distribution. Thus, we shall base point estimates of p on random samples consisting of n trials that satisfy the assumptions underlying the binomial distribution given on page 80. Then, we know that the mean and the standard deviation of x, the number of successes, are given by np and $\sqrt{np(1-p)}$, respectively. If we divide both of these quantities by n we find that the mean and the standard deviation of the **sample proportion**, $\frac{x}{n}$, are

$$\frac{np}{n} = p \quad \text{and} \quad \frac{\sqrt{np(1-p)}}{n} = \sqrt{\frac{p(1-p)}{n}}$$

Thus, it can be concluded that $\frac{x}{n}$ is an unbiased estimate of p and, like the sample mean, its standard deviation is inversely proportional to the square root of the sample size, n.

    Finding exact confidence limits for a population proportion directly from the binomial distribution is difficult. However, approximate large-sample confidence limits can be obtained by making use of the fact that the normal distribution provides an excellent approximation to the binomial distribution for large values of n and values of p that are not very close to 0 or to 1. As we shall discover, confidence intervals for p usually are unacceptably wide unless n is very large; thus use of the normal approximation is practical for most applications.

The mean and standard deviation of the approximating normal distribution are the same as the corresponding quantities for the binomial distribution, namely np and $\sqrt{np\,(1-p)}$, respectively. Thus, approximate large-sample confidence limits can be found for p by making use of the inequality

$$-z_{\alpha/2} < \frac{x - np}{\sqrt{np\,(1-p)}} < z_{\alpha/2}$$

which is true with the approximate probability $1 - \alpha$. Instead of solving this quadratic inequality for p, a further approximation will be made, substituting the sample value $\frac{x}{n}$ for p in the expression $\sqrt{np\,(1-p)}$. (Since the value of the expression $\sqrt{p\,(1-p)}$ is relatively insensitive to the value of p—this expression takes on the value 0.4 for $p = 0.2$ and 0.8, and the value 0.5 for $p = 0.5$—this further approximation is not severe as long as p is not too close to 0 or 1.) Solving the resulting linear inequality for p yields

$$\frac{x}{n} - z_{\alpha/2}\sqrt{\frac{\frac{x}{n}\left(1-\frac{x}{n}\right)}{n}} < p < \frac{x}{n} + z_{\alpha/2}\sqrt{\frac{\frac{x}{n}\left(1-\frac{x}{n}\right)}{n}}$$

with probability approximately equal to $1 - \alpha$.

Thus, for large samples, approximate $1 - \alpha$ confidence limits for a proportion are given by

---

Confidence Limits for p (large samples): $\dfrac{x}{n} \pm z_{\alpha/2}\sqrt{\dfrac{\frac{x}{n}\left(1-\frac{x}{n}\right)}{n}}$

Example.   Records of the last 250 shipments of final assemblies show that 85 were shipped overseas, and the remainder were domestic shipments. Assuming that these shipments represent longstanding experience, find 99% confidence limits for the proportion of shipments sent overseas.

Solution.   In this example, $x = 85$ and $n = 250$. The proportion of overseas shipments in the sample is $x/n = 0.34$. Since $z_{.005} = 2.576$, 99% confidence limits for the true proportion of overseas shipments are

$$0.34 \pm 2.58 \sqrt{\frac{(0.34)\,(0.66)}{250}}$$

or 0.26 and 0.42.

Note the rather wide confidence limits in this example. It is difficult to estimate a proportion with great precision, even for a sample as large as 250. For this reason, very large sample sizes are used to estimate proportions whenever possible. (Firms in the business of taking polls often use samples of 1,000 or more observations.) For such large sample sizes, the use of the normal approximation and substitution of x/n for p introduce negligible errors.

It should be stressed that this method of calculating confidence limits for proportions provides a poor approximation if the value of p is very close to 0 or to 1. In high reliability applications or in sampling inspection where a very small proportion of defectives is found, p usually takes on values very close to 0. In such cases, a one-sided confidence limit is used, as illustrated in Exercise 7.21 below.

EXERCISES

7.17 If a telephone survey of 1,000 registered voters in the United States finds that 21% are registered "Independent," find 95% confidence limits for the true proportion of independent voters in the United States.

7.18 If 65 of 100 marketing executives surveyed stated they were "very satisfied" with their job, find 90% confidence limits for the true proportion of "very satisfied" marketing executives.

7.19 The expense of the survey mentioned in Exercise 7.17 would be doubled if 2,000 voters were interviewed. Suppose this were done, with the same result—21% "Independent." Show that this increase in sample size produces only marginally better results by finding the new 95% confidence limits for the true proportion of independent voters.

7.20 What change would result in the confidence interval requested in Exercise 7.18 if a new sample of 500 executives also showed 65% of the executives "very satisfied" with their job?

The following exercises introduce new material, expanding on some of the topics covered in the text.

7.21 *Upper confidence limits for p.* If p is very close to 0, as often occurs in sampling inspection and applications involving high reliability, upper confidence intervals of the form $p < C$ are useful. For a random sample of size n having x failures, the one-sided interval

$$p < \frac{1}{2n} \chi_\alpha^2$$

for the proportion of failures has a confidence level of $1 - \alpha$. In this formula, χ_α^2 is the value of a random variable having the chi-squared distribution with $2(x + 1)$ degrees of freedom that corresponds to a right-hand tail under this distribution of area α. Find a 95% upper confidence limit for the true proportion defective if a sample of 100 units contains 2 defectives.

7.22 *Continuation.* Two thousand high-reliability electronic components are tested for 1,000 hours under accelerated operating conditions. If 4 of the components failed during this test, find a 99% upper confidence limit for the true proportion of test failures under these conditions.

7.23 *Sample size for estimating* p. Use a method similar to that used for the mean to find a formula for the sample size necessary to estimate p to within a maximum error E with a confidence level of $1 - \alpha$. *Hint*: Since $p(1 - p)$ is a maximum when $p = .5$, use .5 in place of p in this formula.

7.24 *Continuation.* Find the sample size necessary to estimate the proportion of items in inventory for over 30 days to within a maximum error of .05 with 90% confidence.

7.5   Significance Tests Involving Means

Statistical Hypotheses and Significance Tests

The main ideas of testing a hypothesis about a population parameter can be stated quite simply, although application of these ideas sometimes can become complex. A **statistical hypothesis** is a statement about the value or values taken on by a population parameter, or a set of parameters. Thus, the statements: H: "The population mean lies between 2 and 6," or H: "The population mean equals twice its variance," are examples of such hypotheses. A test of a statistical hypothesis involves taking a random sample from the given population and deciding, on the basis of the sample data, whether to accept or reject the statement.

To facilitate the testing of a statistical hypothesis, H, it is useful to select an appropriate *sample statistic*, as well as a *decision rule* which states which sample values of the statistic will lead to acceptance of H and which will lead to rejection of H. For example, suppose we are testing the hypothesis H: "The mean of a given population is less than 10." Then it seems reasonable to use the sample mean as the test statistic. A possible decision rule, but not necessarily the most useful one, might be to accept H if the sample mean is less than 10 and to reject H if the value of this statistic equals or exceeds 10.

Recognizing that random error occurs whenever a sample is taken, it is important to measure and control the errors that can be made when testing a given hypothesis. Two kinds of errors can occur. A **Type I error** occurs when the hypothesis is true, but the decision rule, or test, erroneously rejects it. The Greek letter α is used to designate the probability that a Type I error will be made. A **Type II error** occurs when the hypothesis is false, but the test erroneously accepts this hypothesis. The probability of a Type II error is denoted by β. The kinds of results that can occur when using a sample statistic to test an hypothesis are summarized in the following table:

	Accept H	*Reject H*
H is true	Correct decision	Type I error
H is false	Type II error	Correct decision

Let us suppose that the hypothesis of interest involves the mean of a population, and we wish to test the hypothesis H: $\mu = \mu_0$; that is, the mean of a population equals some given value, μ_0. To perform the test, we must first give a decision rule, stating what sample results will lead to rejection of this hypothesis and what results will lead to acceptance. The decision rule usually takes the form of a statement about the value of a relevant sample statistic. Since the hypothesis in question deals with the mean of a population, it seems reasonable to base our decision about H on the mean of a random sample drawn from that population.

If we wish the probability of a Type I error to equal α, and the probability of a Type II error to equal β, it is necessary to specify exactly what values of the sample mean will lead to rejection of H. To illustrate, suppose we want to test whether the mean content of cans of motor oil produced by a given manufacturer is *at least* equal to one quart, as claimed on the label. To perform this test, let us specify the hypothesis under test to be H: "The mean oil content equals exactly one quart." Perhaps, we wish to reject H when the sample mean is "too large"; that is, there is evidence that the mean content of a can of motor oil exceeds one quart. This procedure is analogous to what we did on page 95 when we discussed sampling inspection. There, we chose a number C so that a lot will be accepted whenever x, the number of defectives, is less than or equal to C. Analogously, in this example we shall choose a number C so that H will be rejected whenever the sample data are such that $\bar{x} > C$.

Now, we are in a position to choose C so that the probability of rejecting H when it is true will equal α. If H is true, then $\mu = \mu_0$, and, by the central limit theorem, the statistic

$$z = \frac{\bar{x} - \mu_0}{\sigma / \sqrt{n}}$$

has approximately the standard normal distribution for large samples. Thus, the probability of rejecting H when it is true is equivalent to the probability that $z > C$. If this probability is to be equal to α, we can choose C so that $C = z_\alpha$, as shown in Figure 7.2.

FIGURE 7.2 Choice of Rejection Criterion

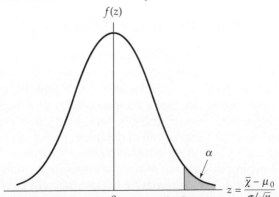

Having assured that the probability of a Type I error will be α, now we can attempt to calculate the probability of a Type II error. β is the probability of accepting H when it is false. Since H is false for any value of μ other than 1, β is not known uniquely; it is a function of the unknown population mean, μ. This suggests that we can draw a curve, analogous to the operating characteristic curve defined on page 95, which gives values of β for different values of μ. This task becomes somewhat complex, and we shall not undertake it here. It is important to understand, however, that the probability of a Type II error is not uniquely defined in most hypothesis tests of practical interest.

We can avoid the problem of defining the Type II error uniquely if we set up a second, competing hypothesis. The decision rule will not allow acceptance of the original hypothesis, H, but only potential rejection of this hypothesis in favor of the competing hypothesis. To distinguish between the two hypotheses, we shall now denote the original hypothesis, H, as H_0, calling it the **null hypothesis**. The competing hypothesis will be denoted as H_1, and it is called the **alternative hypothesis**. Instead of accepting the null hypothesis, we shall ask only whether the data are sufficiently inconsistent with H_0 that the alternative hypothesis must be preferred. In order to be assured that the probability of a Type I error is α, we shall insist that H_0 specifies an *exact value* for a population parameter. (The alternative hypothesis does not need to specify the exact value of the population parameter.)

The resulting test is called a **significance test**. In a significance test for a mean, the null hypothesis always takes the form H: $\mu = \mu_0$. It is customary to restrict the choice of alternative hypothesis in a significance test to H_1: $\mu < \mu_0$, H_1: $\mu > \mu_0$, or H_1: $\mu \neq \mu_0$. When we reject the null hypothesis in favor of the alternative H_1: $\mu \neq \mu_0$, we have shown that the sample mean differs from μ_0 by more than can be expected by chance alone, and it is said that $\bar{x}$ is **significantly different** from μ_0. If the alternative hypothesis is $\mu > \mu_0$ or $\mu < \mu_0$, and the null hypothesis is rejected, it is said that $\bar{x}$ is **significantly greater** or **significantly less** than μ_0, respectively.

In a significance test, the probability of making a Type I error, α, is called the **level of significance**, and its value is fixed in advance. The probability of a Type II error is not known, but β can be plotted for various assumed values of the parameter under test.

Deciding Which Alternative Hypothesis to Use

The decision rule for testing the null hypothesis H_0: $\mu = \mu_0$ depends upon which of the three alternative hypotheses is employed. The null hypothesis has been set up as a "straw man" to see if the alternative hypothesis is more plausible on the basis of a sample result. Since the probability of a Type II error, β, is not stated, the null hypothesis never really is "accepted" in a significance test. The test is conducted only for the purpose of determining whether or not the null hypothesis can be *rejected in favor of the stated alternative hypothesis*.

Keeping this idea in mind, the alternative hypothesis is chosen to bear the "burden of proof." The alternative hypothesis will be accepted only if the sample evidence makes it clear that it is preferable to the null hypothesis. Otherwise, judgment is suspended, and it is customary to act as if the null hypothesis is true. For example, in testing a claim that the mean strength of a material exceeds a given limit L, the null hypothesis H_0: $\mu = L$ will be tested against the alternative hypothesis H_1: $\mu > L$. This choice of alternative hypothesis places the "burden of proof" on the claim; that is, the null hypothesis will be rejected only if the sample mean is *greater* than L by more than can be expected by chance alone, or "significantly greater" than L. In this case, *rejection of the null hypothesis* is equivalent to *acceptance of the claim*; that is, proof that the claim is true. It is for this reason that H_0 is called the "null" hypothesis. It is set up for the explicit purpose of being believed *unless* the sample evidence leads to its rejection.

On the other hand, if the burden of proof were to be placed on those who reject the claim, the alternative hypothesis H: $\mu < L$ would be used. Then, H_0 is rejected only when the sample mean is *less* than L by more than chance alone would allow, or "significantly less" than L. Now, rejection of the null hypothesis is equivalent to proof that the claim is false.

In some cases it is important to test whether a population mean equals a given value, and the test needs to be sensitive to departures from the mean in either direction. For example, suppose an inventory policy requires that a monthly average of 500 parts be kept in inventory; more than 500 would generate excess inventory costs, and fewer than 500 would create potential production delays. In this case, it is appropriate to state the alternative hypothesis in the form H_1: $\mu \neq 500$, rejecting the null hypothesis if the sample mean is "significantly different" from 500 in either direction.

The following example illustrates some of the issues involved in the choice of an alternative hypothesis.

> *Example.* In the example on page 19 concerning the purchase of a new sheet-metal cutter, it was concluded that new cutter would increase the cost of quality. To help make a final decision whether to purchase the new cutter, a test sample of 25 washer-door panels is cut using the new machine. The mean number of burrs on such panels is 3.45 when the old cutting method is used.
>
> a. What null hypothesis should be used?
>
> b. What alternative hypothesis should be used?
>
> c. What alternative hypothesis would have been used if the cost of quality had been less for the new cutter?
>
> *Solution.* a. The null hypothesis should state that the mean number of burrs per panel is that of the old cutting method, since the old method will be retained unless the new method is proved to have fewer burrs; thus, the null hypothesis is H_0: $\mu = 3.45$.
>
> b. Since the cost of quality is increased with the new cutter, it should be purchased only if there is a demonstrable improvement in the burring problem. Thus, the alternative hypothesis should be H_1: $\mu < 3.45$ so that rejection of the null hypothesis will offer proof of fewer burrs and lead to purchase of the new cutter.

c. If the cost of quality had decreased, the new cutter would have been purchased unless it caused a demonstrable decrease in quality. Thus, the alternative hypothesis should be $H_1: \mu > 3.45$ so that rejection of H_0 would lead to a decision not to purchase the new cutter.

Significance Tests for the Sample Mean

To perform a significance test of the null hypothesis $H_0: \mu = \mu_0$, a random sample of size n is taken from the population whose mean is to be tested. Then, we follow the following steps.

Step 1. Decide on what **test statistic** is to be used on which to base the test.

That is, we determine what quantity is to be calculated from the sample data, whose value will determine whether or not the null hypothesis should be rejected. If it can be assumed that the null hypothesis is true, the mean of the population from which the sample is taken equals μ_0. On this assumption, the statistic

$$\textit{Test Statistic for a Mean:} \qquad t = \frac{\bar{x} - \mu_0}{s / \sqrt{n}}$$

will have approximately the standard normal distribution for large samples, or it will have the t distribution for any size sample if the population is normally distributed.

Having decided on the test statistic, we must now go to

Step 2. Specify α, the probability of a Type I error.

Type I errors often are chosen in the range .01 – .10. If the consequences of false rejection of the null hypothesis are serious, α should be chosen to be smaller, and vice versa.

Step 3. Determine an appropriate alternative hypothesis.

The alternative hypothesis can be determined by applying reasoning similar to that of the example on page 169.

Now, we must do

Step 4. Decide for which values of the test statistic the null hypothesis will be rejected.

This will depend upon the choice of alternative hypothesis. If the alternative hypothesis is $H_1: \mu \neq \mu_0$, rejection of H_0 in favor of H_1 takes place in both "tails" of the t distribution, as shown in Figure 7.3. That is, we reject the null hypothesis when the sample mean is "too far away" from μ_0. In this figure, equal tails having areas $\alpha/2$ were chosen. Any combination of tail areas that total α would have assured that the probability of a Type I error is α, but it can be shown that it usually is best to use equal tails. Such a test is called a **two-tail test**.

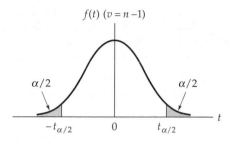

FIGURE 7.3   Two-Tail Test for Testing H_0:
$\mu = \mu_0$ Against H_1: $\mu \neq \mu_0$

If the alternative hypothesis is H_1: $\mu > \mu_0$, rejection of H_0 in favor of H_1 takes place in the right-hand tail of the t distribution, as shown in Figure 7.4. That is, we reject the null hypothesis when the sample mean is "too large" compared with μ_0. Conversely, if the alternative hypothesis is H_1: $\mu < \mu_0$, rejection of H_0 in favor of H_1 takes place in the left-hand tail of the t distribution, as shown in Figure 7.5. That is, we reject the null hypothesis when the sample mean is "too small" compared with μ_0. Such tests are called **one-tail tests**.

Finally,

Step 5. Take a random sample from the appropriate population and decide whether or not to reject the null hypothesis in favor of the chosen alternative.

The criteria for testing the significance of a sample mean are summarized in the following table:

	TESTING $\mu = \mu_0$	
Alternative Hypothesis	*Statement:* $\bar{x}$ *Is Significantly*	*When*
$\mu \neq \mu_0$	Different From μ_0	$t < -t_{\alpha/2}$ or $t > t_{\alpha/2}$
$\mu > \mu_0$	Greater Than μ_0	$t > t_\alpha$
$\mu < \mu_0$	Less Than μ_0	$t < -t_\alpha$

FIGURE 7.4   One-Tail Test for Testing H_0:
$\mu = \mu_0$ Against H_1: $\mu > \mu_0$

FIGURE 7.5   One-Tail Test for Testing H_0:
$\mu = \mu_0$ Against H_1: $\mu < \mu_0$

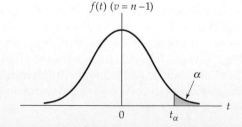

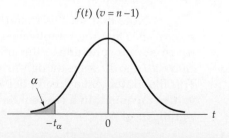

Example. Test at the .05 level of significance whether the mean of a random sample of size $n = 16$ differs significantly from 10 if the distribution from which the sample was taken is approximately normal, and the sample results are $\bar{x} = 8.4$, and $s = 3.2$.

Solution. Step 1. Since we can assume that the population is approximately normally distributed, we can use the t test statistic given on page 170.

Step 2. α is given to be .05.

Step 3. Since we are asked to determine if the sample mean is significantly *different* from a given value, the alternative hypothesis is $H_1: \mu \neq 10$.

Step 4. From the table for testing $\mu = \mu_0$, with the alternative hypothesis $\mu \neq \mu_0$, we shall reject H_0 if $t < -t_{.025}$ or if $t > t_{.025}$. From Table 4, $t_{.025} = 2.131$ with $n - 1 = 15$ degrees of freedom.

Step 5. The standardized sample mean, t, is calculated to be

$$t = \frac{8.4 - 10}{3.2 / \sqrt{16}} = -2.00$$

Since $t = -2.00$ falls between -2.131 and 2.131, we cannot reject the null hypothesis; thus, it cannot be said that the sample mean is significantly different from 10 at the .05 level of significance.

The following example illustrates a test of whether a sample mean is significantly less than a stated value:

Example. Test at the .05 level of significance whether the mean of a random sample of size $n = 16$ is significantly less than 10 if the distribution from which the sample was taken is approximately normal, $\bar{x} = 8.4$ and $s = 3.2$.

Solution. Steps 1 and 2 are identical to those of the preceding example.

Step 3. Since we are asked to determine if the sample mean is significantly *less* than a given value, the alternative hypothesis is $H_1: \mu < 10$.

Step 4. From the table for testing $\mu = \mu_0$, with the alternative hypothesis $\mu < \mu_0$, we shall reject H_0 if $t < -t_{.05}$. From Table 4, $t_{.05} = 1.753$ with $n - 1 = 15$ degrees of freedom.

Step 5. The value of the standardized sample mean is -2.00, identical to that of the preceding example. However, now we compare -2.00 with $-t_{.05} = -1.753$, with 15 degrees of freedom. Since $-2.00 < -1.753$, it can be concluded at the level of significance .05 that the sample mean is significantly less than 10.

Note that, in this example, we were able to conclude at the .05 level of significance that the sample mean was *less* than 10. Yet, in the preceding example, we could not conclude at the .05 level of significance that the sample mean was *different* from 10 even though results of a sample of the same size were identical. This illustrates that a greater departure of the same mean from the null hypothesis value is required in a two-tail test than in a one-tail test having the same level of significance. Why?

When testing the significance of a sample mean, it is important to assure that the population from which the sample has been taken is approximately normally distributed, unless the sample size is large ($n > 29$). In either event, the standardized sample mean will be a value of a random variable having the t distribution. For large samples, ∞ degrees of freedom can be used for t, automatically giving values of z.

The language used in connection with significance tests may seem awkward, but it is necessary! The example on page 171 asks us to determine if a sample mean is significantly different from 10 at the .05 level of significance. Since the sample mean actually has the value 8.4, we know that it is *different* from 10. In making the test of significance, we are asking whether the difference between 8.4 and 10 is greater than can be accounted for by chance alone. The phrase *significantly different* should be used only when using significance tests to compare sample means with population means, or with each other. The word *different* is used to compare population means.

Computer Example

The statistical software package *MINITAB* may be used to perform tests of significance. To test whether the mean of the coking-time data given in Exercise 3.10 on page 36 is significantly *less* than 8 at the .10 level of significance, the command TTEST OF MU = 8 ALTERNATIVE K = −1 IN COL C1 is given. (To test whether the mean is significantly *greater* than 8, use $K = +1$ in place of $K = −1$. To test whether the population mean is significantly *different* from 8, use $K = 0$ or omit the "Alternative" portion of the command entirely). *MINITAB* then prints the results:

TEST OF MU = 8.00 VS MU L.T. 8.00

	N	MEAN	STDEV	SEMEAN	T	PVALUE
C1	30	7.76	1.18	0.22	−1.11	0.14

The P value gives the smallest level of significance at which the sample mean would be found to be significantly less than 8. Since this P value is greater than .10, the level of significance for the test, it cannot be concluded that the sample mean is significantly less than 8.

EXERCISES

7.25 Using the data of Exercise 7.4 on page 160, test at the .01 level of significance whether the mean is significantly greater than 10.

7.26 To test whether the mean of a population is less than 16.8, a random sample of 100 observations was taken. If the sample mean is 16.5 and the sample standard deviation is 1.1, can it be said that the sample mean is significantly less than 16.8 at the .10 level of significance?

7.27 A random sample of size 16, taken from a normally distributed population, has a mean of 0.15 and a standard deviation of 0.20. Is the sample mean significantly greater than 0 at the .05 level of significance?

7.28 A random sample of size 100 has a mean of 3.8 and a standard deviation of 11.5. Is the sample mean significantly different from 6.0 at the .01 level of significance?

7.29 The mean burst strength of the cardboard cartons mentioned in the example on page 127 is specified to be 138 pounds.

  a. What null and alternative hypotheses would you use to test whether the mean burst strength is less than expected?

  b. Upon whom does your choice place the "burden of proof"?

  c. If a random sample of 100 observations has a mean of 135 pounds and a standard deviation of 8 pounds, perform the test using the .05 level of significance.

7.30 a. What null and alternative hypotheses would you use to test whether mean fill weights in the example on page 129 are less than expected.

  b. Upon whom does your choice place the "burden of proof"?

7.31 Use *MINITAB* or other computer software to test whether the mean failure time of the sample of 38 light bulbs given in Exercise 3.25 on page 42 is significantly less than 300 hours. Use the .01 level of significance.

7.32 Use *MINITAB* or other computer software to test whether the mean breaking strength of the sample of 40 solder bonds given on page 43 is significantly greater than 14 grams. Use the .01 level of significance.

The following exercises introduce new material, expanding on some of the topics covered in the text.

7.33 *Significance tests for the difference between two means.* Significance tests for the difference between two means can be conducted using the analysis-of-variance methods to be introduced in Chapter 10. In most cases, we wish to test whether two population means are equal; that is, whether the means of random samples taken from these populations are significantly different. In such cases, we can use the test statistic

$$z = \frac{\bar{x}_1 - \bar{x}_2}{\sqrt{\dfrac{s_1^2}{n_1} + \dfrac{s_2^2}{n_2}}}$$

provided the sample sizes n_1 and n_2 are large enough so we can apply the central limit theorem and the population standard deviations σ_1 and σ_2 can be approximated by the sample standard deviations s_1 and s_2, respectively. (Usually, $n_1 + n_2 > 29$ will suffice.) Given two samples with

	Sample 1	Sample 2
n	25	30
x	119	98
s	28	16

test whether the two sample means are significantly different. Use the .05 level of significance.

7.34 *Continuation.* In a random sample of 50 middle managers, the mean number of years of education was 17.3, with a standard deviation of 2.4. A random sample of 35 research scientists had a mean of 18.4 years of education, with a standard deviation of 3.6 years. Test at the .01 level of significance whether the scientists had significantly more education than the managers.

7.6 Significance Tests Involving Variances

As previously discussed, control of variability is one of the greatest problems of quality improvement. Thus, significance tests involving a single variance are used in the control of the uniformity of manufactured products or operations. Sometimes it is necessary to compare the variability of two different processes, suppliers, or materials. Tests involving the ratio of two variances may be used to make such comparisons. Such tests also are used in "analysis of variance," the method for analyzing the experimental designs described in Chapter 10.

Tests Involving a Single Variance

We begin by describing tests involving a single variance. The chi-square statistic given on page 147 can be used as a basis for such tests. According to the sampling theory so described, if the null hypothesis $H_0: \sigma^2 = \sigma_0^2$ is true, and a random sample is drawn from a normal population whose variance is s^2, then

$$\textit{Test Statistic for a Variance:} \quad \chi^2 = \frac{(n-1)\, s^2}{\sigma_0^2}$$

will have the chi-square distribution with $v = n - 1$ degrees of freedom. The criteria for testing this null hypothesis against a given alternative hypothesis, analogous to those for tests involving a mean, are as follows:

TESTING $\sigma^2 = \sigma_0^2$

Alternative Hypothesis	*Statement: s^2 Is Significantly*	*When*
$\sigma^2 \neq \sigma_0^2$	Different From σ_0^2	$\chi^2 < \chi^2_{1-\alpha/2}$ or $\chi^2 > \chi^2_{\alpha/2}$
$\sigma^2 > \sigma_0^2$	Greater Than σ_0^2	$\chi^2 > \chi^2_{\alpha}$
$\sigma^2 < \sigma_0^2$	Less Than σ_0^2	$\chi^2 < \chi^2_{1-\alpha}$

The tests are conducted following the same five steps that were used to perform significance tests on a mean, as illustrated in the following example.

Example. The intensity of light is regarded as being uniform throughout a production floor if its standard deviation is less than 3.0 foot-candles. To check on a given facility, measurements of light intensity are made at 12 different locations, obtaining a standard deviation of 3.8 foot-candles. Test, at the .05 level of significance, if the light intensity at this facility can be regarded as being uniform.

Solution. *Step 1.* If we can assume that the population is approximately normally distributed, we can use the chi-square test statistic given on page 175.

Step 2. α is given to be .05.

Step 3. We shall choose to place the burden of proof on the statement: "light intensity is not uniform." Thus, we will not reject the null hypothesis unless the alternative hypothesis $H_1: \sigma > 3.0$ is favored by the sample result.

Step 4. From the table for testing $\sigma^2 = \sigma_0^2$, with the alternative hypothesis $\sigma^2 > \sigma_0^2$, we shall reject H_0 if, $\chi^2 > \chi^2_{.05}$, as illustrated in Figure 7.6. From Table 5, $\chi^2_{.05} = 19.675$ with $n - 1 = 11$ degrees of freedom.

Step 5. With $s^2 = 3.8$, chi-square is calculated to be

$$\chi^2 = \frac{(12 - 1)\,(3.8)^2}{3.0^2} = 17.65$$

Since chi-square = 17.65 is not greater than 19.675, we cannot reject the null hypothesis; thus, it cannot be said that the sample variance is significantly greater than $3.0^2 = 9.00$ at the .05 level of significance.

Tests Involving Two Variances

The *F* statistic given on page 149 can be used as the basis for tests involving two variances. Thus, if two independent samples of sizes n_1 and n_2 are taken from normal populations having the variances σ_1^2 and σ_2^2, respectively,

$$F = \frac{s_1^2}{s_2^2}$$

FIGURE 7.6   Chi-Square Test for Example Above

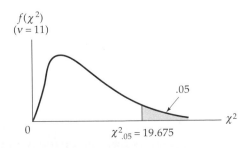

$\chi^2_{.05} = 19.675$

will have the F distribution with $n_1 - 1$ and $n_2 - 1$ degrees of freedom. The exact form of the F statistic to be used in testing hypotheses involving two variances, however, depends upon the alternative hypothesis chosen.

The null hypothesis to be tested is H_0: $\sigma_1^2 = \sigma_2^2$, which can be restated as H_0: $\sigma_1^2/\sigma_2^2 = 1$. To understand how the exact form of the test statistic depends upon the null hypothesis, we state the three potential alternative hypotheses as H_1: $\sigma_1^2/\sigma_2^2 \neq 1$, H_1: $\sigma_1^2/\sigma_2^2 > 1$, and H_1: $\sigma_1^2/\sigma_2^2 < 1$. Table 5 gives only right-hand tails of the F distribution, as shown in Figure 7.7; thus, we must be careful to define the test statistic somewhat differently for each of these alternative hypotheses. Using Table 5, we can reject H_0 only when F is "too large." Thus, to test the null hypothesis against the two-tail alternative H_1: $\sigma_1^2/\sigma_2^2 \neq 1$, we can place the larger of the two variances, s_M^2, in the numerator of the F ratio, and the smaller variance, s_m^2, in the denominator, testing as if the level of significance were $\alpha/2$ instead of α. For the one-tail alternative H_1: $\sigma_1^2/\sigma_2^2 > 1$, we place s_1^2 in the numerator, and for the alternative H_1: $\sigma_1^2/\sigma_2^2 < 1$, we place s_2^2 in the numerator of the F ratio.

The tests for hypotheses involving two variances can be summarized as follows:

TESTING $\sigma_1^2/\sigma_2^2 = 1$

Alternative Hypothesis	Statement: s_1^2 Is Significantly	When
$\sigma_1^2/\sigma_2^2 \neq 1$	Different From s_2^2	$F = \dfrac{s_M^2}{s_m^2} > F_{\alpha/2}$ (d.f. $= n_M - 1, n_m - 1$)
$\sigma_1^2/\sigma_2^2 > 1$	Greater Than s_2^2	$F = \dfrac{s_1^2}{s_2^2} > F_{\alpha}$ (d.f. $= n_1 - 1, n_2 - 1$)
$\sigma_1^2/\sigma_2^2 < 1$	Less Than s_2^2	$F = \dfrac{s_2^2}{s_1^2} > F_{\alpha}$ (d.f. $= n_2 - 1, n_1 - 1$)

The following example illustrates the performance of these tests.

FIGURE 7.7   Rejection Regions for All F-Tests

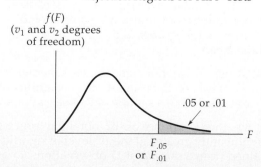

Example. We wish to test whether the variability is the same in the thickness of cold-rolled steel made by two different mills. If random samples of 16 measurements from Mill 1 and 25 measurements from Mill 2 had variances of 2.4 mils and 5.2 mils, respectively, can we conclude at the .10 level of significance that the variability of one mill is different from that of the other?

Solution. *Step 1.* We must assume that the populations are approximately normally distributed; then we can use the F statistic appropriate for the chosen alternative hypothesis.

Step 2. α is given to be .10.

Step 3. Since we are testing for a significant difference, without regard to which variance is larger, we wish to use the alternative hypothesis H_1: $\sigma_1^2/\sigma_2^2 \neq 1$.

Step 4. The larger variance (3.6) arose from Mill 2. Thus, the test statistic is the value of F obtained by placing the larger sample variance in the numerator, and using the level of significance $\alpha/2 = .05$. From Table 6(a), with 15 and 24 degrees of freedom, $F_{.05} = 2.11$.

Step 5. F is calculated from the sample result to be $F = \frac{5.2}{2.4} = 2.17$. Since F is larger than 2.11, we reject the null hypothesis and conclude, at the .10 level of significance, that $\sigma_1^2 \neq \sigma_2^2$.

EXERCISES

7.35 A random sample of size 29, taken from a normal population, has a standard deviation of 12.5. Test, at the .05 level of significance, if this standard deviation is significantly greater than 10.

7.36 If a random sample of size 18, taken from a normal population, has a variance of 93, can it be said at the .01 level of significance that the population variance is less than 85?

7.37 The specific heat of 14 iron samples was measured in a laboratory. If these measurements had a variance of $6.8 \cdot 10^{-5}$, can it be said, at the .01 level of significance, that the variance of the normal population which underlies this sample equals $5.0 \cdot 10^{-5}$?

7.38 In an effort to improve the quality of a polymer mold, a breaking-strength variability standard was set; the standard deviation of breaking strengths was not to exceed 1 gram. If a sample of 20 such molds has a standard deviation of 1.35 grams, is this standard being met? (Use the .05 level of significance.)

7.39 Test, at the .01 level of significance, whether a variance of 12.8, obtained from a sample of 11 observations, is significantly less than a variance of 35.5, obtained from a sample of 9 observations.

7.40 Test whether a variance of 102, obtained from a sample of 25 observations, is significantly greater, at the .05 level of significance, than a variance of 46, obtained from a sample of 17 observations.

7.7 Significance Tests Involving Proportions

Tests Involving a Single Proportion

The proportion of failures in a specified lot of manufactured goods is of interest in sampling inspection (Section 4.6). The proportion of measurements made on product produced by a given production process that do not satisfy tolerances is important in process control (Chapter 8). The probability that the service life of a product will exceed a given limit is an element of reliability measurement (Chapter 11). Many of the methods required for these applications are based on testing the null hypothesis that a proportion equals a given constant.

In Section 7.4 it was observed that a population proportion, p, can be regarded as a parameter of the binomial distribution, and that the binomial distribution can be approximated by the normal distribution for large values of n. Thus, if the null hypothesis $H_0: p = p_0$ is true, the quantity

$$\textit{Test Statistic for a Proportion:} \quad z = \frac{x - np_0}{\sqrt{np_0\,(1 - p_0)}}$$

can be used to perform significance tests for proportions. The test criteria are shown in the following table.

	TESTING $p = p_0$	
Alternative Hypothesis	*Statement:* $\dfrac{x}{n}$ *Is Significantly*	*When*
$p \neq p_0$	Different From p_0	$z < -z_{\alpha/2}$ or $z > z_{\alpha/2}$
$p > p_0$	Greater Than p_0	$z > z_{\alpha}$
$p < p_0$	Less Than p_0	$z < -z_{\alpha}$

The following example illustrates tests of hypotheses involving a single proportion.

Example. To see if at least 50% of the compressors manufactured by a firm can withstand 5 years of continuous operation without failure, 100 compressors were put on test. If 38 of them were still running after 5 years, can it be said, at the .05 level of significance, that this standard has not been met?

Solution. *Step 1.* With $n = 100$, we can use the z test statistic given above.

Step 2. α is given to be .05.

Step 3. We shall choose the alternative hypothesis $H_1: p < p_0$, since we do not wish to reject the null hypothesis unless there is strong evidence that p is "too small."

Step 4. We shall reject H_0 if $z < -z_{.05}$. From Table 3, $z_{.05} = 1.645$.

Step 5. z is calculated to be

$$z = \frac{38 - 50}{\sqrt{100 \; (.5) \; (1 - .5)}} = -2.4$$

Since $z = -2.4$ is less than -1.645, we reject the null hypothesis; thus, it cannot be said that the proportion of compressors surviving 5 years is significantly less than 50%, at the .05 level of significance.

Tests Involving Several Proportions

The preceding tests were restricted to a comparison of a single proportion to a given value. However, it may be desired to test the equality of two or more proportions when we make comparisons of the quality of several different brands, when we determine whether several different production stations are producing to the same level of defectives, and so forth.

In testing whether two or more proportions are equal, the null hypothesis states that k binomial populations, having the parameters $p_1, p_2, \ldots, p_k$ are such that

$$H_0: p_1 = p_2 = \ldots = p_k = p$$

The alternative hypothesis states that at least two of these population proportions are not equal. The parameter p is the common value of the population proportions when the null hypothesis is true.

The following algorithm is used for testing whether two or more proportions are equal. First, the data are arranged in a table as follows:

	Sample 1	Sample 2	...	Sample k	Total
Successes	x_1	x_2	...	x_k	x
Failures	$n_1 - x_1$	$n_2 - x_2$	...	$n_k - x_k$	$n - x$
Total	n_1	n_2	...	n_k	n

The $2k$ numbers at the intersection of each row and column in this table, giving the observed number of successes or failures for each sample, are called **observed cell frequencies**, and they are denoted by o_{ij}.

Next, the **expected cell frequencies** are calculated under the assumption that the null hypothesis is true. Under this assumption, the number of observations in any cell can be expected to be proportional to the number of observations in its row and its column. Thus, if e_{ij} represents the expected cell frequency in row i and column j ($i = 1, 2$ and $j = 1, 2, \ldots, k$), then

$$e_{1j} = C \cdot x n_j \quad \text{and} \quad e_{2j} = C \cdot (n - x) n_j$$

where C is a proportionality constant. Since the expected cell frequencies must sum to n, the total of the observed frequencies, it follows that $C = 1/n$. Thus, *the expected frequency of any cell can be found by multiplying the totals of the row and the column to which the cell belongs, and dividing by the grand total* n.

Now, if we let o_{ij} represent the observed frequency of any cell in the table, and e_{ij} represent the expected frequency of that cell, then it can be shown that the sampling distribution of the statistic

$$\textbf{Test Statistic for Several Proportions:} \qquad \chi^2 = \sum_{i=1}^{2} \sum_{i=1}^{k} \frac{(o_{ij} - e_{ij})^2}{e_{ij}}$$

is approximated by the chi-square distribution with $k - 1$ degrees of freedom. This approximation is reasonably good for large samples, usually defined to be large enough that no expected frequency is less than 5. If one or more of the e's is less than 5, columns can be combined so that each of the combined columns has an expected frequency of 5 or greater. Since any departure from the null hypothesis (any population proportion unequal to the others) will tend to *increase* the value of chi-square, the null hypothesis is rejected for values of chi-square that are "too large"; that is, we say that the population proportions are significantly different at the level of significance α if chi-square takes on a value greater than χ^2_α with $k - 1$ degrees of freedom.

> *Example.* Samples of three materials, under consideration for the housing of machinery on a seagoing vessel, are tested by means of a salt-spray test. Any sample that leaks when subjected to a power spray is considered to have failed. The following are the test results:

	Material A	Material B	Material C	Total
Leaked	36	22	18	76
Not Leaked	63	45	29	137
Total	99	67	47	213

Test, at the .05 level of significance, if the three materials have the same probability of leaking as a result of this test.

Solution. *Step 1.* We use the chi-square test statistic given above.

Step 2. α is given to be .05.

Step 3. In testing the difference among several proportions, the alternative hypothesis always is "the proportions are not all equal."

Step 4. From Table 5, $\chi^2_{.05} = 5.991$ with $k - 1 = 2$ degrees of freedom.

Step 5. The expected frequencies are as follows:

$e_{11} = 76 \times 99/213 = 35.3 \quad e_{12} = 76 \times 67/213 = 23.9 \quad e_{13} = 76 \times 47/213 = 16.8$

$e_{21} = 137 \times 99/213 = 63.7$ $e_{22} = 137 \times 67/213 = 43.1$ $e_{23} = 137 \times 47/213 = 30.2$

Thus,

$$\chi^2 = \frac{(36 - 35.3)^2}{35.3} + \frac{(22 - 23.9)^2}{23.9} + \frac{(18 - 16.8)^2}{16.8}$$

$$+ \frac{(63 - 63.7)^2}{63.7} + \frac{(45 - 43.1)^2}{43.1} + \frac{(29 - 30.2)^2}{30.2}$$

$$= 0.390$$

Since 0.390 is less than $\chi^2_{.05}$ (2 *d.f.*) = 5.991, we cannot reject the null hypothesis; thus, we do not conclude, at the .05 level of significance, that the three materials have different probabilities of leaking as a result of the test.

Note in the foregoing example that we followed the customary practice of rounding all expected frequencies to one decimal place. Note, also, that the expected frequencies in any row or column have the same sum as the observed frequencies in that row or column.

EXERCISES

7.41 In a random survey of 1,000 households in the United States, it is found that 29% of the households contained at least one member with a college degree. Does this finding refute the statement that the proportion of all such United States households is at least 35%? (Use the .05 level of significance.)

7.42 If 61 out of 100 tosses of a coin result in "heads," is there evidence, at the .01 level of significance, that the coin is not balanced?

7.43 A lot consisting of a very large number of integrated circuits is labeled "premium" if at least 50% of the circuits in the lot perform to certain high-performance standards. A random sample of size 50 is taken from a given lot, and 28 circuits in the sample perform to these standards. Can we be sure, at the .05 level of significance, that the lot can be labeled "premium"?

7.44 Show that the proportionality constant C, mentioned on pages 180 and 181 must equal $1/n$.

7.45 Randomly chosen test samples were sent to four different medical laboratories to determine if they were obtaining consistent results. Each laboratory received 30 samples, and the resulting numbers of "positives" are as follows:

Lab 1	Lab 2	Lab 3	Lab 4
8	6	11	9

Test, at the .025 level of significance, whether the laboratories have the same probability of a "positive" on a given sample.

7.46 Samples of an experimental material are produced by three different prototype processes, and tested for compliance to a strength standard. If the tests showed the following results, can it be said, at the .01 level of significance, that the three processes have the same probability of passing this strength standard?

	Process A	Process B	Process C
No. "Pass"	45	58	49
No. "Fail"	21	15	35

The following exercise introduces new material, expanding on some of the topics covered in the text.

7.47 *Contingency tables.* A method closely related to that for testing for differences among several proportions also can be used to test whether two classifications are independent. For example, one classification might involve a rating given to products in final inspection, and the other might be categories describing the severity of warranty complaints on these products.

To perform such a test, we represent each of the r categories of one classification as one of the r rows of a table, and the c categories of the other as one of its c its columns. Thus, the samples are arranged in an $r \times c$ table; the number in the "cell" formed by the intersection of row i and column j is the number of observations falling in category i of the first classification and category j of the second. To test the null hypothesis that the two classifications are independent against the alternative they they are not, a chi-square test statistic is defined, analogous to that used for testing the equality of several proportions, as follows:

$$\chi^2 = \sum_{i=1}^{r} \sum_{i=1}^{c} \frac{(o_{ij} - e_{ij})^2}{e_{ij}}$$

In this formula, o_{ij} is the number of observations in row i and column j, and the quantities e_{ij} are calculated by multiplying the sum of the observations in row i by the sum of those in column j, and dividing this product by the grand total of all the observations.

Tests of the fidelity and selectivity of radio receivers coming off a production line give the following results:

		Fidelity		
		Low	Medium	High
	Low	16	22	41
Selectivity	Medium	44	75	26
	High	20	26	2

Test, at the .01 level of significance, whether there is a relationship between fidelity and selectivity.

7.8 Application to Quality

It is not unusual for an engineer (or a manager, for that matter) to guess about the value of an important parameter, or to make a decision without information, hiding behind his or her "experience and expertise" to give validity to this guess or this decision. Unfortunately, such guesses have no discernable properties. It is nearly impossible to judge how accurate or how precise the guesses or how beneficial the decisions. While estimates and hypothesis tests based on statistical science may seem to some to be pedantic and fussy, they are much better than "guesses." Their properties are known, and they are objective—two "experts" will arrive at the same estimates if they use the same basic theory.

It is an important doctrine of modern quality theory that actions be based on *data*. Scientific methods must be applied to data to make estimates and to verify the validity of theories. The old regime of working "by the seat of the pants" is replaced in an organization dedicated to quality by a commitment to making objective decisions based on valid analyses of data. This means that we must bear with some of the rigor and some of the fussiness of statistical inference if we are going to create and improve quality.

But it doesn't mean that we must have a blind dedication to the methods outlined in this chapter. Often, the assumptions underlying the methods given here simply are not met. Often, we can transform the data to near normality to satisfy the assumptions underlying the use of the chi-square and the *F* statistics, but sometimes we forget to do so. This may not stop us from claiming that our estimates are "good" or that our decisions are "valid" even when they seem to violate common sense. This sort of thing not only gives statistics a bad name but also leads to a false justification of the old slipshod methods.

There are two frequent misuses of statistical inference that should be mentioned. The first misuse involves a misunderstanding of what *statistical significance* really means. As diligent students of statistics, we now know it means that a result is not what we could have expected by chance alone if the null hypothesis were true. It does *not* necessarily mean that the result is useful or important. When we say, "It's statistically significant," we may not be saying any more than, "The sample is large enough so we can detect small and unimportant differences." There is no substitute for judgment and common sense in conducting quality-improvement investigations.

This last statement becomes more vivid when we discuss what some have called "errors of Type III in statistical application." The Type III error is not-so-whimsically defined as the error of doing a masterful job of solving the wrong problem. It should be remembered that statistical theory tells us only how to test a null hypothesis against a given alternative hypothesis in certain circumstances. It doesn't choose these hypotheses, and it doesn't decide the level of significance. These decisions are left to you—and they are very important. Also, it should make you feel better that statistical science does not require you to be a robot, blindly substituting numbers into equations. The really important choices must come out of your training and experience.

It takes a while for these ideas to sink in. No single example that we can give here really will convince you of the truth of what we have said. Perhaps the following brief story about a statistician working in industry—who was trained as an engineer—will help a bit.

The Ace Manufacturing Company has a good reputation with its customers for the quality of its products. Ace does all the right things. It continually surveys its markets for information about customer needs and satisfaction. Ace's new-product designs, based on this information, are market-driven and developed by teams that included production, marketing, and management personnel as well as design engineers. Pilot operations, optimized with the aid of experimental designs (Chapter 10), are utilized extensively to gain a better understanding of the effects of variation on their process parameters and for "trouble-shooting." Ace's production lines make heavy use of statistical process controls (Chapter 8). Reliability (Chapter 11) is "designed in" to their products through careful selection of components and vendors, and life tests are conducted routinely. The Ace Manufacturing Company's deserved reputation for quality has made it very profitable.

Not only does the company try to do the many things needed to assure quality, but it tries to foster attitudes at every level that make its people think about quality. The chief statistician of the Product Assurance Department started out as an engineer. As interest developed in quality, he decided to do graduate study in statistics. He's been around engineers, managers, and statisticians long enough to know how to tell fact from fiction, and to recognize when professional jargon is used to impress, rather than to explain. Above the chief statistician's desk is the following poster:

DON'T BOTHER ME WITH FACTS

VARIABILITY IS EVERYWHERE.
  ACCOUNT FOR IT!

ORACLES GIVE FACTS.
  STATISTICIANS GIVE ESTIMATES WITH CONFIDENCE INTERVALS!

DON'T JUST TELL ME: " IT'S SIGNIFICANT."
  IT MAY NOT BE IMPORTANT!

EXERCISES

7.48 What is the chief statistician trying to say in the first line of the poster: "Variability Is Everywhere."

   a. Nothing is certain.

   b. No number is accurate.

   c. Variability must be recognized and measured.

   d. I don't believe anything you say.

7.49 To what does the word *facts* refer in the poster?

    a. Statements of absolute truth.

    b. Data.

    c. Information believed to be true.

    d. Estimates given with confidence intervals.

7.50 The word *significant* has a different meaning in statistics than it does in everyday English. Distinguish between these two meanings.

7.51 Give an example of a statement that may have "statistical significance" but little or no "importance."

REVIEW EXERCISES

These review exercises can be used for informal review of the chapter, or as two practice examinations, designed to be taken over a time period of about one hour. The odd-numbered exercises comprise one examination; the even-numbered exercises comprise the other.

7.52 If a random sample of size 16, taken from a normally distributed population, has a mean of 11.6 and a standard deviation of 4.2, find a 95% confidence interval for the population mean.

7.53 A random sample of size $n = 25$ was taken from a normal population with the following results: $\bar{x} = 121.8$, $s = 13.7$. Find a 99% confidence interval for the population mean.

7.54 How large a sample should be taken from a population whose variance is assumed to be 24.9 so that the maximum error in using the sample mean to estimate the population mean will be 1.5 with 95% confidence?

7.55 A random sample of 35 observations from an unknown population has a mean of 2.4 and a standard deviation of 1.7.

    a. What is the maximum error that can be made with 99% confidence in using the value 2.4 to estimate the population mean?

    b. If the maximum error at this level of confidence is to be 0.5, would an additional sample be required? If so, how large should it be?

7.56 Find a 99% confidence interval for the standard deviation of the normal population from which a random sample of size 51 was taken, if the sample standard deviation is 11.6.

7.57 A random sample of 30 observations, taken from a normal population, has a variance of 316. Find a 95% confidence interval for the population variance.

7.58 A telephone survey of 568 middle managers in American manufacturing corporations shows that 68% were "satisfied with their jobs." Find a 90% confidence interval for the percentage of all such managers who are so satisfied.

7.59 A clinical trial involving 1,524 patients shows that 27.5% of them had unwanted side effects from a certain drug. Find a 99% confidence interval for the probability that a patient taking this drug will have such side effects.

7.60 To determine if a process has been centered correctly, a sample of 50 units was measured, with a mean value of 29.6 cm and a standard deviation of 1.4 cm. If the process specification for this measurement is 30.0 cm, should corrective action be taken? (Use the .05 level of significance.)

7.61 If the mean service life of a certain kind of light bulb is to be at least 1,000 hours and a random sample of 84 bulbs had a mean life of 1,144 hours, with a standard deviation of 316 hours, can it be said that this criterion has been satisfied at the .01 level of significance?

7.62 Test, at the .01 level of significance, whether the variance of a normal population equals 40 against the alternative that the variance does not equal 40 if the standard deviation of a random sample of size 25 taken from this population is 5.95.

7.63 A random sample of 16 observations is taken from a normal population to test whether its variance is 159 against the alternative that the variance exceeds 159. What conclusion can be drawn at the .05 level of significance if the sample variance is 181?

7.64 Random samples of sizes 16 and 20, taken from two normal populations, have variances of 38.4 and 16.1, respectively. Test, at the .10 level of significance, whether the two population variances are equal against the alternative that they are unequal.

7.65 Two random samples of sizes 25 are taken from normal populations to test the null hypothesis that their variances are equal. If the sample variances are 16.5 and 50.9, respectively, can it be said, at the .01 level of significance, that the second population has a larger variance?

7.66 It has been claimed that corporate purchases will decline in the coming quarter. A survey of 400 corporate purchasing managers shows that 285 plan on decreasing their purchases in the coming quarter. Does this result support the claim (at the .10 level of significance)?

7.67 Would you suspect (at the .05 level of significance) that a coin is unbalanced if there were 310 heads in 500 tosses?

7.68 Test, at the .01 level of significance, whether the proportions of foreign sales contacts are the same in each of three departments if a sample taken from the records for the past year yields the following data:

	Dept. A	Dept. B	Dept. C
Foreign Sales Contacts:	39	16	27
Domestic Sales Contacts:	115	50	32

7.69 Can it be said (at the .05 level of significance) that two urns contain the same proportion of red balls if 100 random drawings from each bowl (with replacement) produce 37 red balls and 54 red balls, respectively?

GLOSSARY

	page		page
Alternative Hypothesis	168	Random Error	154
Bias	154	Sample Proportion	163
Confidence Interval	154	Sample Size (Mean)	158
Confidence Limits		Significance Test	168
for Means (Large Samples)	156	Significantly Different	168
for Means (Small Samples)	157	Significantly Greater	168
for Proportions	164	Significantly Less	168
for Variances	162	Statistical Hypothesis	166
Estimation	154	Statistical Inference	154
Expected Cell Frequencies	180	Test Statistic	170
Level of Significance	168	Tests of Hypotheses	154
Null Hypothesis	168	Two-Tail Test	170
Observed Cell Frequencies	180	Type I Error	166
One-Tail Test	171	Type II Error	166
Point Estimate	154	Unbiased Estimate	154

STATISTICAL PROCESS CONTROL

Where have we been? We have completed our brief tour of the basic ideas of statistics. Although this "fly-over" does not qualify us as statisticians, perhaps now we can better understand how statistics can help us in the pursuit of better quality. Hopefully, we will better understand how the professional statistician approaches such problems so that we can become better partners with statistical science in quality improvement.

Where are we going? Now, we're ready to apply some of what we've learned to some of the most important methods for quality improvement. In this chapter, we learn about one of them—Statistical Process Control (SPC). SPC allows us to discover the "natural" tolerances of a process, to compare them with its design specifications, and to control the process so it continues to produce the best quality of which it is capable.

Section 8.1, *Introduction to SPC*, shows how some of the statistical principles already studied can be applied to the economic control of a manufacturing process and outlines the steps to be taken in initiating a program of statistical process control.

Section 8.2, *Control Charts for Variables*, gives methods for the construction and interpretation of control charts for the means and the variability of continuous data.

Section 8.3, *Control Charts for Attributes*, gives methods for the construction and interpretation of np charts, p charts, and c charts.

Section 8.4, *Cusum Charts*, introduces a variation on standard control-chart techniques.

Section 8.5, *Process Capability*, shows how measurements made on the output of a process can be compared with specification limits to determine the capability of the process to produce to these limits.

Section 8.6, *Maintenance of SPC Programs*, discusses the use of warning limits, and gives rules for shutting down a process to make repairs or adjustments, for re-calculation of control limits, and for adding or deleting control points.

8.1 Introduction to SPC

Comparisons of histograms with specification limits were discussed in Section 3.6. As pointed out in that section, the quality of a product can be determined on the basis of the center and the variation of measurements taken on the products it produces. A production line can be expected to produce a quality product with respect to a given measurement if the measurements are centered between the specification limits, and if their variation is small enough so that the probability of obtaining a measurement outside of specifications is very low.

Many different events can occur during a production run to cause the distribution of a given measurement to change over time. Assignable causes of variability such as tool wear, deterioration, or contamination of chemical solutions can cause a process to "go out of center." Random events, such as inadvertent changes in machine settings, changes in ambient conditions, and so forth, can cause increases in the variability of the process. Any changes in the distribution of a measurement will tend to give rise to out-of-specification products.

It is the function of **Statistical Process Control (SPC)** to detect changes in the distribution of critical measurements as early as possible, preferably *before* discrepant products are produced. Early detection can lead to timely corrective action, higher quality, and lower costs of scrap and rework. As pointed out in Section 4.6, detection of poor quality based entirely on final inspection is uncertain as well as costly. By the time poor quality has been detected by inspection, discrepant product has already been produced. Implementation of an effective SPC program saves time and money by avoiding unacceptable product, while it contributes to higher quality. For this reason, SPC is a major component of total quality management.

The basic ideas of what has come to be known as SPC were given by Walter Shewhart in his 1931 book entitled *Economic Control of Quality of Manufactured Product* (see Bibliography, Appendix B). The cornerstone of an SPC program is the **control chart,** used for detecting changes in the distribution of a measurement taken on a product or a process over a period of time. The control chart makes use of samples taken repeatedly over a period of time. In a dynamic environment, such as can be expected on a production line, the effects of assignable causes are more likely to become evident over time than in a given sample. Thus, the measurements comprising a given sample deliberately are taken over a short space of time to assure that their parent distribution is the same; that is, to minimize variability arising from assignable causes. However, different samples are taken at widely differing times to maximize the effects of assignable causes between samples. Shewhart used the term **rational subgroups** to denote groups of measurements chosen to minimize the effects of assignable causes within groups and to maximize such effects between groups.

The ultimate goal of quality has been described by Taguchi (Section 2.6) as producing a product or rendering a service with measurements as close as possible to their target value. In this spirit, the control chart does not ordinarily include specification limits, but rather it makes use of **control limits** that are based on the demonstrated ability of the process to hold a measurement close to its target. Control-chart measurements are made on samples taken from successive rational subgroups. A process is said to be **in control** if the distribution of these measurements does not appear to have shifted over time. Otherwise, the process is said to be **out of control**. For example, the control chart might consist of the means of the samples taken from each rational subgroup, and the process is regarded as being out of control if a sample mean is outside one of the control limits. The procedure for a given sample mean is similar to the conduct of a two-tail significance test for a sample mean. However, a sequence of many such tests is conducted in a control chart.

Most processes are capable of giving rise to many different measurements, some taken directly from the process and others made on the product it produces. The choice of which measurements to control can be a difficult one. Excessive use of control charts, especially when many of them have shown no process drifts over a long period of time, can trivialize an SPC program to the point of undermining its effectiveness. Variables chosen for measurement and control should satisfy at least one of two criteria. They should be critical in determining the ultimate quality of the product, and they should be chosen to have probative value in determining the source of assignable causes of variability.

It is a good policy to start an SPC program by choosing a limited number of variables to measure in an area where serious quality problems have existed in the past. If the variables have been chosen well, out-of-control events will allow the rapid identification and elimination of causes, reducing rework and scrap and motivating effective continuation of the program. Use of Pareto charts to identify the most frequently occurring defects, and Ishikawa cause-and-effect diagrams to determine relationships between assignable causes and product or process measurements can aid greatly in determining which variables to measure in the beginning. Experimental programs, such as the one illustrated in Section 10.6, can help to determine which process variables exert the most influence on variations in product quality and, thus, should be carefully controlled. Getting started in a limited but important area of a process and making maximum use of the quality-improvement methods previously introduced to help decide which variables to control, will go a long way to ensure the success of an SPC program.

It makes little sense to compare measurements made on a product to the product specification limits if the process producing that product is unstable. Only after a process has been brought into control—that is, the process is producing to its maximum capability—should important measurements be checked against their specification limits. (Methods for determining the capability of a process to meet specification limits will be discussed in Section 8.5.) Thus, we shall begin our introduction to SPC by giving methods for the construction of control charts for the control of various kinds of process measurements.

8.2   Control Charts for Variables

Two general kinds of measurement can be utilized for the control of a process. Measurements of continuous variables, such as diameters, electrical properties, chemical compositions, and so forth, are called **variables measures** in SPC. Measurements of discrete variables, such as proportions of defective product or numbers of defects in a product, are called **attributes measures.**

Control charts for variables usually are maintained in pairs. One member of the pair, called an $\bar{x}$ **chart,** plots the means of the measurements taken in rational subgroups. The second member is an **R chart** or **s chart**, which plots the ranges or the standard deviations of these measurements.

Small sample sizes often must be used to represent a rational subgroup for reasons of economy as well as to minimize the possibility of assignable causes operating within the subgroup. The assumption that the sample measurements, or even the sample means will have a normal distribution may not be reasonable for such small samples. For this reason, "probability limits," such as are used for tests of significance, seldom are used in control charts. Instead, it is common to use **three-sigma limits** obtained from a history of measurements made on the process during a period when it has been in good control; that is, when there is reason to believe that assignable causes of variability have not been present.

We shall illustrate the construction of three-sigma limits by finding these limits for a control chart of sample means. We begin with a set of at least 15 or 20 samples, each of size n, *taken when the process is believed to be in good control.* The grand mean of these samples,

$$\bar{\bar{x}} = \frac{1}{k} \sum_{i=1}^{k} \bar{x}_i$$

will provide a good estimate of μ, the mean of the population from which these samples has been taken. Here, $\bar{x}_i$ is the mean of sample i, and k is the number of samples taken.

Since the sample size is assumed to be small, we shall use the sample range R to estimate the variability of the sample. The quantity

$$\bar{R} = \frac{1}{k} \sum_{i=1}^{k} R_i$$

is the mean range of the k samples. Using the theory about the distribution of a sample range discussed on page 161, it can be seen that three-sigma limits for the sample means can be based on the quantity $A_2\bar{R}$, which provides an estimate of $\frac{3\sigma}{\sqrt{n}}$ where σ is the standard deviation of the distribution of an individual measurement. The quantity A_2 can be found in Table 7 (Appendix A) for values of n from 2 to 20. An $\bar{x}$ control chart consists of values of the sample means, $\bar{x}_i$, plotted against sample numbers, a central line, and three-sigma control limits. The lower control limit is denoted by LCL and the upper control limit by UCL. The resulting control-chart values for an $\bar{x}$ chart are as follows:

Control-Chart Values for an $\bar{x}$ chart based on ranges:

central line $= \bar{\bar{x}}$

$$LCL = \bar{\bar{x}} - A_2\bar{R}$$

$$UCL = \bar{\bar{x}} + A_2\bar{R}$$

The following example illustrates the construction of an $\bar{x}$ chart based on sample ranges.

> *Example.* Fifteen samples of five bearing deflectors were taken during a period when the process for manufacturing these deflectors was believed to be in control, and eccentricities were measured. The means and ranges of these samples, given in thousandths of an inch, were:

Mean	17	14	8	17	12	13	15	16	13	14	16	9	11	9	12
Range	6	11	4	8	9	14	12	15	10	7	12	6	9	11	13

Subsequently, samples of size 5 were taken from the production line each hour and the mean eccentricity was measured, with the following results over the next 32 hours of production:

Mean Eccentricities															
11	14	9	15	17	19	13	22	20	18	21	16	16	20	22	18
21	9	17	19	20	22	4	11	23	22	19	11	9	20	22	21

a. Construct an $\bar{x}$ chart showing the 32 mean eccentricities, and comment on the state of control of the process.

b. Are the original 15 samples suitable for calculating control-chart values? Why?

> *Solution.* a. The control-chart values are based on the initial 15 observations, using the following calculations:

$$\text{central line: } \bar{\bar{x}} = (17 + 14 + \ldots + 12)/15 = 13.1$$
$$\overline{R} = (6 + 11 + \ldots + 13)/15 = 9.8$$
$$LCL = 13.1 - (0.58)(9.8) = 7.4$$
$$UCL = 13.1 + (0.58)(9.8) = 18.8$$

The value of A_2 (0.58) was obtained from Table 7 for $n = 5$. The resulting control chart is shown in Figure 8.1(b), along with the initial 15 means. From this control chart, it is evident that the process has gone out of control.

b. The original 15 means, shown on the control chart in Figure 8.1(a), show an excellent state of control, with no trends toward either control limit. Thus, the process gives evidence of having been in control when the data were taken for computation of the control-chart values.

In the preceding example, we know that the process has gone out of control since the initial 15 samples were taken. When such a conclusion has been reached, it becomes critical to search for a cause or causes for the process to have gone out of control. In making such a search, it is important to discover whether the means have drifted, whether there has been a change in the variability of the measurements, or both. In anticipation of such a contingency, it is customary to keep a companion **R chart** along with each $\bar{x}$ chart. The control-chart values for an R chart are as follows:

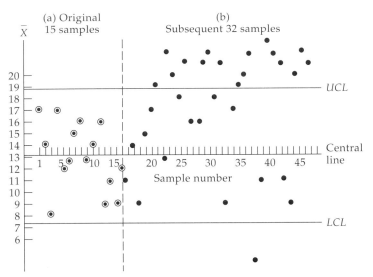

FIGURE 8.1 $\bar{x}$ Chart for Bearing-Deflector Eccentricities

---

Control-Chart Values for an R Chart:

central line $= \bar{R}$

$$LCL = D_3\bar{R}$$

$$UCL = D_4\bar{R}$$

The values of D_3 and D_4 are obtained from Table 7. The following continues the preceding example, constructing the companion R chart and drawing conclusions about the state of the process.

Example. The following are the ranges of the 32 samples of size 5 whose means were given in the preceding example:

Sample Ranges															
7	11	6	4	12	14	11	10	8	6	9	13	12	8	6	7
9	7	11	6	10	8	14	5	10	4	6	9	11	12	8	6

Construct an R chart using the original values of R given in the example on page 193, and determine whether the cause of the out-of-control state of the process can be attributed to changes in the mean eccentricities, their variation, or both.

Solution. The following calculations are made:

$$\text{central line} = \bar{R} = 9.8$$

$$LCL = 0$$

$$UCL = (2.11)(9.8) = 20.7$$

The *R* chart is shown in Figure 8.2, and it is apparent that the variability of the process is in a good state of control.

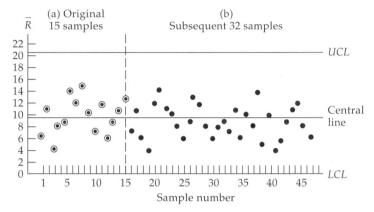

FIGURE 8.2   *R* Chart for Bearing-Deflector Eccentricities

A search for assignable causes now can be narrowed to causes associated with drifts in means, since the variability of the underlying distribution of eccentricities does not appear to have increased. Such a search concentrated on tool wear, wear or mislocation of jigs and fixtures, and so forth. The search disclosed that the fixture used for locating the outer diameter of the bearing deflector needed re-grinding. When this was done, the process came back into control.

Control Charts Based on Standard Deviations

Control charts for variables have been based on the range as a measure of the variability of samples from rational subgroups for many years. Two reasons have been given for the use of the range, rather than the standard deviation. First, the computations are much simpler than if the sample standard deviation had been used. Thus, production-line operators can be taught the mechanics of the computations needed for the control charts, and the charts can be maintained in "real time" on the production line. Second, sample sizes usually are quite small; we have seen that, for small samples, the sample range provides nearly as good an estimate of the population standard deviation σ as does the sample standard deviation (page 161).

Nowadays, it is no longer necessary to perform control-chart computations by hand. An SPC program can be built around microcomputers or computer terminals placed strategically on the production line, together with readily available specialized software designed to perform all the control-chart computations and draw the charts. In some cases, automatic gauges are utilized, which perform the required measuring tasks and feed the resulting data directly to computers that, in turn, produce the finished control charts and even sound an alarm when the process goes out of control.

It is inadvisable to use the range in constructing control charts for samples of sizes greater than 5. As computation of the standard deviation is facilitated by modern computer methods, it is best to get into the habit of using control charts for standard deviations no matter how small the sample size.

The construction of such control charts is similar to that for $\bar{x}$ and R charts, except that the sample standard deviation s is used in place of R and different constants are obtained from Table 7 to produce the following control chart values:

Control Chart Values for $\bar{x}$ and s Based on Standard Deviations:		
	$\bar{x}$ Chart	s Chart
central line	$\bar{\bar{x}}$	$\bar{s}$
LCL	$\bar{\bar{x}} - A_3\bar{s}$	$B_3\bar{s}$
UCL	$\bar{\bar{x}} + A_3\bar{s}$	$B_4\bar{s}$

The construction of an $\bar{x}$ chart based on standard deviations will be illustrated by referring to the eccentricities given in the example on page 193. Suppose the standard deviations, s_i, of the original 15 samples are as follows:

Standard Deviations of the Original 15 Samples														
2.5	3.9	1.7	3.1	3.2	3.7	3.3	4.1	3.0	2.4	3.7	2.3	2.6	3.4	3.7

Thus, $\bar{s} = (2.5 + 3.9 + \ldots + 3.7)/15 = 3.1$. From Table 7, we obtain $A_3 = 1.43$, $B_3 = 0$, and $B_4 = 2.09$ for $n = 5$. Thus, the control-chart values are

	$\bar{x}$ Chart	s Chart
central line	13.1	3.1
LCL	8.7	0.0
UCL	17.5	6.5

Suppose, further, that the standard deviations of the subsequent 32 samples are as follows:

Sample Standard Deviations															
2.1	3.9	1.9	1.4	3.8	4.7	3.7	3.5	2.7	2.0	3.2	4.5	4.1	2.6	2.0	2.4
3.2	3.3	2.9	1.9	3.4	3.0	4.7	1.8	3.2	1.4	2.1	3.5	3.8	4.2	2.7	1.8

The resulting s chart is shown in Figure 8.3. Although similar conclusions can be drawn from the control charts based on standard deviations as were drawn from the charts based on ranges, the narrower control limits for an $\bar{x}$ chart based on standard deviations makes this chart more sensitive to changes in the sample mean than is the $\bar{x}$ chart based on ranges.

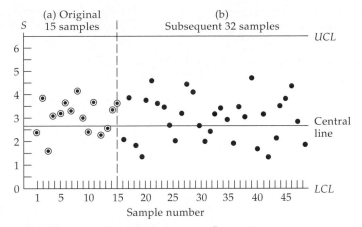

FIGURE 8.3 *s* Chart for Bearing-Deflector Eccentricities

Interpretation of Control Charts

A major advantage of control charts is their capability of detecting imminent out-of-control situations *before the process actually goes out of control*. Thus, corrective action can be taken before discrepant product is produced, saving the time and cost of scrap and rework. A control chart is capable of giving several different kinds of clues that a process may be heading out of control. Three of the most helpful clues are summarized as follows:

RUNS

When a consecutive sequence of points in a control chart all lie on one side of the central line, it is said that there is a **run** above or below the central line. It is customary to regard a run of seven or more points, even if they are all within the control limits, as evidence that the chart will soon go out of control. Runs are illustrated in Figure 8.4.

FIGURE 8.4 Control Chart Exhibiting a Run

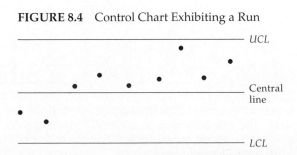

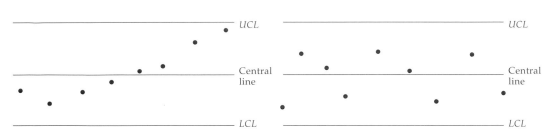

FIGURE 8.5   Control Chart Exhibiting a Trend   **FIGURE 8.6**   Control Chart Exhibiting Periodicity

TRENDS

A continuous rise or fall of consecutive points in a control chart is called a **trend**. Trends consisting of seven or more points are regarded as evidence of an imminent out-of-control situation. Trends are illustrated in Figure 8.5.

PERIODICITY

If the points on a control chart exhibit a periodic or cyclical pattern, such as that illustrated in Figure 8.6, there is evidence that the process variability can be reduced. **Periodicity** often is caused by excessive adjustments made on the production-line equipment in a misguided attempt to keep measurements close to the central line.

To further illustrate the use of control charts for variables in detecting and diagnosing quality problems, we consider the following example involving service quality.

Example.   In an effort to control the quality of counter service at a fast-food restaurant, the franchiser hires a survey research firm to interview a random sample of four customers every hour. One of the questions asked is: "On a scale of 1 to 10, with 1 being least courteous and 10 being most courteous, rate the courtesy of the counter service you received." On the day after a training seminar on counter service, during which representatives of the franchiser were present observing counter service, the following customer ratings were obtained:

				A.M.							P.M.					
Sample	6	7	8	9	10	11	12	1	2	3	4	5	6	7	8	9
1	8	8	9	7	7	8	8	9	8	7	7	9	10	9	8	7
2	7	9	8	8	10	9	9	8	10	10	8	9	9	9	7	8
3	9	10	7	8	9	9	8	10	10	8	9	10	9	8	10	9
4	8	7	8	9	8	10	9	9	9	10	10	8	10	7	9	10

The recency of the training course and the presence of the observers was taken as evidence that the servers were doing as well as they could, and that "the process was in control." After a period of several days had elapsed, routine measurements were taken over the course of one day, with the following results:

	A.M.							P.M.								
	6	7	8	9	10	11	12	1	2	3	4	5	6	7	8	9
Mean	7.8	6.9	6.2	6.4	7.8	7.4	5.9	6.8	7.4	8.1	8.5	7.3	6.2	6.5	7.8	8.5
St. Dev.	0.7	0.8	0.7	0.5	0.6	0.9	1.1	0.8	0.7	0.6	0.7	0.9	0.7	0.8	0.6	0.4

Sketch and interpret the appropriate control charts. What action(s) would you recommend?

Solution. The means and standard deviations of the 16 samples taken during the control period are calculated with the following results:

	A.M.						
	6	7	8	9	10	11	12
Mean	8.0	8.5	8.0	8.0	8.5	9.0	8.5
St. Dev.	0.82	1.29	0.82	0.82	1.29	0.82	0.58

	P.M.								
	1	2	3	4	5	6	7	8	9
Mean	9.0	9.25	8.75	8.5	9.0	9.5	8.25	8.5	8.5
St. Dev.	0.82	0.96	1.50	1.29	0.82	0.58	0.96	1.29	1.29

We shall construct control charts for the mean and the standard deviation of the service ratings. To obtain the control-chart values for these charts, first we calculate

$$\bar{\bar{x}} = 8.609 \quad \text{and} \quad \bar{s} = 0.997$$

and the following control-chart values are found, using $A_3 = 1.63, B_3 = 0$, and $B_4 = 2.27$ found from Table 7 for $n = 4$:

	$\bar{x}$ Chart	s Chart
central line	8.61	1.00
LCL	6.98	0
UCL	10.24	2.08

Control charts for the data obtained on the second day are shown, together with these control-chart values, in Figure 8.7. While the chart for standard deviations is well within control, showing no abnormalities such as runs, trends, or periodicity, the chart for means is badly out of control. Five of the sixteen data points are below the *LCL*, and *a run of all sixteen data points lies below the central line*. Clearly, significant deterioration in service has taken place. There is some evidence of periodicity in this control chart, and a closer look reveals a reduction in service quality during each of the three normal mealtime periods.

It can be conjectured that there is insufficient service staff to accommodate customers during the "rush hours," with the result that service quality deteriorates markedly during these periods. Even if this problem were corrected, however, the overall level of service quality is too low—the mean rating of non-rush periods remains below the central line. Perhaps general morale has suffered as a result of inadequate staffing during the rush periods. After reviewing staffing requirements, and hiring and training the additional staff required, it can be hoped that a new set of control charts will show a serving process that is under good control.

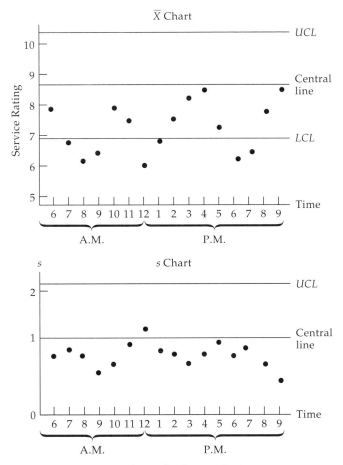

FIGURE 8.7   Control Charts for Service Ratings

Computer Applications

As mentioned earlier, many different software companies have produced computer software packages to aid in the construction of control charts. We shall produce control charts for the preceding example with the use of *MINITAB* to illustrate the use of computers in SPC.

All data may be entered in a single column if the four observations comprising the first sample are entered first, then the four observations of the second sample, and so forth. The following commands will produce the $\bar{x}$ chart:

```
MTB > SET C1
DATA > 8 7 9 8 8 9 10 7 .... 7 8 9 10
DATA > END
MTB > XBARCHART C1 4
```

The last command asks for an *x*-bar chart, using data in C1, and grouping the data into subgroups of size 4. The resulting chart is shown in Figure 8.8(a). The corresponding *s* chart also can be obtained, by giving the command SCHART C1 4. This chart is shown in Figure 8.8(b) along with the *x*-bar chart. A comparison of the control limits in Figure 8.8 with the limits calculated on page 199 shows slight differences. These differences are due in part to roundoff, and also to the fact that *MINITAB* corrects $\bar{s}$ for bias.

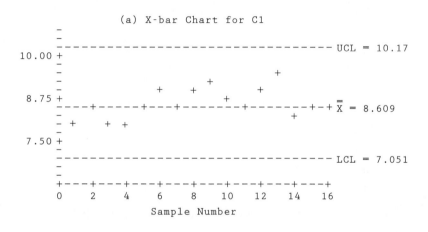

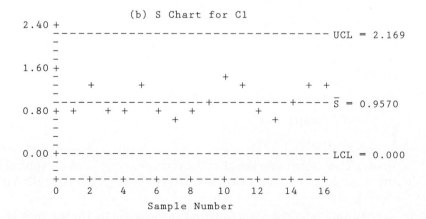

FIGURE 8.8 Control Charts from *MINITAB*

MINITAB also produces control charts based on ranges, and other kinds of control charts. In addition, this software can produce up to eight different tests on control charts for anomalies similar to runs, trends, and periodicities.

EXERCISES

8.1   The following are the results of 15 samples, each consisting of 4 successive measurements of the internal viscosity of styrene (Mooneys):

Sample	1	2	3	4	5	6	7	8	9	10	11	12	13	14	15
Mean	31.6	32.8	30.9	29.6	32.7	32.5	33.7	30.5	31.0	32.9	31.8	29.5	30.7	32.6	30.7
Range	5.5	7.7	3.5	9.6	4.9	1.8	2.6	4.7	3.1	4.0	3.2	3.0	6.4	5.8	3.6
St. Dev.	1.8	2.7	1.2	3.5	1.9	0.7	0.9	2.0	1.2	1.3	1.1	1.0	2.1	1.9	1.5

Use these results to find the control-chart values for an $\bar{x}$ and an R chart.

8.2   The following are the results of making 10 reverse-current readings (nanoamperes) on 20 hourly samples of transistors:

Mean	11.6	13.4	11.9	14.2	11.6	14.3	12.9	10.5	12.6	13.1
Range	4.7	6.1	6.2	11.0	8.2	9.6	4.7	8.5	2.7	5.2
St. Dev.	1.6	1.9	1.4	2.8	1.5	1.9	1.3	2.4	1.9	2.0

Mean	12.2	10.7	11.9	12.6	12.8	12.1	10.8	13.4	11.0	14.2
Range	6.7	2.7	7.7	10.9	7.0	4.7	3.9	4.7	5.7	3.8
St. Dev.	1.4	1.3	1.6	1.9	2.4	2.5	1.6	1.4	1.7	2.4

Use these results to find control-chart values for an $\bar{x}$ and an R chart.

8.3   Find the control-chart values for $\bar{x}$ and s charts, using the data of Exercise 8.1.

8.4   Find the control-chart values for $\bar{x}$ and s charts, using the data of Exercise 8.2.

8.5   Plot the data points given in Exercise 8.1 on

   a. $\bar{x}$ and R control charts

   b. $\bar{x}$ and s control charts

   In each case, determine whether the process is in sufficient control to use the control-chart values calculated from the data for subsequent process control.

8.6   Plot the data points given in Exercise 8.2 on

   a. $\bar{x}$ and R control charts

   b. $\bar{x}$ and s control charts

   In each case, determine whether the process is in sufficient control to use the control-chart values calculated from the data for subsequent process control.

8.7   a.   Plot the following 24 additional sample results for internal viscosity on control charts using the central line and control limits found in Exercise 8.3.

Mean	30.2	31.1	33.6	29.9	33.8	33.1	30.3	33.4	30.6	33.8	32.6	30.5
St. Dev.	1.7	2.4	1.8	2.1	2.5	3.1	1.4	2.3	1.5	1.2	2.5	0.9

Mean	31.6	34.2	33.8	31.0	32.5	32.6	33.9	32.2	33.4	34.0	33.5	35.2
St. Dev.	1.4	1.6	1.1	1.4	1.6	2.0	2.9	3.1	2.8	1.3	2.0	2.7

    b. Has the process gone out of control?

    c. Are there any other indications (runs, trends, etc.) that corrective action may be required?

8.8 a. Plot the following 20 additional sample results for reverse current on control charts using the central line and control limits found in Exercise 8.4.

Mean	11.7	13.4	12.5	14.1	13.0	10.7	12.6	14.0	13.2	11.9
St. Dev.	1.1	1.6	3.1	2.8	1.0	3.0	2.6	1.4	2.9	2.5

Mean	14.1	13.6	11.9	13.8	13.0	12.6	11.8	11.0	10.8	10.6
St.Dev.	1.7	1.5	1.8	2.4	0.9	2.0	0.6	1.2	2.5	2.7

    b. Has the process gone out of control?

    c. Are there any other indications (runs, trends, etc.) that corrective action may be required?

8.9 A control chart for means has the central line 10.6 and lower and upper control limits of 7.1 and 14.1, respectively. Plot the following means on this control chart and check to see if corrective action is needed: 10.9, 11.7, 9.4, 8.6, 12.1, 14.0, 13.1, 11.7, 9.5, 7.8, 10.3, 12.1, 13.8, 13.1, 11.7, 9.4, 7.3, 8.4, 10.6, 12.0.

8.10 A control chart for means has the central line 0.6 and upper and lower control limits of 5.9 and −4.7, respectively. Plot the following means on this control chart and check to see if corrective action is needed: −3.7, −4.1, 5.0, 5.7, 0.1, −2.6, 3.2, 5.4, 5.8, 0.2, −4.0, −2.8, 0.3, 2.8, 3.7, 5.1, 5.5.

8.11 The following measurements were made of the diameters of hourly samples of three consecutive turned rotor shafts (in inches) produced by an automatic lathe.

					Sample					
Shaft	1	2	3	4	5	6	7	8	9	10
1	0.495	0.502	0.504	0.498	0.502	0.504	0.497	0.499	0.498	0.501
2	0.499	0.500	0.499	0.498	0.499	0.503	0.499	0.502	0.502	0.503
3	0.501	0.503	0.498	0.499	0.501	0.501	0.496	0.501	0.499	0.499

					Sample					
Shaft	11	12	13	14	15	16	17	18	19	20
1	0.505	0.502	0.498	0.499	0.502	0.502	0.499	0.500	0.499	0.503
2	0.501	0.502	0.497	0.501	0.503	0.496	0.501	0.501	0.500	0.498
3	0.501	0.499	0.499	0.499	0.497	0.501	0.497	0.502	0.498	0.500

Use *MINITAB* or some other statistical software package to produce $\bar{x}$ and R charts and determine whether the process is in control.

8.12 The following measurements were made of the thicknesses of 4 periodic samples cut from cold-rolled steel plate (in mm).

					Period					
Sample	1	2	3	4	5	6	7	8	9	10
1	25.1	26.4	24.9	25.7	25.6	24.2	26.0	25.5	25.8	24.7
2	24.9	25.3	26.1	27.0	23.9	24.8	25.6	25.4	24.7	26.5
3	25.0	25.6	24.2	26.1	26.3	24.8	25.5	27.0	26.1	25.8
4	24.4	24.9	25.0	25.3	25.4	25.1	26.1	25.8	24.7	25.5

					Period					
Sample	11	12	13	14	15	16	17	18	19	20
1	24.8	25.3	26.1	24.9	26.0	25.6	24.7	26.2	25.3	25.6
2	25.2	25.1	25.8	24.8	25.9	24.9	25.2	24.6	24.9	25.2
3	25.4	26.0	24.9	25.0	26.1	25.2	24.7	24.9	25.2	26.2
4	24.7	25.5	26.1	24.4	25.5	26.1	24.8	25.5	26.0	25.1

Use *MINITAB* or some other statistical software package to produce $\bar{x}$ and s charts and determine whether the process is in control.

The following exercises introduce new material, expanding on some of the topics covered in the text.

8.13 *Unequal sample sizes.* If the samples drawn for a control chart are not of the same size, the control limits can be adjusted for each point on the chart to reflect this difference. For an $\bar{x}$ chart, $\bar{\bar{x}}$ is defined to be the weighted mean of the $\bar{x}_i$ values (see Exercise 3.75 on page 59) with the weights n_i. Similarly, $\bar{R}$ or $\bar{s}$ is the weighted mean of the R_i or s_i values, using the weights n_i. For an $\bar{x}$ chart, new control limits are calculated for each mean by using the appropriate value of n to find A_1 or A_2, as sketched in Figure 8.9. Similarly, for an R chart or an s chart, new control-chart values are calculated for each point on the chart by entering Table 7 with the appropriate value of n. Find the control-chart values for an $\bar{x}$ and an R chart from the following data, and plot the data on control charts:

n	4	5	4	3	5	4	6	4	4	5	4	3
mean	3.1	5.6	2.8	3.7	4.9	3.8	4.2	6.0	4.3	2.9	4.6	4.2
range	0.9	1.2	1.6	1.4	0.8	2.3	1.7	1.4	2.4	0.9	1.0	2.1

FIGURE 8.9 Control Chart for Unequal Sample Sizes

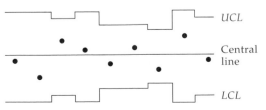

8.14 *Continuation.* Plot an $\bar{x}$ and an s chart for the following data:

n	3	3	4	3	4	4	5	2	4	3	3	5
mean	0.6	0.8	0.4	1.2	0.7	0.9	0.3	1.3	1.4	0.8	0.5	1.0
st. dev.	0.2	0.3	0.1	0.4	0.3	0.4	0.3	0.0	0.2	0.5	0.2	0.3

8.3 Control Charts for Attributes

Control charts for variables are preferable for use in mass production. Data consisting of continuous measurements usually give information that is much more sensitive to process changes than attribute data, such as fractions of defective products. However, control charts for attributes sometimes are useful in "job shops," where production runs are short, and in the control of errors involved in providing a service. It should be noted that control limits for attribute control charts tend to be very wide unless large samples are used. Thus, whenever control charts for variables can be used, they are preferable to control charts for attributes.

In connection with control charts for attributes, it is necessary to distinguish between defectives and defects. A product is said to be **defective** if it has one or more **defects**. Three kinds of control charts are in frequent use for attribute data; two of them are used to control defectives, and one for defects.

A **fraction-defective chart**, or *p* **chart** is used to control the proportion of defective products or services produced. Unfortunately, control of a production line only with the use of *p* charts is "after the fact"; that is, by the time inspection results underlying a *p* chart have shown that a process has gone out of control, many defective products may already have been produced.

A **number-of-defectives** chart, or *np* **chart,** is similar to a *p* chart, except that the *number* of defective products in a sample consisting of *n* products is charted, instead of the *proportion* of defectives. The control-chart values for a number-of-defectives chart, as well as those for a *p* chart, are based on the theory developed in Section 4.4 for the binomial distribution.

Sometimes it is necessary to control the number of defects in a unit of product. For example, in the production of tin plate, a continuous coil of plated steel is generated. This coil may have defects, such as bubbles or pits, at any point on its plated surface, and it is desired to control the number of defects per hundred feet of coil. As a further example, it may be desirable to control the number of defective solder joints on a circuit board. The control chart used in such situations is called a **number-of-defects chart**, or *c* **chart**, and its control-chart values are based on the Poisson distribution, described in Section 4.5.

p Charts and *np* Charts

The control-chart values for a *p* chart are based on theory which states that, if a random sample of size *n* is taken from the binomial population $b(x;n,p)$, then x/n is an unbiased estimate of the binomial parameter *p*. If *k* random samples are taken from the same binomial population, n_i is the sample size of sample *i*, and x_i is the number of defectives in sample *i*, then

$$\textbf{\textit{Estimate of Proportion Defectives:}} \quad \bar{p} = \frac{x_1 + x_2 + \ldots + x_k}{n_1 + n_2 + \ldots + n_k}$$

is an unbiased estimate of p, the proportion of defectives in the population.

To obtain control-chart values that do not vary from sample to sample, we shall assume that $n_1 = n_2 = \ldots = n_k = n$. Suppose a series of samples of size n are taken. It is assumed that these samples come from the binomial distribution whose parameter equals p. If p is estimated by $\bar{p}$, then the standard deviation of x/n, the fraction defective in any one of these samples, is given by the formula on page 84, with $\bar{p}$ replacing p, or

$$\textbf{\textit{Standard Deviation of Fraction Defective:}} \quad s_{x/n} = \sqrt{\frac{\bar{p}\,(1-\bar{p})}{n}}$$

Thus, control-chart values for a p chart are

$$\textbf{\textit{Control-Chart Values for a Proportion-of-Defectives Chart:}}$$
$$\text{central line} = \bar{p}$$
$$LCL = \bar{p} - 3\sqrt{\frac{\bar{p}\,(1-\bar{p})}{n}}$$
$$UCL = \bar{p} + 3\sqrt{\frac{\bar{p}\,(1-\bar{p})}{n}}$$

If $\bar{p}$ is small, the formula for the lower control limit may produce a negative number. In this case, the lower control limit is taken to be zero; in effect, there is only an upper control limit.

Example. A high-volume automobile dealer checks on the quality of the service department by keeping a log of customer complaints and incidents of repeat service. The following data give the proportion of such incidents for 20 batches, consisting of 100 consecutive service orders each:

3, 7, 5, 4, 10, 0, 6, 4, 9, 11, 8, 7, 3, 7, 12, 5, 2, 6, 1, 4.

a. Find the control-chart values for a p chart and plot these data on the chart.

b. Comment on what the chart shows.

Solution. a. $\bar{p}$ is calculated as follows:

$$\bar{p} = \frac{\frac{3}{100} + \frac{7}{100} + \ldots + \frac{4}{100}}{20}$$

The control-chart values are as follows:

central line = 0.057

$$LCL = 0.057 - 3 \sqrt{\frac{(0.057)\,(0.943)}{100}} = -0.013$$

$$UCL = 0.057 + 3 \sqrt{\frac{(0.057)\,(0.943)}{100}} = 0.127$$

The lower control limit is taken to be zero, and the control chart is shown in Figure 8.10.

b. This control chart has the appearance typical of an in-control process. No data points are out of control and there are no measurable runs or trends. The question of periodicity may be moot, since it is not known whether or not the batches were arranged in a temporal sequence.

Calculation of the control-chart values for a number-of-defectives chart, or np chart, is analogous to that of a p chart, except that $n\bar{p}$ is used in place of $\bar{p}$, and the control limits, likewise, are multiplied by n. Thus, the control-chart values for an np chart are as follows:

Control-Chart Values for a Number-of-Defectives Chart:

central line $= n\bar{p}$

$$LCL = n\bar{p} - 3 \sqrt{n\bar{p}\,(1-\bar{p})}$$

$$UCL = n\bar{p} + 3 \sqrt{n\bar{p}\,(1-\bar{p})}$$

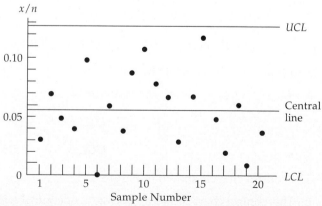

FIGURE 8.10 *p* Chart for Customer Complaints

The *np* chart for the data of the preceding example is identical to the *p* chart shown in Figure 8.10 except that the vertical scale is multiplied by $n = 100$.

c Charts

The control chart for the number of defects is based on the assumption that the distribution of the number of defects per unit of product is approximated by the Poisson distribution. If c_i is the number of defects in the *i*th manufactured unit, and

$$\bar{c} = \frac{1}{k} \sum_{k=1}^{n} c_i$$

is the mean number of defects in *k* units, then $\bar{c}$ is an estimate of the Poisson parameter λ. Thus, the mean of the distribution of the number of defects can be estimated by $\bar{c}$, and its standard deviation by $\sqrt{\bar{c}}$. Thus, the control-chart values for a *c* chart are as follows:

Control-Chart Values for a Number-of-Defects Chart:

central line $= \bar{c}$

$$LCL = \bar{c} - 3\sqrt{\bar{c}}$$

$$UCL = \bar{c} + 3\sqrt{\bar{c}}$$

Example. Suppose the mean number of defects per 1,000 feet of nylon staple used to manufacture carpets has been found to be 5.8. The following are results of an inspection of twenty-five 1,000-foot reels of staple:

4.2	3.5	6.1	7.4	5.0	4.8	3.6	0.0	2.9	6.3	5.4	5.8	4.6
5.3	6.1	4.0	5.2	6.3	4.7	6.7	3.9	4.6	5.9	2.5	4.9	

a. Plot a control chart.   b. Comment on what the chart shows.

Solution.   a. The control-chart values are

central line $= 5.8$

$$LCL = 5.8 - \sqrt{5.8} = 3.4$$

$$UCL = 5.8 + \sqrt{5.8} = 8.2$$

The control chart is shown in Figure 8.11.

b. This control chart shows a good state of control with no trends, runs, or periodicities that might have indicated imminent control problems.

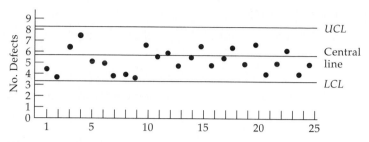

FIGURE 8.11   *c* Chart for Number of Defects in Nylon Staple

Computer Applications

Control charts for *p*, *np*, and *c* can be produced with *MINITAB* software, using the commands PCHART, NPCHART, and CCHART, respectively. We shall illustrate the use of computers in obtaining control charts for attributes by using *MINITAB* to produce the chart shown in Figure 8.12. The data, consisting of the *number* of complaints given in the example on page 206 are entered into C1 first. (In this case, $n = 100$; thus, the *number* of complaints equals the *proportion* of complaints.) Then, the command PCHART C1 100 is given, to indicate that a *p* chart is desired, using the data in C1, and each datum is the number of "nonconformities" in a sample of size 100. The resulting control chart is shown in Figure 8.12.

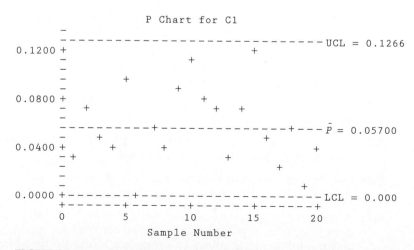

FIGURE 8.12   *p* Chart Produced by *MINITAB*

EXERCISES

8.15 What kind of control chart (p chart, np chart, or c chart) should be used to control each of the following:

    a. The number of defects found in successive lots.

    b. The percentage of painted panels needing touch-up.

    c. The number of missing rivets in an air-frame assembly.

    d. The yield of an integrated-circuit production line.

8.16 If a factory produces rolls of nylon carpeting that are 12 feet wide and 50 feet long, what kind or kinds of control chart should be used to control

    a. the number of defects per roll

    b. the number of defective rolls (rolls having one or more defects) per lot of 50 rolls

8.17 Plot the following proportions on a p chart having the control limits 0 and 0.07 and determine whether corrective action is indicated: 0.02, 0.01, 0.05, 0.04, 0.07, 0.05, 0.03, 0.06, 0.06, 0.05.

8.18 A c chart has the control limits 0.5 and 4.6. Plot the following numbers of defectives per unit on a control chart, and determine whether corrective action is indicated: 3.5, 1.2, 0.9, 2.2, 3.5, 4.4, 3.8, 2.9, 1.6, 0.0, 2.7, 4.1, 3.9, 3.0, 2.4, 0.5.

8.19 The following are the number of carburetors in consecutive lots of 100 found to be defective on final test: 4, 5, 1, 0, 3, 2, 1, 6, 0, 6, 2, 0, 2, 3, 4, 1, 3, 2, 4, 2, 1, 2, 0, 2, 3, 4, 1, 0, 0, 0, 0, 1, 2, 3, 3. Using the results of the first 20 lots, find the control-chart values for a p chart, and plot all 35 test results on this chart.

8.20 The following are the results of inspecting consecutive samples of 50 screws for defective threads: 4, 5, 3, 2, 1, 5, 4, 4, 6, 5, 2, 2, 3, 1, 2, 4, 5, 4, 6, 4, 2, 3, 4, 2, 4, 1, 0, 0, 1, 2, 0, 0, 0, 1, 0, 0, 1, 0, 0, 1, 0, 1, 0, 2, 0, 0, 0, 1, 0, 1. Find the control-chart values for a p chart from the first 20 samples, and plot all data on a control chart.

8.21 Examine the control chart plotted in Exercise 8.19 for evidence of runs or trends. What conclusions can be drawn?

8.22 Examine the control chart plotted in Exercise 8.20 for evidence of runs or trends. What conclusions can be drawn?

8.23 Use the data given in Exercise 8.19 to plot an np chart and compare this chart with the p chart previously plotted.

8.24 Use the data given in Exercise 8.20 to plot an np chart and compare this chart with the p chart previously plotted.

8.25 A company sets type for the publication of technical manuscripts. A random sample of 30 pages of galley proof are read for typographical errors, and the following numbers of errors are found on these pages of galley proof: 4, 7, 9, 3, 0, 4, 2, 1, 6, 3, 5, 8, 2, 1, 4, 6, 5, 2, 2, 1, 6, 3, 4, 6, 5, 2, 1, 0, 4, 0.

Use the first 15 observations to find the control-chart values for a c chart, and plot the chart.

8.26 The following is the number of defective welds on 40 gear-case covers: 0, 2, 0, 0, 5, 0, 5, 1, 0, 1, 0, 2, 3, 0, 0, 5, 0, 1, 4, 0, 0, 0, 3, 1, 4, 5, 1, 0, 0, 0, 1, 0, 0, 0, 0, 0, 2, 0, 0, 0. Use the first 20 observations to find the control-chart values for a c chart, and plot the chart.

8.27 a. Is the type setting process described in Exercise 8.25 in control?

b. Does the control chart exhibit any signs warranting corrective action?

8.28 Comment on the control chart plotted in Exercise 8.26.

8.29 a. Use *MINITAB* or some other statistical computer software package to produce the control chart requested in Exercise 8.19.

b. Use the same data, but now assume that the lot size was 89, to repeat part a. (Note the increased reduction in work when it no longer is necessary to calculate all the proportions.)

8.30 Use *MINITAB* or some other statistical computer software package to produce the control chart requested in Exercise 8.20.

8.4 Cusum Charts

The control charts we have studied so far are known as **Shewhart control charts**. They have the common property that, once the control limits have been established, use is made only of the last data point in determining whether a process has gone out of control. As experience with Shewhart control charts has accumulated, checks for runs, trends, and so forth have been added in an effort to incorporate all the information contained in a given chart.

An alternative has been proposed to the Shewhart control chart that directly incorporates all previous information in the sequence of sample results. Such a chart, called a cumulative-sum, or **cusum chart**, plots the cumulative sum of deviations of the sample values from the target value T. For example, if the means of r samples are taken and $\bar{x}_i$ is the mean of sample i, the quantities

---

Cumulative Sums: $\quad S_k = \sum_{i=1}^{k} (\bar{x}_i - T) \qquad k = 1, 2, \dots, r$

---

are the cumulative sums of the deviations of the sample means from their target, T. Analogous to a Shewhart control chart, the target often is chosen to be the grand mean of a series of samples taken when the process is in control. The cumulative sums are plotted on the y axis of the cusum control chart. If the process remains in control, it can be expected that the cumulative sums will vary randomly about zero. If the process means are increasing, the cumulative sums will trend upward, above zero; if the means are decreasing, the trend will take the cumulative sums below zero. By making use of all data in each control-chart point, the cusum chart sometimes detects out-of-control conditions earlier than the corresponding Shewhart control chart.

A formal decision procedure for determining whether a process is out of control is based on a **V-mask**. A typical V-mask is shown in Figure 8.13. The V-mask is placed on the control chart with the point labeled "O" on a given cumulative sum, and the line OA is placed parallel to the horizontal axis of the chart. The process is said to be in control if the points corresponding to all the previous cumulative sums lie within the V formed by the mask. If one or more such points lie below the lower arm of the V, there is evidence that the process mean has shifted downward; if one or more points lie above the upper arm, the process is considered to have shifted upward. In practice, the V-mask is applied to each point on the cusum chart as the points on the chart are generated.

Construction of V-masks requires careful geometric scaling in both the x and y directions. The geometry of the V-mask is based on the quantity D, defined to be the change in the value of the underlying population mean that must be detected with near certainty. Then, formulas can be given for the quantities

d = distance between the points O and A in Figure 8.13, and
θ = angle between the line OA and either of the "arms" of the V-mask

Using theory developed by E. S. Page in 1954, the formula for d is

$$d = E(\alpha) \left[\frac{\sigma / \sqrt{n}}{D} \right]^2$$

and for θ we have

$$\tan \theta = \frac{D}{2y}$$

In the equation for d, the quantity $E(\alpha)$ depends on the desired probability of a Type I error. $E(\alpha)$ usually is taken to be 13.215 to conform to the 3-sigma limits of the Shewhart control chart. The value of σ usually is estimated by $\bar{s}$; and n is the size of each sample. In the equation for θ, y is a scaling factor used to control the geometry of the control chart. Usually, y is chosen in the interval between $\frac{\sigma}{\sqrt{n}}$ and $2.5 \frac{\sigma}{\sqrt{n}}$

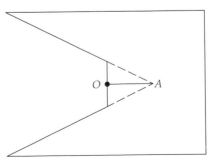

Figure 8.13 *V-Mask*

so that will θ lie between 30 and 60 degrees. The vertical axis of the control chart must be scaled so that there are y units for each unit on the horizontal axis.

Construction of a *V*-mask and its use in a cusum chart will be illustrated by considering the following data, giving the mean fracture strength (in pounds), based on samples of 4 castings each.

Sample Number	Sample Mean	$\bar{x}_i - T$	Cusum $\Sigma(\bar{x}_i - T)$
1	89.2	9.1	9.1
2	83.6	3.5	12.6
3	83.9	3.8	16.4
4	78.1	−2.0	14.4
5	84.2	4.1	18.5
6	78.4	−1.7	16.8
7	82.9	2.8	19.6
8	79.5	−0.6	19.0
9	78.8	−1.3	17.7
10	76.3	−3.8	13.9
11	75.0	−5.1	8.8
12	77.3	−2.8	6.0
13	70.3	−9.8	−3.8
14	71.0	−9.1	−12.9
15	73.6	−6.5	−19.4

In calculating the cumulative sums, the target value, T, was taken to be 80.1, the grand mean of sample means previously obtained when the process was in control. Although the sample standard deviations have not been given in the preceding table, we shall assume that they have been calculated, and that $\bar{s} = 8.8$ pounds. Let us suppose that the minimum detectable change in mean fracture strength is to be $D = 5.0$ pounds. Then, we can calculate the two quantities

$$d = 13.215 \left[\frac{\sigma / \sqrt{n}}{D} \right]^2 = 13.215 \left[\frac{8.8 / \sqrt{4}}{5} \right]^2 = 11.6$$

$$\tan \theta = \frac{D}{2y} = \frac{5}{(2)(4.4)} = 0.57$$

so that $\theta = 30°$, approximately. Here, we took $y = \dfrac{\bar{s}}{\sqrt{n}} = \dfrac{8.8}{\sqrt{4}} = 4.4$. The resulting *V*-mask is shown superimposed on the data in Figure 8.14. Imagine that we make a cutout of the *V*-mask and we slide it along the chart, with the point O placed successively on the point representing each cusum and the line OA parallel to the x axis. Then, we will find that one or more of the cusums is not included in the V of the mask for the first time when we reach sample 13, as shown in Figure 8.14. Thus, it can be concluded that the process has gone out of control with that sample. Examination of the fracture strengths shows that they begin to diminish on or about the tenth sample. This reduction in strength is rapid enough so that the casting process can be said to be out of control by the time sample 13 is reached.

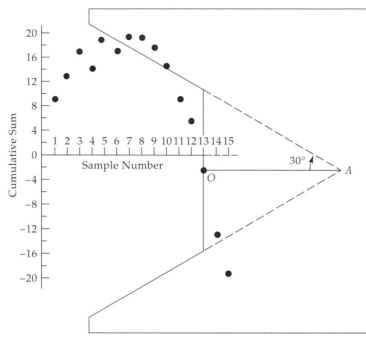

FIGURE 8.14   Cusum Chart with *V*-Mask

EXERCISES

8.31 Calculate the cusums for the data of Exercise 8.1 on page 202, using the grand mean of the data as the target value.

8.32 Calculate the cusums for the data of Exercise 8.2 on page 202, using the grand mean of the data as the target value.

8.33 Using the data of Exercise 8.1 on page 202, make a scale drawing of a *V*-mask appropriate for a cusum chart with $D = 3$ Mooneys.

8.34 Using the data of Exercise 8.2 on page 202, make a scale drawing of a *V*-mask appropriate for a cusum chart with $D = 2$ nanoamperes.

8.35 Make a chart using the cusums calculated in Exercise 8.31 and check this chart for control, using the *V*-mask prepared in Exercise 8.33.

8.36 Make a chart using the cusums calculated in Exercise 8.32 and check this chart for control, using the *V*-mask prepared in Exercise 8.34.

8.37 Using the fracture-strength data given on page 213, construct a Shewhart control chart. (Assume $\overline{R} = 21.7$.) How, if at all, do the conclusions that can be drawn from the Shewhart chart differ from those drawn from the cusum chart?

8.5   Process Capability

The location and spread of a histogram were compared to specification limits in Section 3.6, where it was seen how a distribution could be used to determine the ability of a process to meet specifications. Analogously, control-chart values are based on the location and variability of the distribution of measurements made on a process. Control limits, calculated for a process that has been brought into a state of statistical control, reflect the "natural tolerances" of a process. Comparisons of these limits with process specifications can establish the capability of a process to meet its specifications. Several indices of **process capability** have been formalized in Japanese industry, and application of these indices is becoming increasingly prevalent in the United States.

The first of these indices, which assumes that a process is properly centered and in a state of statistical control, measures the *potential* of the process to meet specifications. This index is based on a comparison of the specification range of a measurement to its natural variability. If U represents the upper specification limit and L represents the lower specification limit, then the specification range is $U - L$. The following index compares the specification range to the six-sigma range of the process.

$$\textbf{C}_p \textbf{ Index:} \qquad C_p = \frac{U - L}{6\hat{\sigma}}$$

Here, $\hat{\sigma}$ is an estimate of the standard deviation of the process measurement. Its value can be estimated from an R chart by $\hat{\sigma} = \frac{\overline{R}}{d_2}$ where d_2 is obtained from the table on page 161 for the appropriate sample size, n. A process is said to be capable if $C_p \geq 1$.

While a process is capable if $C_p = 1$, this value of C_p implies that about 0.3% of the product will fail to meet specifications for the given measurement if it has a normal distribution and if it is centered between the specification limits. If the distribution is not normal, or if it is not centered, the percentage of product failing to meet specifications can be much larger than 0.3%. Also, the quality of a product often depends upon the ability of more than one critical measurement to meet specifications. As will be shown in Chapter 11, the probability of failing to meet specifications increases rapidly as the number of critical measurements increases. Keeping these issues in mind, values of C_p should be sought that are much higher than 1 if overall quality is to be assured. The following classification of product quality in terms of values of C_p has been adopted widely by industry:

Value of C_p	Quality
< 1.00	not capable
1.00 – 1.33	barely manufacturable
1.34 – 3.00	good
> 3.00	excellent

Striving for very high quality, many manufacturing facilities in Japan aim for values of C_p as high as 6, or greater. While this goal may seem impossible to some, it has been achieved often enough to make it credible. When this goal is achieved, the probability of producing an out-of-specification product usually is extremely small, even when distributions are not normal and many different measurements are critical to the product's performance.

The C_p index assumes that the process has been centered; also, it is applicable only to two-sided specification limits. Several additional indices have been defined to measure actual process performance, rather than its potential capability, and to be applicable to one-sided specification limits. First, we introduce an index expressing the degree to which a process is centered. Such an index can be based on a comparison of the difference between an estimate of the process mean and its target value to the distance between the target value and the closest specification limit. If T represents the target value and $\hat{\mu}$ represents an estimate of the process mean, then $|T - \hat{\mu}|$ is the absolute value of the difference between the target value and the actual process center. ($\hat{\mu}$ can be obtained from the central line of an appropriate control chart.) The distance between the target value and the *closest* specification limit is the smaller of the two numbers, $(T - L)$ and $(U - T)$. Denoting this latter quantity by $min\ (T - L, U - T)$, the index

$$\textbf{\textit{Process\,--\,Centering Index:}} \qquad k = \frac{|T - \hat{\mu}|}{min\ (T - L, U - T)}$$

measures the degree to which the process is centered between the specification limits. This index always will have a value greater than or equal to 0, and its most desirable value is zero. In the special case where the target value, T, is centered between the specification limits,

$$min\ (T - L, U - T) = T - L = U - T = (U - L)/2$$

and the index becomes

$$k = \frac{2\,|T - \hat{\mu}|}{U - L}$$

The index k measures the degree to which a process is centered between two specification limits. When there is only one specification limit, it is desired to have the process mean as far from the specification limit as possible. Thus, indices that compare the distance between the estimated process mean and the given limit to 3σ are indicated. When there is a one-sided specification limit, the index

$$\textbf{\textit{Index for Lower Specification Limit:}} \qquad C_{pL} = \frac{\hat{\mu} - L}{3\hat{\sigma}}$$

compares the distance between the estimated process mean and the *lower* specification limit to the three-sigma limit. Similarly, the index

$$\textbf{\textit{Index for Upper Specification Limit:}} \qquad C_{pU} = \frac{U - \hat{\mu}}{3\hat{\sigma}}$$

compares the distance between the estimated process mean and the *upper* specification limit to the three-sigma limit. In these formulas $\hat{\sigma}$ is an estimate of the process standard deviation.

The smaller of these two one-sided indices, C_{pL} and C_{pU}, provides an index of process capability that relates to two-sided specification requirements and also takes into account process off-centering. This index can be expressed in terms of C_{pL} and C_{pU}, as follows:

$$C_{pk} \text{ } \textbf{\textit{Index:}} \qquad C_{pk} = min\ (C_{pL},\ C_{pU})$$

The reader will be asked in Exercise 8.38 below to prove that $C_{pk} = C_p(1 - k)$ in the special case where $T = \dfrac{U + L}{2}$.

If the process is centered between its specification limits, then $C_{pL} = C_{pU}$, and $C_{pk} = C_p$. Thus, C_{pk} can be increased by centering the process. Both index values can be increased by taking steps to reduce the process variability, σ. The index C_p measures the *potential* capability of a process, while C_{pk} measures its *actual* capability. Product quality is classified in terms of values of C_{pk} identically to the classifications given on page 215 for C_p.

> **Example.** In the example involving bearing deflectors, given on page 193, it was found that $\bar{\bar{x}} = 13.1$ and $\bar{R} = 9.8$. Suppose specification limits for the eccentricity of the bearing deflectors are $L = 5$ and $U = 25$. Find C_p and C_{pk}, and determine the quality of the process producing these deflectors.
>
> **Solution.** An estimate of σ is afforded by $\dfrac{\bar{R}}{d_2} = \dfrac{9.8}{2.326} = 4.21$, where the value $d_2 = 2.326$ is found from the table on page 161 for $n = 5$. Thus, $C_p = \dfrac{25 - 5}{6\ (4.21)} = 0.79$. To find C_{pk}, we first calculate C_{pL} and C_{pU}. Using $\bar{\bar{x}} = 13.1$ to estimate μ, and $\hat{\sigma} = 4.21$, we obtain $C_{pL} = \dfrac{13.1 - 5}{(3)\ (4.21)} = 0.64$ and $C_{pU} = \dfrac{25 - 13.1}{(3)\ (4.21)} = 0.94$. Thus, $C_{pk} = min\ (0.64, 0.94) = 0.64$. Since both C_p and C_{pk} have values less than 1, the process is not capable of meeting specifications. In order for the process to become capable, its variability needs to be greatly reduced, and it should be better centered between the specification limits.

EXERCISES

8.38 Use the definition $C_{pk} = min(C_{pL}, C_{pU})$ and the definition of k to prove that $C_{pk} = C_p(1 - k)$ in the special case $T = \dfrac{U + L}{2}$.

8.39 a. What is the value of L when there is only an upper specification limit?

   b. What is the value of U when there is only a lower specification limit?

   c. Show that $C_{pk} = C_{pU}$ in part a and $C_{pk} = C_{pL}$ in part b.

8.40 Suppose the specification limits for the viscosity given in Exercise 8.1 on page 202 are 20 and 40 Mooneys.

   a. Find C_p.

   b. Find C_{pk}.

   c. Make a statement about the quality of the product produced by this process.

8.41 Suppose the specification limits for the reverse current given in Exercise 8.2 on page 202 are 5 and 25 nanoamperes.

   a. Find C_p.

   b. Find C_{pk}.

   c. Make a statement about the quality of the product produced by this process.

8.42 Suppose the specification limits for the measurements given in Exercise 8.9 on page 203 are 3.5 and 17.7. (Assume $n = 5$.)

   a. Find C_p.

   b. Find C_{pk}.

   c. Make a statement about the quality of the product produced by this process.

8.43 Suppose the specification limits for the measurements given in Exercise 8.10 on page 203 are -10 and 10. (Assume $n = 4$.)

   a. Find C_p.

   b. Find C_{pk}.

   c. Make a statement about the quality of the product produced by this process.

8.44 A process measurement has a mean value estimated to be 5.8 and a standard deviation estimated to be 0.6. If there is only an upper specification limit of 10.0, find C_{pU}.

8.45 A process measurement has a mean value estimated to be 10.5 and a standard deviation estimated to be 1.3. If there is only a lower specification limit of 5.0, find C_{pL}.

8.6 Maintenance of SPC Programs

Once it has been determined which product or process measurements should be controlled and control charts have been established, the task of statistical process control has just begun. The pursuit of quality is based on *continuous improvement*; thus, we never should be satisfied simply because all control charts show that a process is "in control." Constant maintenance and interpretation of existing control charts, as well as revision of control limits to reflect process improvement, are important aspects of continuous improvement. Addition or deletion of control charts, as experience and new information suggest improvements, are part of an SPC program.

Warning Limits

The ability it affords to take corrective action *before* a process goes out of control is one of the major advantages of SPC. Some methods for detecting potential out-of-control situations, based on control-chart patterns, were introduced in Section 8.2. A different method for "early detection" relies on **warning limits**. It is customary to draw two-sigma limits on control charts, using dotted lines or lines of a different color to distinguish them from the three-sigma control limits. If a point on a control chart is outside of a two-sigma warning limit, special attention is paid to future points on the chart, and investigation is begun into possible assignable causes of variability.

In using two-sigma warning limits, or three-sigma control limits for that matter, it should be recalled that the control chart affords a kind of hypothesis test. To illustrate, a given point on a control chart for means is the value of the mean of a sample taken from a population whose mean is estimated by $\bar{\bar{x}}$ and whose standard deviation is assumed to be $\bar{R}/d_2$ or $\bar{s}/c_2$. If the sample means are approximately normally distributed, the probability that a given mean will be outside either of the two-sigma warning limits is approximately .05, and the probability that it will be outside either of the three-sigma control limits is about .003. Thus, *for a single point on the control chart*, a test is made of whether the given sample mean is significantly different from the assumed population mean, $\bar{\bar{x}}$. For this test, the probability of a Type I error, α, is approximately .05 if warning limits are used, and .003 if control limits are used.

With $\alpha = .003$, the use of three-sigma control limits may seem very conservative. (The conclusion that a sample mean is "out of control," which is equivalent to the statement "$\bar{x}$ is significantly different from $\bar{\bar{x}}$," will be drawn falsely only with probability .003.) However, the significance test is applied repeatedly to a sequence of many sample means. Even if these means were normally and independently distributed and *not out of control* (not significantly different from μ_0) the probability of finding one or more means outside of the three-sigma control limits in 100 control-chart points is given by $1 - .997^{100} = .26$. Under the same conditions, the probability of finding one or more means outside of the two-sigma warning limits exceeds .99.

Three-sigma control limits are used in control charts to prevent excessive "false alarms." However, when two-sigma warning limits are used, it can be expected that many false alarms will occur over a period of time. Thus, when a control-chart point is outside of the warning limits, care must be taken not to overreact. Consideration should be given to shut down a process *only* when at least two closely neighboring points exceed three-sigma control limits. But close scrutiny of the control chart, always a good idea, should be redoubled when warning limits are exceeded, and vigorous action to uncover possible assignable causes of variability should be taken.

Re-Calculation of Control Limits

As more information is gained about a process, and as expertise grows in its ability to produce higher-quality products, it should be expected that the points on its control charts will vary less and less. When it is observed that a control chart has remained in control for a long period of time, especially if its points no longer come close to its warning limits, it is time to re-calculate the control limits. Using the most recent 15 – 30 samples, new control limits are calculated. The new limits should be narrower than the old ones, and a new process-capability study should show that C_{pk} has increased in value.

A program of periodic re-calculation of control limits, at least annually for all control charts, will keep an SPC program from growing stale. Notice should be taken of which charts do not show improvement when control limits are re-calculated and new values of C_{pk} are obtained. Some of the variables measured by these charts may already have been controlled as much as possible or as much as needed. (Over-control should be avoided.) But excess variability revealed by some control charts may be reduced by additional research. Designed experiments to uncover and remove additional sources of variability should be undertaken in these areas. Constant activity to identify areas needing improvement and to reduce the number of potential problem areas is the essence of quality improvement.

Adding or Deleting Control Points

During the startup phase of an SPC program, a limited area of special importance or concern was selected, and decisions were made about which variables or attributes within this area of the process to control. These decisions were based on engineering knowledge, on cost considerations, and on experiments performed to uncover relationships among the process variables and product measurements. As the program expands, additional "control points" are added, until the critical variables determining the process output have been brought under control. At this point, measurable improvements in product quality should have been noted.

From time to time, a process may "crash"; that is, it will unexpectedly begin to produce discrepant product. Close examination of the control charts maintained for the process often uncover reasons for such an event. Maintenance and inspection of all control charts as close as possible to "real time" play an important role in their use as diagnostic tools. If control charts are not maintained in real time, but they are updated and examined only periodically, say weekly or monthly, reasons for the problems they may uncover often will have been forgotten. Thus, taking speedy and efficient corrective action will become difficult. In addition, by the time potential out-of-control problems are discovered, many discrepant products may have been produced.

Even if all charts show a state of control, there may be trends in some charts which, when operating together, can result in poor quality. A comparison of the correlation matrix (page 244) of points on all charts taken during the period of a crash to the correlations obtained during prior periods of successful operation may reveal combinations of process measurements which lead to trouble. A more sophisticated approach, involving experiments designed to study interactions among critical measurements, may be required (see Chapter 10). Perhaps a variable, previously ignored, has changed in value enough to become critical.

A scientific approach to the solution of process difficulties sometimes shows that some of the wrong variables have been controlled. Close examination of control charts may show that some variables have very small variability compared to specification limits and no longer need to be closely controlled. Process capability analyses may show that the process is not capable of meeting specifications with regard to other variables. Perhaps the only way to change an incapable process into one that is capable is to make a design change. Maybe vendors of parts used in the process have changed, or ambient conditions, such as temperatures or humidities, have changed enough to exert an effect on the process. Or, perhaps, an important variable was overlooked when the SPC program was initiated. Its values may not have varied widely at an earlier time, but now greater variability has made its effect on the product dominant. The search for such variables should be ongoing, rather than delayed until a process shows signs of failure.

The conditions for adding or deleting control points are complex, and there is no simple "formula" for their identification. However, the following summary may be helpful in determining these conditions:

Conditions for adding control points

- Experimental work shows that a new variable, not previously controlled, may exert an important influence on the process output.
- New events have occurred, such as the introduction of new operators, new training programs, new vendors for incoming parts or materials, or specification changes. It is very helpful in diagnosing the causes of potential out-of-control situations to record such events on the appropriate control charts as they occur.
- Relationships involving variables not currently controlled may result in control-chart problems. It may not be possible to restore an existing chart to a state of control until these variables are identified and controlled.

Conditions for deleting control points

- A controlled variable has exhibited almost no variability over a long period of time.
- Two or more controlled variables exhibit a very high degree of correlation. Only one of them needs to be controlled, preferably the "causal" variable, if that can be determined.
- Specification or process changes have made measurement of a variable or variables superfluous.

EXERCISES

8.46 Draw warning limits on the control chart plotted in Exercise 8.9 on page 203, and determine what, if any, action needs to be taken.

8.47 Draw warning limits on the control chart plotted in Exercise 8.10 on page 203, and determine what, if any, action needs to be taken.

8.48 Draw warning limits on the control chart plotted in Exercise 8.25 on page 210, and determine what, if any, action needs to be taken.

8.49 Draw warning limits on the control chart plotted in Exercise 8.26 on page 211, and determine what, if any, action needs to be taken.

8.50 What is the probability that at least one of ten consecutive means plotted on a control chart will exceed the control limits even when the process actually remains in control? (Assume that the ten means come from independent samples taken from the same normal population.)

8.51 What is the probability that at least one of ten consecutive means plotted on a control chart will exceed the warning limits even when the process actually remains in control? (Assume that the ten means come from independent samples taken from the same normal population.)

8.52 Which of the following statements is false? Control limits should be re-calculated

　a. at least annually

　b. whenever the process goes out of control

　c. when warning limits are exceeded

　d. when there are clear signs that improvement has occurred

8.53 Which of the following statements is true? A control point should be deleted

　a. when its control chart stays in control for one year

　b. when its chart shows little or no variability over a long period of time

　c. every time specification limits change

REVIEW EXERCISES

These review exercises can be used for informal review of the chapter, or as two practice examinations, designed to be taken over a time period of about one hour. The odd-numbered exercises comprise one examination; the even-numbered exercises comprise the other.

8.54 Twenty samples, each of size 4, were taken from consecutive rational subgroups. It was found that $\bar{\bar{x}} = 9.7$ and $\bar{R} = 4.4$. Find the control-chart values for $\bar{x}$ and R charts.

8.55 Fifteen samples, each of size 5, were taken from consecutive rational subgroups. It was found that $\bar{\bar{x}} = 11.6$ and $\bar{R} = 8.1$. Find the control-chart values for $\bar{x}$ and R charts.

8.56 Twenty-five samples, each of size 6, were taken from consecutive rational subgroups. It was found that $\bar{\bar{x}} = 9.7$ and $s = 1.4$. Find the control-chart values for $\bar{x}$ and s charts.

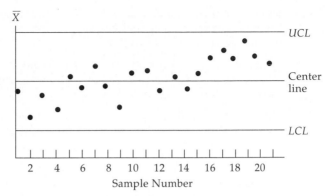

FIGURE 8.15 Control Chart for Exercise 8.58

8.57 Twenty samples, each of size 4, were taken from consecutive rational subgroups. It was found that $\bar{\bar{x}} = 11.6$ and $\bar{s} = 2.1$. Find the control-chart values for $\bar{x}$ and s charts.

8.58 Examine the control chart in Figure 8.15 for indications that it may go out of control and state your findings.

8.59 Examine the control chart in Figure 8.16 for indications that it may go out of control and state your findings.

8.60 The proportion of rejected product on final inspection has been found to be 0.015 under normal operating conditions. Find the control-chart values for a p chart using lots of size 200.

8.61 The proportion of products needing rework has been found to be 0.12 under normal operating conditions. Find the control-chart values for a p chart using lots of size 50.

8.62 Under normal operating conditions, the mean number of defects found in 100 feet of vinyl-coated steel has been found to be 3.6. Find the control-chart values for a c chart.

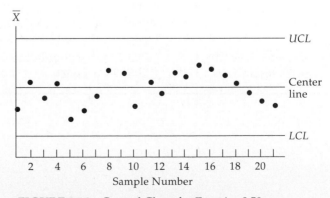

FIGURE 8.16 Control Chart for Exercise 8.59

8.63 Find the control-chart values for a c chart to control soldering quality if the mean number of defective solder joints per circuit board is 1.8 under normal processing conditions.

8.64 Find the constants d and tan θ for a V-mask if the greatest acceptable change in the process mean is $D = 2$, and samples of size $n = 4$ have a mean standard deviation of $\bar{s} = 0.76$.

8.65 Find the constants d and tan θ for a V-mask if the greatest acceptable change in the process mean is 3.5, and samples of size 4 have a mean standard deviation of 1.8.

8.66 Find C_p and C_{pk} for the process in Exercise 8.54 if the specification limits are 2–20.

8.67 Find C_p and C_{pk} for the process in Exercise 8.55 if the specification limits are 6–18.

8.68 Find warning limits for the $\bar{x}$ chart in Exercise 8.54.

8.69 Find warning limits for the $\bar{x}$ chart in Exercise 8.55.

GLOSSARY

9

REGRESSION ANALYSIS

Where have we been?   So far, all our statistical methods have dealt with observations involving a single variable. Many problems of quality improvement involve *relations* among several variables. Thus, we need to add to our armamentarium statistical methods that apply to simultaneous observations made on several variables.

Where are we going?   In this chapter, we begin with the statistics of relations between two variables, and then we generalize the methods discovered to include many variables. We conclude by discussing the usefulness as well as the pitfalls of applying these "multivariate" methods to problems involving quality.

Section 9.1, *Simple Linear Regression*, uses the method of least squares to fit a straight line to data involving two variables and gives a test for the significance of a linear regression.

Section 9.2, *Correlation*, defines the coefficient of correlation and gives a test for significance and confidence intervals for a correlation.

Section 9.3, *Multiple Regression*, extends curve fitting to functions of several independent variables and shows how to use statistical computer software for multiple-regression analysis.

Section 9.4, *Analysis of Residuals*, checks the adequacy of a regression analysis by analyzing patterns of residuals.

Section 9.5, *Application to Quality*, discusses the applicability of the methods of this chapter to the solution of quality problems.

9.1 Simple Linear Regression

Often it becomes necessary to study the relationships among several variables on the basis of unplanned data. An example might be the relationship between the price of a product and its sales, using data collected from historical records. Another example might involve a study of the relationships linking four or five critical production-line settings and a measurement made on the resulting product.

A widely used method for the analysis of such relationships, involving two or more variables, is called **regression analysis**. We shall begin our study of regression analysis by deriving and applying the theoretical basis for regression analyses involving simple relationships between two variables.

Method of Least Squares

The use of statistical methods to find "best-fitting" linear relationships between two variables is called **simple linear regression**. In this kind of problem, we assume that the value taken on by a given random variable may depend upon the value taken on by a second, fixed (non-random) variable. For example, the length of a steel rod may depend upon its temperature. The length of the rod is regarded as a value of a random variable because it can be influenced by measurement errors as well as variability associated with slightly differing steel composition from one rod to another. However, we shall assume that the temperature of the rod can be fixed, without error, at each of several given values. In simple linear regression, we further assume that the mean length of many different rods is a linear function of the temperature to which the rods are heated.

To state these assumptions symbolically, let us suppose that y is a value of a random variable whose mean is a linear function of a fixed (non-random) variable x. It will be assumed further that the standard deviation of y is a constant, independent of x. The random variable having the value y is called the **response variable** or the **dependent variable**, and x represents the **independent variable**. These assumptions can be summarized by the following linear-regression equation:

$$y = \alpha + \beta x + \epsilon$$

where ϵ is a value of a random variable whose mean is zero and whose standard deviation is σ.

The purpose of a regression analysis is to estimate the parameters α and β by finding the straight line of "best fit" to a set of n paired observations, (x_i, y_i), with $i = 1, 2, \ldots, n$. Suppose the fitted line has the equation

$$\hat{y} = a + bx$$

where $\hat{y}$ is the predicted value of y when the independent variable takes on the value x, and a and b are estimates of α and β. Then the error made when $\hat{y}_i$ is used to predict y for a given x_i is

$$e_i = y_i - \hat{y}_i$$

The errors e_i are called **residuals**.

A graphic interpretation of residuals is shown in Figure 9.1. In this figure, a set of data points (x_i, y_i) for $i = 1, 2, \ldots, n$ is graphed on a Cartesian coordinate system, together with a straight line drawn through these points. The residual e_i is the signed distance parallel to the y axis between the data point (x_i, y_i) and the fitted straight line.

The purpose of the regression analysis can be restated as that of finding the straight line—the values of a and b—that minimizes the residuals in some fashion. First, it is observed that, unless all the data points are collinear, no straight line can be found so that all $e_i = 0$. Perhaps a straight line can be sought that minimizes the sum of the residuals. When defining the standard deviation, it was found (page 52) that positive and negative deviations cancel so that the sum of the deviations from the mean is zero. Analogously, many different straight lines can be found so that the sum of the residuals is zero.

An approach can be employed for minimizing residuals that is analogous to the definition of the standard deviation, namely, minimizing the sum of the *squared* residuals. This method for finding the line of best fit is called the **method of least squares.** To find the values of a and b by the method of least squares, first the sum of squared residuals is written as

$$\sum_{i=1}^{n} e_i^2 = \sum_{i=1}^{n} \left[y_i - (a + bx_i) \right]^2$$

This sum of squares can be regarded as a function of a and b, and it is desired to find values of a and b which make it a minimum. A necessary condition for a minimum is that the partial derivatives of the sum of squared residuals equal zero, with respect to both a and b. Differentiating the sum of squared residuals, first with respect to a, and setting the result equal to zero, we obtain

$$2 \sum_{i=1}^{n} \left[y_i - (a + bx_i) \right] \cdot (-1) = 0$$

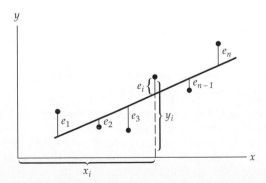

FIGURE 9.1   Residuals in Simple Linear Regression

Then, differentiating with respect to b and setting the result equal to zero, we obtain

$$2 \sum_{i=1}^{n} \left[y_i - (a + bx_i) \right] \cdot (-x_i) = 0$$

If the terms in each of these equations are summed and then rearranged, the following **normal equations** are obtained:

$$\Sigma y = na + b\,\Sigma x$$

$$\Sigma xy = a\,\Sigma x + b\,\Sigma xy$$

(The subscripts and limits of summation have been omitted for simplicity.) It can be seen that the normal equations are linear in a and b. Solving this system of linear equations for a and b, we obtain

> **Least - Squares Solution:** $\quad b = \dfrac{n\Sigma xy - \Sigma x\Sigma y}{n\Sigma x^2 - (\Sigma x)^2} ; \quad a = \bar{y} - b \cdot \bar{x}$

Before fitting a straight line to data, it is important to verify that the relationship does not depart excessively from linearity. Examination of a graph of the data points, called a **scatter plot**, is always recommended before applying the methods of this section.

Example. The following table shows the elongation (in thousandths of an inch) of steel rods of nominally the same composition and diameter, when subjected to various tensile forces (in thousands of pounds):

Force, x	1	5	3	2	4	2	6	8	7	4
Elongation, y	15	80	39	34	58	36	88	111	99	65

a. Draw a scatter plot of these data to see if it is reasonable to fit a straight line.

b. If so, find the slope and the intercept of the least-squares line.

c. Use the equation of the fitted line to estimate the elongation when the tensile force is 5,700 pounds.

Solution. a. The scatter plot is shown in Figure 9.2 and no serious departure from linearity is detected.

b. We shall work our way through the least-squares estimates of the parameters of the best-fitting straight line here to give you a better idea of how the least-squares method works. In practice, you probably will want to do these calculations with the use of computer methods (see page 232). To find the solution to the normal equations, the following quantities are required:

$$n = 10, \quad \Sigma x = 42, \quad \Sigma y = 625, \quad \Sigma x^2 = 224, \quad \Sigma xy = 3{,}273$$

Thus

$$b = \frac{n\Sigma xy - \Sigma x \Sigma y}{n\Sigma x^2 - (\Sigma x)^2} = \frac{(10)\,(3,273) - (42)\,(625)}{(10)\,(224) - (42)^2} = 13.61$$

and

$$a = \bar{y} - b \cdot \bar{x} = \frac{625}{10} - (13.61) \cdot \frac{42}{10} = 5.34$$

c. The estimated elongation (in thousandths of an inch) at a tensile force of 5.7 thousand pounds is

$$\hat{y} = 5.34 + (13.61)(5.7) = 82.9$$

Note that the fitted straight line has been added to the scatter plot shown in Figure 9.2. The value 82.9 is the y value of this line when $x = 5.7$, as shown in the figure.

FIGURE 9.2   Scatter Plot

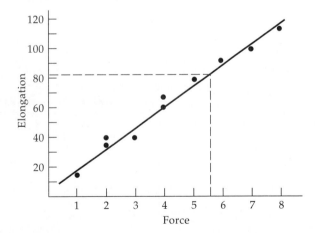

Test for Significance of Regression

The question of whether y is linearly related to x—that is, the slope β of the linear regression equation differs from zero—can be answered by examining the slope of the fitted least-squares line. If the slope of the fitted line is significantly different from zero, we say that there is a *linear* relation between x and y at the chosen level of significance. Note that two variables may have a strong relationship even when there is no *linear* relationship between them. Figure 9.3 shows a graph representing a least-squares line that has a slope of 0 when there is a strong curvilinear relationship between x and y. (The method of least squares can be used to fit curvilinear relationships involving a single independent variable, as shown in Exercise 9.13 on page 235.)

The least-squares estimate, b, of the slope, β, in the linear-regression equation has an error related to the standard deviation of the residuals. The variance of the residuals can be estimated by the mean-squared deviation of the y values of the data points from the least-squares line, or

$$s_e^2 = \frac{\sum_{i=1}^{n}(y_i - \hat{y}_i)^2}{n-2} = \frac{1}{n-2}\sum_{i=1}^{n}\left[y_i - (a + bx_i)\right]^2$$

(Note that the divisor is $n - 2$, rather than $n - 1$ as in the variance of a single variable. Since two constants, a and b, are estimated for a least-squares line, two degrees of freedom are deducted from n.) If it further can be assumed that the e_i are values of a random variable having the normal distribution, then the quantity

---

t *Value for Testing for Linearity:* $\qquad t = \dfrac{b}{s_e}\sqrt{\Sigma x^2 - \dfrac{(\Sigma x)^2}{n}}$

---

has the t distribution with $n - 2$ degrees of freedom when there is no linear relationship, that is, when $\beta = 0$.

A test of the null hypothesis that the linear-regression equation has a slope equal to zero against the hypothesis that the slope does not equal zero can be based on the value of t given in this last equation. After finding a and b by the method of least squares, the residuals and their standard deviation are calculated, and t is

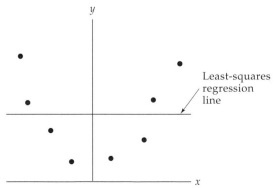

FIGURE 9.3 Curvilinear Relationship With Zero-Slope Linear Regression

calculated. Then, the calculated value of t is compared with the value of $t_{\alpha/2}$ obtained from Table 4 with $n - 2$ degrees of freedom and the level of significance α. If the value of t exceeds $t_{\alpha/2}$, or if it is less than $-t_{\alpha/2}$, it can be said that b is significantly different from zero at the significance level α.

The same formula can be used to construct a $1 - \alpha$ confidence interval for β. Using methods similar to those given in Section 7.2, the following $1 - \alpha$ confidence limits can be found:

$1 - \alpha$ Confidence Limits for β: $\quad b \pm \dfrac{t_{\alpha/2} \cdot s_e}{\sqrt{\Sigma x^2 - \dfrac{(\Sigma x)^2}{n}}}$

Example. a. Test whether the straight line fitted to the data in the example on page 228 has a slope significantly different from zero at the .01 level of significance, and b. find 95% confidence limits for β.

Solution. a. First, it is necessary to find the standard deviation of the residuals. Again, we shall do it once the long way, to give you an appreciation of what is involved. We recommend, however, that you use a computer method, such as the one outlined on page 232. This standard deviation is found by calculating $\hat{y}_i = 5.21 + 13.64\, x_i$ for each value of i, then finding the corresponding residual, $y_i - \hat{y}_i$, then calculating the sum of squares of these residuals, and dividing by $10 - 2 = 8$. This calculation is summarized as follows:

i	1	2	3	4	5	6	7	8	9	10
x_i	1	5	3	2	4	2	6	8	7	4
y_i	15	80	39	34	58	36	88	111	99	65
$\hat{y}_i$	18.9	73.4	46.1	32.5	59.8	32.5	87.1	114.3	100.7	59.8
e_i	−3.9	6.6	−7.1	1.5	−1.8	3.5	0.9	−3.3	−1.7	5.2

The sum of the squared residuals is 168.55, so that $s_e^2 = 168.55/8 = 21.1$, and $s_e = 4.6$. To find the calculated value of t, use is made of the two sums $\Sigma x = 42$, $\Sigma x^2 = 224$ obtained on page 229, yielding

$$t = \frac{13.64}{4.6}\sqrt{224 - \frac{(42)^2}{10}} = 20.5$$

Comparing the calculated value of t with the value of $t_{.005}$ found in Table 4 for 8 degrees of freedom, it is found that the calculated value, 20.5, exceeds the tabular value, 2.306. Thus, it can be concluded at the .01 level of significance that the slope of the least-squares line is significantly different from zero; that is, there is a significant linear relationship between x and y.

b. The 95% confidence limits for β are

$$13.64 \pm \frac{(2.306)\,(4.6)}{\sqrt{224 - \dfrac{(42)^2}{10}}}$$

or 12.10 and 15.18.

It is important to check whether the assumption of normality is approximately met before proceeding with a test for the significance of the slope of a regression line or the construction of a confidence interval. Fortunately, the t value used in testing or constructing a confidence interval for the slope can be used even when the distribution of the residuals departs moderately from normality.

> *Example.*   Check the residuals found in the preceding example to assure that their distribution is approximately normal with a mean of 0.
>
> *Solution.*   Figure 9.4 shows a normal-scores plot of the 10 residuals. Note that they closely follow a straight line, indicating approximate normality, and that their mean value is 0, except for a slight round-off error.

Computer Applications

The calculations involved in linear regression can be lengthy and tedious, often leading to arithmetic errors. Fortunately, many statistical computer software programs are capable of performing these calculations, and it is strongly recommended that one of these programs be used for this purpose. To illustrate these calculations, we shall apply *MINITAB* software to the data of the example on

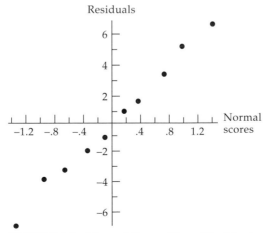

FIGURE 9.4   Normal-Scores Plot of Residuals

page 228. First, the values of x are set in column C1 and the y values are set in C2, in the *same order*. Then, the command REGRESS C2 ON 1 PREDICTOR IN C1 is given. The following is a portion of the resulting output:

THE REGRESSION EQUATION IS

C2 = 5.32 + 13.6 C1

COLUMN	COEFFICIENT	ST. DEV. OF COEF.	T-RATIO COEF/S.D.
	5.324	3.153	1.69
C1	13.6134	0.6662	20.44

S = 4.596

This output gives the values of the estimated regression coefficients, $a = 5.32$ and $b = 13.6$. (Except for small round-off differences, they are the same values found on page 229.) The t ratio given for C1 is the same ratio we use to test whether the slope of the regression line is significantly different from zero, also except for roundoff. *MINITAB* software also provides an option for calculating the residuals, which shall be explored in Section 9.4.

EXERCISES

Use of a computer is recommended for some of the following exercises.

9.1 Use the method of least squares to fit a straight line to the following data:

x	3	7	5	6	4	2	8	4
y	12.8	16.5	13.6	14.1	9.9	7.5	14.3	3.7

9.2 Use the method of least squares to fit a straight line to the following data:

x	5	3	3	4	2	6	1
y	9.8	10.4	17.6	7.1	13.0	7.5	12.3

9.3 The following are the loads (grams) put on the end of like plastic rods with the resulting deflections (cm):

Load	25	30	35	40	45	50
Deflection	1.58	1.39	1.41	1.60	1.78	1.65

a. Would it be appropriate to fit a straight line to these data with deflection as the independent variable and load as the dependent variable? Why?

b. Draw a scatter plot to check the reasonableness of fitting a straight line to these data with deflection as the y variable and load as the x variable.

c. If the scatter plot indicates it is reasonable to do so, find the slope of the least-squares line. State in words what the slope measures.

d. Estimate the deflection when the load is 38 grams.

9.4   The following data gives the sheet resistance (ohm-cm) of silicon wafers that correspond to different diffusion times (hours):

Diffusion time	0.5	1.0	1.5	2.0	2.5
Sheet resistance	89	90	88	92	91

a. Would it be appropriate to fit a straight line to these data with diffusion time as the independent variable and sheet resistance as the dependent variable? Why?

b. Draw a scatter plot to check the reasonableness of fitting a straight line to these data with resistance as the y variable and time as the x variable.

c. If the scatter plot indicates it is reasonable to do so, find the slope of the least-squares line. State in words what the slope measures.

d. Estimate the sheet resistance when the diffusion time is 0.75 minutes.

9.5   Test at the .01 level of significance whether the slope of the line fitted in Exercise 9.1 is significantly different from zero.

9.6   Test at the .05 level of significance whether the slope of the line fitted in Exercise 9.2 is significantly different from zero.

9.7   Find 95% confidence limits for β, the true slope of the line fitted in Exercise 9.1.

9.8   Find 90% confidence limits for β, the true slope of the line fitted in Exercise 9.2.

9.9   Find 95% confidence limits for the true incremental deflection per unit load in Exercise 9.3.

9.10 Find 90% confidence limits for the true increase in sheet resistance per additional hour of diffusion time in Exercise 9.4 and comment on the width of these limits.

The following exercises introduce new material, expanding on some of the topics covered in the text.

9.11 *Limits of prediction.* It can be shown that, with probability $1 - \alpha$, a future value of y corresponding to a given value of x will fall between the limits

$$\hat{y} \pm t_{\alpha/2} \cdot s_e \sqrt{1 + \frac{1}{n} + \frac{(x - \bar{x})^2}{\Sigma x^2 - (\Sigma x)^2 / n}}$$

Using the value of $\hat{y}$ found in the example on page 229, find 95% limits of prediction for y when $x = 5.7$.

9.12 *Extrapolation.* The formula given in Exercise 9.11 shows that the error in **extrapolation**, estimating y for values of x outside the range of the data, can become very large. Demonstrate this statement by calculating 95% limits of prediction when $x = 15$ and comparing these limits with those obtained in Exercise 9.11 for $x = 5.7$.

9.13 *Non-linear regression.* Certain non-linear relations can be transformed to linearity and fitted to data by the method of least squares. For example, $y = ab^x$ can be fitted by taking logarithms, writing $\log y = \log a + x \log b$. The same procedure is followed as before, except that $\log y$ is used in place of y and the results are estimates of $\log a$ and $\log b$. The following data give the curing time of test samples of concrete, x, and their tensile strength, y.

x	1	2	3	4	5	6
y	3.54	8.92	27.5	78.8	225	639

a. Fit a curve of the form $y = ab^x$ to these data.

b. What assumptions must be made to find a confidence interval for the slope of the transformed line?

c. Find 95% confidence limits for β, the true value of b.

9.14 *Continuation.* Find the transformations necessary to fit each of the following using linear regression.

a. $y = \dfrac{1}{a + bx}$ b. $y = ax^b$ c. $y = a + bx^2$

9.2 Correlation

Coefficient of Correlation

The strength of a linear relationship between the fixed (non-random) variable x and the random variable y can be measured by the slope b of the least-squares regression line. This measure, however, is scale dependent; that is, the magnitude of the slope depends on the scale of measurement used for y.

However, a scale-invariant measure of linear relationship can be given by comparing the variance of the residuals to the original variance of the y values. To consider the two extremes, first let us suppose that the residual variance were equal to zero. Thus, there remains no variability after fitting a straight line—*all* the original variability of the y values has been "accounted for" by the regression. Second, suppose the residual variance were equal to the original variance of the y values. In this case, *none* of the original variability has been "explained" by the regression. These observations suggest that a scale-invariant measure of the strength of a linear regression can be constructed by comparing the variability "explained" by the regression to the original variability of y. The variability explained by the regression is $\sigma_y^2 - \sigma_e^2$ and the original variance of y is σ_y^2. The ratio of these quantities,

$$\rho^2 = \frac{\sigma_y^2 - \sigma_e^2}{\sigma_y^2} = 1 - \frac{\sigma_e^2}{\sigma_y^2}$$

measures the proportion of the reduction in variability associated with the regression to the original variability of y. The square root of this ratio is called the **population correlation coefficient**.

Without going into the derivation, we shall simply state that the population correlation coefficient can be estimated by the quantity

$$\textit{Correlation Coefficient:} \quad r = \frac{n\Sigma xy - (\Sigma x)\,(\Sigma y)}{\sqrt{n\Sigma x^2 - (\Sigma x)^2}\,\sqrt{n\Sigma y^2 - (\Sigma y)^2}}$$

called the **sample correlation coefficient**. The sample correlation coefficient can take on any value from −1 to 1. The value of this correlation coefficient reveals a great deal about the least-squares line fitted to the data. If r takes on a *positive value* close to 1, the data points (x_i, y_i) are nearly collinear, and the slope of the corresponding least-squares line is positive, as shown in Figure 9.5(a). Conversely, if r takes on a *negative value* close to −1, the data points also are nearly collinear, but the slope of the least-squares line is negative (Figure 9.5 [b]). If r takes on a value close to 0, there is little or no linear relationship (Figure 9.5[c]).

> *Example.* The following data give the hardness (Rockwell 30-T) of six specimens of cold-reduced steel having different annealing temperatures (°F):
>
> | Hardness | 81.7 | 77.6 | 53.8 | 62.4 | 85.3 | 71.2 |
> |-------------|------|------|------|------|------|------|
> | Temperature | 950 | 1090 | 1370 | 1180 | 900 | 1100 |
>
> a. Find the correlation coefficient between x = annealing temperature and y = hardness.
>
> b. What percentage of the variation in hardness values is explained by the linear relationship with annealing temperature?
>
> *Solution.* a. The calculations for a correlation coefficient are tedious. Although they are illustrated here for completeness, you probably will want to use a computer method, such as the one illustrated on page 239. First, the following quantities are obtained:

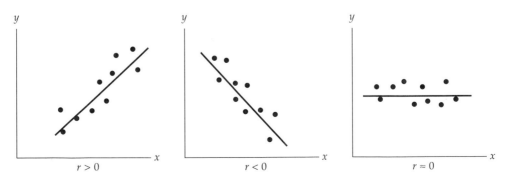

FIGURE 9.5 Sign of the Correlation Coefficient

$$n = 6, \quad \Sigma x = 6{,}590, \quad \Sigma y = 432.0$$
$$\Sigma x^2 = 7{,}379{,}900, \quad \Sigma y^2 = 31{,}830.38, \quad \Sigma xy = 464{,}627$$

Then, the correlation coefficient is calculated as follows:

$$r = \frac{6\,(464{,}627) - (6{,}590)\,(432.0)}{\sqrt{6\,(7{,}379{,}900) - (6{,}590)^2}\ \sqrt{6\,(31{,}830.38) - (432.0)^2}} = -0.97$$

b. The linear relationship with annealing temperature explains $100[(-0.97)^2] = 94\%$ of the variation in hardness values.

The correlation coefficient in this example has a value close to -1, indicating a strong linear relationship such that hardness *decreases* as annealing temperature increases. In this example, it was assumed that x, the annealing temperature, was known without error, while the hardness, y, was a random variable. A least-squares regression line can be calculated in this case, and a test can be made of the significance of its slope if the residuals are approximately normally distributed.

The coefficient of correlation also has meaning when both x and y are random variables. In this case, r^2 also estimates the proportion of the variation of y that is explained by its linear relationship with x. Although a regression line also can be fitted in such cases, the error made when predicting y from a given value of x also must take into account errors in x and this error cannot be found by the methods of the preceding section.

Significance Tests and Confidence Intervals for *r*

Tests for the significance of a coefficient of correlation are made to determine whether the population correlation coefficient differs from zero at a stated level of significance. The purpose of such a test, of course, is to determine whether or not there is a *linear* relationship between x and y. We have already seen how such a test can be made by testing the null hypothesis $\beta = 0$ against the alternative $\beta \neq 0$, where β is the slope of a linear regression line (page 230). That test, however, required the assumption that x was the value of a *fixed* variable. A correlation coefficient is defined both for this case and for the case where both x and y are values of random variables. In this latter case, a test of the null hypothesis $\rho = 0$ against the alternative $\rho \neq 0$ is required to test for a linear relationship between x and y.

The test can be greatly facilitated by using the following **Fisher Z-transformation**:

Fisher Z-transformation: $\quad Z = \dfrac{1}{2} \cdot \ln \dfrac{1+r}{1-r}$

When there is no correlation between x and y, and when both x and y are values of random variables having the normal distribution, it can be shown that the random variable with values given by Z has approximately the normal distribution with the mean 0 and the standard deviation $\dfrac{1}{\sqrt{n-3}}$. Thus, a test of whether a correlation coefficient is significantly different from zero at the level of significance α, can be based on the statistic

$$z = \frac{Z - \mu_Z}{\sigma_Z} = \sqrt{n-3} \cdot Z$$

which, under the stated assumptions, is a value of a standard normal random variable. The correlation coefficient is said to be significant (significantly different from zero) if z is less than $-z_{\alpha/2}$ or greater than $z_{\alpha/2}$. The calculation of Z is facilitated by Table 8 (Appendix A) which gives values of Z for values of r from 0.00 to 0.99, in increments of 0.01. For negative values of r, the sign of Z is changed from plus to minus, and vice versa.

Example. The following data give the carbon content and the permeability of 10 sinter mixtures:

Carbon content	5.5	3.0	4.5	5.0	4.8	4.2	4.7	5.1	4.4	3.6
Permeability	16	31	21	19	16	23	20	11	22	20

a. Calculate the coefficient of correlation for these data.

b. Test, at the .05 level of significance, whether the correlation coefficient is significant.

Solution. a. The correlation coefficient is calculated with the use of MINITAB (see page 239) to be $r = -0.82$.

b. Table 8 yields $Z = 1.157$ with $r = 0.82$, but, since r is negative, a minus sign is put in front of Z. Thus, $z = \sqrt{10-3} \cdot Z = (2.65)(-1.157) = -3.07$. Comparing -3.07 with -1.96, the value found in Table 3 for a left-hand tail of area .025, we conclude that the calculated correlation coefficient is significant at the .05 level of significance; that is, there is a linear relationship between the carbon content and the permeability of the sinter. Since r is negative, either of these variables decreases in value as the other increases.

Confidence limits can be found for a correlation coefficient by first finding confidence limits for μ_Z, the mean of the distribution of the random variable having the value Z, and then using Table 8 to convert Z to r. Since Z is a value of a random variable having the mean 0 and the standard deviation $\dfrac{1}{\sqrt{n-3}}$, confidence limits for the mean of this random variable are given by

---

Confidence Limits for μ_Z: $Z \pm \dfrac{z_{\alpha/2}}{\sqrt{n-3}}$

Example: Find 90% confidence limits for the population correlation co-efficient of the two sinter variables in the preceding example.

Solution: It was found that $r = -0.82$, corresponding to $Z = -1.157$. Thus, 90% confidence limits for μ_Z are $-1.157 \pm \dfrac{1.645}{\sqrt{10 - 3}}$, or -1.779 and -0.535. The values of r closest to $Z = 1.779$ and $Z = 0.535$ in Table 8 are .94 and .49, respectively. Changing to minus signs, the 90% confidence limits for the correlation coefficient are -0.94 and -0.49.

Note the width of the 90% confidence limits for ρ in the preceding example. Usually, it requires samples of a size much larger than 10 to estimate a population correlation coefficient with a reasonable degree of precision.

Computer Applications

To calculate the correlation coefficient in the example on page 238 using *MINITAB* software, first

```
SET C1
5.5  3.0  4.5  5.0  4.8  4.2  4.7  5.1  4.4  3.6
END

SET C2
16   31   21   19   16   23   20   11   22   20
END
```

Then, give the command CORR C1 C2. The correlation coefficient $-.823$ will appear on the screen.

EXERCISES

Use of a computer is recommended for some of the following exercises.

9.15 Calculate the correlation coefficient for the data on the elongation of steel rods given in the example on page 228.

9.16 Calculate the correlation coefficient for the data of Exercise 9.2 on page 233.

9.17 Calculate the correlation coefficient for the data of Exercise 9.1 on page 233.

9.18 Calculate the correlation coefficient for the data of Exercise 9.4 on page 234.

9.19 Calculate the correlation coefficient for the data of Exercise 9.3 on page 233.

9.20 Test for the significance of the correlation coefficient found in Exercise 9.16. Use the .05 level of significance.

9.21 Test for the significance of the correlation coefficient found in Exercise 9.17. Use the .01 level of significance.

9.22 Test for the significance of the correlation coefficient found in Exercise 9.18. Use the .10 level of significance.

9.23 Test for the significance of the correlation coefficient found in Exercise 9.19. Use the .05 level of significance.

9.24 Find 99% confidence limits for the true correlation whose estimated correlation coefficient is calculated in Exercise 9.16.

9.25 Find 95% confidence limits for the true correlation whose estimated correlation coefficient is calculated in Exercise 9.17.

9.26 Find 90% confidence limits for the true correlation whose estimated correlation coefficient is calculated in Exercise 9.18.

9.27 Find 95% confidence limits for the true correlation whose estimated correlation coefficient is calculated in Exercise 9.19.

9.3 Multiple Regression

Multiple Linear Regression

In simple linear regression, we find the line of best fit involving two variables, a random response variable having values given by y and a single fixed (non-random) independent variable, x. In **multiple linear regression**, it continues to be assumed that y is a value of a random variable, but now there are k fixed independent variables, $x_1, x_2, \ldots, x_k$. The purpose of multiple linear-regression analysis is to apply the method of least squares to find a hyperplane of best fit to a set of data points $(x_{1i}, x_{2i}, \ldots, x_{ki}, y_i)$ so that y can be predicted for a given set of values of $x_1, x_2, \ldots, x_k$ with minimum error. The fitted surface has the equation

$$\hat{y} = b_0 + b_1 x_1 + b_2 x_2 + \ldots + b_k x_k$$

where $\hat{y}$ is the predicted value of y when the independent variables take on the values $x_1, x_2, \ldots, x_k$.

Analogous to simple linear regression, the ith residual is the error, $y_i - \hat{y}_i$, made when $\hat{y}_i$ is used to predict the value of y corresponding to a given set of values $x_{1i}, x_{2i}, \ldots, x_{ki}$. Thus, for example, suppose the fitted surface has the equation

$$\hat{y} = 5 + 4x_1 - 3x_2$$

and, for the ith data point, $y_i = 6.1$, $x_{1i} = 0.5$, and $x_{2i} = 0.2$. Then,

$$\hat{y}_i = 5 + 4(0.5) - 3(0.2) = 6.4$$

and the *i*th residual is $y_i - \hat{y}_i = 6.1 - 6.4 = -0.3$. A graphical interpretation of the residuals in multiple regression analysis is illustrated for $k = 2$ in Figure 9.6.

FIGURE 9.6 Regression Plane with Residuals

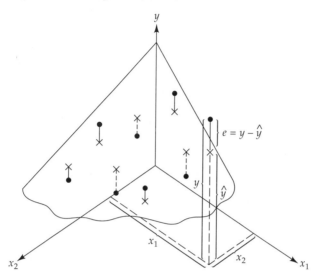

The coefficients $b_0, b_1, b_2, \ldots, b_k$ in the fitted equation can be found by the method of least squares using methods similar to those for simple linear regression. For a multiple linear-regression analysis involving k independent variables, there will be $k + 1$ normal equations. Although these equations are linear in the $k + 1$ coefficients $b_0, b_1, b_2, \ldots, b_k$, the solution of these simultaneous linear equations can pose a formidable computing task, especially when k is large.

Before the advent of modern computers and statistical software written to perform this task, large multiple-regression analyses were rarely attempted. For this reason, the mechanics of computing the coefficients of the least-squares equation and other statistical quantities associated with them will not be emphasized here. Instead, *MINITAB* software will be used to illustrate the kind of computer output produced by many statistical computer software packages. Emphasis will be given to the interpretation of the results of multiple-regression computations, as shown in the following example.

In wave soldering of circuit boards, an entire circuit board is run through the wave-soldering machine, and all solder joints are made. Suppose 5 major variables involved in the machine setup are measured for each run. A total of 25 separate runs of 5 boards each are made. (Each board contains 460 solder joints.) The soldered boards are subjected to visual and electrical inspection, and the number of defective solder joints per 100 joints inspected is recorded, with the following results:

Run No.	Conveyor Angle x_1	Solder Temperature x_2	Flux Concentration x_3	Conveyor Speed x_4	Preheat Temperature x_5	Faults per 100 Solder Joints y
1	6.2	241	0.872	0.74	245	0.201
2	5.6	250	0.860	0.77	229	0.053
3	6.5	258	0.853	0.64	266	0.239
4	6.4	239	0.891	0.68	251	0.242
5	5.7	260	0.888	0.81	262	0.075
6	5.8	254	0.876	0.75	230	0.132
7	5.5	250	0.869	0.71	228	0.053
8	6.1	241	0.860	0.76	234	0.119
9	6.1	256	0.854	0.62	269	0.172
10	6.3	260	0.872	0.64	240	0.171
11	6.6	249	0.877	0.69	250	0.369
12	5.7	255	0.868	0.73	246	0.100
13	5.8	258	0.854	0.80	261	0.105
14	6.1	260	0.879	0.77	270	0.196
15	5.8	262	0.888	0.70	267	0.126
16	6.3	256	0.870	0.81	246	0.216
17	6.4	254	0.862	0.76	233	0.286
18	6.8	247	0.855	0.65	250	0.306
19	6.7	238	0.876	0.69	249	0.403
20	6.3	264	0.884	0.71	265	0.162
21	6.4	260	0.891	0.79	252	0.214
22	5.7	259	0.881	0.80	245	0.287
23	5.8	244	0.863	0.76	238	0.092
24	5.4	259	0.875	0.68	217	0.008
25	5.7	264	0.870	0.64	276	0.102

Using *MINITAB* software to perform a linear multiple-regression analysis, we set the values of x_1 in column C1, x_2 in C2, . . . , x_5 in C5, and y in C6, in the same order as the run numbers shown in the data table. Then, the command REGRESS C6 ON 5 PREDICTORS C1 – C5 is given. The following is a portion of the resulting output:

① THE REGRESSION EQUATION IS
C6 = –1.79 + 0.214 C1 – 0.00096 C2 + 0.90 C3 + 0.122 C4 + 0.000169 C5

② COLUMN	COEFFICIENT	ST. DEV. OF COEF.	TRATIO = COEF/S.D.
	–1.7885	0.9655	–1.85
C1	0.21357	0.03630	5.88
C2	–0.000959	0.001873	–0.51
C3	0.898	1.047	0.86
C4	0.1216	0.2167	0.56
C5	0.0001695	0.0009457	0.18

③ S = 0.05806

④ R-SQUARED = 73.6 PERCENT

⑤

ROW	Y C1	C6	PRED. Y VALUE	ST. DEV. PRED.Y	RESIDUAL	ST.RES.
22	6.70	0.2870	0.1104	0.0220	0.1766	3.29R

R DENOTES AN OBS. WITH A LARGE ST. RES.

We have assigned a number to each element of this computer output. The following is an explanation of each such element:

① The regression equation is the linear equation fitted to the data by the method of least squares. This equation gives the values of the six coefficients $b_0, b_1, b_2, \ldots, b_6$.

② This table gives the value of each coefficient in the regression analysis and its standard deviation. Also, it gives the value of t necessary to perform a t test for the significance of each regression coefficient, on the assumption that the residuals are approximately normally distributed. Each computed value of t should be compared to the value of $t_{\alpha/2}$ given in Table 4 with $n - k - 1$ degrees of freedom, where n is the total number of observations, and k is the number of independent variables. In this example, $n = 25$ and $k = 5$; thus, t has 19 degrees of freedom, and, for tests at the .05 level of significance, the appropriate value of $t_{.025}$ from Table 4 is 2.093. Since only the t value for C1 (the coefficient of x_1, conveyor angle) is less than -2.093 or greater than 2.093, this variable is the only one in this analysis having a coefficient significantly different from zero at the .05 level of significance.

③ S is the standard deviation of the residuals. This statistic is analogous to s_e, defined on page 230.

④ R-squared is the square of the **multiple correlation coefficient**. Its square root, R, is analogous to the correlation coefficient defined in Section 9.2. R^2 measures the strength of the multiple linear relationship of the x's with the dependent variable, y, by comparing the variance of the y values with the variance of the residuals. Expressed as a percentage, $100\,R^2$ measures the percentage of the reduction in the variance of the y values that is attributable to the linear relationship of y with the five predictor variables (x_i) included in the analysis. In this example, the linear relationship reduced the original variance of the dependent variable (faults per 100 solder joints) by 73.6 percent. Had R^2 had the value 0, the variance of the residuals would have been equal to the variance of the y values, and the fitted equation would have contributed no improvement to the ability to predict a value of y. Had R^2

had the value 1 (100%), each data point would have fit the equation perfectly; that is, each residual would have had the value 0.

⑤ Any observation having an unusually large residual (a possible outlier) is identified by this computer analysis. In this case, observation 22, having the y value 0.2870, was predicted by the regression equation to have the value 0.1104. Thus, the residual corresponding to this data point is $0.2870 - 0.1104 = 0.1766$. The standardized value of this residual is $0.1766/0.05806 = 3.04$, or 3.29 with s corrected for degrees of freedom. (This corresponds to a right-hand tail of area only .0012 under the standard normal distribution.)

Pitfalls in the Use of Multiple Regression

It is tempting to conclude that the coefficients in a multiple-regression analysis represent the "effects" of the corresponding predictor variables on the dependent variable. For example, it appears that the coefficient of x_1 in the preceding analysis, having the value 0.214, is the estimated effect on y of increasing x_1 by 1 unit. But it probably is not true that y, the number of faults per 100 solder joints, will increase by 0.214 when x_1, the conveyor angle, is increased by 1 unit. There are several reasons for making this statement.

Any estimate of a coefficient in a regression analysis is subject to random error. A confidence interval can be found for such a coefficient when it can be assumed that the residuals are approximately normally distributed. Thus, this error is relatively easily quantified, and it often plays a small role relative to other sources of error.

A much more serious source of error in interpreting the coefficients of a multiple-regression equation arises from correlations among the independent variables in the multiple-regression equation. When at least some of the independent variables are highly correlated with each other, it is not possible to separate their effects on the dependent variable. Thus, it is said that the effects of the independent variables are **confounded** with each other. To investigate the degree of correlation among the independent variables, the following **correlation matrix** of pairwise correlation coefficients has been computed for the wave-solder data by giving the *MINITAB* command CORRELATE C1 - C5:

	C1	C2	C3	C4
C2	-.328			
C3	-.039	.174		
C4	-.281	.030	.215	
C5	.251	.402	.117	-.207

(Only a portion of the full matrix is shown here, since the matrix is symmetric; for example, the correlation of C1 with C2 equals the correlation of C2 with C1, and the correlation of any column with itself equals 1.) It can be seen that several of the data columns involving independent variables probably are correlated.

The effect of confounding can be observed by performing a multiple linear-regression analysis of y on $x_2, x_3, x_4,$ and x_5 only; that is, by omitting x_1 from the regression equation. The resulting multiple-regression equation is

$$\hat{y} = 0.23 - 0.00617x_2 + 1.18x_3 - 0.150x_4 + 0.00238x_5$$

By comparison, the multiple-regression equation previously obtained when all five independent variables were used in the regression was

$$\hat{y} = -1.79 + 0.214x_1 - 0.0096x_2 + 0.90x_3 + 0.122x_4 + 0.000169x_5$$

It is readily seen that the coefficients of $x_2, x_3, x_4,$ and x_5 have changed by more than trivial amounts when the independent variable x_1 has been omitted from the analysis. For example, the coefficient of x_2, which was -0.0096 when x_1 was included in the regression equation, becomes -0.00617—an increase of 56%—when x_1 is not included, and the coefficient of x_4 changes sign. Evidently, the estimates of the coefficients in these multiple-regression analyses are confounded with each other.

As further evidence of confounding, it is observed that the value of t for the coefficient of x_2 in the regression equation that omitted x_1 is -2.28, which is less than $-t_{.025} = -2.086$ with $25 - 4 - 1 = 20$ degrees of freedom. (The remaining coefficients are associated with values of t less than 2.0.) Thus, when x_1 is omitted from the regression equation, x_2 appears to have a significant effect on the value of y, at the .05 level of significance. No such effect was detected in the analysis that included all five independent variables.

Introduction of Non-Linear Terms

Non-linear terms, such as $x^2, x^3, x_1x_2,$ and so forth, can be introduced into a multiple-regression equation to fit curved surfaces to data. When non-linear terms are added, however, there is a risk of introducing correlations between independent variables in the equation, such as x and x^2, thus increasing any problems of confounding that already may exist. This difficulty may be avoided, or at least minimized, by standardizing the variables used in the regression analysis. (Standardization, in this case, consists of subtracting the mean of each variable from each value of that variable, and dividing the result by its standard deviation.)

The use of large multiple-regression equations, containing many variables in both linear and non-linear forms, can produce an equation with better predictive power than one containing only linear terms. However, this method often creates highly correlated independent variables, even when standardization is employed, thereby making the problem of confounding even worse.

9.4 Analysis of Residuals

The residuals of any multiple-regression analysis should be examined carefully. If tests of significance are to be made, a normal-scores plot of the residuals is needed to check the assumption of normality required for the validity of the t tests. Even if such tests are not planned, an analysis of the residuals is useful in checking if the data are adequately described by the form of the fitted equation, or by the variables included in the equation.

The following three kinds of graphs are helpful in the analysis of residuals:

1. A normal-scores plot of the residuals
2. A plot of the residuals against the values of $\hat{y}$
3. A plot of the residuals against the run numbers

The normal-scores plot checks the assumption that the residuals are approximately normally distributed. While the t tests associated with regression analysis are not highly sensitive to departures from normality, gross departures will invalidate the significance tests associated with the regression. (However, the equation remains useful for estimating values of the coefficients and for obtaining $\hat{y}$, a predicted value of y.)

A plot of the residuals against the predicted values of y can reveal errors in the assumptions leading to the form of the fitted equation. If the chosen equation adequately describes the data, such a plot will show a "random" pattern, without trend or obvious curvilinearity. On the other hand, if a linear equation is fitted to data that are highly non-linear, the residuals will show a curvilinear trend. When the data depart seriously enough from the assumed relationship, excessive errors in prediction will result, and estimates of the coefficients of the independent variables will be relatively meaningless.

A plot of the residuals against integers reflecting the order of taking the observations (or "run number," or the time each observation was taken) also should show a random pattern, without trends. A trend in such a plot can be caused by the presence of one or more variables, not included in the regression analysis, whose values have a measurable influence on the value of y, and which varied over the time period of the experiment. Ambient variables, such as temperature and humidity often exert such effects. A time trend in the residuals may suggest that these (and possibly other) variables should be controlled or their values measured and included in the regression equation when performing further research.

To illustrate these methods for checking residuals, the residuals were computed for the data of the wave-solder regression analysis given on page 242. Standardized residuals can be found directly with *MINITAB* software by giving the command REGRESS C6 ON 5 PREDICTORS C1-C5, PUT RESIDUALS IN

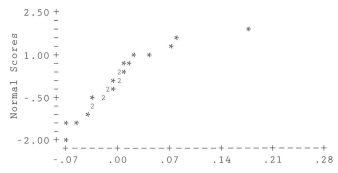

FIGURE 9.7 Normal-Scores Plot of Wave-Solder Regression Residuals

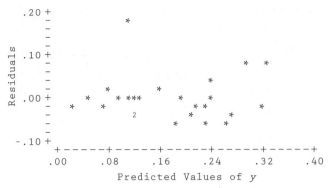

FIGURE 9.8   Plot of Residuals against $\hat{y}$

C7. To obtain the "raw" (non-standardized) residuals, $\hat{y}$ is calculated by giving the command LET C8=-1.7885+.21357*C1-.000959*C2+.898*C3+ 1216*C4 + .0001695*C5, that is, by calculating $\hat{y}$ for each observation, placing the results in column *C*8. Then, giving the command LET C9=C6–C8, places the residuals in column *C*9. Either the standardized or the raw residuals can be used for checking the residuals.

The normal-scores plot of the raw residuals is shown in Figure 9.7. Except for the previously noted outlier, the graph shows good agreement with the assumption that the residuals are normally distributed. (If significance tests are to be performed for the coefficients of the regression, it is recommended that the out-lying observation, run no. 22, be discarded and that the regression be rerun.)

A plot of the raw residuals against $\hat{y}$ is shown in Figure 9.8. Ignoring the outlier, this graph shows a random pattern with no obvious trends or curvilinear-ity. Thus, it appears that the linear multiple-regression equation was adequate to describe the relationship between the dependent variable and the five independent variables over the range of observations.

These residuals are plotted against the run numbers in Figure 9.9. This graph, likewise, shows a random pattern, with no linear or curvilinear trends. It appears that no extraneous variable has affected the value of *y* during the experiment.

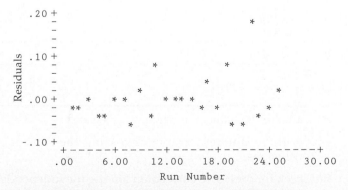

FIGURE 9.9   Plot of Residuals against Run Numbers

EXERCISES

Use of a computer is recommended for some of the following exercises:

9.28 Which of the following are legitimate uses of multiple-regression analysis?

  a. To predict a value of the dependent variable within the range of the data.

  b. To estimate the effect of a given independent variable on a response.

  c. To determine if variation of a given independent variable causes a change in the value of the response variable.

  d. To explore several variables for the purpose of postulating possible relationships for further exploration.

9.29 Which of the following statements are false?

  a. Multiple-regression analysis can be used for prediction.

  b. It is not necessary to check for trends in the residuals if the regression equation contains only linear terms.

  c. The coefficients in a regression equation can be trusted to give the effects of the corresponding independent variables on the dependent variable if the correlation matrix shown no high correlations.

  d. Multiple-regression analysis is a statistical method that has serious limitations.

9.30 a. Fit a linear surface of the form $y = b_0 + b_1x_1 + b_2x_2$ to the following data:

y	27.0	20.4	15.1	16.4	7.7	3.5	1.9	−3.7	−9.4	−12.1
x_1	1.4	2.1	3.9	4.1	5.6	6.0	7.4	8.1	9.3	10.1
x_2	−2	−1	−1	0	2	3	3	5	6	7

  b. Estimate the value of y when $x_1 = 6.5$ and $x_2 = -0.4$.

  c. Plot three different graphs involving the residuals and state your conclusions.

  d. Check for excessive correlation among the independent variables.

9.31 a. Fit a linear surface of the form $y = b_0 + b_1x_1 + b_2x_2$ to the following data:

y	40.1	−34.7	13.2	−27.5	−8.8	−31.5	18.4	8.0	−4.4	44.6
x_1	2	6	3	7	5	9	4	8	10	1
x_2	11	−2	5	0	1	2	6	8	7	9

  b. Estimate the value of y when $x_1 = 4.6$ and $x_2 = 3.8$.

  c. Plot three different graphs involving the residuals and state your conclusions.

  d. Check for excessive correlation among the independent variables.

9.32 a. Fit a linear surface to the following data:

y	x_1	x_2
118	41	−6
38	76	3
156	19	6
45	67	−3
31	62	−1
17	99	−3
109	27	−5
349	43	12
195	25	−8
72	24	2
94	48	5
118	3	4

   b. How good a fit is obtained?

   c. Plot the residuals against $\hat{y}$ and determine whether the pattern is random.

   d. Check for excessive correlation among the independent variables.

9.33 The following data represent measurements of monthly water usage at a plant (gallons), monthly production (tons), mean monthly temperature (°F), and the monthly number of days of plant operation over a period of 20 months:

Water Usage y	Production x_1	Mean Temp. x_2	Days Oper. x_3
2,609	108	70	20
2,228	97	68	19
2,559	113	66	19
2,723	144	58	19
3,088	109	82	21
2,522	64	44	19
2,012	91	61	20
2,254	82	44	18
2,436	126	59	21
2,460	111	62	21
2,147	85	54	18
2,378	101	69	21
2,031	84	58	20
1,717	70	64	21
2,117	107	51	22
1,902	97	36	17
2,251	98	56	22
2,357	96	85	19
1,721	132	49	23
1,980	84	64	19

   a. Fit a linear surface to these data.

   b. How good a fit is obtained?

c. Plot the residuals against $\hat{y}$ and determine whether the pattern is random.

d. Check for excessive correlation among the independent variables.

9.34 Using the data of Exercise 9.32:

a. Create a new variable, x_2^2.

b. Fit a surface of the form

$$y = b_0 + b_1x_1 + b_2x_2 + b_3x_2^2$$

c. Find the correlation matrix of the three independent variables. Is there excessive correlation among any of them?

d. Standardize each of the independent variables, x_1 and x_2, and create a new variable that is the square of the *standardized* value of x_2.

e. Fit a surface of the same form as in part b to the standardized variables. Compare the goodness of fit of this surface to that of the linear surface fitted in Exercise 9.32.

f. Plot the residuals of this regression analysis against the values of $\hat{y}$ and compare this plot to the one obtained in Exercise 9.32.

9.35 Using the data of Exercise 9.33:

a. Create a new variable, x_1x_2.

b. Fit a surface of the form

$$y = b_0 + b_1x_1 + b_2x_2 + b_3x_3 + b_4x_1x_2$$

c. Find the correlation matrix of the four independent variables. Is there excessive correlation among any of them? Why would this result be expected?

d. Standardize each of the three independent variables x_1, x_2, and x_3, and create a new variable that is the product of the *standardized* values of x_1 and x_2.

e. Fit a curved surface of the same form to the standardized variables. Compare the goodness of fit of this surface to that of the linear surface fitted in Exercise 9.33.

f. Find the correlation matrix of the four standardized independent variables and compare with the results of part c.

9.36 When 50 measurements were made of the value of y, its standard deviation was 36.5. After a multiple regression of y on several independent variables, the standard deviation of the residuals was found to be 11.8. What is the value of R^2 for this analysis?

9.5 Application to Quality

Data involving many related variables often present themselves to the manager or the engineer. Issues involving production costs, sales, pricing, and so forth, often involve multivariate relationships. When there is a problem on a production line, the engineer is faced with the need to identify and correct the "cause" of the problem. Such problems often are the result of many variables acting in concert, variables involving production settings, specifications for purchased parts, testing equipment, and so forth. These variables often do not act independently of each other in determining the quality of the final product.

Data collection and data bases have become widely used in this age of communication, aided by computer technology. The temptation to "analyze" a set of data collected for another purpose, or for no reason at all (except, perhaps, that it's easy to read off the machines), becomes overwhelming when one is faced with a problem that has no easy solution in sight. And, how better to analyze the data quickly (and without much preliminary thought) than to perform a regression analysis?

Problems of selection of the best equation to fit to the data, high correlations among the independent variables, and the presence of other, non-measured variables lurking in the background are too easily put off. So, the large multiple regression is run, some apparent relationships are discovered, and maybe this will be the lucky time that the administrative decisions or the process adjustments suggested by these relationships will work. It's a bit like the lure of gambling—you win just often enough to want to do some more.

Notwithstanding this rather severe indictment of multiple regression, the method does have a variety of useful applications. In some cases, a physical law is known; that is, an equation relating some response to one or more "inputs" is accepted as a description of a phenomenon. Then, regression analysis is an excellent tool for estimating the unknown constant or constants in the equation. For example, in heat-transfer theory, the relation between the temperature in a plane wall and the distance from the hot surface is approximately parabolic. Data, consisting of temperatures measured at different points on the wall at the same time, can be used as input to a regression equation to measure the constants defining the parabola. From these, the temperature gradient at any point and the thermal conductivity can be calculated.

Sometimes, a large multiple-regression analysis is useful in "screening," as a first step in reducing the number of variables for further study. If an independent variable is not too highly correlated with another variable in the regression study, and if variation of the values of this variable over its practical range have little effect on the response variable, it is probably safe to exclude this variable from further consideration. If a given independent variable appears to exert a strong influence on the dependent variable in a multiple-regression analysis, it bears further investigation. However, it may turn out that this variable is highly correlated with another variable or variables that drive the response. Thus, while regression analysis is useful as an investigative tool, it can be risky to base *final* conclusions solely on the results of such an analysis.

To illustrate the use of multiple-regression analysis in screening for effective variables, consider a problem of manufacturing plated wire for special electronic applications. The process is a continuous one, involving bath chemistries, bath temperatures, and line speeds. It is nearly impossible to make electrical measurements on the wire until it has been completely processed. In all, 54 processing variables can be identified that could effect the electrical characteristics of the wire, and there is little theory on which to base estimates of their individual or joint effects on the wire. Serious problems were encountered in consistently producing plated wire to acceptable electrical specifications.

It requires a long and expensive run to manufacture a spool of plated wire with a given set of values of the 54 variables. Thus, formal experimentation, in which the values of these variables are deliberately varied, seems precluded. But, it is not difficult to measure the values that these variables take on as the wire proceeds through the process, as well as the final electrical results obtained on the spool. In an effort to reduce the number of variables for further consideration, and to avoid interference with normal production, a multiple-regression analysis involving 80 spools of wire was performed. A simple linear equation was fitted to the values of the 54 process variables and the electrical test result for each spool. It was found that most of the independent variables did not have coefficients significantly different from zero, and they were discarded from further analysis. Some of them did not seem to affect the dependent variable because their values remained relatively constant during the production of these spools. Discarding such variables seems reasonable as long as no fundamental process changes occur in the future that would cause them to vary more extensively.

The regression analysis "boiled down" the number of apparently effective variables to four. A more comprehensive "full quadratic" equation, containing a constant term, 4 linear terms, 4 square terms, and all 6 possible cross-product terms then was fitted to new data on 30 spools. Analysis of the resulting equation suggested some actions that could be taken to improve the process. The resulting improvement in the process doubled the process yield and reduced the electrical variability of the acceptable spools. Since the initial yield was very low, much more work was yet to be done.

The regression analyses gave the process engineers some new ideas about how their process might behave. A few additional variables were suggested, including some that had previously been discarded. The improved process yield provided not only the incentive but also the financial capability to conduct a series of designed experiments (Chapter 10) that eventually led to a better understanding of the process and further improvements in its yield.

REVIEW EXERCISES

These review exercises can be used for informal review of the chapter, or as two practice examinations, designed to be taken over a time period of about one hour. The odd-numbered exercises comprise one examination; the even-numbered exercises comprise the other. The review exercises for this chapter have been chosen to minimize computation.

9.37 In a simple linear-regression analysis, made on 25 observations (x_i, y_i), the following sums were calculated:

$$\Sigma x = 49, \quad \Sigma y = 18.1, \quad \Sigma x^2 = 98, \quad \Sigma y^2 = 18.25, \quad \Sigma xy = 37.5$$

Find the slope of the least-squares regression line.

9.38 In a simple linear-regression analysis, made on 20 observations (x_i, y_i), the following sums were calculated:

$$\Sigma x = 16, \quad \Sigma y = 28, \quad \Sigma x^2 = 21, \quad \Sigma y^2 = 46 \quad \Sigma xy = 19$$

Find the slope of the least-squares regression line.

9.39 Given $s_e = 0.36$, find a 95% confidence interval for the slope of the line fitted in Exercise 9.37.

9.40 Given $s_e = 0.55$, test, at the .10 level of significance, whether the slope of line found in Exercise 9.38 is significantly different from zero.

9.41 Find the coefficient of correlation between x and y for the data of Exercise 9.37.

9.42 Find the coefficient of correlation between x and y for the data of Exercise 9.38.

9.43 Test, at the .05 level of significance, whether the correlation coefficient found in Exercise 9.41 is significantly different from zero.

9.44 Find a 95% confidence interval for the true correlation corresponding to the correlation coefficient found in Exercise 9.42.

9.45 The following multiple-regression equation was fitted to 25 data points:

$$\hat{y} = 125 + 42.6x_1 + 37.5x_2 - 1.5x_1^2$$

a. What is the estimated value of the dependent variable when each independent variable takes on the value 1?

b. The coefficient of the non-linear term x_1^2 is much smaller than the other coefficients. Can it be concluded that this term is "not important," and that a regression using only linear terms would have been adequate? Give reasons for your answer.

9.46 The following multiple-regression equation was fitted to 30 data points:

$$\hat{y} = 12.0 + 11.4x_1 + 3.75x_2 - 97.5x_3$$

a. What is the estimated value of the dependent variable when each independent variable takes on the value 0?

b. Since it has by far the largest coefficient, is it true that x_3 is the "most important" variable in explaining variations in y? Give reasons for your answer.

9.47 The residuals of the regression analysis described in Exercise 9.45 were plotted against the predicted values of the dependent variable. The graph showed a strong linear trend. What does this signify?

9.48 A normal-scores plot was made of the residuals of the regression analysis described in Exercise 9.46. The graph showed a strong linear trend. What does this signify?

9.49 State three potential pitfalls in the interpretation of the results of a multiple-regression analysis.

9.50 State two applications of multiple-regression analysis that are essentially without pitfalls.

DESIGN OF
EXPERIMENTS

Where have we been? The methods so far introduced for handling multivariate data seem powerful, especially when combined with modern computer technology. Yet, we have seen how easily they can lead to incorrect conclusions. We need multivariate methods that eliminate as many of the pitfalls of multiple-regression analysis as possible.

Where are we going? Problems of confounding and causality can be minimized, if not eliminated, by designing the data-collection process in advance. Some of the most widely used experimental designs for studying causal relationships are introduced in this chapter, along with methods for analysis of the data resulting from such designed experiments.

Section 10.1, *The Role of Experimental Design in Quality*, contrasts experimental design with haphazard collection of data and illustrates the steps to be taken in the design of an experimental program.

Section 10.2, *One-Way Designs*, introduces the analysis of variance for experimental designs involving several means.

Section 10.3, *Randomized-Block Designs*, gives designs for the elimination of variability caused by a second variable.

Section 10.4, *Factorial Designs*, shows how to analyze designs involving several variables and introduces the notion of interaction.

Section 10.5, *Fractional Factorial Designs*, discusses the problems involved in running large factorial designs and describes some methods for dealing with these problems.

Section 10.6, *Application to Quality*, presents a case example of a quality-improvement problem and shows how experimental design is used in its solution.

10.1 The Role of Experimental Design in Quality

The collection and analysis of data requires time, effort, and financial resources. Thus, data should be collected only when there is a specific purpose to be served and when it is known, *in advance*, what methods of analysis will be used to draw useful and reliable conclusions. It is essential that a clear purpose be stated when planning data collection. As discussed in Chapter 9, the analysis of data collected without purpose can result in confusion, creating a situation in which several inconsistent conclusions can be drawn. Lost time, misallocation of resources, and poor quality can result.

A planned program of allocating data-collection resources for the purpose of obtaining answers to specific questions is called an **experimental design**. To be complete, an experimental design should include not only the purpose of the experiment, a clear statement of what data are to be collected, and the method of data collection but also a plan of analysis of the resulting data. The data-analysis plan must be capable of obtaining answers that satisfy the purpose of the experiment as well as making probability statements about the validity of the answers. An experimental design without an analysis method that satisfies these criteria is of little practical value.

The steps required for the design of an experiment can be summarized as follows:

DESIGN OF AN EXPERIMENT

1. Clearly and concisely state the purpose of the experiment.
2. State the kind of data to be taken and the method of data collection to be used.
3. Give the sample size, including how the observations are to be allocated with respect to factors that might influence the data.
4. Use randomization whenever possible.
5. Decide on the method of analysis to be used.

These steps are illustrated by the following suggested experimental design for an experiment involving wire-bond yield strengths.

1. *Purpose.* To determine whether the four bonding machines on the production floor are giving similar results.
2. *Kind of data.* Obtain solder-bond strengths in grams, using the yield-strength testing machine.
3. *Allocation of observations.* Take ten observations from each of the four bonding stations.
4. *Randomization.* Select the ten observations from each of the four stations at random time intervals over a period of one week.
5. *Method of analysis.* Test for differences in bonding machines using a one-way analysis of variance (Section 10.2).

In this chapter, several experimental designs are given that are capable of addressing a broad spectrum of questions which frequently occur in quality-improvement programs. Methods also are given for analyzing the data resulting from each design for the purpose of drawing reliable conclusions. The experimental designs in this chapter are capable of eliminating the problem of confounding that plagues the regression method of studying relationships among several variables.

10.2   One-Way Designs

In the proposed wire-bond experiment whose purpose is given on page 256, it is implied that mean yield strengths might vary, depending upon which bonding station is employed. There are many other situations in which one is interested simply in whether or not differences exist among several means. An experimental design for testing such differences is called a **one-way design**. The general problem of testing for differences among several means in a one-way design can be stated as follows. Suppose a random sample of size n is taken from each of k populations. Can it be said, at a given level of significance, that the k population means are different from each other?

Specifically, we wish to test the null hypothesis that several population means are equal against the alternative hypothesis that at least two of them are unequal. To find an appropriate test statistic for this, it is necessary to make several assumptions about the samples as well as the populations from which they were taken. It will be assumed that *the samples are independent*, that *each sample comes from a normally distributed population*, and that *the variances of the k populations are equal*. While these may seem to be very stringent assumptions, as in the case of the t tests described in Chapter 7, the tests that will be performed are relatively insensitive to minor departures from these assumptions. These assumptions are well enough met in sufficiently many applications so that the analyses to be given in this chapter can be undertaken.

A test statistic for testing whether there are differences among the k population means can be found by examining the variance of the data in two different ways. First, the assumed common variance of the k populations, σ^2, can be estimated from the data comprising any given sample, say sample i, by its sample variance, s_i^2. Then, a *pooled within-sample variance* can be calculated by taking the mean of these k variances. This pooled variance estimates σ^2 directly from all the data of the experiment.

Another estimate of σ^2 is afforded by the variance of the k sample means, *provided that the populations from which these samples were taken have the same mean*. On this assumption, the variance of the sample means will estimate $\dfrac{\sigma^2}{n}$. If this assumption is not correct, and the k population means are different from one another, the variance of the sample means (the *between-sample variance*) estimates $\dfrac{\sigma^2}{n}$ plus a component of variability associated with the differences among the means. Thus, n times the between-sample variance can be expected to be larger than σ^2 if the samples come from populations that do not have the same mean.

The foregoing argument suggests that a test of whether the population means are different from one another can be based on a comparison of the between-sample variance with the pooled within-sample variance. If the samples arose from populations having the same mean, then n times the between-sample variance can be expected not to differ significantly from the pooled within-sample variance. If the population means are different, however, the between-sample variance will be increased by the variation associated with this difference, and n times the between-sample variance can be expected to be larger than the pooled within-sample variance.

Thus, the difference among several sample means can be tested by comparing two variances as in performing an F test (page 177). Thus, if the pooled within-sample variance and n times the between-sample variance do not differ, the following variance ratio can be expected to have a value close to 1:

$$ F = \frac{s_B^2}{s_W^2} $$

In this F ratio, s_B^2 is n times the variance of the sample means and s_W^2 is the pooled within-sample variance. If the k means differ, s_B^2 can be expected to exceed s_W^2, and this ratio can be expected to have a value greater than 1. If the null hypothesis of equal means is true, the distribution of the random variable whose values are given by F is the F distribution, first described on page 149. The degrees of freedom for the numerator of F are $k - 1$, since the numerator is a constant times the variance of k sample means. The degrees of freedom for the denominator are $k(n - 1)$, since this variance is the mean of k sample variances, each obtained from n observations.

To summarize, a test of the significance of the difference among the means of k samples, each of size n, is performed by calculating two variances, n times the between-sample variance and the pooled within-sample variance of the k samples. Then, the first variance is divided by the second to form a ratio that has the F distribution when the null hypothesis of no difference among the population means is true. Finally, the value of F so obtained is compared with $F_{.05}$ or $F_{.01}$, with $k - 1$ and $k(n - 1)$ degrees of freedom, depending upon the desired level of significance.

Analysis of Variance

The work of computing the variances required to analyze an experimental design is simplified by means of an algorithm called the **analysis of variance**. With this method of analysis, first one calculates the numerator of a given variance, called a **sum of squares**. Then, each sum of squares is divided by the appropriate degrees of freedom to form estimates of the required variance. For tests involving several means, these computations involve the following sums of squares:

Correction term:

$$ C = (\text{sum of all } nk \text{ observations})^2 / nk $$

Total sum of squares:

$$ SST = \text{sum of the squares of all } nk \text{ observations } - C $$

Sum of squares for means:

$$ SSM = 1/n \text{ times the sum of the squares of the sample totals } - C $$

Sum of squares for error:

$$ SSE = SST - SSM $$

The results are summarized in an analysis-of-variance table, or **ANOVA**, as follows:

Analysis of Variance for One-Way Design

Source of Variability	Degrees of Freedom	Sum of Squares	Mean Square	F
Means Error	$k - 1$ $k(n - 1)$	SSM SSE	$MSM = SSM/(k - 1)$ $MSE = SSE/k(n - 1)$	MSM/MSE
Total	$nk - 1$	SST		

The total sum of squares is the numerator of the variance of all nk observations. The term **mean square** in an analysis of variance refers to a sum of squares divided by its corresponding degrees of freedom; mean squares are estimates of variances. In this *ANOVA*, the mean square for "error" estimates the pooled within-sample variance. The term *error* is used to reflect that this variance represents the background variability against which differences are compared. The mean square for "means" estimates the between-sample variance.

The last column, labeled *F*, gives the value of the test statistic used for testing the equality of the *k* population means. To perform the required significance test, one compares this value of *F* with the value of $F_{.05}$ or $F_{.01}$ with $k - 1$ and $n(k - 1)$ degrees of freedom found in Table 6(a) or Table 6(b), depending upon the desired level of significance. It can be said that "the *k* sample means are significantly different at the desired level of significance" if the value of *F* in the *ANOVA* exceeds the appropriate value obtained from Table 6.

Example. It was decided to perform the wire-bonding experiment described on page 256 by measuring the yield strength of ten wire bonds made by each of the four bonding machines, with the following results:

Bonder	Yield Strength (grams)									
A	10.5	16.7	8.9	9.2	7.4	9.6	9.4	9.8	12.8	10.7
B	9.3	12.3	11.5	14.0	10.1	12.5	10.3	8.1	7.2	8.7
C	11.3	11.8	9.7	12.8	9.4	12.6	12.4	11.0	13.6	10.0
D	8.3	9.0	10.6	4.0	6.7	7.0	5.4	8.0	8.3	11.0

Perform an analysis of variance to test whether the mean bonding strengths associated with the four bonders are significantly different, using the 0.05 level of significance.

Solution. First, the following totals are obtained:

Bonder Totals: (A) 105.0   (B) 104.0   (C) 114.6   (D) 78.3

Then, the correction term is calculated:

$$C = (105.0 + 104.0 + 114.6 + 78.3)^2/40 = 4{,}038.09$$

Finally, the following sums of squares are calculated:

$$SST = (10.5^2 + 16.7^2 + \ldots + 11.0^2) - C = 234.96$$

$$SSM = (105.0^2 + 104.0^2 + 114.6^2 + 78.3^2)/10 - C = 72.42$$

$$SSE = 234.96 - 72.42 = 162.54$$

The results are arranged in the following *ANOVA* table:

Source of Variability	Degrees of Freedom	Sum of Squares	Mean Square	F
Bonders	3	72.42	24.14	5.34
Error	36	162.54	4.52	
Total	39	234.96		

Interpolating in Table 6, with 3 and 36 degrees of freedom, we get $F_{.05} = 2.87$. Since $F = 5.34$ is greater than $F_{.05} = 2.87$, it can be concluded at the .05 level of significance that the four bonder means are significantly different.

We concluded, in the foregoing example, that the four means were "significantly different." Of course, we knew before performing the computations that the sample means were different. As pointed out on page 169, the purpose of the test of significance is to determine *if the observed differences in the sample means are greater than could be accounted for by chance alone*, thus giving credibility to the hypothesis that the population means are not all equal.

The necessary computations for an analysis of variance can be lengthy, even for this simple experimental design. These computations are greatly facilitated with the use of appropriate statistical computer software, which will be relied upon more heavily in this text as the computations for designs to be introduced become more burdensome.

Multiple Comparisons

The analysis of variance in the preceding example did not provide a method for determining whether the differences among the sample means arose because one bonder produced bonds of different strength from that of the other three, or whether two bonders were alike in this respect and the other two bonding machines were different, and so on. A method for answering questions of this kind—a method for making multiple comparisons among means—provides an answer to such questions.

One of the several multiple-comparisons tests available is the **Duncan multiple-range test**, applicable for the comparison of the means of k samples of *equal size*, under the same assumptions underlying the analysis of variance. The following steps are followed in the performance of this test:

1. Calculate the standard error of the means, using the formula:

$$s_{\bar{x}} = \sqrt{\frac{MSE}{n}}$$

where *MSE* is the mean square for error in the analysis of variance and n is the number of observations in each of the k samples.

2. Table 9 (Appendix A) gives values of r_p, for the levels of significance .05 (Table 9[a]) and .01 (Table 9[b]). The value of r_p depends on the degrees of freedom for error in the analysis of variance and on p, the number of adjacent means being compared ($2 \leq p \leq k$).

3. Calculate the **least significant range**, using the formula

$$R_p = r_p \cdot s_{\bar{x}}$$

4. Arrange the means by size, from smallest to largest.

5. Compare the difference of the last and the first mean to R_k. If this difference is greater than R_k, it can be concluded that the k sample means are significantly different at the level of significance used for obtaining r_p from Table 9. Then, similarly compare all adjacent sets of $k - 1$ means, now using R_{k-1} as the criterion for significant differences. Continue this process for sets of $k - 2$ adjacent means, and so forth, down to sets of 2 adjacent means. In making these comparisons, it is helpful to underline the adjacent means in a set whose means are not found to be significantly different. If a later comparison involves a subset of means already connected by an underline, no further comparisons need be made among the means in that subset.

As complex as the procedure for performing step 5 may seem, an organized method for performing this analysis requires much less effort than at first it would appear, as illustrated in the following example.

Example. Perform Duncan's multiple-range test to make multiple comparisons among the four bonder means of the experiment described in the example on page 259. Use the .05 level of significance.

Solution. *Step 1.* From the analysis of variance shown on page 260, $MSE = 4.52$; thus,

$$s_{\bar{x}} = \sqrt{\frac{4.52}{10}} = 0.672$$

Step 2. Interpolating in Table 9(a), for 36 degrees of freedom, find the following values of r_p:

p	2	3	4
r_p	2.87	3.02	3.11

Step 3. Multiply each value of r_p in this table by 0.672 to obtain:

p	2	3	4
R_p	1.93	2.03	2.09

Step 4. Arrange the four means by size, as follows:

Bonding Machine	D	B	A	C
Mean	7.83	10.40	10.50	11.46

Step 5. The difference between the largest and the smallest means is $11.46 - 7.83 = 3.63$. Since this value exceeds $R_4 = 2.09$, these means are significantly different at the .05 level of significance, and no underline is drawn to connect them. (This result was expected, as the analysis of variance showed a difference among the four means at the .05 level of significance.) Next, compare the last three means by calculating their difference, $11.46 - 10.40 = 1.06$, and comparing it to $R_3 = 2.03$. Since $1.06 < 2.03$, these means do not differ significantly, and they can be underlined as follows:

	D	B	A	C
Mean	7.83	10.40	10.50	11.46

Comparing the other set of three adjacent means, the difference $10.50 - 7.83 = 2.67$ is found to be greater than 2.03; thus, no additional underline is drawn to connect them. Now, the only means that need to be compared in sets of two are the first two, since the last three means already are connected by an underline. The difference $10.40 - 7.83 = 2.57$ exceeds $R_2 = 1.93$, and it can be concluded that the first mean (bonder D) is significantly different from the second mean (bonder B). Thus, no more underlines are drawn, and the final conclusion is reached, at the .05 level of significance, that bonder D is producing bonds with smaller pull strengths than the other three bonders.

The result obtained in the foregoing example was simple to interpret. The four means were clearly divided into two non-overlapping groups. Sometimes the result of a multiple-range test is not as clear-cut. To illustrate, suppose the following results had been obtained for a comparison of four means:

$$A \quad B \quad C \quad D$$

This result divides the four means into two overlapping groups. The means labeled A and D are not connected by the same underline; thus, it can be said that they are significantly different from each other. But neither the means in the set A, B, and C nor those in the set B, C, and D differ significantly from one another at the level of significance employed in the test. In general, care must be taken in the interpretation of the results of a multiple-range test to state only that means connected by a common underline are not significantly different, and that only means *not connected by a common underline* differ significantly from each other.

Computer Applications

An analysis of variance for determining significant differences among several sample means can be conducted with *MINITAB* software. To illustrate, suppose seven students from each of five different school districts are given an achievement test, and the following scores result:

District 1:	84	66	71	63	63	69	72
District 2:	97	87	56	82	98	78	85
District 3:	99	77	90	98	91	90	87
District 4:	61	63	59	62	70	64	75
District 5:	86	79	74	61	63	79	72

To test for the significance of the differences among the sample mean scores for the five school districts, first, the data are entered into five columns. All seven observations from District 1 are put into C1, the seven observations from District 2 into C2, and so forth. Then, the command

$$\text{AOVONEWAY C1,C2,C3,C4,C5}$$

is given. The following results appear on the screen:

ANALYSIS OF VARIANCE

SOURCE	DF	SS	MS	F
FACTOR	4	2988.7	747.2	8.93
ERROR	30	2510.7	83.7	
TOTAL	34	5499.5		

```
                                  INDIVIDUAL 95 PCT CI'S FOR MEAN
                                  BASED ON POOLED STDEV
LEVEL  N     MEAN    STDEV     --+---------+---------+---------+-------
C1     7    69.71     7.25        (-----*-----)
C2     7    83.29    14.12                   (-----*-----)
C3     7    90.29     7.34                         (-----*-----)
C4     7    64.86     5.64     (-----*-----)
C5     7    73.43     9.00           (-----*-----)
                                  --+---------+---------+---------+-------
POOLED STDEV = 9.15               60        72        84        96
```

MINITAB refers to the population means as the *factor* of interest. In this problem, the term *factor* refers to the means of the four different bonders. *MINITAB* software gives (pictorially) 95% confidence intervals for each of the population means. While this graph appears similar to the results of a multiple-range test, the same interpretations do not necessarily apply. To obtain valid statements about significant differences among the sample means, it is necessary to perform a multiple-range test, as illustrated in this section.

EXERCISES

Use of a computer is recommended for some of the following exercises.

10.1 Four wafers are chosen at random from each of six different vapor deposition stations used in the production of integrated circuits. The 24 resulting samples are examined microscopically for voids, and the number of voids is counted in a randomly chosen square away from the edge of the wafer, with the following results:

Station	WAFER NUMBER			
	1	2	3	4
1	3	1	2	3
2	4	6	3	2
3	0	2	2	1
4	4	2	3	3
5	1	1	4	2
6	3	3	5	4

Perform an analysis of variance to determine, at the .05 level of significance, whether there are differences in the mean number of voids among the stations.

10.2 To determine the effects of three different lubricants on tool wear, new cutting tools were weighed and then installed on each of 36 lathes, randomly assigning each lubricant to 12 of the lathes. After 20 hours of operation, the tools were removed and measured for wear by taking the difference between the initial and the final weight, with the following results (in grams):

Lubricant A:	1.04	0.70	0.85	1.03	0.67	0.81
	0.85	0.76	0.81	0.92	0.92	0.90
Lubricant B:	0.88	0.96	0.93	0.84	1.16	0.84
	0.91	0.95	0.87	1.02	0.97	0.94
Lubricant C:	0.69	0.81	0.78	0.68	0.83	0.71
	0.77	0.75	0.86	0.64	0.79	0.78

Perform an analysis of variance to determine, at the .01 level of significance, whether there are differences in the mean amount of tool wear associated with the three lubricants.

10.3 Suppose two independent random samples, each of size 30, are taken from normal populations having the same variance, with the following results:

$$\bar{x}_1 = 64.6, \ s_1 = 8.7; \qquad \bar{x}_2 = 50.8, \ s_2 = 7.4$$

Perform an analysis of variance to test whether the two sample means are significantly different at the .05 level of significance. *Hint*: The total of the observations corresponding to each mean can be obtained from the equation on page 50, defining the sample mean. The sums of squares necessary to find *SST* can be found by applying the equation on page 52, the computing formula for the variance, and adding the resulting values of and Σx_1^2 and Σx_2^2.

10.4 For data on 5 observations selected at random from each of 10 groups, it was calculated that *SST* = 3,758 and *SSM* = 1,276. Complete the analysis of variance table. Are the means significantly different from one another at the .05 level of significance?

10.5 If the means of 4 different groups are to be examined by choosing 15 observations at random from each group, and it is calculated that *SSM* = 3.79 and *SST* = 9.48, do the four means differ significantly at the .05 level of significance?

10.6 Use Duncan's multiple-range test at the .05 level of significance to determine what, if any, significant differences exist among the means in Exercise 10.2.

10.7 Apply Duncan's multiple-range test to the results of Exercise 10.1, using the .01 level of significance.

The following exercises introduce new material, expanding on some of the topics covered in the text.

10.8 *Unequal sample sizes.* An analysis of variance can be performed to test the difference among k means using samples of different sizes. If n_i is the number of observations in sample i ($i = 1, 2, \ldots, k$), the total sum of squares and the sum of squares for means are computed from the following formulas:

$$SST = \sum_{i=1}^{k} \sum_{j=1}^{n_i} y_{ij}^2 - C \qquad SSM = \sum_{i=1}^{k} \frac{T_i^2}{n_i} - C$$

Paint stripes are applied to three different kinds of areas on a highway, 55-mph zones, 35-mph zones, and areas just before a traffic signal. Traffic counts are kept, and the number of axles passing over the stripes is recorded until, in the judgment of the highway engineer, the stripes are worn out. Traffic counts obtained from this experiment are as follows:

55-mph zones:	3,485	4,251	2,890	5,116	
35-mph zones:	7,615	5,520	6,135	6,915	5,012
Traffic signals:	2,856	1,199	1,945		

Perform an analysis of variance of these data to determine, at the .05 level of significance, if these data show that highway paint wear depends on the type of traffic.

10.9 *Difference between two means.* In the special case $k = 2$, the test statistic

$$t = \frac{\bar{x}_1 - \bar{x}_2}{\sqrt{(n_1 - 1) s_1^2 + (n_2 - 1) s_2^2}} \sqrt{\frac{n_1 n_2 (n_1 + n_2 - 2)}{n_1 + n_2}}$$

has the t distribution with $n_1 + n_2 - 2$ degrees of freedom. Use the following data to test, at the .01 level of significance, for a difference between the mean heat-producing capacity (in millions of calories per ton) of the two mines.

Mine A:	8.13	8.35	8.34	8.26	8.07	
Mine B:	7.95	7.90	8.14	7.89	7.92	7.84

10.10 *Analysis of proportions.* Data consisting of proportions do not satisfy the assumption of equal variances. However, the variances may be stabilized by making the transformation $y = \arcsin \sqrt{\frac{x}{n}}$. Make use of this transformation to perform an analysis of variance on the following data, giving yields (proportion of products passing electrical tests) on an experimental integrated-circuit line:

Method 1:	.58	.69	.42	.55	.38	.46	.35	.62
Method 2:	.84	.71	.66	.84	.69	.89	.76	.84
Method 3:	.47	.49	.58	.54	.61	.39	.52	.41

10.3   Randomized-Block Designs

After performing the analysis of the wire-bonding experiment described in the example on page 256, further thought was given to the reason for the poor performance of bonder D. A physical examination of this bonder showed no obvious faults, and attention then was turned to the operators of the bonding machines. It was discovered that the same operator made most of the bonds with bonder D, and that the bonds performed with bonders A, B, and C were made mostly by other operators. Thus, it is not clear whether the bonder differences discovered in this experiment can be attributed to the bonders themselves or to the operators. This ambiguity can be clarified by designing an experiment in which an extraneous source of variability—operators, in this case—is "balanced out" of the comparison of bonders.

An extraneous source of variability can be eliminated by holding its value fixed in an experiment. In the wire-bonding experiment, for example, all bonds could be made by the same operator. However, conclusions about bonder differences then would be applicable only to that operator; nothing would be known about what experience could be expected if other operators were to bond wires with these machines. Suppose, instead, that the experiment were repeated in "blocks" of observations, where each block consisted of a different operator making bonds using each bonder. Then, much broader conclusions could be drawn. Such an experiment, in which extraneous sources of variability are removed by repeating a one-way experiment, fixing the value of the extraneous variable at a different level in each repetition, or block, is called a **randomized-block design**. The term *randomized* is used here to reflect that each operator makes bonds with each bonder in a different random order. The purpose of this randomization is to assure that biases associated with fatigue, or other causes, known or unknown, will not effect the experimental results.

For the wire-bonding experiment, a block will consist of the bonds made with each bonder by a given operator. The order of bonders is randomized separately in each block. The analysis of variance for a randomized-block design in which a means are to be compared in each of b blocks has the structure as shown at the top of page 267.

This "dummy" analysis of variance, which omits sums of squares, mean squares, and F, is useful in designing an experiment. It lists all the factors and their degrees of freedom, assuring that nothing has been omitted and providing an opportunity to check if there are sufficient degrees of freedom for error. To find moderate differences among the population means, it is best that the degrees of

Dummy ANOVA for Randomized-Block Design

Source of Variability	Degrees of Freedom
Means	$a - 1$
Blocks	$b - 1$
Error	$(a - 1)(b - 1)$
Total	$ab - 1$

freedom for error equal or exceed 30. If the degrees of freedom for error is judged to be insufficient, the number of blocks can be increased to bring these degrees of freedom to a desirable level. As in the one-way design, the degrees of freedom for means is the number of means compared, minus 1. Similarly, the degrees of freedom for blocks is the number of blocks, minus 1, and the total degrees of freedom is the total number of observations (a means in each of b blocks, or ab), minus 1. The degrees of freedom for error is obtained by subtracting the sum of the degrees of freedom for means and for blocks from the total degrees of freedom.

The total sum of squares is computed as for a one-way design; it is the sum of the squares of all ab observations, minus the correction term. The correction term, again, is the square of the grand total of all observations, divided by the number of observations, ab. To compute the sum of squares for means or for blocks, totals are found for each mean or block, these totals are squared and summed, the sum of squared totals is divided by the number of observations comprising each total, and the correction term is subtracted from this result.

To illustrate, suppose the wire-bonding experiment is rerun as a randomized block experiment, using five operators (blocks), with the following results:

Bond Strength (grams)

	BONDER				
Operator	A	B	C	D	Totals
1	11.8	9.6	12.6	10.2	44.2
2	10.4	12.4	11.0	10.5	44.3
3	9.6	10.2	11.4	3.1	34.3
4	9.8	11.7	10.0	9.7	41.2
5	10.5	10.2	9.8	9.1	39.6
Totals	52.1	54.1	54.8	42.6	203.6

Although the calculations for the *ANOVA* are somewhat lengthy and tedious, they are outlined in the accompanying box, entitled "Calculations for Randomized-Block *ANOVA*," so that the use of a statistical computer package to obtain the *ANOVA* will have less of a "black box" feel. Now, and for the more complicated designs to follow, however, we shall illustrate the analysis of data arising from experimental designs with the use of *MINITAB* statistical computer software.

CALCULATIONS FOR RANDOMIZED-BLOCK ANOVA

Correction term:
$$C = (203.6)^2/20 = 2,072.65$$

Total sum of squares:
$$SST = (11.8^2 + 9.6^2 + \ldots + 9.1^2) - C$$
$$= 2,143.3 - 2,072.65 = 70.65$$

Sum of squares for bonders:
$$SSA = (52.1^2 + 54.1^2 + 54.8^2 + 42.6^2)/5 - C$$
$$= 2,091.80 - 2,072.65 = 19.15$$

Sum of squares for blocks (operators):
$$SSB = (44.2^2 + \ldots + 39.6^2)/4 - C$$
$$= 2,089.56 - 2,072.65 = 16.91$$

Sum of squares for error:
$$SSE = SST - SSA - SSB = 34.59$$

To obtain a randomized-block *ANOVA* with *MINITAB*, we enter three columns of data. The first column, C1, consists of the experimental data, entered in a convenient order. The next two columns denote to which category of mean (which bonder) and to which block (which operator) the corresponding observation in C1 belongs. For example, suppose we construct C1 by entering observations in the following order: first, the four observations corresponding to operator 1 and bonders $A - D$ (the first row of the data given on page 267), then the four observations corresponding to operator 2 (the second row of data), with the bonders in the same order, and so forth. Thus, the data entry in column C1 will be as follows:

C1:   11.8,   9.6,   12.6,   10.2,   10.4,   12.4,   11.0,   10.5,   9.6,   10.2,
      11.4,   3.1,   9.8,   11.7,   10.0,   9.7,   10.5,   10.2,   9.8,   9.1.

Now, we shall identify the bonders (what *MINITAB* calls "factor C2") by placing numbers in C2 so they correspond to the four bonders. Since the first four data entries in C1 are for bonders $A - D$, we first place the *numbers* 1, 2, 3, and 4 in C2 to represent bonders A, B, C, and D, respectively. The next four entries also are 1, 2, 3, and 4, representing the bonders in the second row of the data. We continue placing sets of integers consisting of 1, 2, 3, and 4 in C2 until we have five such sets in all, obtaining:

C2:   1,   2,   3,   4,   1,   2,   3,   4,   1,   2,   3,   4,   1,   2,   3,   4,   1,   2,   3,   4.

We shall use C3 to identify the five levels of "factor C3" (operators). The first four observations in C1 are from operator 1, the next four from operator 2, and so forth. Thus, the column representing operators is as follows:

C3:   1,   1,   1,   1,   2,   2,   2,   2,   3,   3,   3,   3,   4,   4,   4,   4,   5,   5,   5,   5.

Now, we are ready to give the command

$$\text{ANOVA C1} = \text{C2, C3 ;}$$

This command tells *MINITAB* that we wish to perform an analysis of variance with the data in C1 and the factors defined by C2 and C3. It is important to end the command with a semicolon, to inform *MINITAB* that a subcommand is to follow. The subcommand we wish to give is

$$\text{MEANS C2, C3 .}$$

This subcommand will cause a printout of the mean of each level of the factors defined in columns 2 and 3; that is, we will obtain the mean of the observations for each bonder and the mean for each operator. Notice the period placed at the end of this command to inform *MINITAB* that no more subcommands will be given. (When using the command–subcommand structure, each command or subcommand must be followed by a period or a semicolon.)

When these commands are given, we obtain the following output:

Factor	Type	Levels	Values				
C2	fixed	4	1	2	3	4	
C3	fixed	5	1	2	3	4	5

Analysis of Variance for C1

Source	DF	SS	MS	F	P
C2	3	19.156	6.385	2.22	0.139
C3	4	16.907	4.227	1.47	0.273
Error	12	34.589	2.882		
Total	19	70.652			

MEANS

C3	N	C1
1	5	10.420
2	5	10.820
3	5	10.960
4	5	8.520

C2	N	C1
1	4	11.050
2	4	11.075
3	4	8.575
4	4	10.300
5	4	9.900

This output begins by identifying the location of the column representing each factor, and it gives the number of levels of each factor as well as the values assigned to each level. The analysis-of-variance table is similar to that for the one-way design, except now there is a second factor, representing "blocks" (operators, in this example). There are two F values in this table, indicating that now we are making two tests of significance, one for differences among bonder means, and another for differences among operator means. The mean values printed below the *ANOVA* give the mean wire-bond strength for each bonder (C2 means) and for each operator (C3 means), as well as N, the number of observations for each mean. (The reader will be asked, in Exercise 10.11, to verify that substantially the same results are obtained (except for roundoff) when the calculations shown in the table entitled "Calculations for Randomized-Block *ANOVA*" are used.)

The last column in this table, labeled P gives a probability corresponding to the smallest level for which the corresponding value of F would be significant. Thus, for example, a P value of .025 means that the corresponding F value would be significant at the .05 level of significance, but not at the .01 level. (We have encountered P values before, on page 173.) Since neither of these P values is less than .05, we cannot conclude that either the bonder means or the operator means are significantly different at the .05 level of significance.

Note, in the analysis-of-variance table, that the mean square for error (2.882) is less than it was for the one-way design ($MSE = 4.52$ in the *ANOVA* on page 260). In general, if blocking is effective in eliminating the variability associated with an extraneous variable, the error mean square will be reduced. A smaller error term makes the F tests more sensitive to small differences among the means. It is curious that no significant difference was found among the bonder means in this analysis, when we found such a difference in the one-way analysis. This result is especially strange when we also consider the reduced error mean square, indicating that the significance test for bonder means performed here is more sensitive to small differences than heretofore. We shall investigate a possible explanation for this phenomenon in the next section.

EXERCISES

Use of a computer is recommended for some of the following exercises.

10.11 Using the calculations given in the table on page 268, complete the calculations for the *ANOVA* and compare with the results obtained with *MINITAB*.

10.12 Find the minimum number of blocks required for a randomized-block design having k means if the number of degrees of freedom for error is to be at least 30.

10.13 Complete the following *ANOVA* and interpret the results:

ANOVA for Randomized-Block Design

Source of Variability	Degrees of Freedom	Sum of Squares	Mean Square	F
Means	6	174.52		
Blocks	4	101.39		
Error				
Total		357.10		

10.14 Complete the following *ANOVA* and interpret the results:

ANOVA for Randomized-Block Design

Source of Variability	Degrees of Freedom	Sum of Squares	Mean Square	F
Means	4	2.008		
Blocks				
Error		6.065		
Total	39	9.101		

10.15 To practice the calculations involved in a randomized-block design with simple numbers, construct an *ANOVA* table for the following hypothetical experimental results:

Blocks	Variables		
	A	B	C
1	2	3	4
2	1	2	3
3	4	1	2

a. Perform the two significance tests (at the .05 level of significance).

b. State, in words, what the results mean.

c. What objection reasonably could be raised to running this experiment?

10.16 To practice the calculations involved in a randomized-block design with simple numbers, construct an *ANOVA* table for the following hypothetical experimental results:

Blocks	Variables		
	A	B	C
1	1	3	2
2	3	2	3
3	4	1	2
4	5	4	3

a. Perform the two significance tests (at the .05 level of significance).

b. State, in words, what the results mean.

c. What objection reasonably could be raised to running this experiment?

10.17 To test two meat tenderizers, each was applied to a steak before broiling and tasted by an expert food taster who rated the tenderness on a 10-point scale from least tender (1) to most tender (10). Ten tasters employed in this test gave the following ratings:

Tenderizer	Taster									
	1	2	3	4	5	6	7	8	9	10
A	7	8	6	7	6	6	6	8	5	7
B	8	7	8	8	7	8	6	9	7	9

Perform a randomized-block analysis of variance and state your conclusions.

10.18 To test the distance three brands of golf ball could be hit with a driver, eight golfers were asked to drive golf balls of each of the three brands in a random order. Distance was measured (in yards) from the tee to the final landing place, orthogonal to the line of tees, with the following results:

	Golfer							
Brand	1	2	3	4	5	6	7	8
A	184	225	199	265	201	240	290	201
B	196	220	206	253	178	235	287	215
C	212	251	210	284	230	263	308	200

Perform a randomized-block analysis of variance and state your conclusions.

10.19 Do a multiple-range test on the tasters in Exercise 10.17 to see which, if any, tasters were giving scores that were significantly different from the others, at the .05 level of significance.

10.20 Do a multiple-range test on the golf-ball brands in Exercise 10.18 to see if any one brand of golf ball is a "clear winner." Use a significance level of .05.

10.4 Factorial Designs

The *ANOVA* of the randomized-block wire-bonding experiment, given on page 269, showed no significant differences among bonders. This is a surprising result when compared with the significant difference shown in the one-way experiment (page 260). A closer examination of the data of the randomized-block experiment, shows that the mean bonding strength for operator 3 was lower than that for the other operators, but this operator had a poor result only with bonder *D*. The small sample size (only one bond per bonder-operator combination) makes it unrealistic to draw general conclusions, but it can be postulated that this operator is having trouble only with bonder *D*, and not with the other bonders.

Whenever two variables have a *joint* effect on the response variable in an experiment, it is said that there is an **interaction** between them. Thus, our postulate can be restated as: "There is a non-zero interaction between the variables 'bonders' and 'operators.'" This postulate has been suggested because it required *both* a particular bonder and a particular operator to obtain an unusually small bonding-strength response.

In the bonder-operator experiment, there are two variables, called "bonder" and "operator." In an experiment with two or more variables, the term **factor** is used to designate each variable studied, and the values that each factor assumes in the design are called the **levels** of the factors. In our experiment, the factor "bonders" has the four levels 1, 2, 3, and 4 and the factor "operators" has the five levels 1, 2, 3, 4, and 5. Such a design is referred to as a **factorial design**.

It is helpful to have some notation and some specialized language when we discuss factorial designs. When discussing arbitrary factorial designs, we shall use capital letters to refer to its factors. For example, the letters *A*, *B*, and *C* will be used to denote the factors in a three-factor design. The overall effects of these factors on the response are called **main effects**, and the experimental estimate of the effect of a given factor also is denoted by the capital letter used to identify that factor. (Only main effects were capable of being studied in the randomized-block designs.) Estimates of interactions among several factors are denoted by "strings" consisting of the letters that identify those factors. For example, the string *AB* denotes the estimate of the interaction between factor *A* and factor *B* in a factorial experiment. To illustrate, three interactions between two factors can be estimated in a three-factor factorial design; the estimates are denoted by *AB*, *AC*, and *BC*. Also, an estimate can be made of the three-factor interaction, denoted by *ABC*.

In small factorial experiments, it is necessary to repeat the experiment at least once to detect interactions among the factors; each repetition of the experiment is called a **replicate**. For an experimental design to be a factorial design, it is necessary for each replicate to be **fully balanced**; that is, each replicate must contain the same number of observations for each combination of values of the variables. For example, if there are five operators available for the four bonding machines, each replicate of the fully balanced factorial experiment involving these two factors would require a minimum of 20 observations, one observation for each of the 5×4 = 20 combinations of levels of the bonder and operator factors. Each additional replicate also would require 20 observations.

Two-Factor Designs

The analysis of variance for a factorial design, and its interpretation, is introduced by considering a two-factor design. A dummy *ANOVA* for this design is as follows:

Dummy *ANOVA* for Two-Factor Design

Source of Variability	Degrees of Freedom
Replicates	$r - 1$
Factor A	$a - 1$
Factor B	$b - 1$
Interaction	$(a - 1)\ (b - 1)$
Error	$(ab - 1)\ (r - 1)$
Total	$rab - 1$

This design is like the randomized-block design with the following important exceptions. The design is replicated; that is, there are two or more balanced repetitions of the experiment. Replication allows the estimation of the interaction between the two factors. The degrees of freedom for this *ANOVA* are obtained as follows. The degrees of freedom for "replicates" (the difference among the means of the replicates) is the number of replicates, minus 1. As for the randomized-block design, the degrees of freedom for the main effect of each of the two factors is the number of levels of that factor, minus 1. The degrees of freedom for the interaction, or joint effect of the two factors, is the *product* of the degrees of freedom for the two factors.

To illustrate the analysis of a two-factor design, a second replicate has been added to the randomized-block design given on page 267, involving bonders and operators. A new randomization pattern is used to assign the order of operators to bonders in the second replicate. The experimental data of the second replicate are shown, along with the data already given on page 267 for the first replicate, as follows:

Bond Strength (grams)

	REPLICATE 1				REPLICATE 2			
	Bonder				Bonder			
Operator	A	B	C	D	A	B	C	D
1	11.8	9.6	12.6	10.2	10.6	11.9	9.8	9.9
2	10.4	12.4	11.0	10.5	12.0	10.3	10.0	11.6
3	9.6	10.2	11.4	3.1	11.8	9.9	9.1	5.8
4	9.8	11.7	10.0	9.7	10.1	12.1	11.6	9.8
5	10.5	10.2	9.8	9.1	9.4	10.2	9.7	12.1

We shall develop the analysis of variance using *MINITAB* as we did before for the unreplicated, randomized-block design. The presence of the additional replicate adds 20 observations to the experiment; thus C1 now contains 40 observations. We can arrange the 20 observations comprising the first replicate as we did previously (page 268), and then complete C1 by adding the observations of the second replicate in the same order.

Since the experiment includes two experimental factors, bonders and operators, plus a third "factor," replicates, we will need three additional columns to define the order of the levels of these factors. We shall use column C2 to define the levels of the factor bonders much the same as we did previously (page 268), only now we must repeat the sequence of 20 integers involving 5 groups of the levels 1, 2, 3, and 4 to account for the second replicate. Thus, C2 will appear as follows:

C2:   1, 2, 3, 4, 1, 2, 3, 4, 1, 2, 3, 4, 1, 2, 3, 4, 1, 2, 3, 4,
        1, 2, 3, 4, 1, 2, 3, 4, 1, 2, 3, 4, 1, 2, 3, 4, 1, 2, 3, 4.

Using C3 to define the levels of operators, we will have

C3:   1, 1, 1, 1, 2, 2, 2, 2, 3, 3, 3, 3, 4, 4, 4, 4, 5, 5, 5, 5,
        1, 1, 1, 1, 2, 2, 2, 2, 3, 3, 3, 3, 4, 4, 4, 4, 5, 5, 5, 5.

The next column will be used to identify the replicates. Since the first 20 observations in C1 came from the first replicate and the next 20 observations were from the second replicate, C4 will be:

C4:   1, 1, 1, 1, 1, 1, 1, 1, 1, 1, 1, 1, 1, 1, 1, 1, 1, 1, 1, 1,
        2, 2, 2, 2, 2, 2, 2, 2, 2, 2, 2, 2, 2, 2, 2, 2, 2, 2, 2, 2.

After the data have been entered into C1 – C4, the command and subcommand

$$\text{ANOVA C1 = C2 C3 C4 C2*C3;}$$

$$\text{MEANS C2 C3 C4 C2*C3.}$$

will produce the required *ANOVA* and the tables of means. Note that each main effect, C2 and C3, the replicate effect, C4, and the interaction effect, C2*C3 all must be specified in the *ANOVA* command. The *MEANS* subcommand specifies the same effects so that a mean value will be given for each level of the factors C1 and C2, for each of the two replicates defined in C4, and for each of the $4 \times 5 = 20$ combinations of levels of C2 and C3.

After these commands are given, the following output is produced:

Factor	Type	Levels	Values				
C2	fixed	4	1	2	3	4	
C3	fixed	5	1	2	3	4	5
C4	fixed	2	1	2			

Analysis of Variance for C1

Source	DF	SS	MS	F	P
C2	3	16.857	5.619	3.99	0.023
C3	4	23.689	5.922	4.20	0.013
C4	1	0.420	0.420	0.30	0.591
C2*C3	12	44.657	3.721	2.64	0.028
Error	19	26.775	1.409		
Total	39	112.398			

MEANS

C2	N	C1
1	10	10.600
2	10	10.850
3	10	10.500
4	10	9.180

C3	N	C1
1	8	10.800
2	8	11.025
3	8	8.863
4	8	10.600
5	8	10.125

C4	N	C1
1	20	10.180
2	20	10.385

C2	C3	N	C1
1	1	2	11.200
1	2	2	11.200
1	3	2	10.700
1	4	2	9.950
1	5	2	9.950
2	1	2	10.750
2	2	2	11.350
2	3	2	10.500
2	4	2	11.900
2	5	2	10.200
3	1	2	11.200
3	2	2	10.500
3	3	2	10.250

continued

C2	C3	N	C1
3	4	2	10.800
3	5	2	9.750
4	1	2	10.050
4	2	2	11.050
4	3	2	4.450
4	4	2	9.750
4	5	2	10.600

Since the *P* values for *C2* (bonders), *C3* (operators), and *C2*C3* (bonder-operator interaction) all are less than .05, it can be concluded, at the .05 level of significance, that there are differences among bonders and among operators as well as an interaction between operators and bonders.

      To gain a better understanding of the interaction, a plot is made of the mean values for each combination of operator and bonder. (The data points for this plot are the 20 means in the last table of means of the computer output.) It is helpful to make a separate graph for each bonder, with the five operators represented on the horizontal axis and the mean bonding strength for each operator on the vertical axis. The resulting graph is shown in Figure 10.1.

      This graph reveals that, except when operator 3 uses bonder D, there is little difference between the mean bond strengths. Evidently, the difficulty operator 3 is having with this bonder reduces the mean of her overall bond strengths enough to account both for the difference among operators and the difference among bonders. Thus, it would be misleading, for example, to report the main effect of operators by making the unqualified statement: "A difference exists among the operators." In the presence of the operator-bonder interaction, a correct description of the results found in this experiment is "Operator 3 is having trouble using bonder D."

      Interactions between two factors often occur, and experimental programs should be designed that are capable of detecting their presence. When the bond-strength experiment was not replicated, so it was impossible to measure the interaction, no useful conclusions were found. Adding a second replicate to the

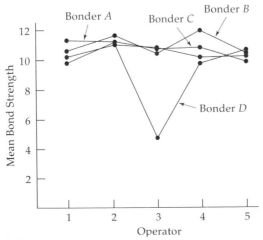

FIGURE 10.1 Bonder-Operator Interaction

experiment allowed the detection of a potential interaction between the two factors, and it provided a large enough sample size to allow the detection of smaller differences among bonders and operators. Of much greater importance, knowledge of the nature of the interaction allowed appropriate action to be taken. Operator 3 was instructed on the use of bonder D, a new type of bonder with which she was unfamiliar. As a result, this operator's performance was improved to put her on a par with the other operators. An otherwise good operator was not discharged or transferred, and overall quality was improved with a minimal investment.

Large Factorial Experiments

The analysis of a multi-factor experiment having n factors will include n main effects and $2^n - 1$ interactions. (There will be $n(n - 1)/2$ two-factor interactions, $n(n - 1)(n - 2)/6$ interactions involving three factors, and so forth.) The methods for performing an *ANOVA* for a multi-factor experiment are a straightforward extension of those for analyzing a two-factor experiment, but the computational work becomes very complex and tedious when the number of factors is large. Fortunately, statistical software is available to solve the computational problem. But no amount of analytical software can solve the problems of managing, performing, and justifying the expense of an excessively large experiment.

  In this section, we shall illustrate the power of factorial experiments to sort out complex interrelationships among many variables by analyzing an experiment having four factors, each at two levels. An experiment of this size often lies within the realm of "do-able" experiments. In the next section, we shall discuss experimental designs that help to reduce the size of excessively large factorial experiments.

  An experiment was designed to control the final phosphorous content of steel by gaining a better understanding of how four factors, pouring temperature, lime ratio, oxygen, and original phosphorous interact to determine final phosphorous. The practical operating ranges of the experimental factors are narrow; thus, it was found reasonable to vary each experimental factor over only two levels. This decision also has kept the number of experimental combinations to

A Pouring Temp. (°F)	B Lime Ratio	C Oxygen	D Original Phosphorous	y FinalPhosphorous Rep 1	y FinalPhosphorous Rep 2
2,400	3	5%	.15%	0.003%	0.001%
2,400	3	5	.30	0.002	0.008
2,400	3	15	.15	0.002	0.005
2,400	3	15	.30	0.004	0.001
2,400	4	5	.15	0.000	0.003
2,400	4	5	.30	0.003	0.007
2,400	4	15	.15	0.010	0.006
2,400	4	15	.30	0.006	0.014
2,600	3	5	.15	0.003	0.007
2,600	3	5	.30	0.011	0.005
2,600	3	15	.15	0.004	0.008
2,600	3	15	.30	0.004	0.008
2,600	4	5	.15	0.007	0.004
2,600	4	5	.30	0.004	0.011
2,600	4	15	.15	0.017	0.011
2,600	4	15	.30	0.014	0.009

within moderate limits. Since this is only a preliminary study, and since it is expensive to manufacture heats of steel, it was decided to include only two replicates of the fully balanced factorial experiment. Complete randomization of the order of all 16 experimental combinations in each replicate would have been inordinately expensive; however, the order of making the heats was randomized as much as possible. The results on page 277 are given, not in the original randomized order of experimentation, but in an arrangement that is more suitable for analysis.

Data entry follows the same concept previously illustrated. It is convenient to enter the 32 experimental results in C1 by going down the column of experimental results for the first replicate, and then appending the results for the second replicate. In entering the data, we multiplied each observation by 1,000 to make the data entry easier and to avoid very small numbers in the analysis-of-variance table. (This change of scale will not change the F values, since they are a *ratio* of two variances, both of which have been scaled up equally.) After entering the data in C1, we create five additional columns to identify the levels of the four experimental factors as well as the replicates. Letting C2 represent factor A, pouring temperature, C3 represent factor B, C4 represent factor C, C5 represent factor D, and C6 represent replicates, we have

```
C2:  1,  1,  1,  1,  1,  1,  1,  1,  2,  2,  2,  2,  2,  2,  2,  2,
     1,  1,  1,  1,  1,  1,  1,  1,  2,  2,  2,  2,  2,  2,  2,  2.

C3:  1,  1,  1,  1,  2,  2,  2,  2,  1,  1,  1,  1,  2,  2,  2,  2,
     1,  1,  1,  1,  2,  2,  2,  2,  1,  1,  1,  1,  2,  2,  2,  2.

C4:  1,  1,  2,  2,  1,  1,  2,  2,  1,  1,  2,  2,  1,  1,  2,  2,
     1,  1,  2,  2,  1,  1,  2,  2,  1,  1,  2,  2,  1,  1,  2,  2.

C5:  1,  2,  1,  2,  1,  2,  1,  2,  1,  2,  1,  2,  1,  2,  1,  2,
     1,  2,  1,  2,  1,  2,  1,  2,  1,  2,  1,  2,  1,  2,  1,  2.

C6:  1,  1,  1,  1,  1,  1,  1,  1,  1,  1,  1,  1,  1,  1,  1,  1,
     2,  2,  2,  2,  2,  2,  2,  2,  2,  2,  2,  2,  2,  2,  2,  2.
```

In these columns, the number 1 represents the first ("low") level of the given factor, and the number 2 represents the second ("high") level.

This experiment has four factors, each at two levels. The response variable is the final phosphorous, and the experiment is performed in two replicates. This experiment provides not only the estimates AB, AC, AD, BC, BD, and CD of the 6 two-factor interactions but also the estimates ABC, ABD, ACD, and BCD, of the 4 three-factor interactions, and $ABCD$, an estimate of the four-factor interaction. Experience has shown that interactions involving three or more factors, the so-called "higher-order" interactions, rarely differ from zero. Thus, it is not unusual to perform analyses of factorial experiments that test for the significance of main effects and two-factor interactions only. If it can be assumed that there are no higher-order interactions in the population, the associated degrees of freedom and sums of squares can be pooled with the error degrees of freedom and sum of squares for error, respectively. This confers the analytic advantage of increasing the degrees of freedom for error, thereby making the analysis more sensitive to main effects and interactions that might otherwise not have been found to be significant.

Making the assumption that higher-order interactions do not exist in the population, or, at least, that they are negligible, we shall ask *MINITAB* to include only the 4 main effects and the 6 two-factor interactions in the analysis. Thus, the 32 experimental observations will produce one degree of freedom for each mean effect and interaction and one degree of freedom for the two replicates, or 11 degrees of freedom. Since there are a total of $32 - 1 = 31$ degrees of freedom, there are $31 - 11 = 20$ degrees of freedom for error. To ask for the inclusion of only the main effects, the effect of replicates, and the 6 two-factor interactions, and to include all relevant tables of means, we give the command and the subcommand

ANOVA C1 = C2 C3 C4 C5 C6 C2*C3 C2*C4 C2*C5 C3*C4 C3*C5 C4*C5;

    MEANS C2 C3 C4 C5 C6 C2*C3 C2*C4 C2*C5 C3*C4 C3*C5 C4*C5.

We shall display fully only the *ANOVA* output, omitting the table giving the factors and their levels as well as the tables of means. However, we shall have occasion to refer to some of the means as we analyze the results.

Analysis of Variance for C1

Source	DF	SS	MS	F	P
C2	1	84.500	84.500	9.10	0.007
C3	1	78.125	78.125	8.41	0.009
C4	1	60.500	60.500	6.51	0.019
C5	1	12.500	12.500	1.35	0.260
C6	1	6.125	6.125	0.66	0.426
C2*C3	1	0.500	0.500	0.05	0.819
C2*C4	1	0.125	0.125	0.01	0.909
C2*C5	1	3.125	3.125	0.34	0.568
C3*C4	1	84.500	84.500	9.10	0.007
C3*C5	1	0.000	0.000	0.00	1.000
C4*C5	1	21.125	21.125	2.27	0.147
Error	20	185.75	9.288		
Total	31	536.88			

This analysis of variance shows P values that are less than .01 for the estimates of the main effects of factor A (pouring temperature), factor B (lime ratio) and the estimate of the BC (lime-ratio – oxygen) interaction, as well as a P value of .019 for the estimate of the main effect of factor C (oxygen). Since factors B and C appear to be involved in an interaction, however, the main effect of, say, factor B will depend upon the level of factor C; thus, it makes no sense to report estimates of such main effects separately. The estimated main effect of factor A, however, is significantly different from 0 at the .01 level of significance. This factor is not involved in an interaction; thus, it makes sense to estimate its main effect with the aid of the following table of means:

Main Effect of Pouring Temperature
(Mean Values of Final Phosphorous)

Level of x_1	Mean of y
2,400	0.0047
2,600	0.0079

Note that we have undone the scaling by multiplying each mean produced by the computer program by 0.001, reporting these means in their original units. The difference between these means, 0.0079 − 0.0047 = 0.0032, is the required estimate of the main effect of factor A. Thus, it is estimated that the effect of an increase of pouring temperature from 2,400°F to 2,600°F will be to increase the final phosphorous of the steel by an estimated 0.0032%.

Since it has been concluded that an interaction exists between Factor B, lime ratio, and Factor C, oxygen, the main effects found for these two variables are of little practical importance. However, their interaction is important. To display this interaction, the following two-way table is constructed:

Interaction Between Lime Ration and Oxygen
(Mean Values of Final Phosphorous)

		Lime Ratio		
		3	4	Means
Oxygen	*5%*	0.0050	0.0049	0.00494
	15%	0.0045	0.0109	0.00769
	Means	0.00475	0.00790	0.00633

This table shows that, while increases in the value of either variable are responsible for increasing y, final phosphorous, there is an estimated additional combined effect. This combined effect is seen when we note that the mean final phosphorous when both oxygen and lime ratio are at their high levels is about twice that of the other three means. The interaction becomes clearer when we construct a graph like that shown in Figure 10.2. In this figure, the line representing the relation of final phosphorous to lime ratio has a slope of nearly zero when the oxygen level is 5%, but the line representing this relationship has a large positive slope when the oxygen level is 15%. Thus, there is a large estimated effect on final phosphorous of increasing the lime ratio (0.0109 − 0.0045 = 0.0064), but this effect is estimated to operate *only when the oxygen level is 15%.*

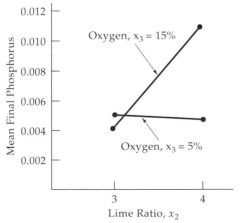

FIGURE 10.2 Lime Ratio—Oxygen Interaction

The function of the analysis of variance is to identify significant effects by performing appropriate significance tests. However, as we have attempted to illustrate in the analysis of the steel-making experiment, the analysis of the results of a designed experiment does not stop at this point. Estimation of the values of the main effects and graphs of the interaction effects found by the analysis of variance are at least as important as the significance tests. If an estimated effect has no statistical significance, all we can say is that the experiment was not capable of proving (at the given level of significance) that the effect exists. However, even if an effect has been estimated to have statistical significance, *it may not be important*. It requires engineering judgment or managerial "know-how" to determine whether results having "statistical significance" also have practical significance.

EXERCISES

Use of a computer is recommended for some of the following exaercises.

10.21 How many degrees of freedom would there be for error in the steel-making experiment described in this section if all of the interactions were included in the *ANOVA*?

10.22 Design a factorial experiment whose three factors have 2, 3, and 4 levels, respectively.

a. List the factors and their levels.

b. Assuming that all interactions are to be included in the *ANOVA*, what is the minimum number of replicates required for the degrees of freedom for error to be at least 30?

c. What would the number of replicates be in part b if only two-factor interactions are included?

d. Give a dummy *ANOVA* for this experiment, including all interactions.

10.23 Use an *ANOVA* to analyze the following data arising from a two-factor experiment. (The three numbers appearing in each cell represent successive replicates.)

		Levels of Factor A		
		1	2	3
Levels of Factor B	1	4 6 3	3 2 1	9 5 7
	2	5 4 6	6 5 4	8 6 7

10.24 Use an *ANOVA* to analyze the following data arising from a two-factor experiment. (The two numbers appearing in each cell represent successive replicates.)

| | Levels of Factor A | | | | |
		1	2	3	4
Levels of	1	6 5	2 0	5 7	4 4
Factor B	2	4 5	5 6	7 8	8 7

10.25 An index of flavor was used to evaluate the effect of adding dioctyl sodium sulfosuccinate (DSS) to milk to stabilize its flavor. Four DSS levels were used including no DSS, and the milk was stored for 7 weeks and 28 weeks to observe the effects of DSS level on storage time. Milk from 5 sources (replicates) was used to make up each set of 8 combinations of DSS and storage time. Perform an analysis of variance on the following experimental results.

| | | Level of DSS (ppm) | | | | | | | |
| | | 0 | | 50 | | 100 | | 150 | |
Time (wks):		7	28	7	28	7	28	7	28
	1	34.6	28.2	35.0	31.1	35.6	33.2	35.4	33.5
	2	33.8	29.0	35.8	30.9	35.8	32.4	35.4	33.9
Reps.	3	34.7	27.2	34.4	29.8	34.6	33.0	36.3	32.5
	4	35.0	28.4	35.1	31.6	35.9	32.9	37.0	34.7
	5	34.2	28.1	34.7	30.6	35.4	33.6	34.9	33.4

10.26 To improve the quality of air conditioners, four compressor designs, *A*, *B*, *C*, and *D*, were tested in each of four geographic regions. The experiment was repeated for two successive cooling seasons (replicates 1 and 2). Two compressors of each design were tested in each region. The following results give the time to failure of each compressor to the nearest month:

| | Replicate 1 | | | | Replicate 2 | | | |
Design:	A	B	C	D	A	B	C	D
Northeast	58	35	72	61	49	24	60	64
Southeast	40	18	54	38	38	22	64	50
Northwest	63	44	81	52	59	16	60	48
Southwest	36	9	47	30	29	13	52	41

Analyze these results to determine if there are any compressor designs that exhibit superior life, and whether such conclusions depend upon the geographic region of use.

10.27 Prepare a brief report on the results of analyzing the experiment in Exercise 10.25. Include the purpose of the experiment, the data-collection method, the method of analysis, and illustrate the results by means of graphs. State your conclusions in words and recommend action(s).

10.28 Prepare a brief report on the results of analyzing the experiment in Exercise 10.26. Include the purpose of the experiment, the data-collection method, the method of analysis, and illustrate the results by means of graphs. State your conclusions in words and recommend action(s).

Exercises 10.29–10.34 are based on data from an experiment described as follows:

Gain of a Semiconductor

Factor:	A Temperature	B Partial Pressure	C Relative Humidity	D Aging Time	E Location of Assembly
Level 1	68°F	10^{-15}	1%	72 hours	Production Line
Level 2	74°F	10^{-4}	30%	144 hours	Laboratory

The following are the resulting observations:

Run No.	A	B	Level of: C	D	E	Response (Gain)
12	1	1	1	1	1	39
22	2	1	1	1	1	32
13	1	2	1	1	1	47
6	2	2	1	1	1	41
28	1	1	2	1	1	38
19	2	1	2	1	1	22
31	1	2	2	1	1	35
24	2	2	2	1	1	31
9	1	1	1	2	1	40
16	2	1	1	2	1	42
3	1	2	1	2	1	55
32	2	2	1	2	1	40
1	1	1	2	2	1	43
30	2	1	2	2	1	30
5	1	2	2	2	1	36
26	2	2	2	2	1	34
25	1	1	1	1	2	43
4	2	1	1	1	2	44
11	1	2	1	1	2	51
14	2	2	1	1	2	40
2	1	1	2	1	2	41
17	2	1	2	1	2	43
7	1	2	2	1	2	48
18	2	2	2	1	2	50
29	1	1	1	2	2	42
20	2	1	1	2	2	41
8	1	2	1	2	2	53
15	2	2	1	2	2	40
21	1	1	2	2	2	40
10	2	1	2	2	2	38
27	1	2	2	2	2	54
23	2	2	2	2	2	44

10.29 Use statistical computer software to test for the significance of all estimated main effects and two-factor interactions at the .05 level of significance.

10.30 This experiment is not replicated. What assumptions are implicit in obtaining the error term in the *ANOVA*?

10.31 Estimate the values of any main effects found to be significantly different from zero in Exercise 10.29.

10.32 Is it appropriate to report only the main effect of a factor involved in an interaction with another factor? Why?

10.33 Sketch a graph to illustrate each interaction estimated in Exercise 10.29 to be non-zero, if any.

10.34 State the conclusions of the experiment in words.

10.5   Fractional Factorial Designs

Multi-factor experiments involving many factors require very large numbers of observations if they are to be fully balanced; that is, if they include all possible combinations of the levels of the factors. A fully balanced experiment requires a number of observations in each replicate equal to the *product* of the levels of all the factors. To illustrate, suppose an experiment contains five factors, A, B, C, D, and E with 3, 4, 2, 3, and 3 levels, respectively. This experiment requires $3 \times 4 \times 2 \times 3 \times 3 = 216$ observations in each replicate. Experiments of this magnitude rarely are performed because, in addition to the expense involved, it is difficult to manage the randomization, record-keeping, and data collection required for a program of such complexity to be successful.

It has been said that the presence of many degrees of freedom for higher-order interactions (interactions involving more than two factors) provides "hidden replication" in a large factorial experiment. In the steel-making example on page 277, where it was assumed that there were no interactions involving more than two factors, it was possible to obtain additional degrees of freedom for error by pooling the effects of the higher-order interactions. Very large factorial experiments involve large numbers of potential higher-order interactions. Rarely is there danger in assuming that such interactions do not exist; thus, estimates of error often can be obtained without the necessity of running large experiments with replication, as illustrated in Exercises 10.29–10.34.

Elimination of replication alone will not always produce a fully balanced experimental design small enough to be practical. The experiment previously described, having five factors with 3, 4, 2, 3, and 3 levels, respectively, requires 216 experimental observations in a single replicate. Drastic reductions in the total experimental size are possible if the number of levels of the each factor in the experiment can be reduced. For example, a five-factor experiment with only two levels per factor has only $2^5 = 32$ observations. Of course, in running only two levels of a factor, only linear effects can be observed. It would require at least three levels to detect possible curvilinearity. Use of four levels assumes the possibility of a complicated curvilinear relationship between the given factor and the experimental response. Thus, it is rare that more than three levels are required for any factor in a factorial experiment.

Several methods have been developed for dealing with the problem of experimental size in addition to elimination of replication. As we have seen, restriction of the number of levels of each factor to 2 provides some help. Such an experiment, having n factors each at two levels, is called a **2^n factorial experiment.** In many situations involving large numbers of potentially important factors, it is useful first to perform a 2^n factorial experiment as a "screening experiment," deciding only on which factors to retain for the next experiment and which to study more closely. As we have noted, only the linear component of the main effects can be studied with a two-level factor. Thus, in screening with a 2^n factorial, it must be assumed that any factor with a large curvilinear effect also will have a large enough linear component to be detected by the experiment.

But, even a 2^n factorial experiment requires a large number of observations if the number of factors, n, is large—over 1,000 observations if $n = 10$. A frequently employed method for further reducing the number of observations required in a balanced experiment involves running only a fraction of the combinations required to achieve full balance. Such an experiment is called a **fractional replicate.**

When a fraction of a fully balanced replicate is performed, estimates of certain main effects and interactions become "tangled" or confounded with one another so that these effects and interactions cannot be estimated separately. Fortunately, a fraction of the experimental combinations often can be chosen so that main effects and two-factor interactions can be estimated separately from each other. In the remainder of this section, we shall introduce the elements of the design of fractional factorials for 2^n experiments, showing how the confounding of effects can be controlled by careful design.

Designing Half Replicates of 2^n Factorial Experiments

The design of a fractional replicate of a 2^n factorial experiment is based on a careful selection of the confounding pattern. To illustrate, let us construct a half replicate of a 2^4 experiment, consisting of $2^{4-1} = 8$ observations. The following is a list of all 15 possible main-effect and interaction estimates in this experiment:

Main Effects:	*A*	*B*	*C*	*D*		
Two-Factor Interactions:	*AB*	*AC*	*AD*	*BC*	*BD*	*CD*
Three-Factor Interactions:	*ABC*	*ABD*	*ACD*	*BCD*		
Four-Factor Interaction:	*ABCD*					

To construct a half replicate, first one of the main effects or interactions is chosen as a **defining contrast.** The effect chosen to be the defining contrast cannot be estimated by the experiment; thus, it is customary to choose a higher-order interaction as the defining contrast.

To illustrate the construction of a fractional replicate, suppose the defining contrast in a half replicate of a 2^4 factorial is the four-factor interaction *ABCD*. This interaction cannot be estimated by the experiment. Furthermore, each estimate of the remaining 14 main effects and interactions in the experiment will be confounded, or **aliased** with its "generalized interaction" with this defining contrast.

The **generalized interaction** of any two factors or interactions is obtained by taking their "product" and canceling like letters. For example, the generalized interaction of ABC with AC is $\cancel{AB}C\cancel{AC} = B$. Thus, if the defining contrast is $ABCD$, the two-factor interaction AB is aliased with CD, its generalized interaction with $ABCD$. The full list of **alias sets** in a half replicate of a 2^4 experiment having the defining contrast $ABCD$ is as follows:

$$A = BCD \quad B = ACD \quad C = ABD \quad D = ABC$$
$$AB = CD \quad AC = BD \quad AD = BC$$

It can be seen from this alias pattern that each main effect is aliased only with a three-factor interaction, and each two-factor interaction is aliased only with another two-factor interaction. If it can be assumed that three-factor and higher-order interactions do not exist, then the main effects of all factors can be estimated in this experiment. Two-factor interactions, however, are aliased in pairs. Thus, if the experimental results show that an estimated two-factor interaction is significant, one or both members of the alias pair may exist, and further experimental work is required to separate such interaction effects.

Having observed how the alias patterns in a half replicate of a 2^4 experiment are determined by the defining contrast, now we turn our attention to the selection of which experimental combinations are to be included in the half replicate. Again, it is helpful to have some notation. We shall denote each of the 16 possible experimental combinations in the 2^4 experiment by a string of lower-case letters. If a given factor is included at level 2, called the *high level*, the corresponding letter appears in the string; if it is included at level 1, the *low level*, the letter is omitted. Thus, for example, *ab* represents the combination where the factors A and B are taken at their high level and factors B and C are taken at their low level. The combination in which all four factors are taken at their low level is denoted by the symbol "1."

Selection of the experimental combinations to be included in a one-half replicate of a 2^4 experiment having a given defining contrast can be made by applying a rule, called the **odds–evens rule**. If "odds" is chosen, for example, all experimental combinations having an odd number of letters in common with the defining contrast are included in the experiment. To illustrate, we shall write down the eight experimental combinations to be included in the "evens" half of the 2^4 factorial experiment having the defining contrast $ABCD$:

$$1 \quad ab \quad ac \quad ad \quad bc \quad bd \quad cd \quad abcd$$

Note that each string of lower case letters denoting an experimental combination has an even number of letters (counting 0 as an even number) in common with the defining contrast $ABCD$.

Although we shall not emphasize the analysis of fractional factorials, the analysis of this experiment proceeds as if it were a 2^3 experiment, involving three factors. Ignoring, say, the factor D, the experiment can be considered to be a full replicate of the 2^3 factorial experiment having the combinations:

$$1 \quad ab \quad ac \quad a(d) \quad bc \quad b(d) \quad c(d) \quad abc(d)$$

Note that the letter *d* is retained (in parentheses) to remind us that the full combination continues to include the factor *D* at its high level. (When this letter is absent, it is understood that the factor *D* is included at its low level.)

Unfortunately, since there is no replication, there are 0 degrees of freedom for error, and no meaningful analysis of variance can be performed. It would be inadvisable to create 1 degree of freedom for error by using a three-factor interaction such as *ABC* for error, inasmuch as it is aliased with the main effect, *D*. (It is risky to assume, without strong evidence obtained from sources outside this experiment, that a main effect does not differ from zero.) The best that can be done is to assume that all two-factor interactions are zero. Since there are six such interactions, aliased with each other in pairs, this assumption can create 3 degrees of freedom for error. The experiment, then, will be capable of estimating main effects only. Since a fully balanced replicate of a 2^4 experiment contains only 16 observations, half replicates of such small experiments normally are not run, and problems of this kind rarely arise in practice.

More realistically, we might wish to run a half replicate of, say, a 2^6 experiment, reducing the number of experimental combinations from $2^6 = 64$ to $2^{6-1} = 32$. If we design such an experiment by using the six-factor interaction, *ABCDEF*, as the defining contrast, it can be seen that each main effect is aliased with a five-factor interaction, and each two-factor interaction is aliased with a four-factor interaction. Thus, the mild assumption that four- and five-factor interactions are zero will guarantee that all main effects and two-factor interactions can be estimated by this experiment without confounding. In addition, we observe that there are 20 three-factor interactions, and that they are aliased with each other, forming 10 pairs. If we further assume that three-factor interactions do not exist, then the 10 pairs of three-factor interactions provide 10 degrees of freedom for error.

Example. Design a half replicate of a 2^5 factorial experiment. (a) Specify the defining contrast. (b) Show the alias pattern. (c) List the combinations to be included in the experiment.

Solution. (a) To minimize aliasing of main effects and two-factor interactions, we shall use the defining contrast *ABCDE*. (b) The alias pattern is as follows:

A = BCDE	B = ACDE	C = ABDE	D = ABCE	E = ABCD
AB = CDE	AC = BDE	AD = BCE	AE = BCD	BC = ADE
BD = ACE	BE = ACD	CD = ABE	CE = ABD	DE = ABC

(c) Using "odds," the 16 experimental combinations are as follows:

a b c d e abc abd abe acd ace ade bcd bce bde cde abcde

The large number of experimental combinations in large factorial experiments can make the selection of a fractional replicate confusing, such as in the preceding example. An order of writing down all the combinations in a factorial experiment so that none of them will be missed is called **standard order.** To write down all the 32 combinations in a 2^n factorial experiment in standard order, we start with 1, the combination having all factors at their low levels. Then, we "multiply" by *a*. Then we introduce the next factor by multiplying 1 and *a* by *b* to obtain b and ab. (So far, we have the combinations $1, a, b, ab$.) To introduce the next factor, we multiply the four preceding combinations by c, obtaining c, ac, bc, and abc. We proceed in this manner until all required letters have been introduced, depending on the value of n. For example, for a 2^5 factorial, we would proceed in this manner until we have multiplied all preceding combinations by the letter e. The 32 experimental combinations are listed in standard order as follows:

$$1 \quad a \quad b \quad ab \quad c \quad ac \quad bc \quad abc \quad d \quad ad \quad bd \quad abd \quad cd \quad acd \quad bcd \quad abcd$$
$$e \quad ae \quad be \quad abe \quad ce \quad ace \quad bce \quad abce \quad de \quad ade \quad bde \quad abde \quad cde \quad acde \quad bcde \quad abcde$$

Designing Other Fractional Factorial Experiments

The design of any fractional replicate of a 2^n experiment follows essentially the same rules that we used in designing half replicates. We shall restrict ourselves to fractions of the form 2^{-p}, where p is an integer; that is, we shall consider only half replicates ($p = 1$), quarter replicates ($p = 2$), eighth replicates ($p = 3$), and so forth, of 2^n factorial experiments. We shall refer to such a fractional replicate as a 2^{n-p} experiment.

To design a 2^{n-p} experiment, first we select p defining contrasts. Then, additional defining contrasts are automatically defined as a consequence of this choice. These additional contrasts consist of all the possible generalized interactions of the original p defining contrasts with each other. For example, to design a quarter replicate of a 2^n experiment, we initially specify two defining contrasts. Then, the generalized interaction of these contrasts also becomes a defining contrast. Thus, there are three defining contrasts in a quarter replicate. (The three defining contrasts cannot be estimated by the experiment.) Each alias set in a quarter replicate consists of four factors or interactions, the factor or interaction itself, plus its generalized interaction with each of the three defining contrasts.

Selection of the experimental combinations to be used in the quarter replicate can be made by applying the odds-evens rule in turn to each *original* defining contrast. Application of this rule to a given defining contrast creates a half replicate. Then, half of the chosen half replicate is found by again applying the odds–evens rule to the second defining contrast. It is preferable to choose odds or evens at random for each selection (a coin toss will do).

The analysis is performed by analyzing a 2^{n-p} design as if it were a full factorial experiment having $n - p$ factors. It is important to identify each main effect or interaction in the analysis with its full alias set. Thus, for example, suppose the interaction AB were found to be significant in a quarter replicate having the defining contrasts $ABCDE$, $CDEFG$, and their generalized interaction $ABFG$. Then AB is aliased with CDE, $ABCDEFG$, and FG, and it is not clear which of the four effects—AB, CDE, $ABCDEFG$, or FG—or some combination of them, is significant. Since there is no replication in a fractional factorial, degrees of freedom for error are created by making assumptions about which higher-order interactions equal zero. We shall not emphasize the analysis of these designs in this text.

Example. Design a quarter replicate of a 2^6 factorial experiment. (a) Specify the defining contrasts. (b) Show the alias pattern. (c) List the experimental combinations to be used.

Solution. (a) We must select the initial two defining contrasts carefully so as to alias as few as possible main effects and two-factor interactions with each other. Suppose we choose the two four-factor interactions $ABCD$ and $CDEF$. Then, $ABEF$ also is a defining contrast. This choice will alias some two-factor interactions with each other, but no main effect will be aliased with an interaction smaller than three factors. (We could have made other choices, perhaps using a three-factor or a four-factor interaction as one of the defining contrasts. You are urged to try some. If you do, you will find that not all main effects can be kept clear of each other or of two-factor interactions with such choices.)

(b) The alias patterns are as follows:

$A = BCD = ACDEF = BEF$ $B = ACD = BCDEF = AEF$ $C = ABD = DEF = ABCEF$
$D = ABC = CEF = ABDEF$ $E = ABCDE = CDF = ABF$ $F = ABCDF = CDE = ABE$
$AB = CD = ABCDEF = EF$ $AC = BD = ADEF = BCEF$ $AD = BC = ACEF = BDEF$
$AE = BCDE = ACDF = BF$ $AF = BCDF = ACDE = BE$ $CE = ABDE = DF = ABCF$
$DE = ABCE = CF = ABDF$ $ACE = BDE = ADF = BCF$ $ACF = BDF = ADE = BCE$

(c) We tossed a coin three times in a row, obtaining T, H. Arbitrarily assigning "odds" to "tails" and "evens" to "heads", we use odds in conjunction with the first defining contrast, $ABCD$, applying this rule to the experimental combinations written down in standard order, to obtain the following half replicate:

a b c abc d abd acd bcd ae be ce $abce$ de $abde$ $acde$
$bcde$ f bf cf $abcf$ df $abdf$ $acdf$ $bcdf$ aef bef cef $abcef$
def $abdef$ $acdef$ $bcdef$

Then, applying the evens rule to these combinations in conjunction with the defining contrast $CDEF$, we obtain the following quarter replicate:

a b acd bcd ce $abce$ de $abde$ cf $abcf$ df $abdf$ aef bef $acdef$ $bcdef$

In the preceding example, no main effect is aliased with another main effect or with a two-factor interaction. However, all two-factor interactions are aliased with one other two-factor interaction. Thus, if a two-factor interaction is found to be significant in the analysis, it will not be known which two-factor interaction in the alias set is affecting the response, or whether both are. Additional experimentation will be needed to resolve such issues if they arise.

A classification scheme has been developed to describe the extent to which a fractional factorial design aliases main effects and two-factor interactions. The "resolution" of such a design describes the extent to which main effects and interactions are aliased with one another. A design is said to be of **Resolution R** if no interaction involving k factors is aliased with any other interaction containing less than $R - k$ factors. Thus, for example,

- A design of resolution $R = $ III does not alias main effects with other main effects, but at least some main effects are aliased with two-factor interactions.
- A design of resolution $R = $ IV does not alias main effects with each other or with two-factor interactions, but it does alias at least some two-factor interactions with other two-factor interactions.

• A design of Resolution $R = V$ does not alias main effects and two-factor interactions with each another, but it does alias two-factor interactions with higher-order interactions.

Designs of Resolution V or higher are said to be "two-factor interaction-clear" designs. For such a design, it is possible to estimate the effects of all factors and all two-factor interactions, on the assumption that higher-order interactions do not exist. Designs of Resolution IV keep main effects "clear," but at least some aliasing of two-factor interactions with each other occurs. The design developed in the example on page 288 is an example of a Resolution IV design. Resolution III designs are used for preliminary studies, where only large effects are sought, and where it is expected that further experimentation will be required. Designs of resolutions less than III are rarely employed, as they alias main effects with each other.

The resolutions of some frequently used 2^{n-p} fractional-factorial designs are listed in the following table:

RESOLUTION OF TWO-LEVEL FRACTIONAL FACTORIAL DESIGNS									
	Number of Factors								
Fraction	*3*	*4*	*5*	*6*	*7*	*8*	*9*	*10*	*11*
1/2	III	IV	V	VI	VII	VIII			
1/4		III	IV	IV	IV	V	VI		
1/8			III	IV	IV	IV	IV	V	
1/16				III	IV	IV	IV	IV	V
1/32						III	III	IV	IV
1/64								III	IV
1/128									III

From the table of resolutions, it can be seen that two-factor–interaction-clear, fractional-factorial designs can be found for 5 to 11 factors. For 10 factors, for example, a one-eighth replicate can be found, having $2^{10-3} = 128$ observations, which does not alias any main effects or two-factor interactions with each other.

Fractional factorials play an important role in experimental designs for engineering applications, where data collection can be difficult and expensive. Resolution III and IV designs are used for screening experiments, where the purpose is to expend the least possible cost to rule out variables having little or no effect on the response. Then, additional experiments of higher resolution are performed with fewer variables. Experimental programs involving a series of experiments frequently are conducted. Such programs gradually narrow down the field of possible effective variables, gaining increasing information about their effects. Such programs of **sequential experimentation** are especially helpful when many potentially effective variables are involved, and their individual and joint contributions to the response initially are unknown.

EXERCISES

10.35 If the defining contrast for a half replicate of a 2^5 factorial experiment is *ABCDE*, find

a. the 16 experimental combinations to be used in the evens half replicate

  b. the alias sets

  c. the resolution of this design

10.36 If the defining contrast for a half replicate of a 2^6 factorial experiment is *ABCDEF*, find

  a. the 32 experimental combinations to be used in the odds half replicate

  b. the alias sets

  c. the resolution of this design

10.37 In constructing a one-quarter replicate of a 2^5 factorial design:

  a. Specify two defining contrasts, and find the third.

  b. Find one of the four sets of 32 experimental combinations with these defining contrasts.

  c. Write down the alias sets.

10.38 In constructing a one-quarter replicate of a 2^7 factorial design:

  a. Specify two defining contrasts and find the third.

  b. Find one of the four sets of 32 experimental combinations with these defining contrasts.

  c. Write down the alias sets.

10.39 a. If it can be assumed that no interactions exist that involve three or more factors, but main effects and two-factor interactions could be present, what 2^{n-p} fractional factorial designs would you choose for $n = 5, 6, 7, 8, 9, 10$, and 11?

  b. Does there exist a 2^{n-p} design for $n = 3$ or 4 and $p > 1$ that allows independent estimation of main effects and two-factor interactions?

10.40 What is the smallest fraction of a 2^{10} factorial design that allows main effects to be estimated independently of each other?

10.41 How many observations are required in a 2^{n-p} experiment if

  a. $n = 8, p = 3$ **b.** $n = 7, p = 2$ **c.** $n = 11, p = 5$

10.42 If the number of experimental combinations is limited to 32 and the number of factors can be 7, 8, or 9, what is the highest resolution 2^{n-p} factorial experiment that can be run?

10.43 What assumptions are implicit in running a Resolution III experiment?

10.44 What assumptions are implicit in running a Resolution IV experiment?

10.6 Application to Quality

One of the major concerns in the production of electronic products is **yield.** The yield of a production line is the ratio of the number of completed units that pass acceptance tests to the number of units started. Ordinarily, yield problems are associated with cost or productivity and not directly with quality. As the yield of a production facility increases, the cost to manufacture a unit of product decreases, and a fixed investment in the facility will produce a larger number of marketable products. Experience shows that low-yield lines tend to ship more "marginal" product, devices that are more likely to drift out of specification limits in storage, shipment, or use. A high-yield production line makes it more economically feasible to produce to higher-quality standards and to hold to specification limits that are well within operating constraints.

An integrated-circuit manufacturing facility experienced difficulty with low yields. That is, the number of silicon chips passing final electrical tests after encapsulation was too low compared to the number of "good" chips on the silicon wafers going into the process. In an effort to come to grips with the problem, a representative of each production area was chosen to participate in a quality-improvement team (QIT).

During one of the early meetings of the QIT, it was decided to see if the problem could be isolated to a single production area. The production areas under consideration were

Wafer fabrication
Photo-resist
Etching
Encapsulation
Testing

A thorough engineering analysis of the most recent of 50 failed circuits was performed in an effort to determine to which area failures could be attributed. The results of this analysis are displayed in the following Pareto chart:

ProductionArea	Number of Failures
Photo-resist	27
Wafer fabrication	12
Etching	7
Encapsulation	3
Testing	1

On the basis of this chart, it was decided first to study the factors effecting failures in the photo-resist area.

A detailed examination of the results of the engineering analyses of the 50 failed circuits, along with a comment made by the head of the photo-resist department, strongly suggested to the group that a major problem in the photo-resist area was line width. As part of the process of manufacturing an integrated circuit, a photo-resistive material is deposited on a properly prepared silicon wafer in a pattern that will become a part of the final circuit. This phase of the process is critical in determining the final quality of the circuit. Many of the patterns laid down during the deposition process are parallel lines whose widths must be carefully controlled. They must be wide enough to carry the required current without excess resistance, and they must not be so wide that the distances between adjacent lines will be small enough to cause electro-magnetic problems, or even short circuits.

A brainstorming session was held in an effort to list all of the factors that might affect line width. More than 20 factors were named. One member of the QIT suggested that these factors could be studied "one at a time" with an economically feasible number of runs. He was quickly squelched by other members who pointed out that this approach would need to be done at arbitrary fixed levels of the other factors. Nothing would be learned about the effect of any factor when the levels of the other factors were changed, and there would be no opportunity to study interactions.

Of course, there is always someone in every group who believes he or she knows the "solution" to the problem, and this QIT was no exception. One member proposed a set of levels of factors to be put immediately into the production line and promised that yields would increase immediately. As tempting as this proposal seemed to some, others argued forcefully against it. One of them said: "If we really knew what was going on, we wouldn't have the problem in the first place." Another said: "We've tried this sort of thing in the past, things got better for a while and we didn't know why; then, things got worse and we still didn't know why."

Another member suggested a fractional replicate of the 2^{21} experiment that would be required to study all the factors simultaneously. While her suggestion was given more serious consideration, it was soon realized that finding a small enough fraction was nearly impossible if it was to be capable of yielding useful information. Following this general approach, however, it was decided to attempt to boil down the large number of factors on the basis of purely engineering and manufacturing considerations. If a small enough set of critical factors could be found, on which there was general agreement about their importance, perhaps a useful screening experiment then could be designed. It was pointed out by another member of the QIT that, even if this approach failed, at least the screening experiment probably would eliminate enough factors so that some previously overlooked factors could be studied in the next experiment.

It was decided to do a cause-and-effects analysis in an effort to gain a better understanding of the contributions of the 21 factors to line width. After several iterations, it began to become clear that the levels of some of the 21 factors were dictated by the levels chosen for some other "controlling" factors. Thus, only the so-called controlling factors were retained. Also, engineering considerations identified other factors that could not play a major role in determining line width. As a result, the number of factors of interest was reduced from 21 to 9. A one-eighth replicate of a 2^9 factorial experiment, requiring 64 wafers, could be designed with Resolution IV. Thus, at least main effects would be "clear" and, if any two-factor interactions were found to be significant as a result of the experimental analysis, further experiments of much smaller size could be designed to "disentangle" the interactions. At the minimum, this experiment would screen out the remaining factors that did not appear to play an important role in determining line width, reducing the problem to something more manageable.

This experiment was performed, and it identified four photo-resist factors as playing a major role in determining line width, namely mask dimension, viscosity, spin speed, and aperture. It was decided to design a second experiment to investigate these factors more thoroughly. Since mask dimension is the controlling factor in determining line width, it was decided to vary it only over a very narrow range to find out how sensitive line width is to errors in this dimension. Also, viscosity practically could be varied only over a narrow range. Thus, it was decided to include these factors only at two levels. It was believed that the other two factors might have a curvilinear effect on line width; thus, they were run at three levels. A member of the QIT was assigned to prepare a brief report, outlining the new experiment for management approval, with the result outlined in the box shown on page 295, entitled "Proposed Line-Width Experiment."

After some grumbling about the cost, it was pointed out that the yield losses in less than one week of production exceed the experimental cost, and management approved.

The proposed experiment was performed, with the following results:

Run	Factor A	Factor B	Factor C	Factor D	Line Width
34	2.4	200	1	low	3.005
22	2.4	200	1	normal	3.181
3	2.4	200	1	high	3.165
14	2.4	200	2	low	2.479
11	2.4	200	2	normal	2.317
29	2.4	200	2	high	3.076
17	2.4	200	3	low	2.542
28	2.4	200	3	normal	3.774
19	2.4	200	3	high	3.310
35	2.4	205	1	low	3.062
26	2.4	205	1	normal	3.176
12	2.4	205	1	high	3.160
25	2.4	205	2	low	2.507
23	2.4	205	2	normal	2.299
30	2.4	205	2	high	3.101
1	2.4	205	3	low	2.498
7	2.4	205	3	normal	3.810
27	2.4	205	3	high	3.333
18	2.6	200	1	low	3.401
4	2.6	200	1	normal	3.220
5	2.6	200	1	high	3.683
31	2.6	200	2	low	3.323
10	2.6	200	2	normal	2.789
8	2.6	200	2	high	2.964
36	2.6	200	3	low	3.380
32	2.6	200	3	normal	3.709
6	2.6	200	3	high	4.015
9	2.6	205	1	low	3.115
20	2.6	205	1	normal	3.264
24	2.6	205	1	high	3.099
16	2.6	205	2	low	2.614
2	2.6	205	2	normal	2.304
13	2.6	205	2	high	3.220
33	2.6	205	3	low	2.501
21	2.6	205	3	normal	3.913
17	2.6	205	3	high	3.376

PROPOSED LINE-WIDTH EXPERIMENT

Purpose:	To study the effects of four factors on line width.
Data:	Microscopic measurements of pre-etch line width (μM) taken from 36 wafers.
Experimental Design:	A single replicate of a $2 \times 2 \times 3 \times 3$ factorial design, including the following factors:

Factor	Level 1	Level 2	Level 3
A. Mask dimension (μM)	2.4	2.6	
B. Viscosity	200	205	
C. Aperture	1	2	3
D. Spin Speed	Low	Normal	High

Method of Analysis:	Analysis of variance to determine existing main effects and two-factor interactions. Graphical method to be used to display and interpret results.
Cost:	20 hours of operator time, 8 hours of engineering time. It is expected that 50% of the experimental wafers will be commercially usable.

The order of the runs in this experiment was randomized, but the results are tabulated in an order that simplifies the analysis. The analysis of variance was performed using a statistical computer program with the following results:

Analysis of Variance of Line-Width Experiment

Source of Variability	Degrees of Freedom	Sum of Squares	Mean Square	F
A Mask Dimension	1	0.4658	0.4658	10.48**
B Viscosity	1	0.2469	0.2469	5.66*
C Aperture	2	2.3530	1.1765	26.98**
D Spin Speed	2	1.1080	0.5540	12.71**
AB	1	0.2800	0.2800	6.42*
AC	2	0.0153	0.0765	1.75
AD	2	0.1094	0.0547	1.25
BC	2	0.0123	0.0062	0.14
BD	2	0.1085	0.0543	1.25
CD	4	2.2598	0.5650	12.96*
Error	16	0.6983	0.0436	
Total	35	7.6571		

The error term for this *ANOVA* was obtained by pooling the degrees of freedom and the sums of squares for all three-factor interactions and the four-factor interaction. (It was not necessary for the computer program to calculate these sums of squares directly, as the error sum of squares was obtained by subtraction once the sums of squares for all main effects and two-factor interactions were calculated.) The asterisks following certain *F* values in this table represent significance at the .05 level (*) or the .01 level (**). It is not surprising, in view of the selection of these factors on the basis of an earlier screening experiment, that all main effects and two of the interactions were found to be significant.

Some members of the QIT were ready to prepare a report on the basis of this *ANOVA* table, believing that they now knew which factors affect line width. At this point, a consulting statistician pointed out that there was more work to be done. "Can the differences revealed by the analysis be quantified?" he asked. "What engineering interpretations can be given to these results? Especially, how does one interpret the interactions? What steps should be recommended next?"

Recognizing that this analysis of variance had served only to confirm which factors played a consistent role in determining line width, the group set out to answer these questions. The first step was to graph the responses associated with each significant factor. Since all four factors were involved in interactions, it was appropriate to graph only the *AB* and *CD* interactions. The graphs they sketched are shown in Figures 10.3 and 10.4.

The graph of the *AB* interaction (Figure 10.3) shows that the two lines are not parallel, a result consistent with an interaction. Examination of this graph supports the conclusion that the line width is less sensitive to a small change in the mask dimension when the viscosity is 205 than when it is set at

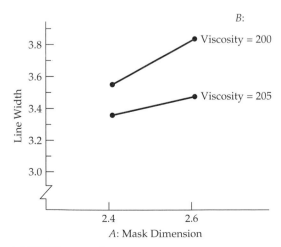

FIGURE 10.3   *AB* Interaction

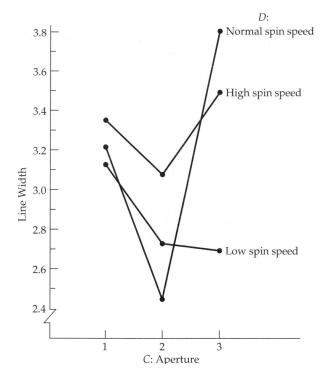

FIGURE 10.4 *CD* Interaction

200. (The absolute value of the difference between the two sample means was 0.444 when the viscosity 200 was used, but it was only 0.273 with a viscosity value of 205.)

The graph shown in Figure 10.4 shows the *CD* interaction. Each of these factors were included at three levels; thus, it was possible to examine curvilinear effects. The curves show that the response, line width, is least sensitive to variations in aperture when the low spin speed is used together with apertures in the range 2 to 3. Also, line-width variation with changes in aperture is greatest at the normal spin speed. It appears from this result that the worst possible choice of spin speed is the speed that has been used on the line—"normal" spin speed. With this spin speed, even small errors in setting the aperture width can result in unacceptable variation of line widths from a given standard.

Upon examination of these results, several changes were made in the settings on the production line. It was decided to increase the viscosity to 205 from the currently used value, 200, and to increase the spin speed to the higher level. Also, it was decided to use a new aperture level midway between 2 and 3. After these changes were made, the yield increased, but not yet to an acceptable level.

This was a first step in quality improvement, but more work was yet to be done.

EXERCISES

10.45 The experimental program described in this section failed to bring the yield to desirable levels. Was the program a failure? Why?

10.46 Select the specific purpose of the experimental program described in this section from among the following alternatives:

a. to find conditions for maximum yield

b. to study factors effecting line width

c. to produce a consistent line width

d. to improve the quality of the product

10.47 Choose between performing further experimental work with the variables studied in the final experiment or studying the effect on yield of variables involved in some other production area. Give reasons for your choice.

10.48 If it is desired to explore line width further, what would you do?

a. Repeat the last experiment.

b. Experiment with additional levels of the four variables studied in the last experiment.

c. Study the combined effects of the variables studied in the last experiment and a limited number of other variables in the screening experiment chosen for engineering reasons and because their effects in the screening experiment, although not significant at the .05 level, were potentially important.

Give reasons for your choice.

REVIEW EXERCISES

These review exercises can be used for informal review of the chapter, or as two practice examinations, designed to be taken over a time period of about one hour. The odd-numbered exercises comprise one examination; the even-numbered exercises comprise the other. Use of a computer is not required, as the exercises have been chosen to minimize computation.

10.49 An experiment was performed with the following results:

	FACTOR A			
Factor B	Level 1	Level 2	Level 3	Level 4
Level 1	1	3	2	0
	2	3	3	2
Level 2	3	4	5	1
	5	4	6	3

Perform a one-way analysis of variance to determine if the means of the four levels of factor A are significantly different at the .05 level of significance. (Use all data, but ignore the presence of factor B.)

10.50 An experiment was performed with the following results:

Factor B	FACTOR A		
	Level 1	Level 2	Level 3
Level 1	3	5	4
	4	4	5
Level 2	4	6	5
	6	6	7
Level 3	2	3	3
	4	4	5

Perform a one-way analysis of variance to determine if the means of the three levels of factor B are significantly different at the .05 level of significance. (Use all data, but ignore the presence of factor A.)

10.51 Regarding the four rows of data in Exercise 10.49 as blocks, perform a randomized-block analysis of variance.

10.52 Regarding the three columns of data in Exercise 10.50 as blocks, perform a randomized-block analysis of variance.

10.53 Perform a two-factor analysis of variance of the data given in Exercise 10.49, assuming that the second row of numbers at each level of factor B represents a second replicate.

10.54 Perform a two-factor analysis of variance of the data given in Exercise 10.50, assuming that the second row of numbers at each level of factor B represents a second replicate.

10.55 Do a multiple-comparisons test of the means of the four levels of factor A in Exercise 10.49. (Use the $ANOVA$ obtained in Exercise 10.53 and the .05 level of significance.)

10.56 Do a multiple-comparisons test of the means of the three levels of factor B in Exercise 10.50. (Use the $ANOVA$ obtained in Exercise 10.52 and the .05 level of significance.)

10.57 Write down the 16 experimental combinations in a 2^4 factorial experiment in standard order.

10.58 Write down the 32 experimental combinations in a 2^5 factorial experiment in standard order.

10.59 Use the defining contrast $ABCD$ to find a half replicate of a 2^4 factorial experiment.

a. Write down the experimental combinations.

b. Write down the alias sets.

10.60 Use the defining contrast $ABCDE$ to find a half-replicate of a 2^5 factorial experiment.

a. Write down the experimental combinations.

b. Write down the alias sets.

10.61 What is the resolution of the experiment designed in Exercise 10.59?

10.62 What is the resolution of the experiment designed in Exercise 10.60?

GLOSSARY

11

RELIABILITY

Where have we been? Statistical process control allows us to control the quality of products produced by a mass-production process. But, once the product has been produced, how do we know that it will continue to perform its task as expected?

Where are we going? The study of how long a product will operate within specifications under a given set of environmental conditions is called "reliability." The theory and applications of reliability have become increasingly important as products have become more sophisticated and have been required to operate in more hostile environments, such as space. This chapter introduces the main ideas of reliability, applying them to frequently met problems of quality.

Section 11.1, *Systems Reliability*, defines *reliability*, giving and applying the product laws of reliabilities and unreliabilities.

Section 11.2, *Failure-Time Distributions*, develops the general distribution of the time to failure of a component, defines failure rates, and applies these concepts to failure times having the exponential distribution and the Weibull distribution.

Section 11.3, *System Reliability under the Exponential Assumption*, derives formulas for calculating mean times to failure for systems whose components have exponential failure-time distributions.

Section 11.4, *Life Testing*, gives methods for estimating the mean life of a component under the exponential and Weibull failure-time assumptions.

11.1 Systems Reliability

Reliability has been characterized as an aspect of quality (page 3), where it was defined to be the ability of a product to deliver its promised performance consistently, with a minimum of breakdowns, over a sufficiently long period of use. Experience has shown that a product may function reliably under one set of conditions but not under another. For example, acceleration on launching and heat on re-entry may render an airframe, which is very reliable for a commercial jetliner, totally unreliable as a space vehicle. Thus, a more complete definition of reliability must take into account the environment in which the product is designed to operate, as well as the time period for which it operates within specified performance limits. Accordingly, we now sharpen our definition as follows: **Reliability** *is the probability that a product will perform within specified limits for at least a specified period of time under given environmental conditions.*

The definition of *reliability* as a probability makes it possible to give numerical values to reliabilities. Furthermore, the rules of probability (see Section 4.2) can be applied to the calculation of reliabilities. Thus, it is possible to calculate the reliability of a product by regarding it as a system, made up of several individual components whose individual reliabilities are known. The reliability of two kinds of systems will be discussed here. A **series system** is one in which failure of *any one* of its components results in failure of the entire system. A **parallel system** is one that will fail only if *all* of its components fail.

Beginning with series systems, suppose such a system consists of n components, and the reliability of component i is given by R_i. Usually, it can be assumed that the components will fail independently of each other. In such cases, repeated application of the special law of multiplication of probabilities, given on page 72, states that the reliability of a series system is the *product* of the reliabilities of its individual components, or

$$\text{\textit{Product Law of Reliabilities}}: \quad R_s = R_1 \cdot R_2 \cdot \ldots \cdot R_n$$

This law demonstrates the effect of increasing the complexity of a system on its reliability. A two-component series system, in which each component has the reliability .95, will have the reliability $.95^2 = .9025$, rounded to .90. Again rounding to two decimal places, if a series system has five such components, its reliability is decreased to $.95^5 = .77$, and the reliability is only $.95^{10} = .60$ with 10 components. The effect on reliability of increasing product complexity can be looked at another way. In order for a product, consisting of 10 components in series, to have a reliability of .95, each component must have a reliability of $\sqrt[10]{.95} = .995$ to three decimal places.

A method for increasing the reliability of a complex product involves the use of parallel redundancy. If several similar components are connected in parallel, failure of one or more of them will not affect the operation of the product, as long as they do not all fail. As attractive as parallel redundancy may appear, it should be recognized that this method of increasing reliability adds cost and weight to the system. In addition, if switching is required to transfer operation from a failed component to another similar component connected with it in series, the reliability of the switch must be taken into account.

The reliability of a parallel system can be calculated by observing that it will fail to function only if all components fail. Since the reliability R_i is the probability that component i will *not* fail, the probability that it *will* fail is obtained by applying the rule for the complement of an event given on page 72. Thus, the *unreliability* of the component is the probability $1 - R_i$. Again, assuming that the components act independently and repeatedly applying the special law of multiplication of probabilities, we obtain $(1 - R_1)(1 - R_2) \cdot \ldots \cdot (1 - R_n)$ for the unreliability of all n components, or the probability that all of them will fail. The system reliability is the probability that the parallel system does not fail, or

---

Product Law of Unreliabilities: $\quad R_p = 1 - (1 - R_1) \cdot (1 - R_2) \cdot \ldots \cdot (1 - R_n)$

To illustrate this law, if a parallel system consists of five components, each having the reliability .7, the system reliability is given by $1 - (1 - .7)^5 = .99757$.

The reliability of a system consisting of components connected both in series and in parallel can be obtained by calculating first the reliabilities of the parallel subsystems, and then by applying the product law of reliabilities to calculate the reliability of the resulting series system. The following example illustrates such a calculation.

Example. A system containing both series and parallel elements is diagrammed in Figure 11.1. Find the reliability of the system.

Solution. The parallel subsystem *BC* can be replaced with an equivalent component having the reliability $1 - (1 - .90)^2 = .99$. Similarly, the parallel subsystem *EFGH* can be replaced with an equivalent component having the reliability $1 - (1 - .75)^4 = .996$, to three decimal places. Thus, the reliability of the entire series system is given by $(.95)(.99)(.97)(.996)(.98) = .890$, rounding to three decimal places.

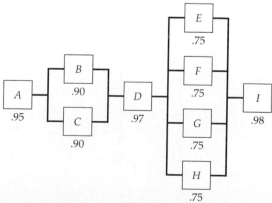

FIGURE 11.1 System Containing Series and Parallel Elements

EXERCISES

11.1 A system consists of five components connected in series with reliabilities of .995, .990, .992, .995, and .998. Find the reliability of the system.

11.2 A series system consists of six components, three components having the reliability .95 each, and three components having the reliability .99 each. Find the reliability of the system.

11.3 Six identical components are connected in series to form a system that must have a reliability of .95. What must be the reliability of each component?

11.4 Ten identical components are connected in series to form a system that must have a reliability of .90. What must be the reliability of each component?

11.5 A system consists of four components connected in parallel with reliabilities of .80, .70, .70, and .65. Find the reliability of the system.

11.6 A system consists of five components connected in parallel with reliabilities of .85, .80, .65, .60, and .70. Find the reliability of the system.

11.7 Five identical components are connected in parallel to form a system that must have a reliability of .99. What must be the reliability of each component?

11.8 Four identical components are connected in parallel to form a system that must have a reliability of .999. What must be the reliability of each component?

11.9 Find the reliability of the system depicted in Figure 11.2.

11.10 Find the reliability of the system depicted in Figure 11.3.

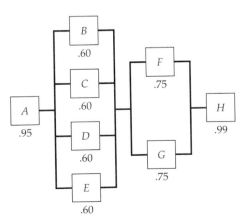

FIGURE 11.2 System for Exercise 11.9

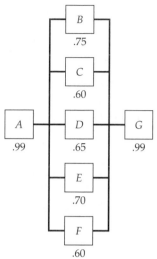

FIGURE 11.3 System for
Exercise 11.10

11.2   Failure-Time Distributions

The definition of reliability implies that the reliability of a component or a system will depend upon the length of time it has been in service. Thus, we shall be interested in the probability of failure during the time interval from 0 to t, denoted by $F(t)$. $F(t)$ can be regarded as a value of a **failure-time distribution** having the probability density function, with the values $f(t)$, and related to $F(t)$ by

$$F(t) = \int_0^t f(x)dx$$

The corresponding **reliability function**, whose values are the probabilities of survival to time t, is given by

$$R(t) = 1 - F(t)$$

The **failure-rate function** of a component or a system at time t is defined to be the rate of change of its failure-time distribution at time t, divided by its reliability at time t. A common kind of failure-rate curve, or plot of failure rates against time, is shared by many systems, whether they be manufactured systems or biological systems. For the early stages in the life of the system, the failure rate tends to decrease; during the middle stages, it tends to remain constant, and for the later stages, it tends to increase. This "bathtub curve" of failure rates is shown in Figure 11.4. When applied to biological systems, the period of early failures often is referred to as "infant mortality." Failures during the period of constant failure rate are caused by accidents, whether they result from genetic deficiencies, automobile accidents, or other kinds of random events. This period, then, is referred to as the

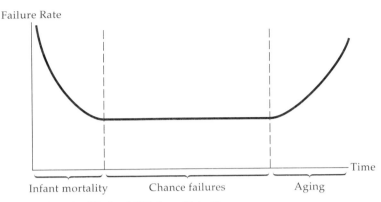

Failure Rate

Infant mortality   Chance failures   Aging

FIGURE 11.4   "Bathtub" Failure-Rate Curve

period of "chance failures." Finally, failures that occur during the period of increasing failure rates are associated with aging, or wearout.

The failure rate, $Z(t)$, of a component or system can be expressed in terms of its reliability function as follows:

$$\textbf{\textit{Failure Rate:}} \quad Z(t) = \frac{f(t)}{R(t)} = \frac{f(t)}{1 - F(t)}$$

Conversely, the reliability, $R(t)$, can be expressed in terms of the failure rate, $Z(t)$, as follows:

$$R(t) = e^{-\int_0^t Z(x)\,dx}$$

Since $Z(t) = f(t)/R(t)$, this formula can be written in terms of the probability density of failure times, as follows:

$$\textbf{\textit{Failure-Time Density Function:}} \quad f(t) = Z(t) \cdot e^{-\int_0^t Z(x)\,dx}$$

Early failures usually occur for a manufactured product as a result of failure to meet specifications. Manufacturers of some products, especially electronic products, subject their products to a "burn-in" period of operation for the express purpose of weeding out early failures. One purpose of a product warranty is to protect the consumer from failures that occur during the period of early failures. Wearout failures occur during the latter stages of operation, and the onset of the wearout-failure period generally defines the normal life of a product.

Exponential Failure Times

The length of the period of chance failures is much greater than that of the early-failure or wearout periods for most products. Thus, the period of chance failures can be described as the normal operating period for the product. During this period, the failure rate is a constant, independent of t. Letting $Z(t) = \lambda$, and substituting this value into the formula for $f(t)$, we obtain

$$f(t) = \lambda \cdot e^{-\lambda t} \quad \text{for} \quad \lambda > 0,\ t > 0$$

This is the equation for the density function of the exponential distribution with parameter λ. (This distribution was defined on page 124.) The exponential reliability function is given by

$$R(t) = 1 - F(t) = 1 - \lambda \int_0^t e^{-\lambda x} dx = e^{-\lambda t}$$

The mean of the exponential distribution is $\mu = 1/\lambda$; thus, the mean time to failure under the assumption of an exponential failure-time distribution is $1/\lambda$. The failure rate of such a system is λ, independent of its age. If the system is repaired after each failure, the mean time between *successive* failures also is given by $1/\lambda$. Thus, in the exponential case, it makes sense to refer to $1/\lambda$ as the mean time *between* failures, abbreviated **MTBF**.

The Weibull Distribution of Failure Times

During the periods of early failures and wearout failures, the failure rate is decreasing or increasing, respectively, as shown in Figure 11.4. Thus, the exponential distribution cannot be used to describe failure times during these periods. A distribution that has been applied successfully to the description of failure times during periods of non-constant failure rates is a generalization of the exponential distribution, called the **Weibull distribution**. The density function of this distribution is given by

$$f(t) = \alpha \beta t^{\beta-1} e^{-\alpha t^\beta} \quad \text{for} \quad \alpha > 0,\ \beta > 0,\ t > 0$$

Graphs of the Weibull distribution are shown in Figure 11.5 for $\alpha = 1$ and several values of β.

The Weibull reliability function is given by

$$R(t) = \alpha \beta \int_0^t x^{\beta-1} e^{-\alpha x^\beta} dx = e^{-\alpha t^\beta}$$

Thus, the Weibull failure rate is

$$Z(t) = \frac{f(t)}{R(t)} = \frac{\alpha \beta t^{\beta-1} e^{-\alpha t^\beta}}{e^{-\alpha t^\beta}} = \alpha \beta t^{\beta-1}$$

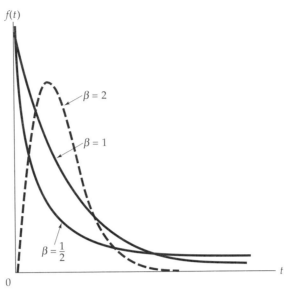

FIGURE 11.5 Weibull Density Functions ($\alpha = 1$)

It can be seen that the Weibull failure-rate function depends on t, increasing as t increases when $\beta > 1$, and decreasing as t increases when $\beta < 1$. If $\beta = 1$, the failure-rate function becomes $Z(t) = \alpha$, corresponding to the exponential failure-time distribution having the parameter α. Thus, the exponential failure-time distribution is the special case of the Weibull distribution having the parameter $\beta = 1$.

The mean time to failure for the Weibull distribution can be obtained by integrating

$$\mu = \int_0^\infty x \cdot \alpha\beta x^{\beta-1} e^{-\alpha x^\beta} dx$$

The resulting mean is

$$\mu = \alpha^{-1/\beta} \Gamma(1 + \frac{1}{\beta})$$

where the Gamma function is defined by

$$\Gamma(a) = \int_0^\infty x^{a-1} e^{-x} dx$$

Note that this mean should not be referred to as the mean time *between* failures, or *MTBF*, as the failure-rate function of the Weibull distribution generally depends on t, the age of the product having the Weibull failure-time distribution. Instead, it is reasonable to refer to this mean as the mean time to *first* failure.

EXERCISES

11.11 A component has an exponential failure-time distribution with a failure rate of 0.0015 failures per hour. Find the probability that the component will fail during the first 500 hours of operation.

11.12 A component has an exponential failure-time distribution with a failure rate of 0.0020 failures per hour. Find the probability that the component will fail during the first 450 hours of operation.

11.13 Find the reliability function corresponding to the failure-time distribution whose density function is $f(t) = \lambda t^2$.

11.14 Find the reliability function corresponding to the failure-time distribution whose density function is $f(t) = \lambda t$.

11.15 Find the failure-rate function corresponding to the failure-time distribution given in Exercise 11.13.

11.16 Find the failure-rate function corresponding to the failure-time distribution given in Exercise 11.14.

11.17 Given the failure-rate function $Z(t) = t^\lambda$, with $\lambda > -1$, find the corresponding reliability function.

11.18 Given the failure-rate function $Z(t) = \alpha t^{\beta-1}$, with $\beta > 0$, find the corresponding reliability function.

11.19 a. Can the Weibull failure-time distribution be applied to each of the three components of the bathtub curve of failure rates?

b. If so, what portion of the curve would correspond to

    i. $\beta = 1$ ii. $\beta < 1$ iii. $\beta > 1$

11.20 Find the mean time to failure corresponding to the Weibull failure-time distribution having $\alpha = 10^{-4}$ and $\beta = 1$.

11.3 System Reliability under the Exponential Assumption

In this section we shall be concerned with finding the times to failure for series and parallel systems when it can be assumed that each component has the exponential failure-time distribution. First, we shall consider a system consisting only of components connected in series. It will be assumed that n components have the exponential failure-time distribution, with failure rates $\lambda_1, \lambda_2, \ldots, \lambda_n$. The product law of reliabilities, given on page 302, states that the reliability of the system is the product of the reliabilities of its individual components. The reliability function corresponding to the exponential failure-time distribution is given on page 307 to be $R(t) = e^{-\lambda t}$. Thus, in the exponential case, we obtain the product law of reliabilities

$$R(t) = \prod_{i=1}^{n} e^{-\lambda_i t} = e^{-t\sum_{i=1}^{n}\lambda_i}$$

It can be seen that the system reliability function also is of the form $R(t) = e^{-\lambda t}$, where $\lambda = \lambda_1 + \lambda_2 + \ldots + \lambda_n$. Thus, the system has an exponential reliability function with a failure rate equal to the *sum* of the failure rates of its components. Now, let us suppose that each component in a series system is replaced immediately when it fails by another component having the same failure rate. Since the *MTBF* is the reciprocal of the failure rate (page 307), the mean time between failures of a series system is given by

$$MTBF = \frac{1}{\lambda_1 + \lambda_2 + \ldots + \lambda_n}$$

Expressing each failure rate, λ_i, in terms of its *MTBF*, $1/\lambda_i$, we obtain the following formula, giving the system *MTBF* in terms of the *MTBF*'s of its components:

MTBF (Series System): $MTBF_s = \dfrac{1}{\dfrac{1}{MTBF_1} + \dfrac{1}{MTBF_2} + \ldots + \dfrac{1}{MTBF_n}}$

In the special case where all n components have the same *MTBF*, the system *MTBF* becomes simply $MTBF_s = MTBF/n$.

The system unreliability for a parallel system is given by

$$F_p(t) = \prod_{i=1}^{n} (1 - e^{-\lambda_i t})$$

Thus, even if all components have the exponential failure-time distribution, the system failure-time distribution is not exponential. However, while the *MTBF* of a parallel system has no meaning, the mean time to first failure of the system can be calculated. The general result is quite complicated, but a very useful result can be obtained in the special case where all components have the same failure rate. It can be shown that the mean time to failure of a parallel system whose components have the exponential failure-time distribution with the same failure rate λ is given by

Mean System Failure Time: $\mu_p = \dfrac{1}{\lambda}(1 + \dfrac{1}{2} + \ldots + \dfrac{1}{n})$

This law for parallel systems shows the futility of relying too heavily on parallel redundancy to increase reliability. In addition to reliability losses that might be caused by switching from a failed component to another one in parallel, each additional component added in parallel contributes less to increasing the system mean time to failure than the last. To illustrate, suppose a parallel system has two components with failure rates of 0.001 failures per hour. The mean time to failure of this system is $\dfrac{1}{0.001}(1 + \dfrac{1}{2}) = 1{,}500$ hours. Addition of a third component in parallel, having the same failure rate, increases the system mean time to failure to $\dfrac{1}{0.001}(1 + \dfrac{1}{2} + \dfrac{1}{3}) = 1{,}833$ hours. Thus, a 50% increase in the number of components increases the mean time to failure of the system by only 22%.

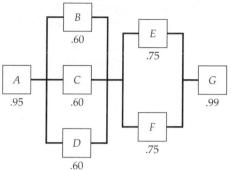

FIGURE 11.6 System for Example below

Example. Find the mean time to failure of the system diagrammed in Figure 11.6. (The numbers given in this figure are reliabilities for 100 hours of operation, and it is assumed that each component has an exponential failure-time distribution.)

Solution. First, we must calculate the failure rate corresponding to each reliability. Since, for each component, $R(100) = e^{-100\lambda}$, the corresponding failure rate can be found by solving for λ. For example, for component A, the failure rate is $-\ln(.95)/100 = 5.13 \cdot 10^{-4}$, or 0.51 failures per thousand hours. Similar calculations give the following failure rates (in failures per thousand hours) for all seven components:

Component	A	B	C	D	E	F	G
Failures per 1000 hrs	0.51	5.11	5.11	5.11	2.88	2.88	0.101

First, the mean time to failure is computed for each of the two parallel subsystems. For the subsystem BCD we obtain the mean failure time $\frac{1}{5.11}(1 + \frac{1}{2} + \frac{1}{3}) = 0.36$ thousand hours. Thus, the subsystem failure rate is $\frac{1}{0.36} = 2.78$ thousand hours. For the subsystem EF we obtain the mean failure time $\frac{1}{2.88}(1 + \frac{1}{2}) = 0.52$ thousand hours, and the failure rate of this parallel sub-system is $\frac{1}{0.52} = 1.92$ thousand hours. Thus, the sum of the failure rates for the series system $A(BCD)(EF)G$ is $0.51 + 2.78 + 1.92 + 0.10 = 5.31$. The system mean time to failure is $1/5.31 = 0.19$ thousand hours, or 190 hours.

EXERCISES

11.21 Find the *MTBF* of a system consisting of 10 components connected in series, each of which has a failure rate of 0.005 failures per hour.

11.22 Find the *MTBF* of a system consisting of 8 components connected in series, each of which has a failure rate of 0.01 failures per hour.

11.23 Find the mean time to failure of a system consisting of 4 components connected in parallel, each of which has a failure rate of 0.10 failures per hour.

11.24 Find the mean time to failure of a system consisting of 5 components connected in parallel, each of which has a failure rate of 0.07 failures per hour.

11.25 A system consists of 3 components, connected in series, whose reliabilities over a 10-hour period are .995, .990, and .985, respectively. Find the *MTBF* of the system.

11.26 A system consists of 2 components, connected in series, whose reliabilities over a 100-hour period are .85 and .90, respectively. Find the mean time to failure of the system.

11.27 Find the mean time to failure of the system diagrammed in Figure 11.2 on page 304 if the reliabilities are given for a 1,000-hour period of operation.

11.28 Find the mean time to failure of a parallel system consisting of 10 components, each having a reliability of .50 for 1,000 hours of operation.

11.4   Life Testing

It has been shown how the reliability of certain kinds of systems can be found if the reliabilities of its components are known. The reliabilities of components or subsystems usually are estimated by **life testing;** that is, by subjecting like components to the same environmental conditions and measuring their times to failure. A life test is said to be a **replacement test** if each component on test that fails is replaced immediately by a new one. Otherwise, the life test is called a **nonreplacement test.** Life tests often are **truncated;** that is, they are terminated before all components on test have failed.

Life Tests under the Exponential Assumption

Let us assume first that the failure-time distribution of each component on test satisfies the exponential assumption. This assumption can be checked by graphing T_i/T_r against i/r; that is, graphing the ratio of the time on test until the ith failure occurs to the time on test until the rth failure occurs against the ratio of i to the total number of failures that have occurred in the test. If the exponential assumption holds, the points on this graph should closely follow a straight line with a 45-degree slope. If the points tend to curve upward, above the 45-degree line, this suggests that the failure rate increases with time; if the points curve downward, this suggests that the failure rate decreases with time.

   To estimate the mean life of a component, $\mu = 1/\lambda$, n components are put on test, and the test is truncated after r of these components have failed. The observed failure times are $t_1 \le t_2 \le \ldots \le t_r$. If T_r is defined to be the accumulated life of all components on test until the rth failure occurs, then for replacement tests,

$$T_r = nt_r$$

and for nonreplacement tests,

$$T_r = (n - r)t_r + \sum_{i=1}^{r} t_i$$

It can be shown that, for both replacement and nonreplacement tests, an unbiased estimate of the mean life is given by

> ***Estimated Mean Life:*** $\hat{\mu} = \dfrac{T_r}{r}$

Confidence limits for the mean life can be based on the chi-square distribution, introduced on page 147. It can be shown that $2T_r/\mu$ is a value of a random variable that has the chi-square distribution with $2r$ degrees of freedom. This statement is true, whether the test is conducted with or without replacement, as long as the appropriate expression is substituted for T_r. Thus, a $(1 - \alpha)$ 100% confidence interval for the mean life, μ, is given by

> ***Confidence Interval for Mean Life:*** $\dfrac{2T_r}{\chi_2^2} < \mu < \dfrac{2T_r}{\chi_1^2}$

In this formula χ_1^2 and χ_2^2 cut off left- and right-hand tails of area $\alpha/2$ under the Chi-square distribution having $2r$ degrees of freedom.

Example. Suppose that 40 components are placed on life test without replacement, and the test is discontinued after 10 of them have failed. The 10 failure times (in thousands of hours) are assumed to be as follows:

0.35   0.78   0.97   1.44   1.56   2.01   2.57   2.68   3.24   3.82

a. Check to see if the exponential assumption is reasonable.

b. Estimate the mean life of these components.

c. Find a 95% confidence interval for the mean life.

Solution.   a. To check the exponential assumption, we first find the values of T_i for $i = 1, 2, \ldots, 10$. Using the nonreplacement-test formula for accumulated life, we calculate $T_1 = 0.35 + (40 - 1)0.35 = 14.00$, $T_2 = 0.35 + 0.78 + (40 - 2)0.78 = 30.77$, etc. All values of T_i are shown in the following table, along with the calculations required to plot the graph.

i	1	2	3	4	5	6	7	8	9	10
T_i	14.00	30.77	38.36	55.38	59.70	75.45	94.50	98.12	116.04	134.02
T_i/T_{10}	0.104	0.229	0.286	0.413	0.445	0.563	0.705	0.732	0.866	1.000
$i/10$	0.1	0.2	0.3	0.4	0.5	0.6	0.7	0.8	0.9	1.0

The resulting graph of T_i/T_r plotted against i/r is shown in Figure 11.7, and it appears that the exponential assumption is reasonable.

b. The estimated mean life is $T_{10}/10 = 134.03/10 = 13.403$ thousand hours.

c. To find a 95% confidence interval for the mean life, first the values of χ^2 that cut off left- and right-hand tails of area .025 under the chi-square distribution are found from Table 5, with $2r = 20$ degrees of freedom. Obtaining $\chi_1^2 = 9.591$ and $\chi_2^2 = 34.170$, we find that the 95% confidence interval is given by

$$\frac{2\,(134.02)}{34.170} < \mu < \frac{2\,(134.02)}{9.591}$$

or $7.844 < \mu < 27.947$, expressed in thousands of hours.

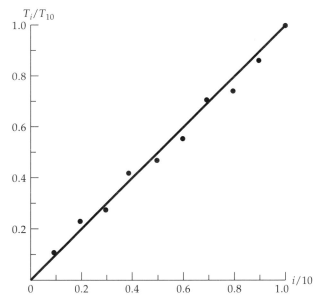

FIGURE 11.7 Plot of T_i/T_r against i/r

Life Tests under the Weibull Assumption

The appropriateness of the assumption of a Weibull failure-time distribution can be checked graphically by transforming its reliability function into a linear relationship. Taking the natural logarithm of the Weibull reliability function

$$R(t) = e^{-\alpha t^\beta}$$

we obtain

$$\ln R(t) = -\alpha t^\beta$$

or

$$\ln \left[\frac{1}{R(t)}\right] = \alpha t^{\beta}$$

Again taking natural logarithms, we obtain

$$\ln \ln \left[\frac{1}{R(t)}\right] = \ln \alpha + \beta \cdot \ln t$$

Then, letting $y = \ln \ln \dfrac{1}{R(t)}$, $A = \ln \alpha$, and $x = \ln t$, this equation can be written as

$$y = A + \beta x$$

which expresses a linear relationship between $x = \ln t$ and $y = \ln \ln \dfrac{1}{R(t)}$.

If n units are put on life test and we observe the failure times $t_1, t_2, \ldots,$ t_r, with $r \leq n$, we can estimate $F(t_i)$ by

$$\hat{F}(t_i) = \frac{i}{n+1}$$

following the same method used to obtain normal scores (page 115). Using the relationship $R(t) = 1 - F(t)$, we can graph

$$y_i = \ln \ln \frac{1}{1 - \hat{F}(t_i)} = \ln \ln \frac{n+1}{n+1-i}$$

against $x_i = \ln t_i$ to check whether the relationship is approximately linear. If so, the assumption of a Weibull failure-time distribution is reasonable.

An estimate of the mean time to failure requires that the Weibull parameters α and β be estimated. While a method of estimation called the method of maximum likelihood produces estimates of the parameters of a distribution with excellent properties, rather complicated equations are obtained when this method is applied to the Weibull distribution. Least-squares estimates for these parameters can be obtained, however, by fitting a straight line to the data points (x_i, y_i). The intercept of the fitted line provides an estimate of $\ln \alpha$, and the slope estimates β directly.

> *Example.* Suppose 50 components are put on life test and the times to failure of the first 10 components (in hours) are
>
> | 28 | 57 | 89 | 95 | 111 | 142 | 165 | 189 | 204 | 220 |
> |----|----|----|----|-----|-----|-----|-----|-----|-----|
>
> a. Check whether these data are consistent with a Weibull failure-time distribution.
>
> b. If so, obtain least-squares estimates of the parameters of the failure-time distribution.
>
> c. Use these estimates to obtain an estimate of the mean time to failure.

Solution. a. To check these data, we first obtain values of x_i and y_i as follows:

i	t_i	$x_i = \ln t_i$	$\dfrac{n+1}{n+1-i}$	$y_i = \ln \ln \dfrac{n+1}{n+1-i}$
1	28	3.33	1.020	−3.92
2	57	4.04	1.041	−3.21
3	89	4.49	1.063	−2.80
4	95	4.55	1.085	−2.51
5	111	4.71	1.109	−2.27
6	142	4.96	1.133	−2.08
7	165	5.11	1.159	−1.91
8	189	5.24	1.186	−1.77
9	204	5.32	1.214	−1.64
10	220	5.39	1.244	−1.52

A plot of y_i against x_i is shown in Figure 11.8, and it appears to be linear, indicating good agreement with the Weibull failure-time assumption.

b. Least-squares estimates of $\ln \alpha$ and β can be obtained by using the methods of Section 9.1. Then the antilog of $\ln \alpha$ is taken to obtain an estimate of α. The resulting estimates are $\hat{\alpha} = 0.0000374$ and $\hat{\beta} = 1.173$. Since $\hat{\beta}$ is estimated to exceed 1, the data indicate that the failure rate is increasing with time.

c. The mean time to failure is estimated by substituting the estimated values of α and β into the formula for the mean of the Weibull distribution given on page 308. This yields (in hours):

$$\hat{\mu} = (0.0000374)^{-1/1.173}\, \Gamma(1 + \frac{1}{1.173}) = 5,627$$

Life tests of components expected to have a long life often are conducted under accelerated environmental conditions. **Accelerated life tests** select con-

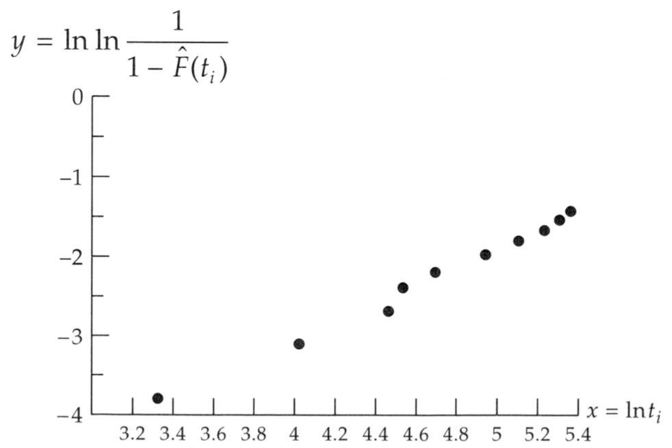

FIGURE 11.8   Weibull Failure-Time Plot

ditions designed to shorten lifetimes by increasing temperature, humidity, shock and vibration, and so forth, well beyond the limits normally expected to be encountered. To facilitate the conduct of an accelerated life test, it is necessary to have a *derating curve* that gives the relationship between lifetime and temperature, for example. If a derating curve has been found, either by performing research to obtain an estimate of the curve or by application of known physical laws, life tests of much shorter duration can be performed. Then the derating curve can be used to translate the estimated mean life calculated under the accelerated environment to an estimated mean life under "normal" conditions.

EXERCISES

11.29 Suppose that 40 units have been placed on life test without replacement, and the test is truncated after 8 units have failed. Their failure times are 78, 132, 456, 504, 601, 695, 780, 994 hours.

a. Check whether it can reasonably be assumed that the failure times are exponentially distributed.

b. If the answer to part a is yes, estimate the mean life, μ of such units.

c. Find a 90% confidence interval for μ.

11.30 To estimate the mean failure time of structural sheet metal, 100 samples were tested by vibrating them with specified frequencies and amplitudes. Failed samples were not replaced. The test was truncated after seven samples failed, with the following times to failure: 2.5, 3.8, 6.1, 9.8, 12.5, 15.3, and 18.2 thousand hours.

a. Check whether it reasonably can be assumed that the failure times are exponentially distributed.

b. If the answer to part a is yes, estimate the mean life, μ of such units.

c. Find a 95% confidence interval for μ.

11.31 In a life test of 100 light bulbs, the bulbs were turned on and off until failure occurred. Failed bulbs were replaced immediately. The numbers of cycles until the first five failures occurred were 265, 390, 615, 945, and 1260. Assume that the number of cycles to failure follows the general pattern of an exponential distribution.

a. Estimate μ, the mean number of cycles to failure.

b. Construct a 90% confidence interval for μ.

11.32 Fifty samples of wire were bent and then straightened until they broke. Broken wires were replaced immediately. The number of bending cycles required for the first six of these samples to be broken was 15, 18, 20, 24, 30, and 42. Assume that the number of bending cycles to failure follows the general pattern of an exponential distribution.

a. Estimate μ, the mean number of cycles to failure.

b. Construct a 95% confidence interval for μ.

11.33 One hundred components were put on life test until eight of them failed. The failure times were 7, 20, 53, 89, 97, 125, 210, and 254 hours.

   a. Check whether it reasonably can be assumed that the failure times have a Weibull distribution.

   b. If the answer to part a is yes, use regression analysis to estimate the parameters of this distribution.

11.34 A sample of 60 cutting tools for a lathe was put on a continuous life test until the first 9 of them failed. The observed failure times were 2.8, 6.7, 9.7, 15.4, 28.1, 40.6, 82.5, 164.6, and 230.0 hours.

   a. Check whether it reasonably can be assumed that the failure times have a Weibull distribution.

   b. If the answer to part a is yes use regression analysis to estimate the parameters of this distribution.

REVIEW EXERCISES

These review exercises can be used for informal review of the chapter, or as two practice examinations, designed to be taken over a time period of about one hour. The odd-numbered exercises comprise one examination; the even-numbered exercises comprise the other.

11.35 If a system consists of four components connected in series with reliabilities of .99, .95, .98, and .99, find the reliability of the system.

11.36 If a system consists of five identical components connected in series, each having the reliability .99, find the reliability of the system.

11.37 If a system consists of three identical components connected in parallel, each having the reliability .75, find the reliability of the system.

11.38 If a system consists of four components connected in parallel with reliabilities .7, .8, .6, and .6, find the reliability of the system.

11.39 A system consists of two components connected in series, each having the reliability .98, and a series-connected subsystem having two components connected in parallel, each having the reliability .8. What is the reliability of the system?

11.40 A system consists of two components connected in parallel, each having the reliability .99, and a series-connected subsystem having three components connected in parallel, each having the reliability .7. What is the reliability of the system?

11.41 Certain components are assumed to have an exponential failure-time distribution with a failure rate of 0.005 failures per hour. What is the probability that a randomly chosen component will fail before 200 hours?

11.42 Certain components are assumed to have an exponential failure-time distribution with a failure rate of 0.001 failures per hour. What is the probability that a randomly chosen component will survive 1,000 hours of operation?

11.43 Certain components are assumed to have a Weibull failure-time distribution with the parameters $\alpha = 0.005$ and $\beta = 0.8$. What is the probability that a randomly chosen component will survive 1,000 hours of operation?

11.44 Certain components are assumed to have a Weibull failure-time distribution with the parameters $\alpha = 0.0001$ and $\beta = 1.5$. What is the probability that a randomly chosen component will fail before 200 hours of operation?

11.45 A series system has four components. If it can be assumed that the components have exponential failure-time distributions with *MTBF*'s of 50, 100, 75, and 80 hours, what is the *MTBF* of the system?

11.46 Three components are connected in series. If it can be assumed that they have exponential failure-time distributions with *MTBF*'s of 500, 750, and 1000 hours, what is the *MTBF* of the system?

11.47 Five components are connected in parallel. If it can be assumed that they have exponential failure-time distributions with the common failure rate 0.0005 failures/hour, what is the mean system failure time?

11.48 Four components, having the identical failure rate of 0.001 failures per hour, are connected in parallel. If it can be assumed that they have exponential failure-time distributions, find the mean system failure time.

11.49 A nonreplacement life test of 100 components was terminated after 5 of them had failed. The failure times were 51, 85, 93, 250, and 512 hours. Assuming the exponential failure-time distribution, find 90% confidence limits for the mean life of the component.

11.50 A life test of 50 components was terminated after 6 of them had failed. (Each component that failed was replaced immediately.) The failure times were 11, 42, 95, 150, 231, and 380 hours. Assuming the exponential failure-time distribution, find 95% confidence limits for the mean life of the component.

GLOSSARY

Appendix: A
STATISTICAL TABLES

TABLE 1 Binomial Distribution Function

$$B(x; n, p) = \sum_{k=0}^{x} \binom{n}{k} p^k (1-p)^{n-k}$$

n	x	.05	.10	.15	.20	.25	.30	.35	.40	.45	.50	.55	.60	.65	.70	.75	.80	.85	.90	.95
2	0	.9025	.8100	.7225	.6400	.5625	.4900	.4225	.3600	.3025	.2500	.2025	.1600	.1225	.0900	.0625	.0400	.0225	.0100	.0025
	1	.9975	.9900	.9775	.9600	.9375	.9100	.8775	.8400	.7975	.7500	.6975	.6400	.5775	.5100	.4375	.3600	.2775	.1900	.0975
3	0	.8574	.7290	.6141	.5120	.4219	.3430	.2746	.2160	.1664	.1250	.0911	.0640	.0429	.0270	.0156	.0080	.0034	.0010	.0001
	1	.9927	.9720	.9393	.8960	.8438	.7840	.7183	.6480	.5748	.5000	.4252	.3520	.2818	.2160	.1563	.1040	.0607	.0280	.0073
	2	.9999	.9990	.9966	.9920	.9844	.9730	.9571	.9360	.9089	.8750	.8336	.7840	.7254	.6570	.5781	.4880	.3859	.2710	.1426
4	0	.8145	.6561	.5220	.4096	.3164	.2401	.1785	.1296	.0915	.0625	.0410	.0256	.0150	.0081	.0039	.0016	.0005	.0001	.0000
	1	.9860	.9477	.8905	.8192	.7383	.6517	.5630	.4752	.3910	.3125	.2415	.1792	.1265	.0837	.0508	.0272	.0120	.0037	.0005
	2	.9995	.9963	.9880	.9728	.9492	.9163	.8735	.8208	.7585	.6875	.6090	.5248	.4370	.3483	.2617	.1808	.1095	.0523	.0140
	3	1.0000	.9999	.9995	.9984	.9961	.9919	.9850	.9744	.9590	.9375	.9085	.8704	.8215	.7599	.6836	.5904	.4780	.3439	.1855
5	0	.7738	.5905	.4437	.3277	.2373	.1681	.1160	.0778	.0503	.0313	.0185	.0102	.0053	.0024	.0010	.0003	.0001	.0000	.0000
	1	.9774	.9185	.8352	.7373	.6328	.5282	.4284	.3370	.2562	.1875	.1312	.0870	.0540	.0308	.0156	.0067	.0022	.0005	.0000
	2	.9988	.9914	.9734	.9421	.8965	.8369	.7648	.6826	.5931	.5000	.4069	.3174	.2352	.1631	.1035	.0579	.0266	.0086	.0012
	3	1.0000	.9995	.9978	.9933	.9844	.9692	.9460	.9130	.8688	.8125	.7438	.6630	.5716	.4718	.3672	.2627	.1648	.0815	.0226
	4	1.0000	1.0000	.9999	.9997	.9990	.9976	.9947	.9898	.9815	.9688	.9497	.9222	.8840	.8319	.7627	.6723	.5563	.4095	.2262
6	0	.7351	.5314	.3771	.2621	.1780	.1176	.0754	.0467	.0277	.0156	.0083	.0041	.0018	.0007	.0002	.0001	.0000	.0000	.0000
	1	.9672	.8857	.7765	.6554	.5339	.4202	.3191	.2333	.1636	.1094	.0692	.0410	.0223	.0109	.0046	.0016	.0004	.0001	.0000
	2	.9978	.9841	.9527	.9011	.8306	.7443	.6471	.5443	.4415	.3438	.2553	.1792	.1174	.0705	.0376	.0170	.0059	.0013	.0001
	3	.9999	.9987	.9941	.9830	.9624	.9295	.8826	.8208	.7447	.6563	.5585	.4557	.3529	.2557	.1694	.0989	.0473	.0158	.0022
	4	1.0000	.9999	.9996	.9984	.9954	.9891	.9777	.9590	.9308	.8906	.8364	.7667	.6809	.5798	.4661	.3446	.2235	.1143	.0328
	5	1.0000	1.0000	1.0000	.9999	.9998	.9993	.9982	.9959	.9917	.9844	.9723	.9533	.9246	.8824	.8220	.7379	.6229	.4686	.2649
7	0	.6983	.4783	.3206	.2097	.1335	.0824	.0490	.0280	.0152	.0078	.0037	.0016	.0006	.0002	.0001	.0000	.0000	.0000	.0000
	1	.9556	.8503	.7166	.5767	.4449	.3294	.2338	.1586	.1024	.0625	.0357	.0188	.0090	.0038	.0013	.0004	.0001	.0000	.0000
	2	.9962	.9743	.9262	.8520	.7564	.6471	.5323	.4199	.3164	.2266	.1529	.0963	.0556	.0288	.0129	.0047	.0012	.0002	.0000
	3	.9998	.9973	.9879	.9667	.9294	.8740	.8002	.7102	.6083	.5000	.3917	.2898	.1998	.1260	.0706	.0333	.0121	.0027	.0002
	4	1.0000	.9998	.9988	.9953	.9871	.9712	.9444	.9037	.8471	.7734	.6836	.5801	.4677	.3529	.2436	.1480	.0738	.0257	.0038
	5	1.0000	1.0000	.9999	.9996	.9987	.9962	.9910	.9812	.9643	.9375	.8976	.8414	.7662	.6706	.5551	.4233	.2834	.1497	.0444
	6	1.0000	1.0000	1.0000	1.0000	.9999	.9998	.9994	.9984	.9963	.9922	.9848	.9720	.9510	.9176	.8665	.7903	.6794	.5217	.3017
8	0	.6634	.4305	.2725	.1678	.1001	.0576	.0319	.0168	.0084	.0039	.0017	.0007	.0002	.0001	.0000	.0000	.0000	.0000	.0000
	1	.9428	.8131	.6572	.5033	.3671	.2553	.1691	.1064	.0632	.0352	.0181	.0085	.0036	.0013	.0004	.0001	.0000	.0000	.0000
	2	.9942	.9619	.8948	.7969	.6785	.5518	.4278	.3154	.2201	.1445	.0885	.0498	.0253	.0113	.0042	.0012	.0002	.0000	.0000
	3	.9996	.9950	.9786	.9437	.8862	.8059	.7064	.5941	.4770	.3633	.2604	.1737	.1061	.0580	.0273	.0104	.0029	.0004	.0000
	4	1.0000	.9996	.9971	.9896	.9727	.9420	.8939	.8263	.7396	.6367	.5230	.4059	.2936	.1941	.1138	.0563	.0214	.0050	.0004
	5	1.0000	1.0000	.9998	.9988	.9958	.9887	.9747	.9502	.9115	.8555	.7799	.6846	.5722	.4482	.3215	.2031	.1052	.0381	.0058
	6	1.0000	1.0000	1.0000	.9999	.9996	.9987	.9964	.9915	.9819	.9648	.9368	.8936	.8309	.7447	.6329	.4967	.3428	.1869	.0572
	7	1.0000	1.0000	1.0000	1.0000	1.0000	.9999	.9998	.9993	.9983	.9961	.9916	.9832	.9681	.9424	.8999	.8322	.7275	.5695	.3366

n	x	.05	.10	.15	.20	.25	.30	.35	.40	.45	.50 (p)	.55	.60	.65	.70	.75	.80	.85	.90	.95
9	0	.6302	.3874	.2316	.1342	.0751	.0404	.0207	.0101	.0046	.0020	.0008	.0003	.0001	.0000	.0000	.0000	.0000	.0000	.0000
	1	.9288	.7748	.5995	.4362	.3003	.1960	.1211	.0705	.0385	.0195	.0091	.0038	.0014	.0004	.0001	.0000	.0000	.0000	.0000
	2	.9916	.9470	.8591	.7382	.6007	.4628	.3373	.2318	.1495	.0898	.0498	.0250	.0112	.0043	.0013	.0003	.0000	.0000	.0000
	3	.9994	.9917	.9661	.9144	.8343	.7297	.6089	.4826	.3614	.2539	.1658	.0994	.0536	.0253	.0100	.0031	.0006	.0001	.0000
	4	1.0000	.9991	.9944	.9804	.9511	.9012	.8283	.7334	.6214	.5000	.3786	.2666	.1717	.0988	.0489	.0196	.0056	.0009	.0000
	5	1.0000	.9999	.9994	.9969	.9900	.9747	.9464	.9006	.8342	.7461	.6386	.5174	.3911	.2703	.1657	.0856	.0339	.0083	.0006
	6	1.0000	1.0000	1.0000	.9997	.9987	.9957	.9888	.9750	.9502	.9102	.8505	.7682	.6627	.5372	.3993	.2618	.1409	.0530	.0084
	7	1.0000	1.0000	1.0000	1.0000	.9999	.9996	.9986	.9962	.9909	.9805	.9615	.9295	.8789	.8040	.6997	.5638	.4005	.2252	.0712
	8	1.0000	1.0000	1.0000	1.0000	1.0000	1.0000	.9999	.9997	.9992	.9980	.9954	.9899	.9793	.9596	.9249	.8658	.7684	.6126	.3698
10	0	.5987	.3487	.1969	.1074	.0563	.0282	.0135	.0060	.0025	.0010	.0003	.0001	.0000	.0000	.0000	.0000	.0000	.0000	.0000
	1	.9139	.7361	.5443	.3758	.2440	.1493	.0860	.0464	.0233	.0107	.0045	.0017	.0005	.0001	.0000	.0000	.0000	.0000	.0000
	2	.9885	.9298	.8202	.6778	.5256	.3828	.2616	.1673	.0996	.0547	.0274	.0123	.0048	.0016	.0004	.0001	.0000	.0000	.0000
	3	.9990	.9872	.9500	.8791	.7759	.6496	.5138	.3823	.2660	.1719	.1020	.0548	.0260	.0106	.0035	.0009	.0001	.0000	.0000
	4	.9999	.9984	.9901	.9672	.9219	.8497	.7515	.6331	.5044	.3770	.2616	.1662	.0949	.0473	.0197	.0064	.0014	.0001	.0000
	5	1.0000	.9999	.9986	.9936	.9803	.9527	.9051	.8338	.7384	.6230	.4956	.3669	.2485	.1503	.0781	.0328	.0099	.0016	.0001
	6	1.0000	1.0000	.9999	.9991	.9965	.9894	.9740	.9452	.8980	.8281	.7340	.6177	.4862	.3504	.2241	.1209	.0500	.0128	.0010
	7	1.0000	1.0000	1.0000	.9999	.9996	.9984	.9952	.9877	.9726	.9453	.9004	.8327	.7384	.6172	.4744	.3222	.1798	.0702	.0115
	8	1.0000	1.0000	1.0000	1.0000	1.0000	.9999	.9995	.9983	.9955	.9893	.9767	.9536	.9140	.8507	.7560	.6242	.4557	.2639	.0861
	9	1.0000	1.0000	1.0000	1.0000	1.0000	1.0000	1.0000	.9999	.9997	.9990	.9975	.9940	.9865	.9718	.9437	.8926	.8031	.6513	.4013
11	0	.5688	.3138	.1673	.0859	.0422	.0198	.0088	.0036	.0014	.0005	.0002	.0000	.0000	.0000	.0000	.0000	.0000	.0000	.0000
	1	.8981	.6974	.4922	.3221	.1971	.1130	.0606	.0302	.0139	.0059	.0022	.0007	.0002	.0000	.0000	.0000	.0000	.0000	.0000
	2	.9848	.9104	.7788	.6174	.4552	.3127	.2001	.1189	.0652	.0327	.0148	.0059	.0020	.0006	.0001	.0000	.0000	.0000	.0000
	3	.9984	.9815	.9306	.8389	.7133	.5696	.4256	.2963	.1911	.1133	.0610	.0293	.0122	.0043	.0012	.0002	.0000	.0000	.0000
	4	.9999	.9972	.9841	.9496	.8854	.7897	.6683	.5328	.3971	.2744	.1738	.0994	.0501	.0216	.0076	.0020	.0003	.0000	.0000
	5	1.0000	.9997	.9973	.9883	.9657	.9218	.8513	.7535	.6331	.5000	.3669	.2465	.1487	.0782	.0343	.0117	.0027	.0003	.0000
	6	1.0000	1.0000	.9997	.9980	.9924	.9784	.9499	.9006	.8262	.7256	.6029	.4672	.3317	.2103	.1146	.0504	.0159	.0028	.0000
	7	1.0000	1.0000	1.0000	.9998	.9988	.9957	.9878	.9707	.9390	.8867	.8089	.7037	.5744	.4304	.2867	.1611	.0694	.0185	.0016
	8	1.0000	1.0000	1.0000	1.0000	.9999	.9994	.9980	.9941	.9852	.9673	.9348	.8811	.7999	.6873	.5448	.3826	.2218	.0896	.0152
	9	1.0000	1.0000	1.0000	1.0000	1.0000	1.0000	.9998	.9993	.9978	.9941	.9861	.9698	.9394	.8870	.8029	.6779	.5078	.3026	.1019
	10	1.0000	1.0000	1.0000	1.0000	1.0000	1.0000	1.0000	1.0000	.9998	.9995	.9986	.9964	.9912	.9802	.9578	.9141	.8327	.6862	.4312
12	0	.5404	.2824	.1422	.0687	.0317	.0138	.0057	.0022	.0008	.0002	.0001	.0000	.0000	.0000	.0000	.0000	.0000	.0000	.0000
	1	.8816	.6590	.4435	.2749	.1584	.0850	.0424	.0196	.0083	.0032	.0011	.0003	.0001	.0000	.0000	.0000	.0000	.0000	.0000
	2	.9804	.8891	.7358	.5583	.3907	.2528	.1513	.0834	.0421	.0193	.0079	.0028	.0008	.0002	.0000	.0000	.0000	.0000	.0000
	3	.9978	.9744	.9078	.7946	.6488	.4925	.3467	.2253	.1345	.0730	.0356	.0153	.0056	.0017	.0004	.0001	.0000	.0000	.0000
	4	.9998	.9957	.9761	.9274	.8424	.7237	.5833	.4382	.3044	.1938	.1117	.0573	.0255	.0095	.0028	.0006	.0001	.0000	.0000
	5	1.0000	.9995	.9954	.9806	.9456	.8822	.7873	.6652	.5269	.3872	.2607	.1582	.0846	.0386	.0143	.0039	.0007	.0001	.0000
	6	1.0000	.9999	.9993	.9961	.9857	.9614	.9154	.8418	.7393	.6128	.4731	.3348	.2127	.1178	.0544	.0194	.0046	.0005	.0000
	7	1.0000	1.0000	.9999	.9994	.9972	.9905	.9745	.9427	.8883	.8062	.6956	.5618	.4167	.2763	.1576	.0726	.0239	.0043	.0002
	8	1.0000	1.0000	1.0000	.9999	.9996	.9983	.9944	.9847	.9644	.9270	.8655	.7747	.6533	.5075	.3512	.2054	.0922	.0256	.0022
	9	1.0000	1.0000	1.0000	1.0000	1.0000	.9998	.9992	.9972	.9921	.9807	.9579	.9166	.8487	.7472	.6093	.4417	.2642	.1109	.0196
	10	1.0000	1.0000	1.0000	1.0000	1.0000	1.0000	.9999	.9997	.9989	.9968	.9917	.9804	.9576	.9150	.8416	.7251	.5565	.3410	.1184
	11	1.0000	1.0000	1.0000	1.0000	1.0000	1.0000	1.0000	1.0000	.9999	.9998	.9992	.9978	.9943	.9862	.9683	.9313	.8578	.7176	.4596

TABLE 1 Continued

n	x	.05	.10	.15	.20	.25	.30	.35	.40	.45	.50	.55	.60	.65	.70	.75	.80	.85	.90	.95
13	0	.5133	.2542	.1209	.0550	.0238	.0097	.0037	.0013	.0004	.0001	.0000	.0000	.0000	.0000	.0000	.0000	.0000	.0000	.0000
	1	.8646	.6213	.3983	.2336	.1267	.0637	.0296	.0126	.0049	.0017	.0005	.0001	.0000	.0000	.0000	.0000	.0000	.0000	.0000
	2	.9755	.8661	.6920	.5017	.3326	.2025	.1132	.0579	.0269	.0112	.0041	.0013	.0003	.0001	.0000	.0000	.0000	.0000	.0000
	3	.9969	.9658	.8820	.7473	.5843	.4206	.2783	.1686	.0929	.0461	.0203	.0078	.0025	.0007	.0001	.0000	.0000	.0000	.0000
	4	.9997	.9935	.9658	.9009	.7940	.6543	.5005	.3530	.2279	.1334	.0698	.0321	.0126	.0040	.0010	.0002	.0000	.0000	.0000
	5	1.0000	.9991	.9925	.9700	.9198	.8346	.7159	.5744	.4268	.2905	.1788	.0977	.0462	.0182	.0056	.0012	.0002	.0000	.0000
	6	1.0000	.9999	.9987	.9930	.9757	.9376	.8705	.7712	.6437	.5000	.3563	.2288	.1295	.0624	.0243	.0070	.0013	.0001	.0000
	7	1.0000	1.0000	.9998	.9988	.9944	.9818	.9538	.9023	.8212	.7095	.5732	.4256	.2841	.1654	.0802	.0300	.0075	.0009	.0000
	8	1.0000	1.0000	1.0000	.9998	.9990	.9960	.9874	.9679	.9302	.8666	.7721	.6470	.4995	.3457	.2060	.0991	.0342	.0065	.0003
	9	1.0000	1.0000	1.0000	1.0000	.9999	.9993	.9975	.9922	.9797	.9539	.9071	.8314	.7217	.5794	.4157	.2527	.1180	.0342	.0031
	10	1.0000	1.0000	1.0000	1.0000	1.0000	.9999	.9997	.9987	.9959	.9888	.9731	.9421	.8868	.7975	.6674	.4983	.3080	.1339	.0245
	11	1.0000	1.0000	1.0000	1.0000	1.0000	1.0000	1.0000	.9999	.9995	.9983	.9951	.9874	.9704	.9363	.8733	.7664	.6017	.3787	.1354
	12	1.0000	1.0000	1.0000	1.0000	1.0000	1.0000	1.0000	1.0000	1.0000	.9999	.9996	.9987	.9963	.9903	.9762	.9450	.8791	.7458	.4867
14	0	.4877	.2288	.1028	.0440	.0178	.0068	.0024	.0008	.0002	.0001	.0000	.0000	.0000	.0000	.0000	.0000	.0000	.0000	.0000
	1	.8470	.5846	.3567	.1979	.1010	.0475	.0205	.0081	.0029	.0009	.0003	.0001	.0000	.0000	.0000	.0000	.0000	.0000	.0000
	2	.9699	.8416	.6479	.4481	.2811	.1608	.0839	.0398	.0170	.0065	.0022	.0006	.0001	.0000	.0000	.0000	.0000	.0000	.0000
	3	.9958	.9559	.8535	.6982	.5213	.3552	.2205	.1243	.0632	.0287	.0114	.0039	.0011	.0002	.0000	.0000	.0000	.0000	.0000
	4	.9996	.9908	.9533	.8702	.7415	.5842	.4227	.2793	.1672	.0898	.0426	.0175	.0060	.0017	.0003	.0000	.0000	.0000	.0000
	5	1.0000	.9985	.9885	.9561	.8883	.7805	.6405	.4859	.3373	.2120	.1189	.0583	.0243	.0083	.0022	.0004	.0000	.0000	.0000
	6	1.0000	.9998	.9978	.9884	.9617	.9067	.8164	.6925	.5461	.3953	.2586	.1501	.0753	.0315	.0103	.0024	.0003	.0000	.0000
	7	1.0000	1.0000	.9997	.9976	.9897	.9685	.9247	.8499	.7414	.6047	.4539	.3075	.1836	.0933	.0383	.0116	.0022	.0002	.0000
	8	1.0000	1.0000	1.0000	.9996	.9978	.9917	.9757	.9417	.8811	.7880	.6627	.5141	.3595	.2195	.1117	.0439	.0115	.0015	.0000
	9	1.0000	1.0000	1.0000	1.0000	.9997	.9983	.9940	.9825	.9574	.9102	.8328	.7207	.5773	.4158	.2585	.1298	.0467	.0092	.0004
	10	1.0000	1.0000	1.0000	1.0000	1.0000	.9998	.9989	.9961	.9886	.9713	.9368	.8757	.7795	.6448	.4787	.3018	.1465	.0441	.0042
	11	1.0000	1.0000	1.0000	1.0000	1.0000	1.0000	.9999	.9994	.9978	.9935	.9830	.9602	.9161	.8392	.7189	.5519	.3521	.1584	.0301
	12	1.0000	1.0000	1.0000	1.0000	1.0000	1.0000	1.0000	.9999	.9997	.9991	.9971	.9919	.9795	.9525	.8990	.8021	.6433	.4154	.1530
	13	1.0000	1.0000	1.0000	1.0000	1.0000	1.0000	1.0000	1.0000	1.0000	.9999	.9998	.9992	.9976	.9932	.9822	.9560	.8972	.7712	.5123
15	0	.4633	.2059	.0874	.0352	.0134	.0047	.0016	.0005	.0001	.0000	.0000	.0000	.0000	.0000	.0000	.0000	.0000	.0000	.0000
	1	.8290	.5490	.3186	.1671	.0802	.0353	.0142	.0052	.0017	.0005	.0001	.0000	.0000	.0000	.0000	.0000	.0000	.0000	.0000
	2	.9638	.8159	.6042	.3980	.2361	.1268	.0617	.0271	.0107	.0037	.0011	.0003	.0001	.0000	.0000	.0000	.0000	.0000	.0000
	3	.9945	.9444	.8227	.6482	.4613	.2969	.1727	.0905	.0424	.0176	.0063	.0019	.0005	.0001	.0000	.0000	.0000	.0000	.0000
	4	.9994	.9873	.9383	.8358	.6865	.5155	.3519	.2173	.1204	.0592	.0255	.0093	.0028	.0007	.0001	.0000	.0000	.0000	.0000
	5	.9999	.9977	.9832	.9389	.8516	.7216	.5643	.4032	.2608	.1509	.0769	.0338	.0124	.0037	.0008	.0001	.0000	.0000	.0000
	6	1.0000	.9997	.9964	.9819	.9434	.8689	.7548	.6098	.4522	.3036	.1818	.0950	.0422	.0152	.0042	.0008	.0001	.0000	.0000
	7	1.000	1.0000	.9994	.9958	.9827	.9500	.8868	.7869	.6535	.5000	.3465	.2131	.1132	.0500	.0173	.0042	.0006	.0000	.0000
	8	1.0000	1.0000	.9999	.9992	.9958	.9848	.9578	.9050	.8182	.6964	.5478	.3902	.2452	.1311	.0566	.0181	.0036	.0003	.0000
	9	1.0000	1.0000	1.0000	.9999	.9992	.9963	.9876	.9662	.9231	.8491	.7392	.5968	.4357	.2784	.1484	.0611	.0168	.0022	.0001
	10	1.0000	1.0000	1.0000	1.0000	.9999	.9993	.9972	.9907	.9745	.9408	.8796	.7827	.6481	.4845	.3135	.1642	.0617	.0127	.0006
	11	1.0000	1.0000	1.0000	1.0000	1.0000	.9999	.9995	.9981	.9937	.9824	.9576	.9095	.8273	.7031	.5387	.3518	.1773	.0556	.0055
	12	1.0000	1.0000	1.0000	1.0000	1.0000	1.0000	.9999	.9997	.9989	.9963	.9893	.9729	.9383	.8732	.7639	.6020	.3958	.1841	.0362
	13	1.0000	1.0000	1.0000	1.0000	1.0000	1.0000	1.0000	1.0000	.9999	.9995	.9983	.9948	.9858	.9647	.9198	.8329	.6814	.4510	.1710
	14	1.0000	1.0000	1.0000	1.0000	1.0000	1.0000	1.0000	1.0000	1.0000	1.0000	.9999	.9995	.9984	.9953	.9866	.9648	.9126	.7941	.5367

.95	.90	.85	.80	.75	.70	.65	.60	.55	.50	.45	.40	.35	.30	.25	.20	.15	.10	.05	x	n
.0000	.0000	.0000	.0000	.0000	.0000	.0000	.0000	.0000	.0000	.0001	.0003	.0010	.0033	.0100	.0281	.0743	.1853	.4401	0	16
.0000	.0000	.0000	.0000	.0000	.0000	.0000	.0000	.0001	.0003	.0010	.0033	.0098	.0261	.0635	.1407	.2839	.5147	.8108	1	
.0000	.0000	.0000	.0000	.0000	.0000	.0000	.0001	.0006	.0021	.0066	.0183	.0451	.0994	.1971	.3518	.5614	.7892	.9571	2	
.0000	.0000	.0000	.0000	.0000	.0000	.0002	.0009	.0035	.0106	.0281	.0651	.1339	.2459	.4050	.5981	.7899	.9316	.9930	3	
.0000	.0000	.0000	.0000	.0000	.0003	.0013	.0049	.0149	.0384	.0853	.1666	.2892	.4499	.6302	.7982	.9209	.9830	.9991	4	
.0000	.0000	.0000	.0000	.0003	.0016	.0062	.0191	.0486	.1051	.1976	.3288	.4900	.6598	.8103	.9183	.9765	.9967	.9999	5	
.0000	.0000	.0000	.0002	.0016	.0071	.0229	.0583	.1241	.2272	.3660	.5272	.6881	.8247	.9204	.9733	.9944	.9995	1.0000	6	
.0000	.0000	.0002	.0015	.0075	.0257	.0671	.1423	.2559	.4018	.5629	.7161	.8406	.9256	.9729	.9930	.9989	.9999	1.0000	7	
.0000	.0001	.0011	.0070	.0271	.0744	.1594	.2839	.4371	.5982	.7441	.8577	.9329	.9743	.9925	.9985	.9998	1.0000	1.0000	8	
.0000	.0005	.0056	.0267	.0796	.1753	.3119	.4728	.6340	.7728	.8759	.9417	.9771	.9929	.9984	.9998	1.0000	1.0000	1.0000	9	
.0001	.0033	.0235	.0817	.1897	.3402	.5100	.6712	.8024	.8949	.9514	.9809	.9938	.9984	.9997	1.0000	1.0000	1.0000	1.0000	10	
.0009	.0170	.0791	.2018	.3698	.5501	.7108	.8334	.9147	.9616	.9851	.9951	.9987	.9997	1.0000	1.0000	1.0000	1.0000	1.0000	11	
.0070	.0684	.2101	.4019	.5950	.7541	.8661	.9349	.9719	.9894	.9965	.9991	.9998	1.0000	1.0000	1.0000	1.0000	1.0000	1.0000	12	
.0429	.2108	.4386	.6482	.8029	.9006	.9549	.9817	.9934	.9979	.9994	.9999	1.0000	1.0000	1.0000	1.0000	1.0000	1.0000	1.0000	13	
.1892	.4853	.7161	.8593	.9365	.9739	.9902	.9967	.9990	.9997	.9999	1.0000	1.0000	1.0000	1.0000	1.0000	1.0000	1.0000	1.0000	14	
.5599	.8147	.9257	.9719	.9900	.9967	.9990	.9997	.9999	1.0000	1.0000	1.0000	1.0000	1.0000	1.0000	1.0000	1.0000	1.0000	1.0000	15	
.0000	.0000	.0000	.0000	.0000	.0000	.0000	.0000	.0000	.0000	.0000	.0002	.0007	.0023	.0075	.0225	.0631	.1668	.4181	0	17
.0000	.0000	.0000	.0000	.0000	.0000	.0000	.0000	.0000	.0001	.0006	.0021	.0067	.0193	.0501	.1182	.2525	.4818	.7922	1	
.0000	.0000	.0000	.0000	.0000	.0000	.0000	.0001	.0003	.0012	.0041	.0123	.0327	.0774	.1637	.3096	.5198	.7618	.9497	2	
.0000	.0000	.0000	.0000	.0000	.0000	.0001	.0005	.0019	.0064	.0184	.0464	.1028	.2019	.3530	.5489	.7556	.9174	.9912	3	
.0000	.0000	.0000	.0000	.0000	.0001	.0006	.0025	.0086	.0245	.0596	.1260	.2348	.3887	.5739	.7582	.9013	.9779	.9988	4	
.0000	.0000	.0000	.0000	.0001	.0007	.0030	.0106	.0301	.0717	.1471	.2639	.4197	.5968	.7653	.8943	.9681	.9953	.9999	5	
.0000	.0000	.0000	.0001	.0006	.0032	.0120	.0348	.0826	.1662	.2902	.4478	.6188	.7752	.8929	.9623	.9917	.9992	1.0000	6	
.0000	.0000	.0000	.0005	.0031	.0127	.0383	.0919	.1834	.3145	.4743	.6405	.7872	.8954	.9598	.9891	.9983	.9999	1.0000	7	
.0000	.0000	.0003	.0026	.0124	.0403	.0994	.1989	.3374	.5000	.6626	.8011	.9006	.9597	.9876	.9974	.9997	1.0000	1.0000	8	
.0000	.0001	.0017	.0109	.0402	.1046	.2128	.3595	.5257	.6855	.8166	.9081	.9617	.9873	.9969	.9995	.9999	1.0000	1.0000	9	
.0000	.0008	.0083	.0377	.1071	.2248	.3812	.5522	.7098	.8338	.9174	.9652	.9880	.9968	.9994	.9999	1.0000	1.0000	1.0000	10	
.0001	.0047	.0319	.1057	.2347	.4032	.5803	.7361	.8529	.9283	.9699	.9894	.9970	.9993	.9999	1.0000	1.0000	1.0000	1.0000	11	
.0012	.0221	.0987	.2418	.4261	.6113	.7652	.8740	.9404	.9755	.9914	.9975	.9994	.9999	1.0000	1.0000	1.0000	1.0000	1.0000	12	
.0088	.0826	.2444	.4511	.6470	.7981	.8972	.9536	.9816	.9936	.9981	.9995	.9999	1.0000	1.0000	1.0000	1.0000	1.0000	1.0000	13	
.0503	.2382	.4802	.6904	.8363	.9226	.9673	.9877	.9959	.9988	.9997	.9999	1.0000	1.0000	1.0000	1.0000	1.0000	1.0000	1.0000	14	
.2078	.5182	.7475	.8818	.9499	.9807	.9933	.9979	.9994	.9999	1.0000	1.0000	1.0000	1.0000	1.0000	1.0000	1.0000	1.0000	1.0000	15	
.5819	.8332	.9369	.9775	.9925	.9977	.9993	.9998	1.0000	1.0000	1.0000	1.0000	1.0000	1.0000	1.0000	1.0000	1.0000	1.0000	1.0000	16	
.0000	.0000	.0000	.0000	.0000	.0000	.0000	.0000	.0000	.0000	.0000	.0001	.0004	.0016	.0056	.0180	.0536	.1501	.3972	0	18
.0000	.0000	.0000	.0000	.0000	.0000	.0000	.0000	.0000	.0001	.0003	.0013	.0046	.0142	.0395	.0991	.2241	.4503	.7735	1	
.0000	.0000	.0000	.0000	.0000	.0000	.0000	.0000	.0001	.0007	.0025	.0082	.0236	.0600	.1353	.2713	.4797	.7338	.9419	2	
.0000	.0000	.0000	.0000	.0000	.0000	.0000	.0002	.0010	.0038	.0120	.0328	.0783	.1646	.3057	.5010	.7202	.9018	.9891	3	
.0000	.0000	.0000	.0000	.0000	.0000	.0003	.0013	.0049	.0154	.0411	.0942	.1886	.3327	.5187	.7164	.8794	.9718	.9985	4	
.0000	.0000	.0000	.0000	.0000	.0003	.0014	.0058	.0183	.0481	.1077	.2088	.3550	.5344	.7175	.8671	.9581	.9936	.9998	5	
.0000	.0000	.0000	.0000	.0002	.0014	.0062	.0203	.0537	.1189	.2258	.3743	.5491	.7217	.8610	.9487	.9882	.9988	1.0000	6	
.0000	.0000	.0000	.0002	.0012	.0061	.0212	.0576	.1280	.2403	.3915	.5634	.7283	.8593	.9431	.9837	.9973	.9998	1.0000	7	
.0000	.0000	.0001	.0009	.0054	.0210	.0597	.1347	.2527	.4073	.5778	.7368	.8609	.9404	.9807	.9957	.9995	1.0000	1.0000	8	
.0000	.0000	.0005	.0043	.0193	.0596	.1391	.2632	.4222	.5927	.7473	.8653	.9403	.9790	.9946	.9991	.9999	1.0000	1.0000	9	
.0000	.0002	.0027	.0163	.0569	.1407	.2717	.4366	.6085	.7597	.8720	.9424	.9788	.9939	.9988	.9998	1.0000	1.0000	1.0000	10	
.0000	.0012	.0118	.0513	.1390	.2783	.4509	.6257	.7742	.8811	.9463	.9797	.9938	.9986	.9998	1.0000	1.0000	1.0000	1.0000	11	
.0002	.0064	.0419	.1329	.2825	.4656	.6450	.7912	.8923	.9519	.9817	.9942	.9986	.9997	1.0000	1.0000	1.0000	1.0000	1.0000	12	

TABLE 1 *Continued*

Values given are cumulative binomial probabilities; columns are values of p.

n	x	.05	.10	.15	.20	.25	.30	.35	.40	.45	.50	.55	.60	.65	.70	.75	.80	.85	.90	.95
18	13	1.0000	1.0000	1.0000	1.0000	1.0000	1.0000	.9997	.9987	.9951	.9846	.9589	.9058	.8114	.6673	.4813	.2836	.1206	.0282	.0015
	14	1.0000	1.0000	1.0000	1.0000	1.0000	1.0000	1.0000	.9998	.9990	.9962	.9880	.9672	.9217	.8354	.6943	.4990	.2798	.0982	.0109
	15	1.0000	1.0000	1.0000	1.0000	1.0000	1.0000	1.0000	1.0000	.9999	.9993	.9975	.9918	.9764	.9400	.8647	.7287	.5203	.2662	.0581
	16	1.0000	1.0000	1.0000	1.0000	1.0000	1.0000	1.0000	1.0000	1.0000	.9999	.9997	.9987	.9954	.9858	.9605	.9009	.7759	.5497	.2265
	17	1.0000	1.0000	1.0000	1.0000	1.0000	1.0000	1.0000	1.0000	1.0000	1.0000	1.0000	.9999	.9996	.9984	.9944	.9820	.9464	.8499	.6028
19	0	.3774	.1351	.0456	.0144	.0042	.0011	.0003	.0001	.0000	.0000	.0000	.0000	.0000	.0000	.0000	.0000	.0000	.0000	.0000
	1	.7547	.4203	.1985	.0829	.0310	.0104	.0031	.0008	.0002	.0000	.0000	.0000	.0000	.0000	.0000	.0000	.0000	.0000	.0000
	2	.9335	.7054	.4413	.2369	.1113	.0462	.0170	.0055	.0015	.0004	.0001	.0000	.0000	.0000	.0000	.0000	.0000	.0000	.0000
	3	.9868	.8850	.6841	.4551	.2631	.1332	.0591	.0230	.0077	.0022	.0005	.0001	.0000	.0000	.0000	.0000	.0000	.0000	.0000
	4	.9980	.9648	.8556	.6733	.4654	.2822	.1500	.0696	.0280	.0096	.0028	.0006	.0001	.0000	.0000	.0000	.0000	.0000	.0000
	5	.9998	.9914	.9463	.8369	.6678	.4739	.2968	.1629	.0777	.0318	.0109	.0031	.0007	.0001	.0000	.0000	.0000	.0000	.0000
	6	1.0000	.9983	.9837	.9324	.8251	.6655	.4812	.3081	.1727	.0835	.0342	.0116	.0031	.0006	.0001	.0000	.0000	.0000	.0000
	7	1.0000	.9997	.9959	.9767	.9225	.8180	.6656	.4878	.3169	.1796	.0871	.0352	.0114	.0028	.0005	.0001	.0000	.0000	.0000
	8	1.0000	1.0000	.9992	.9933	.9713	.9161	.8145	.6675	.4940	.3238	.1841	.0885	.0347	.0105	.0023	.0003	.0000	.0000	.0000
	9	1.0000	1.0000	.9999	.9984	.9911	.9674	.9125	.8139	.6710	.5000	.3290	.1861	.0875	.0326	.0089	.0016	.0001	.0000	.0000
	10	1.0000	1.0000	1.0000	.9997	.9977	.9895	.9653	.9115	.8159	.6762	.5060	.3325	.1855	.0839	.0287	.0067	.0008	.0000	.0000
	11	1.0000	1.0000	1.0000	1.0000	.9995	.9972	.9886	.9648	.9129	.8204	.6831	.5122	.3344	.1820	.0775	.0233	.0041	.0003	.0000
	12	1.0000	1.0000	1.0000	1.0000	.9999	.9994	.9969	.9884	.9658	.9165	.8273	.6919	.5188	.3345	.1749	.0676	.0163	.0017	.0000
	13	1.0000	1.0000	1.0000	1.0000	1.0000	.9999	.9993	.9969	.9891	.9682	.9223	.8371	.7032	.5261	.3322	.1631	.0537	.0086	.0002
	14	1.0000	1.0000	1.0000	1.0000	1.0000	1.0000	.9999	.9994	.9972	.9904	.9720	.9304	.8500	.7178	.5346	.3267	.1444	.0352	.0020
	15	1.0000	1.0000	1.0000	1.0000	1.0000	1.0000	1.0000	.9999	.9995	.9978	.9923	.9770	.9409	.8668	.7369	.5449	.3159	.1150	.0132
	16	1.0000	1.0000	1.0000	1.0000	1.0000	1.0000	1.0000	1.0000	.9999	.9996	.9985	.9945	.9830	.9538	.8887	.7631	.5587	.2946	.0665
	17	1.0000	1.0000	1.0000	1.0000	1.0000	1.0000	1.0000	1.0000	1.0000	.9999	.9998	.9992	.9969	.9896	.9690	.9171	.8015	.5797	.2453
	18	1.0000	1.0000	1.0000	1.0000	1.0000	1.0000	1.0000	1.0000	1.0000	1.0000	1.0000	.9999	.9997	.9989	.9958	.9856	.9544	.8649	.6226
20	0	.3585	.1216	.0388	.0115	.0032	.0008	.0002	.0000	.0000	.0000	.0000	.0000	.0000	.0000	.0000	.0000	.0000	.0000	.0000
	1	.7358	.3917	.1756	.0692	.0243	.0076	.0021	.0005	.0001	.0000	.0000	.0000	.0000	.0000	.0000	.0000	.0000	.0000	.0000
	2	.9245	.6769	.4049	.2061	.0913	.0355	.0121	.0036	.0009	.0002	.0000	.0000	.0000	.0000	.0000	.0000	.0000	.0000	.0000
	3	.9841	.8670	.6477	.4114	.2252	.1071	.0444	.0160	.0049	.0013	.0003	.0000	.0000	.0000	.0000	.0000	.0000	.0000	.0000
	4	.9974	.9568	.8298	.6296	.4148	.2375	.1182	.0510	.0189	.0059	.0015	.0003	.0000	.0000	.0000	.0000	.0000	.0000	.0000
	5	.9997	.9887	.9327	.8042	.6172	.4164	.2454	.1256	.0553	.0207	.0064	.0016	.0003	.0000	.0000	.0000	.0000	.0000	.0000
	6	1.0000	.9976	.9781	.9133	.7858	.6080	.4166	.2500	.1299	.0577	.0214	.0065	.0015	.0003	.0000	.0000	.0000	.0000	.0000
	7	1.0000	.9996	.9941	.9679	.8982	.7723	.6010	.4159	.2520	.1316	.0580	.0210	.0060	.0013	.0002	.0000	.0000	.0000	.0000
	8	1.0000	.9999	.9987	.9900	.9591	.8867	.7624	.5956	.4143	.2517	.1308	.0565	.0196	.0051	.0009	.0001	.0000	.0000	.0000
	9	1.0000	1.0000	.9998	.9974	.9861	.9520	.8782	.7553	.5914	.4119	.2493	.1275	.0532	.0171	.0039	.0006	.0001	.0000	.0000
	10	1.0000	1.0000	1.0000	.9994	.9961	.9829	.9468	.8725	.7507	.5881	.4086	.2447	.1218	.0480	.0139	.0026	.0002	.0000	.0000
	11	1.0000	1.0000	1.0000	.9999	.9991	.9949	.9804	.9435	.8692	.7483	.5857	.4044	.2376	.1133	.0409	.0100	.0013	.0001	.0000
	12	1.0000	1.0000	1.0000	1.0000	.9998	.9987	.9940	.9790	.9420	.8684	.7480	.5841	.3990	.2277	.1018	.0321	.0059	.0004	.0000
	13	1.0000	1.0000	1.0000	1.0000	1.0000	.9997	.9985	.9935	.9786	.9423	.8701	.7500	.5834	.3920	.2142	.0867	.0219	.0024	.0000
	14	1.0000	1.0000	1.0000	1.0000	1.0000	1.0000	.9997	.9984	.9936	.9793	.9447	.8744	.7546	.5836	.3828	.1958	.0673	.0113	.0003
	15	1.0000	1.0000	1.0000	1.0000	1.0000	1.0000	1.0000	.9997	.9985	.9941	.9811	.9490	.8818	.7625	.5852	.3704	.1702	.0432	.0026
	16	1.0000	1.0000	1.0000	1.0000	1.0000	1.0000	1.0000	1.0000	.9997	.9987	.9951	.9840	.9556	.8929	.7748	.5886	.3523	.1330	.0159
	17	1.0000	1.0000	1.0000	1.0000	1.0000	1.0000	1.0000	1.0000	1.0000	.9998	.9991	.9964	.9879	.9645	.9087	.7939	.5951	.3231	.0755
	18	1.0000	1.0000	1.0000	1.0000	1.0000	1.0000	1.0000	1.0000	1.0000	1.0000	.9999	.9995	.9979	.9924	.9757	.9308	.8244	.6083	.2642
	19	1.0000	1.0000	1.0000	1.0000	1.0000	1.0000	1.0000	1.0000	1.0000	1.0000	1.0000	1.0000	.9998	.9992	.9968	.9885	.9612	.8784	.6415

TABLE 2   Poisson Distribution Function

$$F(x; \lambda) = \sum_{k=0}^{x} e^{-\lambda} \frac{\lambda^k}{k!}$$

λ \ x	0	1	2	3	4	5	6	7	8	9
0.02	.980	1.000								
0.04	.961	.999	1.000							
0.06	.942	.998	1.000							
0.08	.923	.997	1.000							
0.10	.905	.995	1.000							
0.15	.861	.990	.999	1.000						
0.20	.819	.982	.999	1.000						
0.25	.779	.974	.998	1.000						
0.30	.741	.963	.996	1.000						
0.35	.705	.951	.994	1.000						
0.40	.670	.938	.992	.999	1.000					
0.45	.638	.925	.989	.999	1.000					
0.50	.607	.910	.986	.998	1.000					
0.55	.577	.894	.982	.998	1.000					
0.60	.549	.878	.977	.997	1.000					
0.65	.522	.861	.972	.996	.999	1.000				
0.70	.497	.844	.966	.994	.999	1.000				
0.75	.472	.827	.959	.993	.999	1.000				
0.80	.449	.809	.953	.991	.999	1.000				
0.85	.427	.791	.945	.989	.998	1.000				
0.90	.407	.772	.937	.987	.998	1.000				
0.95	.387	.754	.929	.984	.997	1.000				
1.00	.368	.736	.920	.981	.996	.999	1.000			
1.1	.333	.699	.900	.974	.995	.999	1.000			
1.2	.301	.663	.879	.966	.992	.998	1.000			
1.3	.273	.627	.857	.957	.989	.998	1.000			
1.4	.247	.592	.833	.946	.986	.997	.999	1.000		
1.5	.223	.558	.809	.934	.981	.996	.999	1.000		
1.6	.202	.525	.783	.921	.976	.994	.999	1.000		
1.7	.183	.493	.757	.907	.970	.992	.998	1.000		
1.8	.165	.463	.731	.891	.964	.990	.997	.999	1.000	
1.9	.150	.434	.704	.875	.956	.987	.997	.999	1.000	
2.0	.135	.406	.677	.857	.947	.983	.995	.999	1.000	

TABLE 2   Poisson Distribution Function *(Continued)*

λ \ x	0	1	2	3	4	5	6	7	8	9
2.2	.111	.355	.623	.819	.928	.975	.993	.998	1.000	
2.4	.091	.308	.570	.779	.904	.964	.988	.997	.999	1.000
2.6	.074	.267	.518	.736	.877	.951	.983	.995	.999	1.000
2.8	.061	.231	.469	.692	.848	.935	.976	.992	.998	.999
3.0	.050	.199	.423	.647	.815	.916	.966	.988	.996	.999
3.2	.041	.171	.380	.603	.781	.895	.955	.983	.994	.998
3.4	.033	.147	.340	.558	.744	.871	.942	.977	.992	.997
3.6	.027	.126	.303	.515	.706	.844	.927	.969	.988	.996
3.8	.022	.107	.269	.473	.668	.816	.909	.960	.984	.994
4.0	.018	.092	.238	.433	.629	.785	.889	.949	.979	.992
4.2	.015	.078	.210	.395	.590	.753	.867	.936	.972	.989
4.4	.012	.066	.185	.359	.551	.720	.844	.921	.964	.985
4.6	.010	.056	.163	.326	.513	.686	.818	.905	.955	.980
4.8	.008	.048	.143	.294	.476	.651	.791	.887	.944	.975
5.0	.007	.040	.125	.265	.440	.616	.762	.867	.932	.968
5.2	.006	.034	.109	.238	.406	.581	.732	.845	.918	.960
5.4	.005	.029	.095	.213	.373	.546	.702	.822	.903	.951
5.6	.004	.024	.082	.191	.342	.512	.670	.797	.886	.941
5.8	.003	.021	.072	.170	.313	.478	.638	.771	.867	.929
6.0	.002	.017	.062	.151	.285	.446	.606	.744	.847	.916

λ \ x	10	11	12	13	14	15	16
2.8	1.000						
3.0	1.000						
3.2	1.000						
3.4	.999	1.000					
3.6	.999	1.000					
3.8	.998	.999	1.000				
4.0	.997	.999	1.000				
4.2	.996	.999	1.000				
4.4	.994	.998	.999	1.000			
4.6	.992	.997	.999	1.000			
4.8	.990	.996	.999	1.000			
5.0	.986	.995	.998	.999	1.000		
5.2	.982	.993	.997	.999	1.000		
5.4	.977	.990	.996	.999	1.000		
5.6	.972	.988	.995	.998	.999	1.000	
5.8	.965	.984	.993	.997	.999	1.000	
6.0	.957	.980	.991	.996	.999	.999	1.000

TABLE 2 Poisson Distribution Function *(Continued)*

λ \ x	0	1	2	3	4	5	6	7	8	9
6.2	.002	.015	.054	.134	.259	.414	.574	.716	.826	.902
6.4	.002	.012	.046	.119	.235	.384	.542	.687	.803	.886
6.6	.001	.010	.040	.105	.213	.355	.511	.658	.780	.869
6.8	.001	.009	.034	.093	.192	.327	.480	.628	.755	.850
7.0	.001	.007	.030	.082	.173	.301	.450	.599	.729	.830
7.2	.001	.006	.025	.072	.156	.276	.420	.569	.703	.810
7.4	.001	.005	.022	.063	.140	.253	.392	.539	.676	.788
7.6	.001	.004	.019	.055	.125	.231	.365	.510	.648	.765
7.8	.000	.004	.016	.048	.112	.210	.338	.481	.620	.741
8.0	.000	.003	.014	.042	.100	.191	.313	.453	.593	.717
8.5	.000	.002	.009	.030	.074	.150	.256	.386	.523	.653
9.0	.000	.001	.006	.021	.055	.116	.207	.324	.456	.587
9.5	.000	.001	.004	.015	.040	.089	.165	.269	.392	.522
10.0	.000	.000	.003	.010	.029	.067	.130	.220	.333	.458

λ \ x	10	11	12	13	14	15	16	17	18	19
6.2	.949	.975	.989	.995	.998	.999	1.000			
6.4	.939	.969	.986	.994	.997	.999	1.000			
6.6	.927	.963	.982	.992	.997	.999	.999	1.000		
6.8	.915	.955	.978	.990	.996	.998	.999	1.000		
7.0	.901	.947	.973	.987	.994	.998	.999	1.000		
7.2	.887	.937	.967	.984	.993	.997	.999	.999	1.000	
7.4	.871	.926	.961	.980	.991	.996	.998	.999	1.000	
7.6	.854	.915	.954	.976	.989	.995	.998	.999	1.000	
7.8	.835	.902	.945	.971	.986	.993	.997	.999	1.000	
8.0	.816	.888	.936	.966	.983	.992	.996	.998	.999	1.000
8.5	.763	.849	.909	.949	.973	.986	.993	.997	.999	.999
9.0	.706	.803	.876	.926	.959	.978	.989	.995	.998	.999
9.5	.645	.752	.836	.898	.940	.967	.982	.991	.996	.998
10.0	.583	.697	.792	.864	.917	.951	.973	.986	.993	.997

λ \ x	20	21	22
8.5	1.000		
9.0	1.000		
9.5	.999	1.000	
10.0	.998	.999	1.000

TABLE 2 Poisson Distribution Function *(Continued)*

λ \ x	0	1	2	3	4	5	6	7	8	9
10.5	.000	.000	.002	.007	.021	.050	.102	.179	.279	.397
11.0	.000	.000	.001	.005	.015	.038	.079	.143	.232	.341
11.5	.000	.000	.001	.003	.011	.028	.060	.114	.191	.289
12.0	.000	.000	.001	.002	.008	.020	.046	.090	.155	.242
12.5	.000	.000	.000	.002	.005	.015	.035	.070	.125	.201
13.0	.000	.000	.000	.001	.004	.011	.026	.054	.100	.166
13.5	.000	.000	.000	.001	.003	.008	.019	.041	.079	.135
14.0	.000	.000	.000	.000	.002	.006	.014	.032	.062	.109
14.5	.000	.000	.000	.000	.001	.004	.010	.024	.048	.088
15.0	.000	.000	.000	.000	.001	.003	.008	.018	.037	.070

λ \ x	10	11	12	13	14	15	16	17	18	19
10.5	.521	.639	.742	.825	.888	.932	.960	.978	.988	.994
11.0	.460	.579	.689	.781	.854	.907	.944	.968	.982	.991
11.5	.402	.520	.633	.733	.815	.878	.924	.954	.974	.986
12.0	.347	.462	.576	.682	.772	.844	.899	.937	.963	.979
12.5	.297	.406	.519	.628	.725	.806	.869	.916	.948	.969
13.0	.252	.353	.463	.573	.675	.764	.835	.890	.930	.957
13.5	.211	.304	.409	.518	.623	.718	.798	.861	.908	.942
14.0	.176	.260	.358	.464	.570	.669	.756	.827	.883	.923
14.5	.145	.220	.311	.413	.518	.619	.711	.790	.853	.901
15.0	.118	.185	.268	.363	.466	.568	.664	.749	.819	.875

λ \ x	20	21	22	23	24	25	26	27	28	29
10.5	.997	.999	.999	1.000						
11.0	.995	.998	.999	1.000						
11.5	.992	.996	.998	.999	1.000					
12.0	.988	.994	.997	.999	.999	1.000				
12.5	.983	.991	.995	.998	.999	.999	1.000			
13.0	.975	.986	.992	.996	.998	.999	1.000			
13.5	.965	.980	.989	.994	.997	.998	.999	1.000		
14.0	.952	.971	.983	.991	.995	.997	.999	.999	1.000	
14.5	.936	.960	.976	.986	.992	.996	.998	.999	.999	1.000
15.0	.917	.947	.967	.981	.989	.994	.997	.998	.999	1.000

TABLE 2   Poisson Distribution Function *(Continued)*

λ \ x	4	5	6	7	8	9	10	11	12	13
16	.000	.001	.004	.010	.022	.043	.077	.127	.193	.275
17	.000	.001	.002	.005	.013	.026	.049	.085	.135	.201
18	.000	.000	.001	.003	.007	.015	.030	.055	.092	.143
19	.000	.000	.001	.002	.004	.009	.018	.035	.061	.098
20	.000	.000	.000	.001	.002	.005	.011	.021	.039	.066
21	.000	.000	.000	.000	.001	.003	.006	.013	.025	.043
22	.000	.000	.000	.000	.001	.002	.004	.008	.015	.028
23	.000	.000	.000	.000	.000	.001	.002	.004	.009	.017
24	.000	.000	.000	.000	.000	.000	.001	.003	.005	.011
25	.000	.000	.000	.000	.000	.000	.001	.001	.003	.006

λ \ x	14	15	16	17	18	19	20	21	22	23
16	.368	.467	.566	.659	.742	.812	.868	.911	.942	.963
17	.281	.371	.468	.564	.655	.736	.805	.861	.905	.937
18	.208	.287	.375	.469	.562	.651	.731	.799	.855	.899
19	.150	.215	.292	.378	.469	.561	.647	.725	.793	.849
20	.105	.157	.221	.297	.381	.470	.559	.644	.721	.787
21	.072	.111	.163	.227	.302	.384	.471	.558	.640	.716
22	.048	.077	.117	.169	.232	.306	.387	.472	.556	.637
23	.031	.052	.082	.123	.175	.238	.310	.389	.472	.555
24	.020	.034	.056	.087	.128	.180	.243	.314	.392	.473
25	.012	.022	.038	.060	.092	.134	.185	.247	.318	.394

λ \ x	24	25	26	27	28	29	30	31	32	33
16	.978	.987	.993	.996	.998	.999	.999	1.000		
17	.959	.975	.985	.991	.995	.997	.999	.999	1.000	
18	.932	.955	.972	.983	.990	.994	.997	.998	.999	1.000
19	.893	.927	.951	.969	.980	.988	.993	.996	.998	.999
20	.843	.888	.922	.948	.966	.978	.987	.992	.995	.997
21	.782	.838	.883	.917	.944	.963	.976	.985	.991	.994
22	.712	.777	.832	.877	.913	.940	.959	.973	.983	.989
23	.635	.708	.772	.827	.873	.908	.936	.956	.971	.981
24	.554	.632	.704	.768	.823	.868	.904	.932	.953	.969
25	.473	.553	.629	.700	.763	.818	.863	.900	.929	.950

λ \ x	34	35	36	37	38	39	40	41	42	43
19	.999	1.000								
20	.999	.999	1.000							
21	.997	.998	.999	.999	1.000					
22	.994	.996	.998	.999	.999	1.000				
23	.998	.993	.996	.997	.999	.999	1.000			
24	.979	.987	.992	.995	.997	.998	.999	.999		
25	.966	.978	.985	.991	.994	.997	.998	.999	1.000	

Based on E.C. Molina, Poisson's *Exponential Binomial Limit*, 1973 Reprint, Krieger Publishing Company, Malabar, Florida, by permission of the publisher.

TABLE 3 Normal Distribution Function
$$F(z) = \frac{1}{\sqrt{2\pi}} \int_{-\infty}^{z} e^{-t^2/2} dt$$

z	0.00	0.01	0.02	0.03	0.04	0.05	0.06	0.07	0.08	0.09
0.0	.5000	.5040	.5080	.5120	.5160	.5199	.5239	.5279	.5319	.5359
0.1	.5398	.5438	.5478	.5517	.5557	.5596	.5636	.5675	.5714	.5753
0.2	.5793	.5832	.5871	.5910	.5948	.5987	.6026	.6064	.6103	.6141
0.3	.6179	.6217	.6255	.6293	.6331	.6368	.6406	.6443	.6480	.6517
0.4	.6554	.6591	.6628	.6664	.6700	.6736	.6772	.6808	.6844	.6879
0.5	.6915	.6950	.6985	.7019	.7054	.7088	.7123	.7157	.7190	.7224
0.6	.7257	.7291	.7324	.7357	.7389	.7422	.7454	.7486	.7517	.7549
0.7	.7580	.7611	.7642	.7673	.7704	.7734	.7764	.7794	.7823	.7852
0.8	.7881	.7910	.7939	.7967	.7995	.8023	.8051	.8078	.8106	.8133
0.9	.8159	.8186	.8212	.8238	.8264	.8289	.8315	.8340	.8365	.8389
1.0	.8413	.8438	.8461	.8485	.8508	.8531	.8554	.8577	.8599	.8621
1.1	.8643	.8665	.8686	.8708	.8729	.8749	.8770	.8790	.8810	.8830
1.2	.8849	.8869	.8888	.8907	.8925	.8944	.8962	.8980	.8997	.9015
1.3	.9032	.9049	.9066	.9082	.9099	.9115	.9131	.9147	.9162	.9177
1.4	.9192	.9207	.9222	.9236	.9251	.9265	.9279	.9292	.9306	.9319
1.5	.9332	.9345	.9357	.9370	.9382	.9394	.9406	.9418	.9429	.9441
1.6	.9452	.9463	.9474	.9484	.9495	.9505	.9515	.9525	.9535	.9545
1.7	.9554	.9564	.9573	.9582	.9591	.9599	.9608	.9616	.9625	.9633
1.8	.9641	.9649	.9656	.9664	.9671	.9678	.9686	.9693	.9699	.9706
1.9	.9713	.9719	.9726	.9732	.9738	.9744	.9750	.9756	.9761	.9767
2.0	.9772	.9778	.9783	.9788	.9793	.9798	.9803	.9808	.9812	.9817
2.1	.9821	.9826	.9830	.9834	.9838	.9842	.9846	.9850	.9854	.9857
2.2	.9861	.9864	.9868	.9871	.9875	.9878	.9881	.9884	.9887	.9890
2.3	.9893	.9896	.9898	.9901	.9904	.9906	.9909	.9911	.9913	.9916
2.4	.9918	.9920	.9922	.9925	.9927	.9929	.9931	.9932	.9934	.9936
2.5	.9938	.9940	.9941	.9943	.9945	.9946	.9948	.9949	.9951	.9952
2.6	.9953	.9955	.9956	.9957	.9959	.9960	.9961	.9962	.9963	.9964
2.7	.9965	.9966	.9967	.9968	.9969	.9970	.9971	.9972	.9973	.9974
2.8	.9974	.9975	.9976	.9977	.9977	.9978	.9979	.9979	.9980	.9981
2.9	.9981	.9982	.9982	.9983	.9984	.9984	.9985	.9985	.9986	.9986
3.0	.9987	.9987	.9987	.9988	.9988	.9989	.9989	.9989	.9990	.9990
3.1	.9990	.9991	.9991	.9991	.9992	.9992	.9992	.9992	.9993	.9993
3.2	.9993	.9993	.9994	.9994	.9994	.9994	.9994	.9995	.9995	.9995
3.3	.9995	.9995	.9995	.9996	.9996	.9996	.9996	.9996	.9996	.9997
3.4	.9997	.9997	.9997	.9997	.9997	.9997	.9997	.9997	.9997	.9998
3.5	.9998									
4.0	.99997									
5.0	.9999997									
6.0	.999999999									

TABLE 4   Values of t_α

ν	$\alpha = 0.10$	$\alpha = 0.05$	$\alpha = 0.025$	$\alpha = 0.01$	$\alpha = 0.005$	ν
1	3.078	6.314	12.706	31.821	63.657	1
2	1.886	2.920	4.303	6.963	9.925	2
3	1.638	2.353	3.182	4.541	5.841	3
4	1.533	2.132	2.776	3.747	4.604	4
5	1.476	2.015	2.571	3.365	4.032	5
6	1.440	1.943	2.447	3.143	3.707	6
7	1.415	1.895	2.365	2.998	3.499	7
8	1.397	1.860	2.306	2.896	3.355	8
9	1.383	1.833	2.262	2.821	3.250	9
10	1.372	1.812	2.228	2.764	3.169	10
11	1.363	1.796	2.201	2.718	3.106	11
12	1.356	1.782	2.179	2.681	3.055	12
13	1.350	1.771	2.160	2.650	3.012	13
14	1.345	1.761	2.145	2.624	2.977	14
15	1.341	1.753	2.131	2.602	2.947	15
16	1.337	1.746	2.120	2.583	2.921	16
17	1.333	1.740	2.110	2.567	2.898	17
18	1.330	1.734	2.101	2.552	2.878	18
19	1.328	1.729	2.093	2.539	2.861	19
20	1.325	1.725	2.086	2.528	2.845	20
21	1.323	1.721	2.080	2.518	2.831	21
22	1.321	1.717	2.074	2.508	2.819	22
23	1.319	1.714	2.069	2.500	2.807	23
24	1.318	1.711	2.064	2.492	2.797	24
25	1.316	1.708	2.060	2.485	2.787	25
26	1.315	1.706	2.056	2.479	2.779	26
27	1.314	1.703	2.052	2.473	2.771	27
28	1.313	1.701	2.048	2.467	2.763	28
29	1.311	1.699	2.045	2.462	2.756	29
inf	1.282	1.645	1.960	2.326	2.576	inf

Richard A. Johnson, Dean W. Wichern, *Applied Multivariate Statistical Analysis*, 3e, © 1992, p. 630. Adapted by permission of Prentice Hall, Englewood Cliffs, New Jersey.

TABLE 5   Values of χ_α^2

ν	$\alpha = 0.995$	$\alpha = 0.99$	$\alpha = 0.975$	$\alpha = 0.95$	$\alpha = 0.05$	$\alpha = 0.025$	$\alpha = 0.01$	$\alpha = 0.005$	ν
1	0.0000393	0.000157	0.000982	0.00393	3.841	5.024	6.635	7.879	1
2	0.0100	0.0201	0.0506	0.103	5.991	7.378	9.210	10.597	2
3	0.0717	0.115	0.216	0.352	7.815	9.348	11.345	12.838	3
4	0.207	0.297	0.484	0.711	9.488	11.143	13.277	14.860	4
5	0.412	0.554	0.831	1.145	11.070	12.832	15.056	16.750	5
6	0.676	0.872	1.237	1.635	12.592	14.449	16.812	18.548	6
7	0.989	1.239	1.690	2.167	14.067	16.013	18.475	20.278	7
8	1.344	1.646	2.180	2.733	15.507	17.535	20.090	21.955	8
9	1.735	2.088	2.700	3.325	16.919	19.023	21.666	23.589	9
10	2.156	2.558	3.247	3.940	18.307	20.483	23.209	25.188	10
11	2.603	3.053	3.816	4.575	19.675	21.920	24.725	26.757	11
12	3.074	3.571	4.404	5.226	21.026	23.337	26.217	28.300	12
13	3.565	4.107	5.009	5.892	22.362	24.736	27.688	29.819	13
14	4.075	4.660	5.629	6.571	23.685	26.119	29.141	31.319	14
15	4.601	5.229	6.262	7.261	24.996	27.488	30.578	32.801	15
16	5.142	5.812	6.908	7.962	26.296	28.845	32.000	34.267	16
17	5.697	6.408	7.564	8.672	27.587	30.191	33.409	35.718	17
18	6.265	7.015	8.231	9.390	28.869	31.526	34.805	37.156	18
19	6.844	7.633	8.907	10.117	30.144	32.852	36.191	38.582	19
20	7.434	8.260	9.591	10.851	31.410	34.170	37.566	39.997	20
21	8.034	8.897	10.283	11.591	32.671	35.479	38.932	41.401	21
22	8.643	9.542	10.982	12.338	33.924	36.781	40.289	42.796	22
23	9.260	10.196	11.689	13.091	35.172	38.076	41.638	44.181	23
24	9.886	10.856	12.401	13.484	36.415	39.364	42.980	45.558	24
25	10.520	11.524	13.120	14.611	37.652	40.646	44.314	46.928	25
26	11.160	12.198	13.844	15.379	38.885	41.923	45.642	48.290	26
27	11.808	12.879	14.573	16.151	40.113	43.194	46.963	49.645	27
28	12.461	13.565	15.308	16.928	41.337	44.461	48.278	50.993	28
29	13.121	14.256	16.047	17.708	42.557	45.772	49.588	52.336	29
30	13.787	14.953	16.791	18.493	43.773	46.979	50.892	53.672	30
40	20.706	22.164	24.433	26.509	55.758	59.342	63.691	66.766	40
50	27.991	29.707	32.357	34.764	67.505	71.420	76.154	79.490	50
60	35.535	37.485	40.482	43.118	79.082	83.298	88.379	91.952	60
70	43.275	45.442	48.758	51.739	90.531	95.023	100.425	104.215	70
80	51.172	53.540	57.153	60.391	101.879	106.629	112.329	116.321	80
90	59.196	61.754	65.646	69.126	113.145	118.136	124.116	128.299	90
100	67.328	70.065	74.222	77.929	124.342	129.561	135.807	140.169	100

TABLE 6(a)   Values of $F_{0.05}$

v_1 = Degrees of freedom for numerator

v_2 = Degrees of freedom for denominator	1	2	3	4	5	6	7	8	9	10	12	15	20	24	30	40	60	120	∞
1	161	200	216	225	230	234	237	239	241	242	244	246	248	249	250	251	252	253	254
2	18.50	19.00	19.20	19.20	19.30	19.30	19.40	19.40	19.40	19.40	19.40	19.40	19.40	19.50	19.50	19.50	19.50	19.50	19.50
3	10.10	9.55	9.28	9.12	9.01	8.94	8.89	8.85	8.81	8.79	8.74	8.70	8.66	8.64	8.62	8.59	8.57	8.55	8.53
4	7.71	6.94	6.59	6.39	6.26	6.16	6.09	6.04	6.00	5.96	5.91	5.86	5.80	5.77	5.75	5.72	5.69	5.66	5.63
5	6.61	5.79	5.41	5.19	5.05	4.95	4.88	4.82	4.77	4.74	4.68	4.62	4.56	4.53	4.50	4.46	4.43	4.40	4.37
6	5.99	5.14	4.76	4.53	4.39	4.28	4.21	4.15	4.10	4.06	4.00	3.94	3.87	3.84	3.81	3.77	3.74	3.70	3.67
7	5.59	4.74	4.35	4.12	3.97	3.87	3.79	3.73	3.68	3.64	3.57	3.51	3.44	3.41	3.38	3.34	3.30	3.27	3.23
8	5.32	4.46	4.07	3.84	3.69	3.58	3.50	3.44	3.39	3.35	3.28	3.22	3.15	3.12	3.08	3.04	3.01	2.97	2.93
9	5.12	4.26	3.86	3.63	3.48	3.37	3.29	3.23	3.18	3.14	3.07	3.01	2.94	2.90	2.86	2.83	2.79	2.75	2.71
10	4.96	4.10	3.71	3.48	3.33	3.22	3.14	3.07	3.02	2.98	2.91	2.85	2.77	2.74	2.70	2.66	2.62	2.58	2.54
11	4.84	3.98	3.59	3.36	3.20	3.09	3.01	2.95	2.90	2.85	2.79	2.72	2.65	2.61	2.57	2.53	2.49	2.45	2.40
12	4.75	3.89	3.49	3.26	3.11	3.00	2.91	2.85	2.80	2.75	2.69	2.62	2.54	2.51	2.47	2.43	2.38	2.34	2.30
13	4.67	3.81	3.41	3.18	3.03	2.92	2.83	2.77	2.71	2.67	2.60	2.53	2.46	2.42	2.38	2.34	2.30	2.25	2.21
14	4.60	3.74	3.34	3.11	2.96	2.85	2.76	2.70	2.65	2.60	2.53	2.46	2.39	2.35	2.31	2.27	2.22	2.18	2.13
15	4.54	3.68	3.29	3.06	2.90	2.79	2.71	2.64	2.59	2.54	2.48	2.40	2.33	2.29	2.25	2.20	2.16	2.11	2.07
16	4.49	3.63	3.24	3.01	2.85	2.74	2.66	2.59	2.54	2.49	2.42	2.35	2.28	2.24	2.19	2.15	2.11	2.06	2.01
17	4.45	3.59	3.20	2.96	2.81	2.70	2.61	2.55	2.49	2.45	2.38	2.31	2.23	2.19	2.15	2.10	2.06	2.01	1.96
18	4.41	3.55	3.16	2.93	2.77	2.66	2.58	2.51	2.46	2.41	2.34	2.27	2.19	2.15	2.11	2.06	2.02	1.97	1.93
19	4.38	3.52	3.13	2.90	2.74	2.63	2.54	2.48	2.42	2.38	2.31	2.23	2.16	2.11	2.07	2.03	1.98	1.93	1.88
20	4.35	3.49	3.10	2.87	2.71	2.60	2.51	2.45	2.39	2.35	2.28	2.20	2.12	2.08	2.04	1.99	1.95	1.90	1.84
21	4.32	3.47	3.07	2.84	2.68	2.57	2.49	2.42	2.37	2.32	2.25	2.18	2.10	2.05	2.01	1.96	1.92	1.87	1.81
22	4.30	3.44	3.05	2.82	2.66	2.55	2.46	2.40	2.34	2.30	2.23	2.15	2.07	2.03	1.98	1.94	1.89	1.84	1.78
23	4.28	3.42	3.03	2.80	2.64	2.53	2.44	2.37	2.32	2.27	2.20	2.13	2.05	2.01	1.96	1.91	1.86	1.81	1.76
24	4.26	3.40	3.01	2.78	2.62	2.51	2.42	2.36	2.30	2.25	2.18	2.11	2.03	1.98	1.94	1.89	1.84	1.79	1.73
25	4.24	3.39	2.99	2.76	2.60	2.49	2.40	2.34	2.28	2.24	2.16	2.09	2.01	1.96	1.92	1.87	1.82	1.77	1.71
30	4.17	3.32	2.92	2.69	2.53	2.42	2.33	2.27	2.21	2.16	2.09	2.01	1.93	1.89	1.84	1.79	1.74	1.68	1.62
40	4.08	3.23	2.84	2.61	2.45	2.34	2.25	2.18	2.12	2.08	2.00	1.92	1.84	1.79	1.74	1.69	1.64	1.58	1.51
60	4.00	3.15	2.76	2.53	2.37	2.25	2.17	2.10	2.04	1.99	1.92	1.84	1.75	1.70	1.65	1.59	1.53	1.47	1.39
120	3.92	3.07	2.68	2.45	2.29	2.18	2.09	2.02	1.96	1.91	1.83	1.75	1.66	1.61	1.55	1.50	1.43	1.35	1.25
α	3.84	3.00	2.60	2.37	2.21	2.10	2.01	1.94	1.88	1.83	1.75	1.67	1.57	1.52	1.46	1.39	1.32	1.22	1.00

This table is reproduced from M. Merrington and C. M. Thompson, "Tables of percentage points of the inverted beta (F) distribution," *Biometrika*. Vol. 33 (1943), by permission of the *Biometrika* trustees.

TABLE 6(b) Values of $F_{0.01}$

v_1 = Degrees of freedom for numerator

v_2 = Degrees of freedom for denominator	1	2	3	4	5	6	7	8	9	10	12	15	20	24	30	40	60	120	∞
1	4,052	5,000	5,403	5,625	5,764	5,859	5,928	5,982	6,023	6,056	6,106	6,157	6,209	6,235	6,261	6,287	6,313	6,339	6,366
2	98.50	99.00	99.20	99.20	99.30	99.30	99.40	99.40	99.40	99.40	99.40	99.40	99.40	99.50	99.50	99.50	99.50	99.50	99.50
3	34.10	30.80	29.50	28.70	28.20	27.90	27.70	27.50	27.30	27.20	27.10	26.90	26.70	26.60	26.50	26.40	26.30	26.20	26.10
4	21.20	18.00	16.70	16.00	15.50	15.20	15.00	14.80	14.70	14.50	14.40	14.20	14.00	13.90	13.80	13.70	13.70	13.60	13.50
5	16.30	13.30	12.10	11.40	11.00	10.70	10.50	10.30	10.20	10.10	9.89	9.72	9.55	9.47	9.38	9.29	9.20	9.11	9.02
6	13.70	10.90	9.78	9.15	8.75	8.47	8.26	8.10	7.98	7.87	7.72	7.56	7.40	7.31	7.23	7.14	7.06	6.97	6.88
7	12.20	9.55	8.45	7.85	7.46	7.19	6.99	6.84	6.72	6.62	6.47	6.31	6.16	6.07	5.99	5.91	5.82	5.74	5.65
8	11.30	8.65	7.59	7.01	6.63	6.37	6.18	6.03	5.91	5.81	5.67	5.52	5.36	5.28	5.20	5.12	5.03	4.95	4.83
9	10.60	8.02	6.99	6.42	6.06	5.80	5.61	5.47	5.35	5.26	5.11	4.96	4.81	4.73	4.65	4.57	4.48	4.40	4.31
10	10.00	7.56	6.55	5.99	5.64	5.39	5.20	5.06	4.94	4.85	4.71	4.56	4.41	4.33	4.25	4.17	4.08	4.00	3.91
11	9.65	7.21	6.22	5.67	5.32	5.07	4.89	4.74	4.63	4.54	4.40	4.25	4.10	4.02	3.94	3.86	3.78	3.69	3.60
12	9.33	6.93	5.95	5.41	5.06	4.82	4.64	4.50	4.39	4.30	4.16	4.01	3.86	3.78	3.70	3.62	3.54	3.45	3.36
13	9.07	6.70	5.74	5.21	4.86	4.62	4.44	4.30	4.19	4.10	3.96	3.82	3.66	3.59	3.51	3.43	3.34	3.25	3.17
14	8.86	6.51	5.56	5.04	4.70	4.46	4.28	4.14	4.03	3.94	3.80	3.66	3.51	3.43	3.35	3.27	3.18	3.09	3.00
15	8.68	6.36	5.42	4.89	4.56	4.32	4.14	4.00	3.89	3.80	3.67	3.52	3.37	3.29	3.21	3.13	3.05	2.96	2.87
16	8.53	6.23	5.29	4.77	4.44	4.20	4.03	3.89	3.78	3.69	3.55	3.41	3.26	3.18	3.10	3.02	2.93	2.84	2.75
17	8.40	6.11	5.19	4.67	4.34	4.10	3.93	3.79	3.68	3.59	3.46	3.31	3.16	3.08	3.00	2.92	2.83	2.75	2.65
18	8.29	6.01	5.09	4.58	4.25	4.01	3.84	3.71	3.60	3.51	3.37	3.23	3.08	3.00	2.92	2.84	2.75	2.66	2.57
19	8.19	5.93	5.01	4.50	4.17	3.94	3.77	3.63	3.52	3.43	3.30	3.15	3.00	2.92	2.84	2.76	2.67	2.58	2.49
20	8.10	5.85	4.94	4.43	4.10	3.87	3.70	3.56	3.46	3.37	3.23	3.09	2.94	2.86	2.78	2.69	2.61	2.52	2.42
21	8.02	5.78	4.87	4.37	4.04	3.81	3.64	3.51	3.40	3.31	3.17	3.03	2.88	2.80	2.72	2.64	2.55	2.46	2.36
22	7.95	5.72	4.82	4.31	3.99	3.76	3.59	3.45	3.35	3.26	3.12	2.98	2.83	2.75	2.67	2.58	2.50	2.40	2.31
23	7.88	5.66	4.76	4.26	3.94	3.71	3.54	3.41	3.30	3.21	3.07	2.93	2.78	2.70	2.62	2.54	2.45	2.35	2.26
24	7.82	5.61	4.72	4.22	3.90	3.67	3.50	3.36	3.26	3.17	3.03	2.89	2.74	2.66	2.58	2.49	2.40	2.31	2.21
25	7.77	5.57	4.68	4.18	3.86	3.63	3.46	3.32	3.22	3.13	2.99	2.85	2.70	2.62	2.53	2.45	2.36	2.27	2.17
30	7.56	5.39	4.51	4.02	3.70	3.47	3.30	3.17	3.07	2.98	2.84	2.70	2.55	2.47	2.39	2.30	2.21	2.11	2.01
40	7.31	5.18	4.31	3.83	3.51	3.29	3.12	2.99	2.89	2.80	2.66	2.52	2.37	2.29	2.20	2.11	2.02	1.92	1.80
60	7.08	4.98	4.13	3.65	3.34	3.12	2.95	2.82	2.72	2.63	2.50	2.35	2.20	2.12	2.03	1.94	1.84	1.73	1.60
120	6.85	4.79	3.95	3.48	3.17	2.96	2.79	2.66	2.56	2.47	2.34	2.19	2.03	1.95	1.86	1.76	1.66	1.53	1.38
∞	6.63	4.61	3.78	3.32	3.02	2.80	2.64	2.51	2.41	2.32	2.18	2.04	1.88	1.79	1.70	1.59	1.47	1.32	1.00

TABLE 7 Control-Chart Constants

Number of Observations in Sample, n	CHART FOR AVERAGES		CHART FOR STANDARD DEVIATIONS		CHART FOR RANGES	
	A_2	A_3	B_3	B_4	D_3	D_4
2	1.88	2.66	0	3.27	0	3.27
3	1.02	1.95	0	2.57	0	2.57
4	0.73	1.63	0	2.27	0	2.28
5	0.58	1.43	0	2.09	0	2.11
6	0.48	1.29	0.03	1.97	0	2.00
7	0.42	1.18	0.12	1.88	0.08	1.92
8	0.37	1.10	0.19	1.81	0.14	1.86
9	0.34	1.03	0.24	1.76	0.18	1.82
10	0.31	0.98	0.28	1.72	0.22	1.78
11	0.29	0.93	0.32	1.68	0.26	1.74
12	0.27	0.89	0.35	1.65	0.28	1.72
13	0.25	0.85	0.38	1.62	0.31	1.69
14	0.24	0.82	0.41	1.59	0.33	1.67
15	0.22	0.79	0.43	1.57	0.35	1.65
16	0.21	0.76	0.45	1.55	0.36	1.64
17	0.20	0.74	0.47	1.53	0.38	1.62
18	0.19	0.72	0.48	1.52	0.39	1.61
19	0.19	0.70	0.50	1.50	0.40	1.60
20	0.18	0.68	0.51	1.49	0.41	1.59

TABLE 8* Values of $Z = \dfrac{1}{2}\ln\dfrac{1+r}{1-r}$

r	0.00	0.01	0.02	0.03	0.04	0.05	0.06	0.07	0.08	0.09
0.0	.000	.010	.020	.030	.040	.050	.060	.070	.080	.090
0.1	.100	.110	.121	.131	.141	.151	.161	.172	.182	.192
0.2	.203	.213	.224	.234	.245	.255	.266	.277	.288	.299
0.3	.310	.321	.332	.343	.354	.365	.377	.388	.400	.412
0.4	.424	.436	.448	.460	.472	.485	.497	.510	.523	.536
0.5	.549	.563	.576	.590	.604	.618	.633	.648	.662	.678
0.6	.693	.709	.725	.741	.758	.775	.793	.811	.829	.848
0.7	.867	.887	.908	.929	.950	.973	.996	1.020	1.045	1.071
0.8	1.099	1.127	1.157	1.188	1.221	1.256	1.293	1.333	1.376	1.422
0.9	1.472	1.528	1.589	1.658	1.738	1.832	1.946	2.092	2.298	2.647

*For negative values of r put a minus sign in front of the corresponding Zs, and vice versa.

TABLE 9(a)   Values of r_p for $\alpha = 0.05$

d.f. \ p	2	3	4	5	6	7	8	9	10
1	17.97								
2	6.09	6.09							
3	4.50	4.52	4.52						
4	3.93	4.01	4.03	4.03					
5	3.64	3.75	3.80	3.81	3.81				
6	3.46	3.59	3.65	3.68	3.69	3.70			
7	3.34	3.48	3.55	3.59	3.61	3.62	3.63		
8	3.26	3.40	3.48	3.52	3.55	3.57	3.57	3.58	
9	3.20	3.34	3.42	3.47	3.50	3.52	3.54	3.54	3.55
10	3.15	3.29	3.38	3.43	3.47	3.49	3.51	3.52	3.52
11	3.11	3.26	3.34	3.40	3.44	3.46	3.48	3.49	3.50
12	3.08	3.23	3.31	3.37	3.41	3.44	3.46	3.47	3.48
13	3.06	3.20	3.29	3.35	3.39	3.42	3.46	3.46	3.47
14	3.03	3.18	3.27	3.33	3.37	3.40	3.43	3.44	3.46
15	3.01	3.16	3.25	3.31	3.36	3.39	3.41	3.43	3.45
16	3.00	3.14	3.23	3.30	3.34	3.38	3.40	3.42	3.44
17	2.98	3.13	3.22	3.28	3.33	3.37	3.39	3.41	3.43
18	2.97	3.12	3.21	3.27	3.32	3.36	3.38	3.40	3.42
19	2.96	3.11	3.20	3.26	3.31	3.35	3.38	3.40	3.41
20	2.95	3.10	3.19	3.25	3.30	3.34	3.37	3.39	3.41
24	2.92	3.07	3.16	3.23	3.28	3.31	3.35	3.37	3.39
30	2.89	3.03	3.13	3.20	3.25	3.29	3.32	3.35	3.37
40	2.86	3.01	3.10	3.17	3.22	3.27	3.30	3.33	3.35
60	2.83	2.98	3.07	3.14	3.20	3.24	3.28	3.31	3.33
120	2.80	2.95	3.04	3.12	3.17	3.22	3.25	3.29	3.31
∞	2.77	2.92	3.02	3.09	3.15	3.19	3.23	3.27	3.29

This table is reproduced from H. L. Harter, "Critical Values for Duncan's New Multiple Range Test." It contains some corrected values to replace those given by D. B. Duncan in "Multiple Range and Multiple F Tests," *Biometrics*, Vol. 11 (1955). The above table is reproduced with the permission of the author and the Biometric Society.

TABLE 9(b)   Values of r_p for $\alpha = 0.01$

d.f. \ p	2	3	4	5	6	7	8	9	10
1	90.02								
2	14.04	14.04							
3	8.26	8.32	8.32						
4	6.51	6.68	6.74	6.76					
5	5.70	5.90	5.99	6.04	6.07				
6	5.24	5.44	5.55	5.62	5.66	5.68			
7	4.95	5.15	5.26	5.33	5.38	5.42	5.44		
8	4.74	4.94	5.06	5.13	5.19	5.23	5.26	5.28	
9	4.60	4.79	4.91	4.99	5.04	5.09	5.12	5.14	5.16
10	4.48	4.67	4.79	4.88	4.93	4.98	5.01	5.04	5.06
11	4.39	4.58	4.70	4.78	4.84	4.89	4.92	4.95	4.97
12	4.32	4.50	4.62	4.71	4.77	4.81	4.85	4.88	4.91
13	4.26	4.44	4.56	4.64	4.71	4.75	4.79	4.82	4.85
14	4.21	4.39	4.51	4.59	4.66	4.70	4.74	4.77	4.80
15	4.17	4.34	4.46	4.55	4.61	4.66	4.70	4.73	4.76
16	4.13	4.31	4.43	4.51	4.57	4.62	4.66	4.70	4.72
17	4.10	4.27	4.39	4.47	4.54	4.59	4.63	4.66	4.69
18	4.07	4.25	4.36	4.45	4.51	4.56	4.60	4.64	4.66
19	4.05	4.22	4.33	4.42	4.48	4.53	4.57	4.61	4.64
20	4.02	4.20	4.31	4.40	4.46	4.51	4.55	4.59	4.62
24	3.96	4.13	4.24	4.32	4.39	4.44	4.48	4.52	4.55
30	3.89	4.06	4.17	4.25	4.31	4.36	4.41	4.45	4.48
40	3.82	3.99	4.10	4.18	4.24	4.29	4.33	4.38	4.41
60	3.76	3.92	4.03	4.11	4.18	4.23	4.37	4.31	4.34
120	3.70	3.86	3.97	4.04	4.11	4.16	4.20	4.24	4.27
∞	3.64	3.80	3.90	3.98	4.04	4.09	4.13	4.17	4.21

This table is reproduced from H. L. Harter, "Critical Values for Duncan's New Multiple Range Test." It contains some corrected values to replace those given by D. B. Duncan in "Multiple Range and Multiple *F* Tests," *Biometrics*, Vol. 11 (1955). The above table is reproduced with the permission of the author and the Biometric Society.

Appendix: B
BIBLIOGRAPHY

1. QUALITY

Crosby, P. B., *The Eternally Successful Organization*. New York: McGraw-Hill, 1988.

Crosby, P. B., *Let's Talk Quality*. New York: McGraw-Hill, 1989.

Crosby, P. B., *Quality Is Free*. New York: McGraw-Hill, 1979.

Deming, W. E., *Out of the Crisis*. Cambridge, Mass.: Massachusetts Institute of Technology, Center for Advanced Engineering Study, 1986.

Deming, W. E., *Quality, Productivity, and Competitive Position*. Cambridge, Mass.: Massachusetts Institute for Advanced Engineering Study, 1982.

Feigenbaum, A. V., *Total Quality Control*. New York: McGraw-Hill, 1987.

Gitlow, H., Gitlow, Oppenheim, and Oppenheim, *Tools and Methods for the Improvement of Quality*. Homewood, Ill.: Richard D. Irwin, 1989.

Ishikawa, K., *Guide to Quality Control*. Tokyo: Asia Productivity Organization, 1976.

Ishikawa, K., *What Is Total Quality Control? The Japanese Way*. Englewood Cliffs, N.J.: Prentice Hall, 1985.

Juran, J. M., *Juran on Planning for Quality*. New York: Free Press, 1988.

Juran, J. M., *Managerial Breakthrough*. New York: McGraw-Hill, 1964.

Juran, J. M. and Gryma, F. M., *Quality Planning and Analysis*, 2nd ed. New York: McGraw-Hill, 1980.

Ross, P. J., *Taguchi Techniques for Quality Engineering—Loss Function, Orthogonal Experiments, Parameter and Tolerance Design*. New York: McGraw-Hill, 1988.

Taguchi, G., *Introduction to Quality Engineering*. Tokyo: Asian Productivity Organization, 1986 (English translation: White Plains, N.Y., UNIPUB/Quality Resources).

Taguchi, G., *System of Experimental Design*, Vols. I and II. Co-published White Plains, N.Y.: UNIPUB and Dearborn, Mich.: American Supplier Institute, 1938.

Taguchi, G., Elsayed, A. E., and Hsiang, T., *Quality Engineering in Production Systems*. New York: McGraw-Hill, 1989.

2. ENGINEERING STATISTICS

Devore, J. L., *Probability and Statistics for Engineering and the Sciences*, 3rd ed. Pacific Grove, Calif.: Brooks/Cole, 1991.

Guttman, I., Wilks, S. S., and Hunter, J. S., *Introductory Engineering Statistics*, 3rd ed. New

York: John Wiley, 1982.

Kennedy, J. B., and Neville, A. M., *Basic Statistical Methods for Engineering and Scientists*, 3rd ed. New York: Harper and Row, 1986.

Miller, I., Freund, J. E., and Johnson, R. A., *Probability and Statistics for Engineers*, 4th ed. Englewood Cliffs, N.J.: Prentice Hall, 1990.

Walpole, R. E., and Meyers, R. H., *Probability and Statistics for Engineers and Scientists*, 3rd ed. New York: Macmillan, 1985.

3. THEORETICAL STATISTICS

Freund, J. E., and Walpole, R., *Mathematical Statistics*, 4th ed. Englewood Cliffs, N.J.: Prentice Hall, 1987.

Hoel, P., *Introduction to Mathematical Statistics*, 5th ed. New York: John Wiley, 1984.

Hogg, R. V., and Craig, A. T., *Introduction to Mathematical Statistics*, 4th ed. New York: Macmillan, 1978.

Kendall, M. G., and Stuart, A., *The Advanced Theory of Statistics*, Vol. 1, 4th ed., Vol. 2, 4th ed., Vol. 3, 3rd ed. New York: Hafner Press, 1977, 1979, 1983.

Mood, A. M., and Graybill, F. A., *Introduction to the Theory of Statistics*, 3rd ed. New York: McGraw-Hill, 1973.

4. DESIGN OF EXPERIMENTS

Box, G. E. P., and Draper, N. R., *Empirical Model-Building and Response Surfaces*. New York: John Wiley, 1987.

Box, G. E. P., Hunter, W. G., and Hunter, J. S., *Statistics for Experimenters*. New York: John Wiley, 1978.

Cochran, W. G., and Cox, G. M., *Experimental Designs*, 2nd ed. New York: John Wiley, 1957.

John, P. W. M., *Statistical Analysis and Design of Experiments*. New York: Macmillan, 1971.

5. STATISTICAL PROCESS CONTROL

Duncan, A. J., *Quality Control and Industrial Statistics*, 5th ed. Homewood, Ill.: Richard D. Irwin, Inc., 1986.

Grant, E. L., and Leavenworth, R. S., *Statistical Quality Control*, 6th ed. New York: McGraw-Hill, 1988.

Ishikawa, K., *Guide to Quality Control*. Tokyo: Asian Productivity Organization, 1976.

Messina, W., *Statistical Quality Control for Manufacturing Managers*. New York: John Wiley, 1987.

Taguchi, G., and Wu, Y., *Introduction to Off-Line Quality Control*. Nagoya, Japan: Central Japan Quality Control Association.

6. RELIABILITY AND LIFE TESTING

Bain, L. J., *Statistical Analysis of Reliability and Life Testing Models Theory and Methods*. New York: Marcel Dekker, 1989.

Barlow, R. E., and Proschan, F., *Statistical Theory of Reliability and Life Testing: Probability Models*. New York: Holt, Rinehart and Winston, 1975.

Lawless, J. F., *Statistical Models and Methods for Lifetime Data*. New York: John Wiley, 1981.

Lloyd, D. K., and Lipow, M., *Reliability: Management, Methods, and Mathematics*, 2nd ed. ASQC Quality Press, Milwaukee, Wisconsin, 1984.

Nelson, W., *Applied Life Data Analysis*. New York: John Wiley, 1981.

7. GENERAL REFERENCE WORKS AND TABLES

DoD 5000 XX-G: Total Quality Management: A Guide for Implementation. Washington, D.C.: U.S. Department of Defense, 1989.

Kendall, M. G., and Buckland, W. R., *A Dictionary of Statistical Terms*, 3rd ed. New York, John Wiley, 1983.

Kotz, S., and Johnson, N. L., *Encyclopedia of Statistical Sciences*, Vol 2. New York: John Wiley, 1982.

Military Standard 105D. Washington, D.C.: U.S. Government Printing Office, 1963.

National Bureau of Standards Handbook 91: Experimental Statistics. Washington, D.C.: U.S. Government Printing Office, 1963.

National Bureau of Standards: Tables of the Binomial Distribution. Washington, D.C.: U.S. Government Printing Office, 1963.

Pearson, E. S., and Hartley, H. O., *Biometrika Tables for Statisticians*, 3rd ed. Cambridge: Cambridge University Press, 1966.

Romig, H. G., *50 – 100 Binomial Tables*. New York: John Wiley, 1953.

ANSWERS TO ODD-NUMBERED EXERCISES

CHAPTER 2

2.1 Deming.

2.3 Shewhart.

2.5 a. Motivational programs, new organizational disciplines, quality tools;

b. motivational programs;

c. quality tools;

d. new organizational disciplines.

2.7 Department of Defense.

2.17 a. $5,000; b. $36,000/year; c. $27,000/year; d. −$4,000/year.

2.19 $10 million.

2.23 a. $120/pound; b. $80,000; c. $120/ounce; d. $1.0625; e. $75; f. $80 million.

2.25 a. 0.41 pound; b. 0.32 ounce; c. 0.82%; d. 2,169 psi.

2.27 42.8%.

2.29 8.5 nanoseconds.

2.31 a. $k = C_1 T_1^{-3/2} = C_2 T_2^{-3/2}$; b. $t_2 = (C_2/C_1)^{2/3} T_1$.

2.33 Shewhart — quality control;
Deming — 14 obligations for management;
Juran — four points for quality improvement;
Crosby — "quality is free";
Taguchi — loss function;
Ishikawa — fishbone diagram.

2.35 Vision, principles, practices, and techniques and tools.

2.37 Provides a systematic approach to linking causes to effects.

2.41 $28,400/year.

2.43 $17.90.

2.45 0.04 cm.

CHAPTER 3

3.1 Accuracy is measured in terms of the distance from the "center" of the data to the "target" value; precision is measured by the variability, or "spread," of the data about its center.

3.3 a. 5 | 9 4 5 7 9 9 8
       6 | 1 3 5 0 2 1 7 0 8 4 5 2 0 2 1 3 1

    b. 5f | 4
       5s | 5 9 7 9 9 8
       6f | 1 3 0 2 1 0 4 2 0 2 1 3 1
       6s | 5 7 8 5

    c. The double-stem display better shows the pattern of the data.

3.7 a. ZDI;   b. 41.6%.

3.9 12. 8 9
    13* 1 3
    13. 6 8 9
    14* 0 0 1 2 2 4
    14. 6 7
    15* 1
    15. 6

3.11 11 | 5 9 7
     12 | 4 9 8 8 7 4
     13 | 5 7 6 3 0 2
     14 | 1 5 0 6

3.13

Class Limits	Frequency
40.0 – 44.9	5
45.0 – 49.9	7
50.0 – 54.9	15
55.0 – 59.9	23
60.0 – 64.9	29
65.0 – 69.9	12
70.0 – 74.9	8
75.0 – 79.9	1
	100

3.15 *Class boundaries:* 39.95, 44.95, 49.95, 54.95, 59.95, 64.95, 69.95, 74.95, 79.95;
       Class interval: 5;
       Class marks: 42.45, 47.45, 52.45, 57.45, 62.45, 67.45, 72.45, 77.45.

3.17

Class Limits	Frequency
0 – 1	12
2 – 3	7
4 – 5	4
6 – 7	5
8 – 9	1
10 – 11	0
12 – 13	1
	30

3.19

Class Limits	Percentage
40.0 – 44.9	5
45.0 – 49.9	7
50.0 – 54.9	15
55.0 – 59.9	23
60.0 – 64.9	29
65.0 – 69.9	12
70.0 – 74.9	8
75.0 – 79.9	1
	100

3.21

Upper Class Boundary	Cumulative Frequency
44.95	5
49.95	12
54.95	27
59.95	50
64.95	79
69.95	91
74.95	99
79.95	100

3.23

Upper-Class Boundary	Cumulative Percentage Shipping	Security
1.5	43.3	45.0
3.5	73.3	72.5
5.5	90.0	90.0
7.5	96.7	97.5
9.5	100.0	100.0

3.25 a.

Class Boundaries	Frequency
0 – 99	4
100 – 199	3
200 – 299	4
300 – 324	7
325 – 349	14
350 – 399	6
	38

b. Yes.

3.27 b. Approximately bell-shaped.

3.33 b. The data are categorical.

3.35 b. Positive Skewness;   c. positive skewness.

3.37

Middle of Interval	Number of Observations
115	2 **
120	1 *
125	3 ***
130	5 *****
135	4 ****
140	2 **
145	2 **

3.39

Middle of Interval	Number of Observations
40	1 *
45	7 *******
50	11 ***********
55	21 *********************
60	21 *********************
65	23 ***********************
70	10 **********
75	6 ******

3.47 a. 4,869.5;   b. 4,875 (0.1% error).

3.49 $\bar{x} = 3.0, s = 3.05$.

3.51 $s_A = 0.37, s_B = 0.10$.

3.53 $\bar{x} = 2.4, s = 2.2$.

3.57 Median $= 2$, range $= 12$.

3.59 Bumper A:　range $= 1.3$; bumper B:　range $= 0.3$.

3.63 $\bar{x} = 59.46, s = 7.90$.

3.67 101.3%.

3.69 $Q_1 = 0, Q_2 = 2, Q_3 = 4.5$.

3.71 $Q_1 = 273, Q_2 = 326, Q_3 = 344$.

3.73 a. $P_{.05} = 20, P_{.95} = 376$ (approximately);　b. 90%.

3.75 2.67 ounces.

3.81 a. Distribution is bimodal, suggesting two distinct underlying causes of complaints;

b. identify and correct the cause giving rise to the second mode.

3.83 b. $\bar{x} = 3.8$, median $= 3.5$;　c. $s = 2.5$, range $= 8$.

3.87 $Q_2 =$ median $= 3.5$.

CHAPTER 4

4.1 a. $S = \{HHH, HHT, HTH, THH, TTH, THT, HTT, TTT\}$;

b. "Two or more heads" $= \{HHH, HHT, HTH, THH\}$.

4.3 a. $S = \{CC, CN, CO, NC, NN, NO, OC, ON, OO\}$, where $C =$ "critical," $N =$ "noncritical," and $O =$ "no defect";

b. "neither part has a critical defect" $= \{NN, NO, ON, OO\}$.

4.5 a. Yes;　b. no, the card could be the ace of diamonds.

4.7 1/50.

4.9 .15.

4.11 a. .9;　b. 0;　c. .5.

4.13 a. .37;　b. .58.

4.15 a. Yes;　b. no;　c. yes;　d. yes.

4.17 a. .36;　b. yes.

4.19 5/7.

4.21 .98.

4.23 No.

4.27 $\mu = 2.8, \sigma = 1.3$.

4.29 $\mu = 3.7, \sigma = 1.2$.

4.31 1/55.

4.33 $\mu = 7, \sigma = 2.4$.

4.35 1/10.

4.37 .113.

4.39 .03125.

4.41 No.

4.43 a. Yes;   b. no, the probability applies only to the remainder lot.

4.47 a. .9912;   b. .8021;   c. .1964;   d. .1158.

4.49 a. .0596;   b. .8593;   c. .0600;   d. .4656;   e. .1873;   f. .7758.

4.51 $\mu = 0.66, \sigma = 0.72$.

4.53 $\mu = 2.50, \sigma = 1.12$.

4.55 a. $\mu = 0.5$;   b. $\sigma = 0.071$.

4.57 Reduce n to ¼ the original sample size.

4.59 .3294.

4.61 Required probability is .3410; the result is consistent with the claim $p = .10$ because there is a fair chance (.341) of getting 2 or more such claims in 12 trials.

4.63 a. .0042;   b. yes, probability of such a result is very small (.0042) if $p = .01$.

4.65 a. .8823;   b. yes, probability of 23 or more sevens is only .021.

4.67 .003.

4.69 10/21.

4.71 a. .815;   b. .948;   c. .125;   d. .091.

4.73 .155.

4.75 .710.

4.77 .736.

4.79 .758.

4.83 $\sigma = 4$.

4.85 a. .7157   b. .7149.

4.87 a. .0045;   b. no, the exact binomial probability is .0042.

4.89 .736.

4.91 a. 70;   b. 30.

4.93 Producer's risk $= .10$; consumer's risk $= .05$.

4.95 AQL $= .03$, LTPD $= .26$ (approximately).

4.101 AQL = .07, LTPD = .33 (approximately).

4.103 a. $P(A) + P(B)$; b. 0; c. $P(B) + P(C)$;

d. 1; e. $1 - P(C)$; f. $1 + P(A) - P(B)$.

4.105 a.

No. Spades	Probability
0	27/64
1	27/64
2	9/64
3	1/64

b. $\mu = 3/4, \sigma^2 = 0.5625$.

4.107 a. .9456; b. .0004; c. .0028; d. .9857.

4.109 .5839.

4.111 a. 0.15; b. 0.036.

4.113 a. .238; b. .028; c. .891; d. .232.

4.115 .572.

4.117 AQL = .02 (approximately).

CHAPTER 5

5.1 c. 1/4.

5.3 c. $\dfrac{e-1}{e^2}$.

5.5 a. .8413; b. .8413; c. .1587; d. .1359; e. .1587; f. .6826.

5.7 a. .1151; b. .3446; c. .2620; d. .6449.

5.9 .1836.

5.11 a. 2.33; b. 1.96.

5.13 $Q_1 = -0.675, Q_2 = 0, Q_3 = 0.675$.

5.15 a. .8394; b. .8394 (same result).

5.17 Normal scores are $-0.95, -0.44, 0, 0.44, 0.95$.

5.19 $y_i = \dfrac{2i-1}{38}, i = 1, 2, \ldots, 19$.

5.21 No.

5.23 No.

5.29 a. 2/15; b. $\mu = 12.5, \sigma = 4.33$.

5.33 .6849.

5.35 .287.

5.37 a. 1.645; b. 1.96; c. 2.575.

5.39

k	1	2	3	4	5	6
Prob.	.3174	.0456	.0026	.00006	.0000006	.000000002

5.41   0.089.

5.43   a. 1 inch;   b. .00006 (.006%);   c. .0013 (.13%).

5.45   0.0039.

5.47   Only mildly, if at all; probability of getting a score this far away from the mean, or farther, is .1528.

5.49   .0392; yes.

5.51   b. .4095.

5.53   a. .7734;   b. .1056;   c. .4288.

5.55   a. −0.68;   b. 1.04.

5.57   −0.97, −.43, 0, 0.43, 0.97.

5.59   .0498 (4.98%).

5.61   Yes, probability of watching 1.5 hours or less is only .0183.

5.63   a. .607;   b. 10.

CHAPTER 6

6.1   The top can might be less prone to damage than the others.

6.3   Longer sections take more time to pass the inspection station; thus, they are more likely to be included in the sample than are shorter sections.

6.5   a. 20 feet;   b. 2 inches.

6.7   0.17 inch.

6.9   .0228.

6.11   .1096.

6.13   Yes; probability that the sample mean will differ from 40 by more than 0.5 is only .0026.

6.15   No. Probability of getting a sample mean of 5.9 or less is .0007.

6.17 $\mu = 47$ feet, $\sigma^2 = 6.8$ inches.

6.19   1.39.

6.21   2.528.

6.23   .025.

6.25   a. No;   b. population of breaking strengths is normal.

6.27   .025.

6.29   94.

6.31   a. .05;   b. .10.

6.33 No; student bodies are not all of the same size.

6.35 .0018.

6.37 No; probability of a sample mean as small as 4.5 or smaller is only .0124.

6.39 No; probability of a sample mean as small as 4.5 or smaller increases to .1446.

6.41 .01.

6.43 .01.

CHAPTER 7

7.1 a. Mean of sampling distribution is η.   b. Prefer g_2; it will have smaller sampling variability.

7.3 (131.1, 142.9).

7.5 (1.67, 2.03).

7.7 97.

7.9 (127.31, 134.37).

7.11 a. 4.9;   b. 5.9.

7.13 (2.3, 10.7).

7.15 (6.06, 16.10).

7.17 (.185, .235).

7.19 (.192, .228).

7.21 0.063.

7.23 $n = \left(\dfrac{z_{\alpha/2}}{2E} \right)^2$.

7.25 Since $z = 2.89 > z_{.01} = 2.326$, the sample mean is significantly greater than 10 at the .01 level of significance.

7.27 Yes, since $t = 3 > t_{.05}$ (15 *d.f.*) = 1.753.

7.29 a. $H_0: \mu = 138, H_1 : \mu < 138$;

b. the user of the cartons;

c. since $z = -3.75 < -z_{.05} = -1.645$, the null hypothesis can be rejected at the .05 level of significance, and we conclude that the mean bond strength is less than the specification.

7.31 Not significant, since $t = -0.91$ with a P value of .18 > .01.

7.33 Since $z = 3.32 > z_{.025} = 1.96$, we conclude that the two means are significantly different at the .05 level of significance.

7.35 Since $\chi^2 = 43.75 > \chi^2_{.05}$ (28 *d.f.*) = 41.33, we conclude that the sample standard deviation is significantly greater than 10 at the .05 level of significance.

7.37 Since $\chi^2_{.995}$ (13 *d.f.*) = 3.565 < χ^2 = 17.68 < $\chi^2_{.005}$ (13 *d.f.*) = 29.189; we conclude that the sample standard deviation is not significantly different from 10 at the .05 level of significance.

7.39 Since F = 2.77 < $F_{.01}$ (8, 10 *d.f.*) = 5.06, we conclude that the first variance is not significantly less than the second at the .01 level of significance.

7.41 Yes, since z = −3.98 < −$z_{.05}$ = −1.645.

7.43 No, since z = 0.85 < $z_{.05}$ = 1.645.

7.45 Since χ^2 = 2.13 < $\chi^2_{.025}$ (3 *d.f.*) = 9.348, we cannot conclude at the .025 level of significance that the laboratories are producing different results.

7.47 Since χ^2 = 45.9 > $\chi^2_{.01}$ (4 *d.f.*) = 13.277, we can conclude at the .01 level of significance that Fidelity and Selectivity are related.

7.49 a.

7.53 (114.1, 129.5).

7.55 a. 0.74;   b. yes, 42 additional observations.

7.57 (200.2, 571.1).

7.59 (.246, .304).

7.61 Yes, since z = 4.18 > $z_{.01}$ = 2.326.

7.63 Since χ^2 = 17.08 < $\chi^2_{.05}$ (15 *d.f.*) = 24.996, we cannot reject H_0: σ^2 = 159 in favor of H_1: σ^2 > 159.

7.65 Yes, since F = 3.085 > $F_{.01}$ (24, 24 *d.f.*) = 2.66.

7.67 Yes, since z = 5.37 > $z_{.025}$ = 1.96.

7.69 Yes, since χ^2 = 5.827 > $\chi^2_{.05}$(1 *d.f.*) = 3.841.

CHAPTER 8

8.1

	$\bar{x}$ *Chart*	*R Chart*
Central line	31.6	4.6
LCL	28.2	0
UCL	34.9	10.6

8.3

	$\bar{x}$ *Chart*	*S Chart*
Central line	31.6	1.7
LCL	28.9	0
UCL	34.3	3.7

8.5 Yes, the process is in control.

8.7 b. Yes (see run 24 on $\bar{x}$ chart);

c. yes, a run of 7 points above the central line.

8.9 Although the process is in control (barely), there appears to be strong periodicity.

8.13

	$\bar{x}$ Chart		R Chart	
n	(Central line = 4.20)		(Central line = 1.44)	
	LCL	UCL	LCL	UCL
3	2.7	5.7	0	3.7
4	3.2	5.2	0	3.3
5	3.4	5.0	0	3.0
6	3.5	4.9	0	2.9

8.15 a. np chart;   b. p chart ;   c. c chart;   d. p chart.

8.17 No corrective action yet, but monitor the apparent upward trend for potential future problems.

8.19 Central line = 0.0255, LCL = 0, UCL = 0.074.

8.21 There is a run of 7 points below the central line (runs 27 – 33).

8.23 The np chart is identical to the p chart except that the left-hand scale is multiplied by 100.

8.25 Central line = 3.93, LCL = 0, UCL = 9.87.

8.27 a. Yes;   b. no.

8.31

k:	1	2	3	4	5	6	7	8	9	10	11	12	13	14	15
Cusum:	0	1.2	0.5	–1.5	–0.4	0.5	2.6	1.5	0.9	2.2	2.4	0.3	–0.6	0.4	–0.5

8.33 d = 0.999, if we take y = 1.5, θ = 45°.

8.39 a. $-\infty$;   b. ∞.

8.41 a. 1.66;   b. 1.23;

c. product is barely manufacturable; since C_p = 1.66, product quality can be improved to "good" by centering the process.

8.43 a. 0.94;   b. 0.89;

c. process is not capable, thus product quality will be very poor.

8.45 1.41.

8.47 LWL = –2.92, UWL = 4.13; since warning limits are exceeded by 9 points, the causes of excess variability should be discovered and corrected.

8.49 LWL = 0, UWL = 3.85; since the upper warning limit is regularly exceeded, the causes of excess defects should be discovered and corrected.

8.51 .373.

8.53 b.

8.55

	$\bar{x}$ Chart	R Chart
Central line	11.6	8.1
LCL	6.9	0
UCL	16.3	17.1

8.57

	$\bar{x}$ *Chart*	*R Chart*
Central line	11.6	2.1
LCL	8.2	0
UCL	15.0	4.8

8.59 The last 7 points are trending down, indicating a potential out-of-control situation.

8.61 Central line $= 0.12$, $LCL = 0$, $UCL = 0.26$.

8.63 Central line $= 1.8$, $LCL = 0$, $UCL = 5.8$.

8.65 $d = 0.874$; if we take $y = 1.75$, $\theta = 45°$.

8.67 $C_p = 0.57$, $C_{pk} = 0.54$.

8.69 $LWL = 8.5$, $UWL = 14.7$.

CHAPTER 9

9.1 $\hat{y} = 4.82 + 1.38x$.

9.3 a. No, deflection is measured with error;   b. data appear to be linear;   c. since $b = 0.0098$, the increase in deflection per unit increase in load is estimated to be 0.0098 cm;   d. 1.57 cm.

9.5 Since $t = 2.17 < t_{.005}$ (6 *d.f.*) $= 3.707$, the slope is not significantly different from zero at the .01 level of significance.

9.7 $(-0.18, 2.94)$.

9.9 $(-0.0075, 0.0271)$.

9.11 $(71.5, 94.3)$.

9.13 a. $\hat{y} = 1.18 + 2.86^x$;   b. assume log y is linear in x and the residuals, log $\hat{y} -$ log y, are approximately normally distributed;   c. $(2.77, 2.94)$.

9.15 0.99.

9.17 0.66.

9.19 0.62.

9.21 Since $z = 1.77 < z_{.005} = 2.576$, r is not significantly different from zero at the .01 level of significance.

9.23 Since $z = 1.26 < z_{.025} = 1.96$, r is not significantly different from zero at the .05 level of significance.

9.25 $(-0.69, 0.94)$.

9.27 $(-0.39, 0.95)$.

9.29 b. and c.

9.31 a. $\hat{y} = 3.42 - 4.36x_1 + 4.74x_2$;   b. 1.4;   c. there is a slight trend with run number suggesting that another variable, correlated with time, is affecting the value of y;   d. no correlations are significantly different from zero at the .05 level of significance.

9.33 a. $\hat{y} = 2{,}097 + 6.34x_1 + 12.9x_2 - 61.5x_3$;   b. $R^2 = .346$, the regression explains 34.6% of the variability of the y values;

    c. the pattern appears random;

    d. $r_{x_1x_2} = 0.133, r_{x_1x_3} = 0.344, r_{x_2x_3} = 0.192$.

9.35 b. $\hat{y} = 11{,}024 - 98.2x_1 - 170x_2 + 2.7x_3 + 1.85x_1x_2$;

    c.

	x_1	x_2	x_3
x_2	0.133		
x_3	0.344	0.192	
x_1x_2	0.729	0.769	0.325

    e. $\hat{y} = 2{,}218 + 261x'_1 + 192x'_2 + 4.2x'_3 + 446x'_1x'_2$, where the prime (′) denotes a standardized value;

    f.

	x'_1	x'_2	x'_3
x'_2	0.133		
x'_3	0.344	0.192	
$x'_1x'_2$	−0.515	−0.218	−0.452

9.37 1.03.

9.39 (0.50, 1.56).

9.41 0.64.

9.43 Since $z = 3.56 > z_{.025} = 1.96$, r is significantly different from zero at the .05 level of significance.

9.45 a. 203.6;   b. no, the actual value of x_1^2 may be large enough so that the term $b_3x_1^2$ could have a value as great as or greater than the value of the other terms in the equation.

9.47 A probable non-linear relationship.

9.49 Coefficients cannot be interpreted as "effects," potential confounding, and possible incorrect choice of the form of the equation to fit to the data.

CHAPTER 10

10.1 Since $F = 0.31 < F_{.05}$ (3, 20 d.f.) = 3.10, there are no significant differences among the wafers at the .05 level of significance.

10.3 Since $F = 43.80 > F_{.01}$ (1, 58 d.f.) = 7.09, it can be concluded that the sample means differ significantly at the .01 level of significance.

10.5 Since $F = 12.6 > F_{.05}$ (3, 56 d.f.) = 2.80, it can be concluded that the four means differ significantly at the .05 level of significance.

10.7

C_1	C_2	C_4	C_3
2.500	2.500	2.500	3.167

10.9 Since $t = 4.19 > t_{.005}$ (9 *d.f.*) = 3.250, it can be concluded that the mean heat-producing capacity of the mines differs significantly at the .01 level of significance.

10.11

Source	D.F.	SS	MS	F
Bonders	3	19.15	6.38	2.22
Operators	4	16.91	4.23	1.47
Error	12	34.59	2.88	
Total	19	70.65		

10.13

Source	D.F.	SS	MS	F
Means	6	174.52	29.09	8.95
Blocks	4	101.39	25.35	7.80
Error	24	81.19	3.25	
Total	34	357.10		

10.15 a. $F_{\text{variables}} = 0.44 < F_{.05}$ (2, 4 *d.f.*) = 6.54, and $F_{\text{blocks}} = 0.44 < F_{.05}$ (2, 4 *d.f.*) = 6.54;

b. neither the differences among the block means or those among the variables means are significant at the .05 level of significance;

c. there are insufficient degrees of freedom for error.

10.17 $F_{\text{tenderizers}} = 12.3 > F_{.01}$ (1, 9 *d.f.*) = 10.6, and $F_{\text{tasters}} = 2.73 < F_{.05}$ (9, 9 *d.f.*) = 3.18; thus, the two meat tenderizers give significantly different tenderness results at the .01 level of significance, but it cannot be concluded at the .05 level of significance or less that the tenderness results differ among the 10 tasters.

10.19

Taster:	7	9	5	3	6	1	2	4	10	8
Mean Rating:	6.0	6.0	6.5	7.0	7.0	7.5	7.5	7.5	8.0	8.5

10.21 15.

10.23 $F_A = 12.5 > F_{.01}$ (2, 10 *d.f.*) = 7.56, thus we can conclude that the levels of Factor A give significantly different experimental results at the .01 level of significance; $F_B = 4.42 < F_{.05}$ (1, 10 *d.f.*) = 4.96 and $F_{AB} = 2.45 < F_{.05}$ (2, 10 *d.f.*) = 4.10, thus neither the effect of Factor B nor the AB interaction is significant at the .05 level of significance or less.

10.25 $F_{\text{DSS}} = 78.55 > F_{.01}$ (3, 28 *d.f.*) = 4.54, $F_{\text{Time}} = 486.33 > F_{.01}$ (1, 28 *d.f.*) = 7.60, $F_{\text{Interaction}} = 30.40 > F_{.01}$ (3, 28 *d.f.*) = 4.54, and $F_{\text{Replicates}} = 3.77 > F_{.01}$ (4, 28 *d.f.*) = 2.63; thus, we conclude, at the .01 level of significance, that *DSS* levels, storage time, and their interaction give significantly different flavor-index values and that the replicates have significantly different mean values at the .05 levels.

10.27 *Purpose*: To evaluate the effect on flavor of stored milk when *DSS* is added at four different levels.

Data-Collection Method: Milk was obtained from five different sources, and two different storage times were studied. Milk was tasted after storage to assign a flavor-index value to each milk sample.

Analysis Method: *ANOVA* (two-factor experiment with five replicates.)

Results: (See answer to Exercise 10.25.) A graph of the interaction between *DSS* and storage time shows that the reduction in the mean flavor index with time is diminished at the higher levels of *DSS*.

10.29

Significant Factors	MS	F	p value
A	270.28	12.45	.003
B	205.03	9.45	.007
C	124.03	5.71	.029
E	357.78	16.49	.001
CE	157.53	7.26	.016

10.31 The main effect of *A* is −5.81, and for *B* it is 5.06; the main effects of *C* and *E* are not reported because there is a significant *CE* interaction:

10.33 The following table shows the *CE* interaction:

	Level of C	
Level of E	1	2
1	42.00	33.63
2	44.25	44.75

The response (semiconductor gain) decreases sharply when relative humidity increases from 1% to 30%, but only when assembly is performed on the production line.

10.35 a. (1) *ab ac ad ae bc bd be cd ce de abcd abce abde acde bcde*;

b. *A = BCDE　B = ACDE　C = ABDE　D = ABCE　E = ABCD*
AB = CDE　AC = BDE　AD = BCE　AE = BCD　BC = ADE
BD = ACE　BE = ACD　CD = ABE　CE = ABD　DE = ABC;

c. Resolution V.

10.37 a. *ABC, CDE, ABDE*;

b. using odds for *ABC* and for *CDE: c ad ae bd be abc cde abcde*;

c. *A = BC = ACDE = BDE　　B = AC = BCDE = ADE*
C = AB = DE = ABCDE　　D = ABCD = CE = ABE
E = ABCE = CD = ABD　　AD = BCD = ACE = BE　　AE = BCE = ACD = BD

10.39 a.

n	5	6	7	8	9	10	11
Fraction	1/2	1/2	1/2	1/4	1/4	1/8	1/16

b. no.

10.41 a. 32;　b. 32;　c. 64.

10.43 All interactions are zero.

10.45 No; the yield levels have been increased somewhat and several variables have been eliminated from further consideration so that the continuous search for future yield increases will be more efficient.

10.47 While not all is known about the variables affecting line width, further major improvements in yields probably will not come from this area; it would be best to look at variables affecting fabrication next.

10.49 $F = 2.31 < F_{.05}$ (3, 12 *d.f.*) = 3.49; thus, factor A is not significant at the .05 level of significance.

10.51 $F_A = 7.8$ and $F_{Blocks} = 10.5$; since both F values exceed $F_{.01}$ (3, 9 *d.f.*) = 6.22, both factor A and blocks are significant at the .01 level of significance.

10.53 $F_A = 5.03 > F_{.05}$ (3, 8 *d.f.*) = 4.07, $F_B = 14.96 > F_{.01}$ (1, 8 *d.f.*) = 11.30, $F_{AB} = 1.13 < F_{.05}$ (3, 8 *d.f.*) = 4.07; thus factor A is significant at the .05 level of significance, and factor B is significant at the .01 level of significance.

10.55

	Level 4	*Level 1*	*Level 2*	*Level 3*
Means	1.50	2.75	3.50	4.00

10.57 1 *a b ab c ac bc abc d ad bd abd cd acd bcd abcd*

10.59 a. odds: *a b c d abc abd acd bcd*

    b. $A = BCD$ $B = ACD$ $C = ABD$ $D = ABC$ $AB = CD$ $AC = BD$ $AD = BC$

10.61 Resolution IV.

CHAPTER 11

11.1 .970.

11.3 .991.

11.5 .994.

11.7 .602.

11.9 .859.

11.11 .528.

11.13 $R(t) = 1 - \dfrac{\lambda}{3} t^3.$

11.15 $Z(t) = \dfrac{3\lambda t^2}{3 - \lambda t^3}.$

11.17 $R(t) = \dfrac{e^{-t^{\lambda+1}}}{\lambda + 1}.$

11.19 a. Yes; b. i. random failures, ii. early failures or "infant mortality," iii. wearout or "aging."

11.21 20 hours.

11.23 20.83 hours.

11.25 331 hours between failures.

11.27 2,008 hours (mean time to failure).

11.29 a. Yes; b. 4,506 hours; c. (2,742, 9,055).

11.31 a. 25,200 cycles; b. (13,765, 63,959).

11.33 a. Yes; b. $\hat{\alpha} = 0.032$, $\hat{\beta} = 0.59$.

11.35 .912.

11.37 .984.

11.39 .922.

11.41 .632.

11.43 .715.

11.45 17.9 hours.

11.47 4,567 hours.

11.49 (5,433; 25,244).

INDEX

—

Y0-BCF-232

Clough

Rise and fall of
civilization

Oakland Community College
Highland Lakes Library
7350 Cooley Lake Road
Union Lake, Michigan

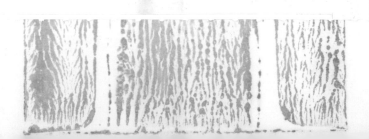

THE RISE AND FALL OF CIVILIZATION

THE RISE AND FALL OF CIVILIZATION • An Inquiry into the Relationship between Economic Development and Civilization

by SHEPARD B. CLOUGH

PROFESSOR OF HISTORY, COLUMBIA UNIVERSITY

Columbia University Press, New York and London

To TONY AND PETE

PREFACE

TO EXPLAIN the rise and fall of civilization in several cultures, as I have attempted to do in this book, is a formidable task, and in retrospect I wonder that I ever had the courage to undertake it. I believed, however, that I could throw light on what in Western Europe and America is meant by civilization, that I could clarify the major factors in the rises and falls of civilization, and that I might discover some pattern in these fluctuations. I felt that the Western World was anxiously groping for a fuller understanding of these questions, but that this understanding could not be obtained, at least not by the general public, from existing literature. Indeed, some of this literature, particularly the work of Arnold Toynbee, seemed to me to obscure human experience through time rather than to make it clear.

From my study of history, I find that the chief ideology of Western man, that is, the most common and most fundamental goal toward which Western men strive, is the attainment of a higher level of civilization. Secondly, the evidence seems to show that in Western culture civilization is considered to be measured by attainments in the arts and sciences, by the establishment of human relations based on known rules of equity and justice, and by the extent of human control over physical environment. Thirdly, I have come to the conclusion that heights of civilization in the chief cultures of which the West is heir were always reached just after peaks of material well-being had been attained. Thus, if my reading of the past is correct, it would seem to follow that one of the necessary conditions for the realization of Western culture's goal is a high level of economic achievement.

In the present analysis I have restricted myself largely to economic matters, realizing full well that they do not constitute a full explanation of my theme. In a future work, I hope to show how ideologies, styles in art, desires for knowledge of the universe, and patterns of social behavior have been differentiated and developed. At the present time I am studying the process of diffusion—the process by which ideas and techniques are spread—for herein are to be found most of the stimuli that lead to change.

For the research which has gone into this book I have made use of the remarkable resources of the Nicholas Murray Butler Library of Columbia University and of the excellent collections of the Dartmouth College Library, which have been opened to me for more summers than I wish to count. To the staffs of both of these institutions I express my thanks for past services and many favors.

SHEPARD B. CLOUGH

CONTENTS

Population and labor
Concentration of wealth
Achievement in the arts
Control over physical and human environment
In retrospect and in prospect

I

A FOREWORD

FEW questions of our times have aroused more curiosity than the rise and fall of what are commonly called civilizations, for modern man is witnessing great changes in the relative economic, political, artistic, and intellectual positions of various peoples of the earth. He is seeing Western Europe lose its place of world dominance which was so clearly established at the time of European expansion overseas in the sixteenth century. He is observing the once-great Eastern cultures bowed down by war, poverty, and overpopulation. And he is witnessing an enormous growth of power and intellectual activity in Russia and in the United States. Being of an inquiring mind, modern man seeks an explanation of what is going on around him—some orderly description and analysis of what is taking place. He wants a comprehensible and usable concept of what civilization is, a yardstick for measuring its rises and falls, and information regarding those factors which contribute to its fluctuations.

⌘ *The problem stated*

Existing literature provides little light on these subjects, at least not in concise form. Many authors who have addressed themselves to the question of civilization have shied away from even an attempt to define civilization and hence have failed to make clear what forces contribute to upward or downward trends of achievement. This is largely true of Arnold J. Toynbee whose *The Study of History* (1934 *ff.*) has received extensive attention since World War II. He employs such clichés as "challenge and response" and "withdrawal and return" to explain what leads to higher or lower stages of civilization, but he fails to indicate what "challenges" under what circumstances have elicited what "responses," or what effect a "withdrawal" of a culture has in determining a "return."

Similarly Oswald Spengler, whose *The Decline of the West* (1926–1928) had a remarkable vogue after World War I, is vague concerning a concept of civilization, and he confuses the issue of civilizing forces by drawing an analogy between the life cycle of a living organism and that of a culture. Both presumably are born, have a youth, mature, grow old, and die, which, if true, would provide no causal explanation of the phenomenon before us. In fact, societies are not organisms and do not behave like them.

Other authors attribute change in degrees of civilization primarily to some single factor, like geography, as does Ellsworth Huntington in his *Man and Civilization* (1945), and thereby oversimplify what is extremely complex. Still others, who avoid many of these pitfalls, like Alfred L. Kroeber in his *Configurations of Culture* (1944), make no serious attempt to analyze the relationships between the economic well-being of a people and gradations of civilization.

Because of these lacunae, and particularly the last, an historical inquiry into the rise and fall of civilization primarily in its economic aspects seems a legitimate enterprise for an economic historian. But before embarking upon the historical phase of this study, it is essential, as in all scientific work, to state as clearly and as precisely as possible the object of the investigation. Thus at the very outset the author must endeavor to explain what is generally meant by the term civilization. Although the word is relatively new, being derived from the concept of people living together according to accepted customs, that is, of acting civilly, it has acquired within the last 150 years a different and broader connotation. It refers to achievements in such aesthetic and intellectual pursuits as architecture, painting, literature, sculpture, music, philosophy, and science and to the success which a people has in establishing control over its human and physical environment. The greater the extent to which a people produces aesthetic and intellectual works of high merit and provides physical and social security for its members, the more civilized that people is; and conversely the less a people realizes either qualitatively or quantitatively any of these tenets of civilization, the less civilized it is.

Various aspects of this definition of civilization require elucidation. First, the term "people" needs to be made more precise, for we are not concerned with any heterogeneous conglomeration of human beings, but with people living in society, that is, with humans who are involved in a complex of relationships which grow out of action to attain common ends, including that of perpetuating the society. More specifically attention will be focused on societies which comprise cultures or, stated differently, with societies which have a body of learnable and transmissible knowledge the fundamental parts of which are homogeneous enough and firmly enough held to effect similarly and sharply the behavior of a large number of persons. Thus we shall speak of French, German, or American societies and of Western culture of which these societies are parts.

Second, this definition of civilization is one that comes from Western culture and from certain other prominent cultures during periods of their most brilliant accomplishments. It involves a value judgment which is not shared by all cultures. In fact, members of some societies have diametrically opposed conceptions of civilization, for they consider the highest form of human existence to be a condition in which man does no work and receives the necessities of life from a bountiful and unfailing nature.

Third, attention should be called to the fact that in the definition of a culture such qualitative and relative words as "homogeneous" and "similar" appear. One may well ask what amount of homogeneity or what degree of similarity must exist among people to justify distinguishing them as a culture. In whatever answer is given to this question an element of arbitrariness must enter, for there are no precise measures or standards of differentiation. All can recognize a profound difference between the behavior of New Yorkers and Eskimos, of English and Chinese, and of French and

Russians, yet one can observe considerable overlappings in the behavior of the last two peoples and a twilight zone between Western and Eastern European cultures.

Fourth, we must realize that in estimating the degree of civilization in any culture at a given time there is a nice problem of weighing different kinds of achievement. For example, is the degree of civilization in a culture higher where great aesthetic works are being produced but where individual security does not exist than where these conditions are completely reversed? This is a difficult matter to resolve, but reasonable conclusions can be obtained by comparing the degree to which all the tenets of civilization are met by a culture at various periods of time. Similarly, the measurement of the greatness of a work of art is extremely difficult, for who can determine the respective beauties of an Egyptian vase and a Grecian urn. In this matter we shall adopt the arbitrary procedure of considering to be of a high order those things which have been most generally and most consistently recognized by competent authorities to be great.

Last, we must recognize that neither cultures nor civilizations are static but on the contrary are ever changing. Thus Greek culture, although it displayed a continuity of certain ideologies over a long period of time, was not a fixed entity, nor was the degree of civilization attained by Athens in the fifth and fourth centuries B.C. attained by all the Greek city-states at the same moment or maintained by Athens for an indefinite space of time. Why such changes took place and what has accounted for shifts in centers of civilization are problems in which we are interested. Our findings should throw light upon what is happening within Western culture today, upon the recent rise of Russia, and upon the predicament in which Eastern cultures find themselves.

✑ Factors in achieving higher degrees of civilization

Having stated the primary theme of this study and having discussed the term civilization and its components, we turn to a consideration of some of the major conditions which have contributed to the achievement of higher levels of civilization. Of all of these conditions none seems to be more fundamental than the possession of economic surplus, that is, of agricultural goods, industrial products, and various services in excess of what is immediately necessary for the maintenance of life. Obviously only when the necessities of physical existence have been provided can energies be devoted to creative enterprise of an aesthetic and intellectual character or can physical security be assured. This is true even in those periods when wealth has been highly concentrated in the hands of a few who supported artists and intellectuals in one way or another and provided at least a modicum of security against want and violence as well as in those times when relatively extensive economic well-being existed throughout society.

The idea that relative economic well-being in society is one of the most necessary conditions to a high stage of civilization leads to an exceedingly appealing hypothesis regarding the fluctuating character of civilization within a culture—a hypothesis which is much more impelling than Toynbee's "challenge and response" or Spengler's analogy. This hypothesis, succinctly stated, is that as a large proportion of human energies in a culture is devoted to the achievement

and enjoyment of great works of art—that is, in achieving
a higher civilization—the smaller the proportion of energies
which will be available for creating economic well-being—that
is, in creating a necessary condition in the process of civilization. This hypothesis, which admittedly points to only one
of many factors that contribute to fluctuations in civilization,
would lead to two further hypotheses: (1) that the peak of
civilization in any culture would be expected at the time
when economic decline begins and (2) that the prolongation
of high stages of civilization can be achieved by a nice balance
in the utilization of human energies between aesthetic and
economic activity.

In any case, the evidence presented in the pages of this
book supports the contention that economic surplus has contributed to advances in civilization, although some of these
contributions have been indirect. For example, one of the
major concomitants of economic well-being has been trade,
and trade has been essential in promoting exchanges of ideas
among peoples and in providing creative stimuli through
economic and cultural competition. Economic surplus pre-
supposes considerable business activity, and great business
enterprise implies the keeping of business records. In fact,
it was in keeping records of business that the earliest known
forms of writing were used, that the first condensed systems
of numerical notation were conceived, and that the most
ancient calendars were developed. Furthermore, business has
employed artists to design its products, architects to plan
buildings to house its operations, and intellectuals of various
kinds to help it produce what people want.

Lastly, economic surplus has made urbanization possible,

which is of great moment because the outstanding achievements of civilization have been products of urban centers. In fact, the aesthetic creations of various cultures have been so highly concentrated in cities that the treasures of places like Athens, Rome, and Paris are regarded as epitomes of their respective cultures.

The role of cities in the process of civilization is to be accounted for in part by the fact that in urban centers a larger number of persons than in rural areas has had time over and above that required for meeting the material needs of life to devote to devising ways to extend man's control over his environment and to engage in cultural activity. Recent studies of cultural influence confirm the historical observation of the dominant role of cities. In fact, they indicate that there is a distinct correlation between cultural influence on the one hand and size of city and commercial activity on the other, that the influence of the city wanes with effective distance,[1] and that the greater the economic dependence upon cities, the greater is the influence of urban centers.[2] Plans for the realization of the most generally recognized triumphs of science, engineering, sculpture, music, architecture, painting, and literature have been produced in cities or by persons who received the stimuli for their work in urban areas. Also the process of acculturation—the borrowing of knowledge, ideas, techniques, forms of organization, art styles, and ideals toward which society should strive—has taken place primarily in cities and has radiated from them.

Essential as economic surplus may have been in providing leisure for the civilizing process, basic as trade has been for the diffusion of ideas, and important as has been the role of cities as centers of cultural achievement, the place of economic factors in the rise and fall of civilization must not be exaggerated. In fact, of equal, if not of greater, importance are the controlling ideologies of a culture. These principles become established chiefly in periods of great and rapid change, as at the very beginning of a culture. At such times goals, intellectual standards, art forms, and technological attitudes become differentiated and specialized. If in this stage the culture embraces the idea that great works of art should be created, that knowledge of the universe should be acquired, and that man should extend his control over human and physical environment, then subsequently more of the energies of the culture will be devoted to the attainment of a high degree of civilization than would otherwise be the case.

What leads to the adoption of such an ideology is difficult to determine. It may be borrowed from some other culture, or it may be developed indigenously. In the latter case climatic and geographic elements may have an influence, for a too bounteous nature may be conducive to sloth, while a too hostile environment may permit of no leisure. Possibly a climate with wide seasonal fluctuations is of importance, for it requires that a surplus be stored during the productive months to tide one over the unproductive months of the year; and this practice provides leisure for other than the obtaining of sustenance. Furthermore natural resources have something to do with the case, because a relatively barren land would not permit the accumulation of surplus. At all events most of the great civilizations which we know have been mainly in temperate regions, and there has been a tendency for the highest civilizations to move toward more invigorating, colder, and stormier areas.

Another ideology which is important for the effective development of civilization is a desire for a social and political organization which will permit individuals to realize their total potential as contributors to civilization. What is implied here is that in a system where social taboos or political restrictions prevent large segments of a culture's population from engaging in types of activity which add most to civilization, the culture cannot attain the highest degree of civilization of which it is capable. Thus the caste system in India, restrictions on choice of occupation in medieval Europe, and the anti-Semitic laws of Nazi Germany curtailed the civilizing process. Also implied in the foregoing statement is freedom from destructive strife, for obviously civilization is retarded if man uses his energies for destroying what has been built rather than for building anew. In general, ravaging warfare has preceded a decline in civilization, for cultures with a high degree of civilization have aroused the envy of, and have become the object of attack by, less favored cultures.

Lastly, among factors conducive to the achievement of higher forms of civilization, mention should be made of a good state of health in the population, the propagation of those elements in society which have the ability to produce offspring with good physiological equipment for creative activity, a balance between production and population so that an increase in people does not absorb all the potential for economic surplus, and the placing of leadership in the hands of those who would advance civilization rather than some other cause.

Of great importance also is the prevention of such a rigid structuring of the culture that the individual has no choice of action. Although there is evidence to indicate that some rigidities, let us say styles in art, tend to be fulfilled, that is, become so stereotyped that they are destroyed, others may last a long time and limit thought and aesthetic expression. Indeed the *range of opportunities for alternative decisions* is a useful concept in analyzing civilizing forces. Everything which contributes to the extension of that range, whether it be economic surplus, the use of leisure, the development of cities, or the absence of stultifying rigidities, is of the utmost significance in furthering the process of civilization.

⚘ *Factors of economic progress*

Now that consideration has been given to what civilization in various cultures is and to what factors have contributed to its development, especial attention is to be focused on economic progress. Indeed economic achievement in relation to degrees of civilization will be a basic concern in the subsequent pages of this study. By economic progress is meant an increase in output of goods and services per capita in a society. Factors of economic progress may be grouped under the following headings: (1) natural resources, (2) technology and techniques of production, (3) labor, (4) capital, (5) business leadership and economic institutions, (6) demand for goods and techniques of distribution. An explanation for the location of economic activity in one area rather than in another is to be found primarily in terms of the extent to which these factors are present at a given place.

Inasmuch as frequent references to these elements of economic progress will be made in the following discussion of the rise and fall of civilization, brief treatment of each of them here will save subsequent digressions from the main story. This treatment begins logically with natural resources. They are products of nature which can be utilized by man to satisfy material wants.

Accordingly a product of nature is not a natural resource until man has the knowledge, skill, means, and desire to utilize it for his own purposes.

Copper was not a natural resource until man learned to work it into useful articles about 3500 B.C. The same can be said of iron (*circa* 1300 B.C.), and of coal before the seventeenth or eighteenth century, when it first began to be used on a relatively large scale. Hence the value of the products of nature has varied widely as man has developed an ability to use one or the other of them.

This has even been true of agricultural land, perhaps the chief natural resource, as man learned to domesticate animals and plants, to drain and irrigate, and to transport bulky goods over long distances. Forests, which may have been barriers to settlement before sharp axes for felling trees were available, became important natural resources as more charcoal was needed for smelting. In fact, the abundance of timber in Western Europe gave that area an advantage over the relatively treeless areas of North Africa and Hither Asia and accounts in part for the shift in the center of civilization from the Mediterranean to the northward.[3] In more recent times power-producing natural resources, especially coal, have had a particularly strategic importance in economic progress, areas with the largest amount of coal available for use in industry having scored the greatest measure of industrialization.[4] In general, it may be said that the economic development of an area is conditioned by the amount and accessibility of its natural exploitable [5] resources per capita of the population.

The second condition of economic progress listed above is technology of production, that is, the body of abilities possessed by mankind to turn out goods and services. A technique, by comparison, is a particular way of accomplishing a stated task. In all cultures some of the existing technology has been inherited or borrowed from other cultures, and some has been developed or discovered indigenously. In this regard it should be noted that no major technical advance has ever been completely lost, even though some have remained dormant for long periods of time.

In the process of borrowing or acculturation the thing that is borrowed and put to use is determined primarily by the borrower's state of technology, by natural resources, by market demand, relative costs of production, available financing, and labor supply. In the spread of mechanized industry at the time of the industrial revolution in England in the eighteenth century, cotton textile machines were frequently the first to be adopted in other countries because the manner of operating them was similar to handicraft methods; raw material was frequently obtained by other areas as easily as by England; the demand for cotton textiles was great; and costs of the equipment were not high. At the same time new methods of iron production moved first to those inhabited areas in which natural resources were abundant and where some knowledge of ironworking already existed.

In invention and discovery the same factors come into play. Most of the great inventions and discoveries have been worked upon simultaneously and often independently by persons in separate but similar environments; they have resulted from the slow accretion of many details from many sources; and they have been applied first where "circumstances were favorable" for their use.[6] Technology and techniques are basic to any understanding of economic progress, for they provide the most important means by which mankind can produce more goods and services with less expenditure of human energy.

The third condition of economic progress is the capacity of the individual worker. Capacity depends upon mastery of techniques, health, morale, and the division of labor, that is, the specialization of workers in the production of some particular part of an object rather than in the production of the whole object. Indeed, cultures with the highest gross product per capita have been those in which a large proportion of the labor force has been engaged in industry, commerce, and transport, which are branches of economic activity permitting a great division of labor among workers and among geographical regions.[7] Still another consideration here is the recompense obtained by workers for their labor, for workers must be paid enough so that their demands for goods will be effective in bringing forth greater supplies.[8] Yet it is essential that there be savings from wages or profits so that investment in additional economic enterprise will continue to take place.[9]

The fourth condition of economic progress is capital. Capital may be defined as valuables in the form of land, tools, machines, buildings, materials, securities, money, or other transferables which can be employed for the production of more goods and services. The more abundant and the more mobile (quickly transferable) capital is, the more readily it can be employed to take advantage of opportunities for production. Consequently the development of symbols which represent wealth, like the deed, mortgage, bond, stock, and money, are of extreme importance in facilitating transfer of ownership.

Of all these symbols of wealth the most important has undoubtedly been money. In fact, money has been classed as one of the most fundamental finds in history, ranking close to the discovery of fire, the domestication of grains and animals, writing, and the use of machine-produced power. As a medium of exchange and as a store of wealth money has been more useful than all other forms in transferring wealth, in accumulating capital, and in permitting a division of labor. As our story progresses we shall have occasion to consider capital formation, growing reliance upon money in the exchange of goods, and the development of organs for the accumulation and investment of capital, like banks and stock exchanges.

The fifth condition of economic progress has to do with business leadership. In general the issue of leadership seems to boil down to this: Men are born with biological equipment which may be extremely poor or exceptionally good; this biological equipment develops and acts primarily in response to stimuli from environment; and leaders are distinguished from other men in the degree to which stimuli effect creative responses. Thus in economic activity leaders are recognized by the extent to which they have been able to combine factors of production to turn out more goods and services.

What they have been able to accomplish is determined in part by the conditions of economic progress here under consideration, but their task has always been easier when cultural ideologies have been favorable to their aims. Thus business leaders have had an easier row to hoe when within society there has been a general conviction that it was well to increase the output of goods and services than when such production, or participation in such production, was scorned as socially and ethically degrading. Similarly economic progress has been enhanced by a generally accepted belief that economic energies should be exerted to produce capital goods (goods which could produce more goods) or consumer goods and services which improve physical well-being.

The sixth and last group of conditions of economic progress embraces the demand for goods and the technology of distribution. We have already touched on this subject in discussing labor and in that connection emphasized the need for sufficient remuneration to permit labor to be a high consumer. Clearly a large demand for goods is a stimulus to the production of more goods. This large demand may come in part from such a distribution of wealth that purchasing power is not concentrated in the hands of a few whose wants are surfeited with supply. It may come from popularizing the desirability of more goods and services like a radio in every room. And it may come, indeed it frequently has, from tapping markets in distant lands never before supplied with certain products. But a large market implies not only effective demand, that is, desire plus the ability to acquire goods, but also the means of effecting exchange and of transport. Money we have listed as particularly important in exchange, and we shall see later what wonders of distribution money made possible as compared to the system of barter. But in commerce we should not minimize the importance of credit, selling organizations like the store, and especially means of transportation. The use of the wheel as against packs and sledges, of hard-surfaced roads and rails as opposed to trails and dirt roads, of the steam vessel compared to the oar-propelled boat, of the railway as against the oxcart—these at once indicate the enormous advances which technological change has wrought in the exchange of goods and the division of labor, and thus in the history of economic development and of civilization itself.

In what has preceded, particular stress has been placed upon factors for advances in civilization and for progress in economic affairs—upon what is in general "environment." This raises the age-old question as to whether important developments in history have really been the result of "social forces," that is, environment, or whether they have been the result of "great men," that is, leaders. This question requires explanation. Obviously all human accomplishments have been realized by human effort; and the greater the individual accomplishment, the greater the man or men who were responsible for it. Yet the creations of great men have been limited by "conditions" of the time and place in which they lived.

Thus one would not expect the invention of calculus to have been made by men of limited intelligence or by a thousand men at any one time and place, yet it is not surprising that two or three "great men" at about the same time and within the same civilization invented calculus quite independently of each other. Environment and "great men" seem to be mutually interdependent, with "great men" working on environment to understand, control, or change it, and with a changing environment stimulating, challenging, and limiting the creative responses of these men.

Still another question arises from the material presented concerning factors for the development of civilization or for economic progress; and that is whether or not a single factor can be isolated as the prime moving force. An affirmative answer to this question is especially tempting because it would enormously simplify analysis and control of social relations. Unfortunately, however, such a simplification would vitiate available evidence. The list of factors which I have made seems to be an irreducible minimum, although in specific cases one or another condition may be predominant. How this was will be seen as we proceed with the story.

For Further Reading

Brooks Adams, *The Law of Civilization and Decay* (1895)
Charles A. Beard, *The Making of American Civilization* (1942)
A. M. Carr-Saunders, *World Population* (1936)
Colin Clark, *The Conditions of Economic Progress* (1951)
S. C. Gilfillan, *The Sociology of Invention* (1935)
Ellsworth Huntington, *Man and Civilization* (1945)
Alfred L. Kroeber, *Configurations of Culture* (1944)
Ralph Linton, *The Science of Man in the Changing World* (1945)

R. M. MacIver, *Society: Its Structure and Changes* (1931)
————, *The Web of Government* (1947)
R. D. McKenzie, *The Metropolitan Community* (1933)
Flinders Petrie, *The Revolutions of Civilization* (1941)
Pitrim A. Sorokin, *Social Philosophies in an Age of Crisis* (1950)
Oswald Spengler, *The Decline of the West* (1926–1928), 2 vols.
Arnold J. Toynbee, *The Study of History* (1934–1939), 6 vols.
  One-volume abridgment by D. C. Somervell (1947).

II

THE EARLIEST CULTURES

―――

THE planet on which we live, it is thought, was created by the congealing of gases trillions of years ago. Over an estimated billion and a half years ago the earth mass cooled and contracted, developing folds and continents. Sometime, perhaps 1,000,000 or 500,000 years ago, man appeared on this globe. In the neighborhood of 3000 B.C. man had developed civilization to a point at which written records were made— records that have come down to us of the present day. Thus the last 5,000 years, or less than 1 per cent of man's existence, is to be the primary concern of this study.

✍ Limitations of our study in time and space

Even in this small fraction of man's total time on earth, human beings have created such a vast and complex body of records about themselves that one man cannot possibly cope with it. Consequently from among the thousands of cultures about which information is available, we shall select for study those which have contributed most directly and in the largest measure to the development of Western culture.

Thus attention will be focused primarily upon the cultures of Sumer, Egypt, Babylon, the Aegean, Greece, Rome, and Western Europe.[1] Furthermore, we shall not treat all phases of these cultures but shall stick to a central theme—to see how various factors of economic progress developed, to judge whether or not certain forces resulted in an increase in the production of goods and services per capita, and to suggest the relationship between economic progress and the achievement of higher levels of civilization. In the present chapter we shall consider certain basic accomplishments which started man on the long road to greater control over his environment and to the creation of objects for intellectual and aesthetic enjoyment. More specifically we shall be concerned with technological changes which permitted the creation of economic surplus, with the division of labor and the growth of trade, with the establishment of cities, and with the development of writing and numerical notations.

Paleolithic civilization and the Neolithic agricultural revolution

When the earliest form of man appeared on this planet, he probably lived much like a wild animal. Although we lack information concerning his earliest economy, archaeological evidence seems to indicate that 140,000 years ago, as the last glacial period was approaching, man was a food gatherer rather than a food producer. He hunted game, sought fruit and nuts, and collected the seeds of plants for grain. He relied similarly on such products of nature as caves for shelter and on animal skins for clothing.

In the course of time man added to the technology by which

he sustained life. He commenced to make use of fire and probably developed ways of producing it. He chipped pieces of stone from rocks in order to get weapons and tools with which to perform his elementary tasks more effectively and more expeditiously. Stone implements, which gave this earliest form of civilization its name of Paleolithic or Old Stone Age, were gradually improved in quality and beauty and became differentiated in design (ax, spear, and knife) for the performance of special jobs. Subsequently Paleolithic man added bones, horns, and ivory to the list of raw materials for tools and weapons. He caught fish with hook and line, used the dog in the chase, employed rotary motion for drilling holes even in stone, utilized the principle of the lever and fulcrum for moving heavy objects, and invented the bow and arrow, the first composite mechanism made by human kind.

Yet the development of these tools and weapons made possible the accumulation of but a very small surplus, probably just enough to provide minimum human needs during the unproductive months of the year. They did not lead to any extensive division of labor, to any important trade, or to any residual body of aesthetic works. When men moved, as they frequently did, they could carry their belongings on their backs or draw them on sledges. Yet, in spite of the low level of accomplishment, one very important ideology became established in many societies—man wanted to increase his control over his physical environment so that he would not suffer from want. This desire is clear from improvements in techniques and from the character of supplications to supernatural powers. In nearly all primitive religions there are prayers, rites, spells, and incantations to the totem to make the hunting good or to assist the supplicant one way or another in overcoming some material difficulty.[2]

About 10,000 years ago important changes took place which were to lead to the creation of economic surplus and to the achievement of higher levels of civilization. With the melting of the northern ice sheets at the end of the last glacial period, the steppes and tundras of Europe were transformed into temperate forests and the prairies south of the Mediterranean and in Hither Asia were converted into deserts with oases. Consequently hunting in these areas became poorer, and man sought a new way to provide himself with sustenance. In the solution of his problem he effected one of the major technological revolutions in history—the domestication of grains and animals. This agricultural revolution came with the beginning of a new age in history—the New Stone or Neolithic Age.[3]

Probably the first grains to be sowed and harvested for human food were related to wheat and barley; and they were selected from a number of wild plant seeds and tubers which had been gathered by Paleolithic man in his quest for food. To wheat and barley were added in various places rice, millet, Indian corn or maize, yams, manioc, and squash. The first animals to be brought under control, cared for, and used to aid in the chase or to supplement its rewards were dogs, sheep, goats, cattle, and pigs. They were chosen either because they had not fled to the north like the reindeer, because they were relatively docile, or because they were particularly productive of the things men wanted.

The domestication of plants and animals had far-reaching effects. Tools were improved and further differentiated.

Polishing of stone was more generally resorted to in order to get a finer edge for harvesting grain, cutting wood, and slaughtering animals. The hoe and sickle, the polished stone ax, and better knives bear testimony of the trend. The cultivation of plant crops involved new problems of food storage and led to the creation of the earthern pot which was far superior for this purpose to bags of skins, horns, or woven baskets. Fermentation was discovered, which allowed Neolithic man to make beverages with which to enliven festive occasions or to relieve the boredom of rainy days. Spinning and weaving were developed, which made possible a substitute for animal skins as clothing. And bodily decoration with shells and beads, which indicated the existence of aesthetic considerations, became more common.

As can readily be imagined, the new agricultural technology placed a premium upon readily arable and fertile land. Oases and naturally drained marsh areas along rivers or around lakes were probably the most desirable, for there few if any trees had to be cleared away. In such areas caves did not exist, so man began to turn to other types of housing—to tents made from hides, to shelters from branches or grass, and to wooden or sod houses. Such structures made possible the establishment of villages of the fixed settlement type, which had the advantages of being convenient to the place of work, of facilitating social intercourse, of accumulating human aesthetic achievements and knowledge, and of providing protection. On the last score, the need was apparently very great, for in times of poor crops and threatened famine people migrated in search of greener pastures which, if need be, they would take by force. Then, too, people who lived on the periphery of settled communities preyed upon their neighbors. These nomadic and seminomadic people were for long a serious problem, for inasmuch as they lived as food gatherers and later as shepherds they had to contend with wild beasts and keep their fighting ability at a high level. They were the "barbarians" who invaded settled areas.

✄ The technological revolution in the Age of Copper

Important as was settled agriculture as a foundation for the production of a surplus which would provide that leisure which man needs for intellectual and artistic pursuits, the Neolithic Age failed to meet some of the basic conditions of economic progress. It did not have sufficiently effective or specialized tools to permit much of a division of labor. It did not employ animal or mechanical power on a large enough scale to supplement human effort to a meaningful degree. It had only the beginnings of a social or political system which could organize the collective efforts of a group for effecting major capital improvements. And it did not have means of transportation adequate for the development of extensive commerce.

The Age of Copper, or, as it is more elaborately called, the Chalcolithic Age, marked an advance in most of these respects. It inaugurated a "technological revolution" during the 1,000 years prior to 3000 B.C. which was, in the opinion of many, more far-reaching in its consequences than any other prior to 1600 A.D. Improvements in techniques made possible important gains in production per capita and laid the foundation for an urban development that was essential to forward strides in intellectual and cultural endeavor.

The geographical area within which the technological revolution of the Copper Age took place is bounded on the west by the Sahara and the Mediterranean Sea, on the east by the Thar Desert and the Himalayas, on the north by the Balkan, Caucasus, Elburz, and Hindu Kush Mountains, and on the south by the Tropic of Cancer. It is a region in which a surplus has to be provided for nonproductive seasons. It has fertile land around oases, along river valleys, and in mountain passes. It had the wild grapevine, dependable crops of dates, figs, and olives with which man could supplement his cultivated crops. It had easily navigable waterways for commerce. It was comparatively well protected from mass migrations from outside. And it had relatively cloudless skies, which facilitated observation of the heavens—observation which led to the acquisition of knowledge useful in guiding the traveler on his way and in keeping track of the seasons.

Settled agriculture within this area was carried on most successfully near oases and along the Nile, at the mouth of the Tigris and Euphrates Rivers, and along the Indus River. Agricultural surpluses were exchanged along the rivers and with the wilder food gatherers on the fringes of settlement. Commerce of this kind made men familiar with new products turned out by their neighbors and whetted a demand for new utilitarian and luxury goods. Furthermore problems of drainage and irrigation in river agriculture contributed to cooperative effort on a relatively large scale. Out of this experience came organization for the pooling of human resources, leadership in productive enterprise, some division of labor, and a recognition of the necessity of "investment" of present energies for the purpose of future gain. Gradually the desire for more products, the possibility of obtaining them in exchange for surplus agricultural goods, and the need for control over the use of water created a situation conducive to technological change.

As the reader has already assumed from the name of the age which we are examining, one of the major technological developments in this period consisted in the discovery and use of copper. This metal had obvious advantages over stone. It was less fragile, could be more easily repaired, took a better edge, made lighter and stronger tools, and could be more readily shaped by hammering or casting into a greater variety of products. Although it was not so universally distributed over the earth's surface as stone, its lightness and consequent ease of transport gave it a preferred position in areas, like the delta of the Tigris and Euphrates, which had no stone. Its cost was high, however, which restricted its use, especially in districts like Egypt where flint was readily available. Nevertheless, we find it being used among the Sumerians in the delta of the Tigris and Euphrates for tools, weapons, vessels, and seals with which to indicate the ownership of goods.

Probably the first copper was found in a relatively pure state in nature, but by 3000 B.C. the metal was being reduced from ore in furnaces fired with charcoal.[4] It was placed in crucibles which would withstand high temperatures; it was handled with tongs; and it was cast in molds. The production of copper provided techniques essential to the manufacture of a large number of metals, and therefore it is not strange that we find silver, lead, and tin being produced by similar methods almost at once. And before long tin began to be added to copper to make bronze.

Important as were these technological advances in metallurgy, they were rivaled in significance by changes in ceramics. Pottery made from clay and fired to obtain greater durability was inherited from earlier times. Within the time span under discussion Egyptians developed ways of glazing and decorating pottery to make the more durable and beautiful faïence. The inhabitants of Hither Asia invented the potter's wheel which made possible extremely rapid manufacture of fine products of perfectly symmetrical design. The invention appears to have reached Egypt by 2700 B.C. and the Indus by 2500 B.C. The principle of the wheel was in itself of greatest moment, for it is essential in the transmission of power and in the reduction of friction. Wheeled vehicles were in use in Mesopotamia, for example, by 3000 B.C., in India before 2500 B.C., in Crete by 2000 B.C., in Egypt by 1600 B.C., and in China and Sweden by 1000 B.C. Curiously enough, they were never developed in the Maya, Inca, or Aztec civilizations, nor was the principle of the wheel used in them for anything except toys!

Among other technological developments mention should be made of the manufacture of brick, at first sun-dried but in India by 2500 B.C. fired in kilns. Wooden plows appeared in Egypt and Mesopotamia about 3000 B.C. and soon afterward in India, although not in China until about 1400 B.C. Plows required animal power for efficient traction, and oxen were equipped with yokes that fitted on the broad shoulders or horns of the beast to pull them.

This union of livestock with the cultivation of crops meant not only the addition of animal energy to human energy in agriculture, but it also meant the production of more humus for the soil and the beginning of a new form of transportation. By 3000 B.C. camels and asses, or horses, were known to the people of Hither Asia and were gradually put to use by others, the horse having been introduced into Egypt between 1800 and 1600 B.C. Unfortunately, the collar for the horse, probably adopted from the ox yoke, was so constructed that it choked the animal when he pulled a heavy load.[5] For this reason and because the ass and horse were much smaller than now, they were not used as draft animals in ancient times except for light vehicles like chariots but were employed as pack and riding animals. It was not until the ninth or tenth century A.D. that the breastplate and later the collar were adjusted to fit on the shoulders of the horse rather than on his throat. For his part, the ox was not very efficient for transport, because he was slow; his load was limited by friction of the wheels on dirt roads; and his hoofs, like the horse's, wore out on hard surfaces. Not until the tenth century A.D. were iron shoes with nails used to overcome this last difficulty.[6]

Finally, the period of which we are speaking witnessed a considerable improvement in water transport. Although Neolithic men had at least rafts and canoes with which they made hazardous journeys, it was probably not until a little before 3000 B.C. that craft capable of carrying substantial cargoes were constructed and that the sail was used to propel them; and it was not until about 2200 B.C. that the rudder came to supplement the steering oar. The sail was a major invention, because it increased the productive efficiency of man by adding to his technology one of the earliest methods of harnessing inorganic energy—a source of power which was in modern times to revolutionize man's economic life.[7]

Technological advances of the Age of Copper had two major economic consequences—they helped bring about a greater amount of trade and a greater division of labor. The latter development is particularly clear in the case of coppersmiths, pottery makers, and to a degree carpenters. The metallurgical process of refining and working copper required specialized knowledge, equipment, and concentration of activity, which took it out of the hands of farmers working at home. Coppersmiths devoted their labors almost exclusively to their craft. By restricting information about their skill to a limited number of apprentices, they retained for themselves enough of the market to keep themselves steadily employed. There is evidence that at least in the making of finished products many of the smiths were itinerants or were employed in noble or priestly households.

Similarly, the potter possessed a skill that was learned only with patience and practice, and he too was frequently an itinerant worker. He carried his wheel to suitable clay near a market, for that was easier than trying to transport finished and fragile products over long distances. Carpentry, masonry, and brickmaking were somewhat less specialized, for farmers could perform in a rough way at least the major tasks of these crafts. As more elaborate tools came into use, as more refined products were demanded, and as the effective demand of the market became larger, a division of labor and a great increase in productivity per worker took place.

The growth of commerce is indicated by a variety of data. Inasmuch as the earliest centers of advanced civilization (delta of the Tigris and Euphrates, the Nile Valley, and the Indus River area) had no copper deposits, it is clear that copper was imported by these areas. Incidentally it would appear that working copper and some of the other new techniques originated in mountainous areas and oases and were borrowed by districts with large agricultural surpluses. In any case, copper first used on the lower Tigris and Euphrates came from Oman on the Persian Gulf, while lead and silver were brought chiefly from the Taurus Mountains in Asia Minor. Copper came into Egypt from Mount Sinai at the northern end of the Red Sea and to the Indus River people from Rajputana and possibly Baluchistan. Similarly wood entered trade both for construction purposes and for charcoal, Egypt obtaining supplies from Syria through the port of Byblos, near the modern city of Beirut, and from Central Africa. Tin, which is not widely found, must have been carried long distances; and we know that Egypt obtained gold from Nubia to the south. Archaeologists have found finished Egyptian products in Sumer, and the products of Sumer and India, far from their place of origin. They believe, moreover, that the spread of such techniques as the potter's wheel indicates a far-flung, if not an intensive, commerce. Finally, a new development was taking place that would have been impossible without commerce—that was the establishment of cities. These new concentrations of population, which were to contribute so much to civilization, were almost completely dependent upon trade for food and for industrial raw materials.

✣ The urban revolution of the early Bronze Age

The first cities [8] of which we have record were established prior to 2500 B.C. in the three river valleys that have been mentioned as the earliest centers of advanced civilization. Somewhat later, cities appeared in other river valleys, like that of the Yellow River in China, at places producing such goods as copper, and at commercial centers around the fringes or along trade routes between the earlier cultures. In all cases, it should be noted, the earliest cities came into being primarily for economic reasons. They were not essentially the creation of political systems to facilitate administration.

That the establishment of cities should be glorified by the term "urban revolution" requires a word by way of explanation. As was stated in Chapter I, cities have been the loci for the creation of the great intellectual and cultural works of man, and hence the establishment of urban centers is a significant part of our story. Second, these early cities witnessed an intensification of economic activity on a large scale, a greater concentration of capital than had ever before been known, the growth of new techniques of production and distribution, a division of labor, and the beginnings of such important economic institutions as money, banking, and a law of contracts. Third, these early cities developed writing and systems of numerical notation which made possible the ready communication of ideas and the accumulation of man's experience for the benefit of posterity.

The geographical environment in which the earliest cities appeared is so strikingly similar that it is worthy of closer examination. The delta of the Tigris and Euphrates Rivers, where the Sumerian culture developed, is an area about the size of Denmark. The soil there is alluvial and extremely fertile.[9] The marshy land had to be drained in order to be cultivated, which necessitated the combined efforts of many men. The rivers stretch through a broad and somewhat arid plain 600 miles from their headwaters in the Armenian plateau and empty into the Persian Gulf, the country thus having water routes both inland and to neighboring lands overseas. The plain was protected by deserts and mountains, but through these barriers were gaps which made the area more susceptible to attack than either Egypt or the Indus River region. In fact, an inordinate amount of the energy of the Tigris-Euphrates peoples was diverted to war, which accounts, in part, for their many contributions to the art of warfare and for the endless destruction of their accomplishments. Finally, the plain was too long to be consolidated successfully into a unit, and it developed into two major parts, Assyria to the north of the Thirty-fourth Parallel and Babylonia to the south.

The valley of the Nile, the home of Egyptian culture, is 550 miles in length from the Delta to the First Cataract but is extremely narrow above the Delta, averaging about 12 miles in width. It comprises an area less than that of modern Switzerland. The annual floods of the river had created a deep alluvial soil with many marshes. Here, too, land had to be reclaimed and floodwaters had to be controlled by drainage and irrigation—a task which was easier than in the Tigris-Euphrates because of the greater regularity of the inundations and their arrival well before the planting season. The Nile Valley was better protected by barren deserts than the Tigris-Euphrates Valley, yet it had equally good facilities for river transportation, and the river emptied into a sea that was easily navigable. It also had stone from the nearby edges of the valley, which the Tigris-Euphrates lacked, but it was equally poor in wood and minerals. Finally, its length is noteworthy; one of the recurring political problems in Egyptian history was how to knit all the valley together rather than to let it break into two major sections or, what was worse, into a multitude of noble states.

In India the earliest cities appeared prior to 2500 B.C. in the flood plains of the Indus River. Here good protection was afforded by deserts and mountains, and water transportation was made possible by the river and its five tributaries. Nor was the Indian Ocean too formidable for the existing techniques of navigation, although it was more repelling than the mild and tideless Mediterranean or the Persian Gulf. Land to be fertile had to be properly watered, while minerals and timber had to be transported from afar.[10]

In places like these, then, rich in soil, made productive by a water supply which had to be controlled by man, poor in minerals and timber which had to be imported, and provided with waterways which could be navigated, the earliest cities came into being. These urban agglomerations were not large by our standards, some having a population of perhaps 40,000, but they were considerably bigger than the agricultural villages which had preceded them. The fundamental techniques of production in them had not advanced much over those of the Age of Copper, except that bronze was supplanting copper as the chief metal, finer goods were being produced, and a greater division of labor was visible in such trades as baking, brewing, and building as a result of the concentrated market.

What impresses us most in these early cities is, however, the concentration of capital, in the broadest sense of the word. In Sumer this concentration was effected primarily by the temples, in Egypt by the pharaohs, and in India probably by a bourgeois class. In Sumer, for which more complete information is available than for the other areas, the temples owned land which they leased to tenants; they granted advances (loans) of seed and implements to tillers of the soil in return for a charge which in essence was interest; they built and maintained drainage and irrigation ditches; they planned production; they employed craftsmen in their own shops and sold the goods thus produced; and they catered to those engaged in trade. From this activity they were able to amass a surplus which could be and was used for expanding production. They were not, however, responsible for the defense of the city, this task being left to the city governor or king.

In Egypt the situation seems to have been somewhat different. Here a political leader was the chief agent in the concentration of wealth, although he assumed godlike prerogatives which fortified his claim to ultimate ownership of land and to goods and services from the people. From payments made him and from profits in the crafts in his extensive household economy, he accumulated great amounts of capital. A relatively large part of it seems to have been employed in nonproductive activity, like the construction of elaborate tombs and the burial of wealth with the dead. The Great Pyramid, for example, constructed between 2420 and 2270 B.C., contains 2,300,000 stone blocks weighing an average of $2\frac{1}{2}$ tons apiece and required, according to Herodotus, the labor of 100,000 men working over a twenty-year period. But some of the wealth of the pharaohs was used to build canals and cities, to exploit the copper deposits at Sinai, to outfit trading expeditions, or to expand production in the royal workshops.

With the appearance of cities, the concentration of capital, a greater division of labor, and increased trade, money came into being. Money, which may be defined as a medium of exchange, a store of wealth, and a measure of value, albeit a fluctuating one, was in the form of metals (gold, silver, or copper) which were weighed out at each transaction. Subsequently bars of metal had the amount of their contents stamped on them. Only much later (*circa* 700 B.C.) were pieces of metal (coins) of fixed weight and fineness issued and guaranteed by political states. Money, although used in a limited degree in this period, was a tremendous convenience, for goods could readily be exchanged for it. Under a system of barter a person with goods to exchange must find someone who has goods which he wants and who wants what he has—an operation which is at best cumbersome. Because money was a boon to trade, it deepened and widened the division of labor. It also facilitated the amassing of surplus, or capital formation, for it could be stored more easily and with less deterioration than, let us say, wheat. Finally it could be brought together from many more sources than could real products, because it was easier to transport, and hence aided in tapping the savings of more people.

Urban development was also accompanied by several other new business practices. We have already mentioned the beginnings of the institution of credit and interest taking in Sumer—a practice which was to become of the utmost importance in economic development because it provided a way of placing capital in the hands of enterprising persons. We have likewise referred to the seal as a means of indicating the ownership of goods, which by its increased use seems to show a growth in the principle of private property. In this period, too, a law of contracts, which recognized the binding force of business agreements, came into being. This law was written into early codes, like the Code of Hammurabi [11] shortly after 1750 B.C., and became a generally accepted concept in business procedure. Employing and employed classes became more clearly delineated, and with them there appeared the wage system. In fact, many of the aspects of modern economic institutions, at least in embryonic form, seem to have been present in the early cities.

The question now remains, did the more adequate combining of many of what we have called factors of economic progress actually result in a larger production of goods and services per capita? Of course we have no direct statistical data as a basis for a firm reply, but all the circumstantial evidence indicates that the answer should be in the affirmative. The cities seem to have brought with them a higher standard of living. People were better housed in the brick and stone buildings than they had been previously in mud and reed huts. In some places there are indications of organized sewage disposal and water supply. City granaries held stores of food which could be, and were, used to alleviate local famines, and the upper classes enjoyed luxuries in greater abundance than ever before. Finally, there was an increase in population which, so long as there is no diminution of goods per capita, is an indication of economic progress.

To be sure, there were germs of decay within the economy

of the urban areas. Wages were too low or too much in the form of goods which had to be consumed at once to permit the development of an ever-expanding market. Wealth was concentrated in the hands of priests, kings, nobles, or traders who had put much of their surplus into nonproductive things, like religious worship, rather than into efforts which would have increased goods available for consumption. Finally, population reached a size at which existing techniques, resources, capital, and trade were unable to provide a further increase in goods and services per capita. While cities were growing, however, there appears to have been greater surpluses than previously for cultural and intellectual activity.

Revolutions in knowledge and culture

The establishment of cities did, in fact, contribute to a "revolution in knowledge" and to a "revolution in culture." The former was of particular importance and may be classed along with fire, the invention of money, and the use of inorganic energy as one of the epoch-making events in man's past.

In the most primitive state of *homo sapiens*, little emphasis is to be placed on the adjective *sapiens*. Man undoubtedly acquired knowledge by simple trial-and-error methods. In learning to smelt copper ore, for example, he probably got copper accidentally, but then, to get more, he endeavored to reproduce the conditions which gave the original result without knowing precisely what the essential conditions were. After many attempts, in which earth with no metallic content was undoubtedly heated many times, man learned that only one kind of earth, that is, copper ore, would produce the desired metal. This "rule of thumb" he passed on to his successors, who followed it slavishly without knowing what actually happened in the process of smelting. The "revolution in knowledge" occurred when man began to understand and was able to explain *why* one event followed another—when man could tell what the result would be if certain conditions were given. This is the essence of science—the ordering of demonstrable or observable data to explain a sequence of events.

The development of science from mere rules of thumb has had an enormous impact on the history of man, especially in that it has given him a greater control over his environment and has allowed him to plan for that control. But the establishment of the scientific ideal has been a long-drawn-out process and is not universally established even today, impeded as it has been by superstition, tradition, false concepts, and vested interests. Consequently a full-blown and generally accepted scientific attitude was not to be expected at the early date of the urban revolution. What happened then was that certain tools necessary for scientific development were created, and meager beginnings in the ordering of demonstrable and observable data were made.

The chief tools invented were writing, simplified systems of numerical notation, and systems of weights and measures. Significantly enough, each one of these tools was developed from economic activity. In Sumer, for example, temple priests who managed elaborate business undertakings needed a way of keeping records which would be intelligible to their successors as well as to people with whom they dealt. The earliest form of writing was, as has been previously stated, in book-

keeping and consisted of symbols which had an agreed upon
meaning. These first symbols were easily understood pictures,
or pictograms. In time these were abbreviated so that they
were conventional signs (hieroglyphs) and were supplemented
by symbols for ideas (ideograms) which could not be easily
rendered pictorially.

Then in Sumer, at least, another step was taken which
was to give a sound value to things and ideas and to render
the sound value by a phonetic symbol, or phonogram. This
last development made easier the adoption of an alphabet,
which appeared in a Semitic language in the nineteenth century B.C. Similarly a system of numerical notation was simply
the adoption of symbols in place of a collection of grouped lines
on a tally stick. Thus in Sumer, while numbers 1 to 9 were expressed by straight lines, 10 was expressed by a circle, 20
by two circles, etc. (a decimal system), except in counting
the volume of certain things like beer, where a semicircle
stood for 60, two semicircles for 120, etc. (a sexagesimal
system).

Concomitantly with these developments more attention was
given to generally agreed upon weights and measures. A man
building his own reed hut could measure his materials according to the size of his hands or the length of his arm and
be accurate enough for his purposes. But a corps of specialized
workers building an elaborate temple could not rely upon
such haphazard methods, for a beam resting on two pillars
had to fit more exactly than would have been possible if each
workman did his measuring with his own anatomy. Similarly,
in trade it was necessary to have generally agreed upon meas-

ures and weights (to be used in a balance) for all transactions involving the transfer of goods.

Certain elementary principles of mathematics, probably the oldest of the sciences, came directly out of economic activity. Thus we find Sumerians using multiplication and multiplication tables in various activities. They multiplied length times breadth in order to determine the size of a field and length times breadth times height to get the volume of a stack of bricks; and they could explain *why* these procedures *always* gave the right answer. They used fractions, and the Egyptians had a formula which gave the correct value to π. The latter were particularly adept at the geometry necessary for constructing buildings connected with their funerary cult. They knew how to determine the batter of a pyramid and were able to execute their plans so exactly that in the case of the Great Pyramid the sides of approximately 755 feet have a mean error of only a little over half an inch.

Even the beginnings of astronomy were closely connected with economic needs. From celestial observation it was learned that the position of the heavenly bodies could measure the passage of time, tell the seasons of the year, and indicate the points of the compass. As a result calendars were developed by which the businessman could decide when a contract should be executed and the farmer could determine when to prepare for floods or when to plant. Similarly diagrams were constructed by which the traveler could find his way across the trackless desert or the sailor his route when out of sight of land. The Sumerians developed a lunar calendar and the Egyptians, a solar one—the lineal ancestor of our own. The latter people divided the year at first into 365 days but came to realize from further observation that the seasonal year is approximately 365¼ days. In the invention of apparatus for keeping track of shorter spaces of time, the exigencies of economic life seem likewise to have played a role, for the businessman needed to "know the time" in order to make and keep appointments, and workingmen engaged upon a cooperative task had to have a measure of time to determine when to begin their labors. To meet these requirements, the Sumerians divided the day into twelve double hours, which has given us our twenty-four hour day, the week into seven days, and the year into twelve months. Both they and the Egyptians employed sundials, movements of the stars, or water clocks, which worked on the principle of the hourglass, for regularizing the activities of life.

In addition to these achievements, there were many others. Some knowledge of the effects of drugs was learned, although medicine was retarded by beliefs in the healing powers of chants and incantations. Surgery made greater strides forward, for, if a stone rolled on a workingman's leg and broke it, the cause of the injury and the need for remedial action minimized recourse to magic cures. Moreover, the Egyptians learned something of human anatomy from the practice of mummification which involved the removal of the viscera. They were able to set broken bones, to hold them in place with splints, to bandage wounds, and to perform certain operations. In engineering, they employed the inclined plane, the water level, and the square. And they were expert enough to build dams and gates whereby they reclaimed land from the Faiyum, which was below sea level, and at the same time provided storage for floodwater that could be released for irrigation in the dry season.

From what has preceded it is clear that in both Sumer and Egypt a beginning was made in the collection of observable and demonstrable data and that some effort was expended in organizing in sequential order those data pertaining particularly to practical questions. Perhaps most of the findings should be classed as "rules of thumb," but in some instances, as in geometry, attempts were made to provide "proofs" which would explain why given conditions always gave the same results. In any case, writing and the use of writing materials, which by comparison with the flimsy paper of today gave great permanency and perhaps imparted to authors an added sense of responsibility, made possible the preservation of data for the analytical scrutiny of generations yet to come.[12]

With the urban revolution there were what have generally been recognized by historians as advances in the arts. The mud-brick temples of Sumer had an austere, if somewhat ponderous, beauty, while Sumerian sculpture, although failing to elicit the admiration of all modern critics, was at least an attempt to render human forms—gods, city governors, and priests—in a realistic manner. In Egypt architecture reflected more refinement, with the fluted columns (copied in stone from the bundles of papyrus reeds which had previously been used as pillars) arranged in colonnades of great beauty. Egyptian sculpture connected with funerary rites is held in high esteem. Paintings in tombs, representing the life on the deceased's estate, are fine in line, show a search for a solution of perspective and are bright if monotonous in color. Even in music, these early urban centers made a contribution, for they assembled drums, flutes, horns, and harps, formed melodies, and employed, if they did not invent, the heptatonic (seven-note) scale. In brief the arts reflect greater refinement, more precise workmanship, and a more extensive development than had been apparent in previous ages.

Thus from Paleolithic times, some half a million years ago to approximately 2700 B.C., men in a few selected places had definitely moved in the direction of what we in Western culture consider civilization. They had improved their techniques of making a living to the point where surplus was available to permit the creation and enjoyment of the "finer things of life." They had displayed an interest in the collection and ordering of observable data—one of the greatest of intellectual pursuits. And they had created works of art that even by our own standards have aesthetic value.

For Further Reading

Gordon Childe, *Man Makes Himself* (1941)
———, *What Happened in History* (1946)
F. M. Feldhaus, *Technik der Vorzeit* (1914)
G. P. Murdock, *Our Primitive Contemporaries* (1939)
G. Renard, *Life and Work in Prehistoric Times* (1929)
M. Rostovtzeff, *A History of the Ancient World,* vol. I (1926)
Ralph Turner, *The Great Cultural Traditions* (1941), 2 vols.

III

EMPIRE CULTURES
OF THE BRONZE AGE

FROM THE THIRD TO THE FIRST
MILLENNIUM B.C.

─────────

FROM approximately the third to the first millennium B.C., Hither Asia, the Nile Valley, and the region of the Aegean Sea passed from the Age of Copper through the Age of Bronze to the Age of Iron. In these areas, which had experienced the agrarian, technological, and urban revolutions of the preceding period, great empires were formed, each with a distinctive culture. During the Age of Bronze these empire cultures attained, in comparison with other cultures, their highest levels of civilization; then in the first millennium B.C. they bowed before the superiority of Hellas and adopted much of Greek culture as their own.

⚜ Economic progress and war

Prior to the attainment of their peaks of civilization, these ancient empires achieved considerable economic progress. This they did without advances in techniques comparable to those of the thousand years prior to 3000 B.C. In fact, the only technological innovations in the millennium and a half from 2700 to 1200 B.C. worth mentioning here were the development of bronze as a substitute for copper, the use of larger ships equipped with the rudder, the making of clear glass in Egypt,[1] and the smelting of iron, which inaugurated the next period. Most economic progress seems to have come from an extension of trade, a greater division of labor, and a deepening of investment. The expansion of commerce increased the supply of natural resources by opening up formerly isolated or undeveloped territories, and it permitted various areas to concentrate on the production of those goods in which they had comparative advantages. Labor became more specialized as commerce was extended, as the size of business operations was enlarged, and as economic goods became more complex. Investment per worker increased, and the craftsman at the end of the Bronze Age had more and better tools than his counterpart of the Age of Copper. Finally, better political organization provided greater security for economic activity and made large-scale undertakings possible.

In the Bronze Age, however, several forces were at work which tended to hinder the functioning of the positive factors of economic progress. Wealth was used unproductively; labor did not provide an expanding market; and population increases resulted in a pressure of people upon the means of subsistence. At least the end of the Bronze Age was marked by mass migrations of many peoples, presumably impelled to move by a need for food.

A decline in economic well-being within a culture was, moreover, frequently accompanied by internal troubles and by attacks from other cultures. In fact, from the beginning of written records, war looms large as a phase of human activity and as a cause of shifts in centers of power and civilization. What often happened was that some backward area would borrow techniques and technology from an advanced culture, just as Russia and Japan have done in recent times, and employ them to develop their own economies and military power. In cases where resources, organizing ability, and the will to succeed were great, backward areas were able to conquer more highly civilized regions and to alter drastically, if not destroy, the older culture.

So often did this happen between 2700 and 1200 B.C. that we cannot refrain from reflecting upon the relation of war to great civilizing achievements. From the history of this period it is clear that a distinction must be made between wars which have been overwhelmingly destructive of material well-being and of civilization and those which have made possible a greater exploitation of nature's resources and a spread of civilized cultures. Second, advanced cultures have seldom taken the precautions necessary to prevent the diffusion of their military techniques and weapons to backward areas and thus to preserve advantages which were the foundation of their superiority. Third, it is remarkable that so-called backward areas have so frequently defeated more civilized areas because of some particular military ascendancy in strategy, tactics, logistics, or morale. Finally, it is worth mentioning that those cultures which looked to war—to preying on others —as a fundamental ideology and as a way of sustaining themselves did not achieve such high levels of civilization as more peaceful cultures. They devoted too large a proportion of their energies to destructive rather than to constructive ends.

⚓ *Rises and falls of civilization in Egypt*

Of all the empire cultures of the Bronze Age that of Egypt permits one of the most instructive case studies of factors which determine *degrees* of civilization. Not only did Egyptian culture have a recorded existence of over three millenniums in which basic ideologies and art forms were amazingly stable, but it experienced rises and falls in its level of civilization which were so regular that they seem almost rhythmic. Furthermore, data regarding Egypt are adequate enough to allow the making of general statements concerning those forces which contributed to the shifting stages of artistic and intellectual achievement and of control over human and physical environment.

The first appreciable economic surplus in Egypt of which we know was amassed as a result of the agricultural, technological, and urban revolutions described in the last chapter. In the course of time, however, this surplus was greatly increased and, in part, as a result of what were fundamentally political forces. As a late tradition has it,[2] one of the many local lords of the land succeeded in bringing Upper and Lower Egypt together in 3200 B.C., and his successors maintained their supremacy and enforced peace until 2270 B.C. This long period of law and order permitted an inordinate amount of human energy to be devoted to economic and cultural ends. Furthermore, centralized authority led to an organization of effort that produced economic results far beyond what could have been realized by the dispersed efforts of individuals.

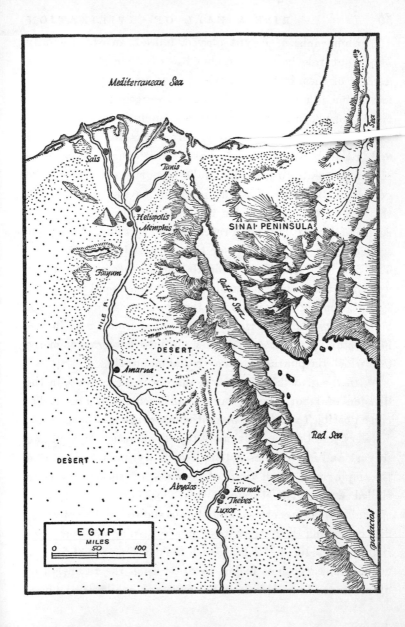

Mediterranean Sea

Sais

Tanis

Heliopolis
Memphis

SINAI PENINSULA

Faiyum

NILE R.

Gulf of Suez

DESERT

Amarna

Red Sea

DESERT

Abydos

Karnak
Thebes
Luxor

Dead Sea

Palestine

EGYPT
MILES
0   50   100

The pharaohs of Egypt played, indeed, an exceptionally important role in the economic life of their country. The earliest of them in the formative period of Egyptian culture had claimed that they were divine and were entitled to special rights, privileges, and prerogatives. Such claims, which are frequent in the pages of history, were successfully maintained and in time became established as institutions. They gave the pharaohs rights to an individual's immortal soul, to all the goods and services of a subject, and to ownership of all the land of the kingdom. On the basis of such principles the rulers of Egypt demanded revenues and offerings from their subjects and in the course of time amassed great fortunes. The pharaohs along with officials and priests constituted a wealthy group which determined the "investment policy" of Egypt. Some of their resources they put into manufacturing, most of Egypt's industrial output coming from lordly household establishments, and some went into foreign commerce over which the pharaohs had a virtual monopoly.

With the growth of industry, there was an increase in the division of labor. Crafts became differentiated, and it became possible to distinguish smiths, carpenters, stonemasons, jewelers, and the like. Obviously the construction of anything so vast and intricate as a temple or a pyramid necessitated the employment of all kinds of specialists from ox drivers to skilled architects. Moreover, tasks within crafts became specialized, pictures taken from tombs showing different workmen combining their separate efforts in turning out standardized products like pots. Then with the development of foreign commerce, Egyptians began to draw upon the resources of other areas for materials which were not available at home.

A few of these goods were obtained by conquest, as in the case of copper from the nearby mines of Mount Sinai, but most were procured through normal channels of trade. Thus Egypt received lumber from Byblos in Syria, gold from Nubia to the south, perfumes and spices from Arabia, and lapis lazuli and other gems from Arabia.

Under conditions of unification and internal peace, of wealth in the hands of lords and kings, of increased investments in industry and commerce, and of a growth in the division of labor, Egypt enjoyed a period of relatively great economic well-being, especially during the Third, Fourth, and Fifth Dynasties, that is, from 2780 to 2420 B.C. Probably the over-all growth in population was not great enough to make the ancient Egyptians wonder, as Thomas Malthus did some four and a half millenniums later, if the goal of humankind was to improve the economic condition of a given number of men or merely to increase the number of men on earth. All that we can say with certainty is that Egypt had enough production per capita to make possible a considerable concentration of economic surplus.

During this period Egypt attained one of its highest peaks of civilization. Political and religious organization, or control over human environment, was such as to provide individuals with a considerable degree of physical security; control over physical environment was sufficient to permit a relatively high degree of certainty that one would have the necessities of life; and leisure was adequate to allow the development of both the sciences and the arts. As we saw in the last chapter, Egyptians devised a solar calendar about this time, constructed a form of writing and numerical notation, and brought their science of mathematics to a point where they could erect with very little error such difficult buildings as pyramids.

In the arts, also, concepts of what to render in inanimate form, methods of work, and styles became differentiated and established. Belief in a physical life after death and the need for providing at least pictorially for a post-mortem existence gave rise to a search for a realistic portrayal of earthly things. Yet a lack of knowledge of perspective and of color made a compromise with realism necessary; and this was done by a stylized representation of actual forms. So firmly were art styles of the Third, Fourth, and Fifth Dynasties established that they persisted without great change to Greek times. Moreover, artists had unusual opportunities to use their talents, for economic surplus and the elevated position of the pharaohs and their officials made possible the construction of extravagant tombs and temples. These edifices represent in the opinion of most scholars one of the highest, if not the highest, attainment in Egyptian art.[3]

For many reasons the happy state of affairs which characterized the years from 2780 to 2420 B.C. did not continue. Not enough surplus was put into investments of those things which would have helped to increase production to keep pace with an expanding population. An extraordinary amount of wealth went into the construction of temples and tombs, and much was buried with the dead, although it was frequently put back in circulation by robbers. In brief, Egypt failed to maintain that nice balance in the use of surplus which would have resulted in an expansion both of goods and services per capita and of cultural accomplishments. The pharaohs insisted, for example, upon using so much labor for ostentatious display that they failed to perform satisfactorily such mundane tasks as irrigating and draining the land or maintaining foreign commerce. Then, to make matters worse, religious restrictions curbed the search for new knowledge, and widespread rigidities in society stifled artistic expression. School instruction became stereotyped, as it so often does, was divorced from the real world and lauded the virtues of "traditional truths." Finally, households of the lords became more economically self-sufficient and independent, and their heads became more reluctant to surrender goods and services to the king or otherwise to recognize his rule. From approximately 2400 B.C. the authority of the pharaohs began to wane, and by 2270 B.C. the *nomes*, or small political units based on villages, became autonomous.

During the ensuing three centuries Egypt underwent a political, economic, and cultural dark age. Rivalry among the lords led to incessant local warfare. Lack of a strong central government prevented the undertaking of important land-reclamation projects, made difficult a rational utilization of the waters of the Nile, deprived the country of a force for protecting its frontiers, upset the most extensive manufacturing establishment—the household of the pharaohs—and disrupted the conduct of both foreign and domestic commerce. Available evidence indicates clearly that economic production deteriorated during this period, and little has remained to give proof of anything but cultural decline. Yet techniques of production were kept alive in the household economy of the lords; astronomical lore appeared on coffin lids so that the deceased could keep track of time; embalming, and incidentally an interest in anatomy, were maintained; and private persons, because of the lack of economic centralization, began to engage in trade and manufacturing for their own accounts.

By about 2000 B.C. the Counts of Thebes managed by war and diplomacy to overcome the anarchy of the preceding years and to unite Egypt once again in a single state.[4] The reestablishment of a well-ordered, centralized government inaugurated a period of prosperity which lasted for over 200 years. Once again surplus was employed in long-range improvements, like irrigation projects and drainage in the Faiyum. Household manufacturing was revived and extended. Foreign commerce was restored with Syria, Palestine, Cyprus, Crete, and the Red Sea area. The use of money as a means of exchanging goods was more widely developed, and with it there was a greater division of labor, more private business for profit, and a development of business law. Foreign wars were few, and those in which Egypt was involved were not exhaustive and were easily won. Semitic tribes, which had penetrated the Delta in the previous period, were expelled, and Nubians to the south, who had frequently invaded Egypt in the past, were punished and were kept out by a series of fortresses at the First Cataract of the Nile.

All the evidence which we have indicates that, as economic activity was intensified in this period of the Twelfth Dynasty, conditions were again propitious for amassing surplus which could be devoted to the arts and sciences. It was then that the famous Labyrinth was constructed near the Faiyum reservoir—a series of buildings which covered an area of nearly a million square feet. Tombs, temples, and obelisks regained their earlier splendor. Jewelry, like that from the tombs in Memphis, was of excellent workmanship and composition, as were also painting and sculpture. This period marked the development of the earliest-known literature of a fictional and nonreligious character. Poetry and prose, which frequently dealt with the preceding time of trouble when emotions ran high, enjoy the distinction of being regarded as the classics of Egyptian literature.

The years, then, from about 2000 to 1788 B.C. were a time when control over human and physical environment was favorable to individual security, to economic well-being, and to artistic achievement. As previously, however, the relatively high state of civilization which was attained came to an end as a result of the lords becoming independent and of ensuing political and economic disintegration. As has so frequently happened in history, foreigners took advantage of unstable conditions to invade a comparatively wealthy area. In this case the newcomers were nomads of inferior civilization—the Hyksos, or Shepherd Kings, who were probably Syrians led by Hittites and Aegeans. In any event they had apparently learned much about Egypt through commerce and had acquired some Egyptian techniques. In military affairs they had distinct advantages over the Egyptians, for not only were they able to consolidate their forces for war, but they employed the horse and light chariot which gave them superior mobility. Yet the Hyksos, although they overran the Nile Valley without great difficulty, were not successful in consolidating their conquests. In fact, they never succeeded in controlling their subjects effectively, and their supremacy was marked by anarchy and Egypt's second dark age.

From the Hyksos the Egyptians did, however, learn something. First, they acquired from their conquerors the horse and the chariot, the latter being the first wheeled vehicle on the banks of the Nile, and second, they had impressed upon their minds more strongly than ever before the economic advantages of political unification and order. Thus a movement was begun to oust the foreign overlords, and by about 1500 B.C. the Count of Thebes succeeded in mastering the Hyksos and once more in unifying the country. The pharaohs of the succeeding dynasty, the Eighteenth (1580–1350 B.C.), were not, however, satisfied with the mere freeing of their country or with political unification. They had visions of grandeur which involved foreign conquests—an idea which they may have taken from the Hyksos' book of statecraft. At first, to be sure, the Egyptians concentrated upon a defense of their frontiers in order to prevent further invasions from Asia, but gradually their campaigns lost the aspect of punitive expeditions against recalcitrant neighbors and became wars of conquest. Palestine, Syria, Phoenicia, Cyprus, the territory of the Euphrates, and Nubia were subjugated and made to pay tribute to the homeland.[5]

For a time, conquest seems to have been a profitable enterprise for Egypt, with the spoils of war adding notably to the economic surplus of the country and with trade following the flags of the victorious armies. An inflow of precious metals and the need for a satisfactory medium of exchange for expanding commercial activity gave rise to a wider use of money and incidentally to a greater division of labor. Laws were further developed to provide more adequate rules for the conduct of business affairs; and the transfer of property, an essential element in economic progress, became more common, even land with its peasants and slaves changing hands with greater facility than hitherto. In fact, Egypt experienced a new wave of economic expansion, and its people enjoyed relative prosperity.

This economic revival of the Eighteenth Dynasty, like the previous ones, was accompanied by a cultural renaissance. Although this reawakening is not considered to have attained a height of achievement commensurate with that of the Third, Fourth, and Fifth Dynasties, or with that of the Twelfth Dynasty, it was during this period of the Great Egyptian Empire that beautiful, if colossal, structures were built, like the Great Temple of Luxor with its famous avenue, which had no rival in the ancient world, the Temple of Amen-Ra at Karnak (the modern name for Thebes), the Temples of Deir-el-Bahari, and the buildings in the new capital at Tell-el-Amarna. Moreover, in the 200 years from 1580 to 1350 B.C. portrait statues of the pharaohs displayed refinement and individuality; painting remained of high quality, although it showed signs of Eastern influence and of becoming over-ornate and even gaudy; tombs, like that of Tutankhamen, continued to be furnished lavishly; and literature flourished once more.

Toward the close of the Eighteenth Dynasty (1350 B.C.) and the end of the Bronze Age signs of Egypt's decline were in evidence. To maintain its supremacy in the foreign parts of the Great Empire, Egypt spent its resources lavishly on campaigns and fleets. Neither diplomacy, subsidies, nor dynastic marriages were able to arrest the endless troubles, frequently fomented by the Hittites and Assyrians whom the Egyptians were unable completely to subdue. To make matters

worse, the diversion of the country's energies into less and less
profitable wars tended to reduce goods and services available
to an enlarged population, while mercenaries returning from
foreign conflicts often indulged their bellicose arts at home.
Then an attempt to establish a single god for the whole
world [6] ran afoul of regional attachments to local gods and
almost precipitated civil war in Egypt itself. With the large-
scale migrations of the "People of the Sea" about 1200 B.C.,
which included migrations from the eastern Adriatic which
pushed the people of Hellas east and southward, Egypt lost
its foreign holdings, was invaded, and was torn by civil
strife.

Subsequently the country was overrun by Nubians from
the south, and still later it became an Assyrian province.
Although during the Twenty-sixth Dynasty (663–609 B.C.)
Egypt regained its independence, endeavored to restore the
boundaries of the Great Empire, and enjoyed a brief economic
and cultural renaissance—the Saitic Revival—the days of
its greatest glory were passed. Egypt came under Persian
domination in 525 B.C., was conquered by Alexander the
Great in 332 B.C., and finally became a Roman possession
in 30 B.C.

Civilization in the cultures of Mesopotamia

Not only along the Nile but also along the Tigris and
Euphrates high levels of civilization were attained during
the Bronze Age. Here, as was seen in the last chapter, the
agrarian, technological, and urban revolutions of the Age of
Copper had made an impact and had inaugurated a movement
which resulted in the creation of economic surplus. And this

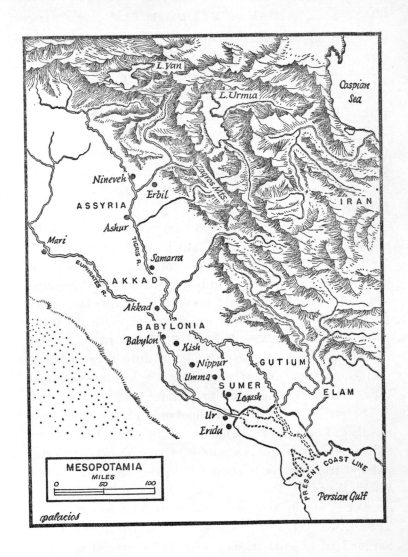

L. Van

L. Urmia

Caspian Sea

Nineveh

Erbil

ASSYRIA

IRAN

Ashur

Zagros Mts

Mari

TIGRIS R.

Samarra

EUPHRATES R.

AKKAD

Akkad

BABYLONIA

Babylon

Kish

Nippur

GUTIUM

Umma

SUMER

Logash

ELAM

Ur

Eridu

PRESENT COAST LINE

MESOPOTAMIA

MILES

0   50   100

Persian Gulf

palacios

economic surplus had accompanied the extension of knowledge of the physical world, of greater control over that world, and of the beginning of writing and numerical notation. In the ensuing two millenniums further economic expansion took place, which again permitted greater civilizing achievements. This expansion did not result from technical innovations. On the contrary it is to be accounted for by the extension of the use of existing techniques, by the establishing of political organization for the maintenance of order, and by the expansion of commerce.

The development of trade was particularly essential in the economic development of the Tigris and Euphrates area, for the only important natural resource of the plain was mud. Although it produced lush crops, bricks, cooking utensils, and clay tablets it could not meet the demands for copper, tin, stone, and timber. The search for these things led to political unification, to a division of labor, and to the spread of the techniques of the dominant culture to neighboring peoples. In the course of time these peoples were to attack their erstwhile teachers and to conquer them.

The earliest culture in Mesopotamia of which we know was that of Sumer near the mouths of the rivers. From it were diffused the ideologies, knowledge, styles of art, and other parts of transmissible learning to a district known as Akkad. Within the areas of Sumer and Akkad independent cities grew up, but gradually they were united to form two kingdoms. Then between 2500 and 2425 B.C. a famous ruler of Akkad, Sargon I, and his illustrious grandson, Naram-Sin, embarked on a policy of conquest. Thanks to superior military organization and equipment, which included the phalanx, the donkey, the chariot, and perhaps the horse, these kings had mobility and striking power beyond that of their enemies. Thus they were able to conquer Sumer, Assyria on the central Tigris, Elam to the east, and Syria on the Mediterranean, and even to conduct campaigns to Cyprus and to the shores of the Black Sea. From their foreign subjects they extracted tribute and with them, as well as with the people in Oman, Arabia, and the Bahrein Islands, they carried on trade. Thus they amassed wealth which inaugurated a golden age of lower Mesopotamian culture.

Soon, however, conditions became less propitious for economic and cultural growth. Less civilized peoples acquired the arts of war practiced by the Akkadians and Sumerians and turned them against the conquerors. About 2400 B.C. a barbarian tribe from Gutium, east of the central Tigris, attacked the Empire, which had been built by Sargon, and obtained dominion over it. Now the direction of the flow of tribute was reversed; commerce languished; and household and temple industry declined. Then there followed a brief dark age, about contemporary with the first dark age in Egypt, in which no important contributions were made to civilization.

Toward the end of the twenty-fourth century B.C. the people of Sumer and Akkad succeeded in expelling the invaders and in reconstituting much of Sargon's empire. In this period and in the twenty-third century B.C. an economic revival took place which was contemporary with the apogee of Sumer's cultural achievements. There was great activity in the collection of data regarding natural phenomena, especially in astronomy, and in mathematics. Architecture also flourished, and many of the great brick structures which were characterized by an unrelieved massiveness and by high towers or ziggurats [7] were constructed. Painting was not extensive, probably because the religion of the Sumerians and the Akkadians prescribed a future life of darkness that could not be rendered pictorially, but sculpturing was widely practiced, artists turning out squat figures representing life on earth.

Toward the beginning of the twenty-first century B.C. civilization in Sumer and Akkad began to decline again as a result of invasions by a less advanced people from Elam in the east. Then the domination of the Elamites was followed by that of the Semitic Amorites,[8] also a people of inferior civilization who had infiltrated from the west. The Amorites founded, about 2050 B.C., the Babylonian Empire with its capital at Babylon. The sixth king in this state was Hammurabi (about 1750 B.C.), the greatest of Babylonian rulers. He brought all Mesopotamia under his sway and established the region's first really centralized government, complete with bureaucracy and military force. He amassed wealth through taxes and other collections from his subjects and devoted that wealth to the expansion of commerce, to land improvements, and to the embellishment of Babylon. He also codified the laws, which constituted a definite advance in the establishment of control over human behavior, and he encouraged the development of the arts and sciences.

The reign of Hammurabi was a high-water mark in Babylonian economic life. The period of prosperity was made possible by the maintenance of order, by drawing upon the natural resources of other areas, and by extending industrial production in royal, lordly, or temple households. Furthermore the high level of economic well-being was attained primarily by private enterprise, the Code of Hammurabi being very specific regarding the contractual rights of individuals and the status of private property.

A portion of the wealth which had been amassed by the time of Hammurabi was devoted to the arts and sciences. The Babylonians continued the astronomical observations of their predecessors and laid the foundations for a genuine science of mathematics upon which Hellenic and Arabic scholars were later to build modern mathematics. They developed place value, that is, a simplified form of numerical notation whereby the value of a sign is determined by its place in relation to other signs, like 1 *over* 3 and second decimal *place;* and by 1800 B.C. they had established certain geometrical relations, like Pythagoras's theorem. In their architecture, they continued the traditions of the Sumerians, although they endeavored to relieve blank walls with sculpture, tiles, and carpets. In sculpture, their artists glorified the living and represented their gods in the form of man or mythical animals which have come down to us in heraldry and mythology—the gryphon, the dragon, the lion, and the bull. And in the field of literature, writers concentrated upon mythological and historical epics, some of which were incorporated in the Bible and hence are familiar to us, like the stories of the Creation, the Fall, and the Flood. In brief, the Babylonians developed those traditions of intellectual and artistic endeavor which had early become differentiated and fixed in Sumerian culture.

Soon after the reign of Hammurabi, which historians regard as the height of Babylonian civilization, the political power and organization of the Amorites was destroyed. Babylonia was torn by internal disorder and soon became the victim of foreign invaders. About 1650 B.C. the Kassites from the east of Babylon endeavored to extend their dominion over the land, but throughout the 570 years of their rule they failed to create a strong central authority or to establish conditions conducive to economic strength and cultural achievement. Furthermore, long-range economic progress was curbed by the lack of technical innovations, the limited range of natural resources, the failure of a mass market to expand steadily, and a social attitude, which came from the institution of slavery, that was hostile to manual labor.

After a series of debilitating wars, Assyria to the north gained control of Babylon (729 B.C.), and Babylonia was not reconstituted as an independent state until 625 B.C. At the latter date consolidation of political power ushered in a new period of economic expansion, which led in the sixth century B.C. to a revival of civilizing activity—to a kind of Neo-Babylonian renaissance. Subsequently Babylonia fell under Persian domination (538 B.C.), and then it was conquered by Alexander the Great (332 B.C.). Its heyday among the cultures of the world was then definitely past.

⪜ Other cultures of Hither Asia

Among the many peoples on the periphery of Babylonia, the most important were the Assyrians, Hittites, Mitanni, Persians, Syrians, Jews, and Phoenicians. Of them all we shall not treat in detail, for information about them is too scanty for purposes of our analysis, and some of them made no important contributions to transmissible knowledge which was to be handed down to Western culture. A word concerning the more prominent of these people will have to suffice.

One of the more important cultures of this group was the Assyrian. It was located in the central Tigris region where agriculture is restricted by a narrow plain and uncertain rainfall. Perhaps this fact accounts in part for the Assyrians' reliance upon war as a way of life. To be sure, they borrowed wholesale from the civilization of the lower Mesopotamian region, taking over from the Sumerians, Akkadians, and Babylonians their techniques, their types of arms, their script, their learning, and to an extent their ideologies and art forms. Yet they remained peculiarly bellicose and ruthless and provide a classical example of a people who attempted to make war pay.[9]

That they had some success from such a policy is attested to by their large and imposing cities of Assur and Nineveh, their far-flung commercial outposts, and their ability to carry on extensive campaigns. Their prowess in arms was derived partly from their use of the horse and their ability to mobilize their resources but also partly from the development of certain military techniques—the heavily armed chariot, the battering-ram, and the assault tower. Their periods of greatest prosperity seem to be correlated with success in wars that gave them tribute or commercial advantages; and their periods of greatest economic decline coincide roughly with their reliance upon their own production or with failure in war.

During most of the second millennium B.C. the Assyrians withstood with only a moderate degree of success attacks from Egyptians, Hittites, Babylonians, and Mitanni. Toward the beginning of the eleventh century B.C., however, they were able to get control of the trade routes of Western Asia and to enjoy considerable prosperity. After a subsequent setback, Assur-nazir-pal II (883–859 B.C.) restored their position and obtained tribute from Phoenician cities. It was during his reign that Assyria experienced its classical period in architecture and art. After the conquest of Babylonia, which included the destruction of the city of Babylon (689 B.C.), and the conquest of Egypt (671 B.C.), Assyria enjoyed another period of economic and cultural success. Then toward the end of the seventh century B.C., exhausted by extensive campaigns in Egypt and other far-distant places, the Assyrian Empire began to disintegrate and finally fell under the combined blows of Medes from Persia and Babylonians.

Of the Hittites, Mitanni, and Persians, we need say very little. The Hittites, located in central Asia Minor, created a feudal empire which through war and trade achieved its highest level of economic well-being in the thirteenth century B.C., which is the time in which the most important Hittite monuments were built. Then with migrations from the Aegean into their territory at the end of the second millennium B.C., their culture was swallowed up and disappeared as an independent entity. Similarly the Mitanni established an empire in the upper Tigris area, preyed upon their neighbors, enjoyed a brief period of economic success, and constructed their best buildings just prior to being brought to submission by Assyria (about 1275 B.C.). Finally the Persians, a people living in the Iranian plateau, developed a distinctive culture which attained its highest point of development in the sixth and fifth centuries B.C. It was in this period that Cyrus the Great (550–529 B.C.) founded an empire that extended from the Indus River to the Mediterranean and from the Caucasus to the Indian Ocean. The wealth drawn from this vast area permitted the construction of great palaces and a development of the arts. The Persian Empire threatened Greece in the fifth century B.C. but was finally demolished by Alexander the Great (336–330 B.C.). This conquest marked a definite decline in the civilization of Persia.

In addition to these cultures of Hither Asia, we should mention briefly those along the coasts of Palestine, Phoenicia, Syria, and Cilicia. Here the land was for the most part mountainous and dry and did not provide important agricultural surpluses, as did Egypt and Mesopotamia. Nor were the people of these regions particularly warlike, so that they made little effort to live off others, as did the Assyrians and Hittites. Their chief economic advantage was that they were located along trade routes between centers of important cultures and could thus reap a profit from acting as middlemen. From the fourth millennium onward we find that they were engaging in commerce and carrying, that they exported what natural resources they had in excess of their own needs, notably lumber and minerals, and that they borrowed industrial techniques from their neighbors for the production of such goods as textiles and pottery. Upon the basis of this economic activity they obtained a surplus which was able to maintain cities—Byblos, Tyre, Sidon, Tarsus, Aleppo, Damascus, and Jerusalem—and to achieve what by our definition was a moderately high degree of civilization.

To determine with any degree of accuracy when peaks of

economic well-being and of civilization were reached in these cultures and what the correspondence between them was is extremely difficult. It is probably correct to say, however, that the Jews attained their greatest wealth following David's victories over the Philistines of the coast, that is, during the reign of Solomon (973–933 B.C.) and that the greatest cultural achievement of the Jews, the writing of the Old Testament, came mostly after this period. Phoenicia, for its part, probably reached the apex of its economic activity at about the beginning of the first millennium B.C., although by 1500 B.C. the people of the area had given a phonetic value to twenty-nine cuneiform characters taught them by the Babylonians and had thus established an alphabet.

ꙮ Aegean culture

Beside the cultures of the Nile and of Hither Asia, the Bronze Age witnessed the development of another culture pertinent to our investigation—that of the Aegean Sea.[10] This region was bounded by Crete, Mycenae in the Peloponnesus, Troy at the western exit of the Hellespont, and the Syrian coast. Although the failure thus far to decipher the hieroglyphic tablets of this culture limits our knowledge of accomplishments, especially in science, literature, and law, archaeological remains give evidence of the attainment of a high degree of economic and cultural achievement. It was from this culture of the Aegean that Greek culture in its formative and most impressionable period borrowed many of its ideologies and patterns of behavior.

The geographical environment of Aegean culture is markedly different from that of the river valleys which we have mentioned in connection with Egypt and Mesopotamia. The islands and coastlands of the Aegean have many arable sections, but they are so effectively divided by mountains and sea that the inhabitants were led, if not driven, to the sea. For its part, the sea is easily navigable, its many islands and mountainous shores providing landmarks for the navigator, and its numerous ports furnishing havens for retreat before the elements and for the loading and unloading of ships. In Aegean civilizations water transport on the sea played a role analogous to that of river transportation in the case of Egypt and Mesopotamia.

Prior to about 3000 B.C. the Aegean region had two types of Neolithic culture with its improved stone tools and weapons, its domesticated animals, and its settled agriculture—that of the mainland and that of the island of Crete. By the early part of the third millennium B.C., however, a wave of invasions struck the Aegean and brought its various parts into contact one with the other. Some uninhabited places, like Cyprus, were settled; trade between Byblos and Egypt was increased; and techniques of copper metallurgy were introduced. Gradually the Aegean people moved out of the Neolithic Age into the Age of Copper, and as they did so the people of Crete acquired a position of primacy.

The island of Crete is relatively mountainous, is only one-third the size of the state of Vermont, and is not rich in natural resources. The economic success which it achieved cannot, therefore, be attributed to an environment richly endowed by nature but rather to a strategic location, to easy access to markets, and to the enterprise of its people. Cretans turned early to manufacturing and to trade in manufactured goods. Soon after 3000 B.C. they were producing a distinctive pottery with bright colors on dark background which appealed to the tastes of the market; they were turning out metal products, particularly the triangular dagger, which were in great demand; and they were making commercially desirable stone and ivory carvings. Some of the raw materials for these products had to be imported, as, for example, copper from Cyprus (the Copper Island) and the Cyclades [11] and ivory from Egypt. These things the Cretans carried mostly in their own ships, for they had become excellent seamen, and they had a central position in the Eastern Mediterranean. In fact, they developed a thriving carrying trade and became the middlemen for the products of the entire Aegean. From this activity, they and the people of the Cyclades, who shared with them part of this development, accumulated a surplus above immediate needs which permitted the establishment of cities and the devotion of energies to cultural things.

About 2500 B.C. changes took place in the Aegean which further enhanced the situation of the Cretans. A new movement of population from Europe and Asia, associated with the Hittites coming into central Asia Minor, a new people into Thessaly, and the building of a second Troy on the ruins of the first, pushed the center of Aegean culture southward toward Crete.

Then the introduction of bronze at about this time gave Crete another advantage, for the island was strategically located for receiving tin mined in Etruria, Spain, Gaul, Cornwall, and the Erz Mountains and for combining it with copper from Cyprus. Finally the proximity of Crete to Egypt made it possible to benefit from trade with that highly developed area. The superiority which Crete achieved in bronze, as, for example, in the long bronze dagger, was soon extended to other trades—to pottery turned on the wheel, finished as faïence, and decorated with polychrome, to goldworking, and to engraving. Reliance upon imported raw materials and exported processed goods made necessary the development of a great merchant marine and a mastery of the sea.

For about a millennium, from 2500 to 1400 B.C., the great sea power, or thalassocracy, which was Crete maintained its economic, cultural, and maritime hegemony in the Aegean. Large palaces were built in cities like Knossos, which were a combination of apartments, workshops, storehouses, and sanctuaries. These palaces were adorned with colonnades and frescoes of profusion and beauty. From the necessity of keeping records of business activity in these beehives, Cretan writing is believed to have come. It passed from the ideographic to the hieroglyphic script stage about 2000 B.C. and was further simplified in the succeeding quarter millennium. Of the political life of the island we know little, although it is thought that unity was achieved and some federation or working arrangement existed between the Cretans and their neighbors. In about 1750 B.C. some misfortune befell the island, probably a revolution or civil war, which resulted in the destruction of the great palaces in the leading cities. Within a short time, however, new palaces, more beautiful and more richly decorated than the first, were constructed, and Cretan industry and art were flourishing once again. Buildings, pottery, jewelry, swords, and frescoes testify to beauty of design and exquisite workmanship. In fact, during the sixteenth and fifteenth centuries B.C. Crete achieved the acme of its artistic production—its classical period.

All was not well, however, in the Cretan scene. During and following the troubles in Egypt attendant upon the domination of the Hyksos (1680–1580 B.C.), the Cretans intensified their activity in the Aegean Sea region itself. In the process of exporting its goods, Crete exported also its institutions, methods of organization, its techniques, its styles, and some of its people. Thus Cretan civilization was literally expanding overseas. In some places this development could be controlled by Crete to its own advantage, but in others the maintenance of a dominant position was impossible. This was particularly true in Mycenae and Tiryns in Argolis. Here the barbarian Achaeans, who had come from the Danube Valley about 3000 B.C. and had subsequently occupied all Hellas, seemed particularly adept at borrowing from Crete and at the same time in maintaining their own independence. Perhaps these towns had certain locational advantages in being able to draw on more resources, to control a larger market, or to obtain cheaper labor. Whatever the reasons, they began the manufacture of "Cretan goods" which they transported in their own ships and protected with their own navy. Their more standardized products appear to have had an advantage over those of Crete, which were becoming more flowery and were destined more exclusively for the luxury trade.

About 1400 B.C. a devastating blow hit Crete. Its cities, no longer fortified because of a consummate confidence in the protection of the sea and in a naval arm, were laid waste. Cretan supremacy was at an end. Whether the destruction of Crete was effected by invading Mycenaeans or whether it was wrought by internal revolution is not certain. Yet the fact remains that for the next two centuries Mycenae was to dominate the Aegean world. Production in Crete deteriorated, as is witnessed by the substitution of clay for stone and metal vessels and the paucity of archaeological finds of Cretan goods in other areas. On the other hand, production in the Hellas of the Achaeans increased considerably. Its cities grew rapidly; its wares were distributed throughout the Aegean; and its supply of hard-to-get raw materials, like tin, seems to have been abundant. Although art forms, workmanship, and a decline in writing all testify to retrogression in the intellectual and cultural aspects of Aegean life, a concentration upon the production of staples and the export of these staples brought wealth to Hellas—to the Hellas described with poetic license in the *Iliad* as being as opulent as Egypt.

Along with this economic development the Achaeans of Mycenae and other important cities embarked upon a policy of expansion—either by peaceful penetration, outright conquest, or a combination of the two. Gradually all the Aegean came under their influence. But the Achaeans overreached themselves, as have so many peoples in the past. Although they were successful against the small states which could not draw upon the organized resources of a large area and even against Troy and its allies in the war celebrated in the *Iliad* and *Odyssey*, they were crushed in their efforts to invade Egypt (about 1229 B.C.). This defeat marked the end of Achaean expansion and together with other expeditions sapped some of the strength of Hellas itself. Gradually bands of barbarians from Albania, the Dorians, closely related to the Achaeans in language, pushed southward. By 1200 B.C.

what had begun as infiltration became invasion. Rapidly the newcomers overran Hellas, pushing the Achaeans before them onto the coasts of Asia Minor and onto Aegean islands, while the inhabitants of these places in turn attacked both Hittites and Egyptians. This was part of the movement of the "People of the Sea." In their wake the Dorians left little but destruction of the Aegean bronze culture, yet out of this destruction was to arise a new and more glorious civilization—that of the Greeks.

As one reflects on this rapid survey of ancient-empire cultures, certain generalizations take shape which are worth careful consideration. When techniques of production became stabilized, efforts to increase output of goods and services consisted primarily (1) in the exploitation of new resources either by land reclamation, commerce, or conquest and (2) in a more efficient organization of production and of distribution by a deepening of investment, a division of labor, the use of money, and better transportation. The success of these efforts depended to a large extent upon political unification of large areas to provide peace at home and victory abroad. In the process of unification, however, rulers frequently attained so much power that they effected a concentration of wealth which, on the one hand, permitted them to support the arts but, on the other, militated against the general sharing of the benefits of civilization and aroused the hostility of subordinate lords.

During the Bronze Age peaks of production appear to have been concomitants of peace or of successful foreign war and to have had a close relationship to the most generally

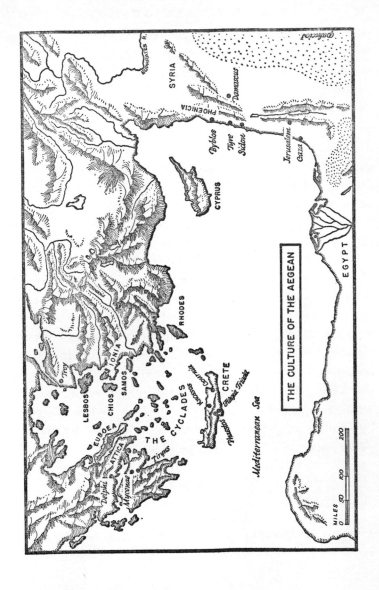

THE CULTURE OF THE AEGEAN

SYRIA

EUPHRATES R.

PHOENICIA

Damascus

Byblos
Tyre
Sidon

Jerusalem
Gaza

CYPRUS

EGYPT

RHODES

IONIA

Tmos

SAMOS

CHIOS

LESBOS

CRETE

Knossos

Phaestos Hagia Triada

THE CYCLADES

EUBOEA

ATTICA

Delphi

Mycenae

Tiryns

Mediterranean Sea

MILES
0 50 100 200

recognized peaks of civilizing achievement. Furthermore, concepts of what a civilization should strive for—whether material gain, a future life, scientific achievement, artistic expression, or dominion over others—took shape at an early date. Similarly the forms which much of this activity was to follow were established toward the beginning of a culture's history and changed, if at all, only after some devastating experience. In the formation of standards and forms all the factors of human existence—economic, social, political, intellectual, and religious—came into play.

One aspect of the process, however, stood out as particularly arresting, that is, the manner in which one culture borrowed from another and built upon the borrowed product. Geographic proximity of one culture to another, degree of intercourse and dependence between them, the adaptability and usefulness of the borrowed part in the environment of the borrower, and some degree of similarity between what was borrowed and the technique or institution which it replaced seem to have been the controlling forces in determining what was acquired. Finally the highest degrees of civilization, as we have defined the word, appear to have been achieved not by cultures which placed emphasis upon war but by the more peace-loving and industrious peoples.

For Further Reading

James Baikie, *A History of Egypt from the Earliest Times to the End of the XVIII Dynasty* (1939), 2 vols.
J. H. Breasted, *A History of the Ancient Egyptians* (1920)
Louis Delaporte, *Mesopotamia: The Babylonian and Assyrian Civilizations* (1925)
F. M. Feldhaus, *Technik der Antike und des Mittelalters* (1931)

Gustave Glotz, *The Aegean Civilization* (1925)

A. T. Olmstead, *History of Assyria* (1923)

————, *A History of Palestine and Syria to the Macedonian Conquest* (1931)

J. D. S. Pendlebury, *The Archaeology of Crete: an Introduction* (1939)

Jacques Pirenne, *Les Grands courants de l'histoire* (1945 *ff.*)

A. P. Usher, *A History of Mechanical Inventions* (1929)

Quincy Wright, *A Study of War* (1944), 2 vols.

IV

ANCIENT GREECE

━━━━━

Dᴜʀɪɴɢ the thousand years prior to the birth of Christ, there was a shift in the location of the most advanced cultures from Hither Asia and Egypt, northwestward to the European continent, first to Greece and then to Rome. These cultures, borrowing heavily from their predecessors, attained higher levels of civilization than had ever been reached before. Greek culture was particularly brilliant; and from it Western culture was eventually to draw many of its concepts, art styles, bodies of knowledge, and ideologies. So great is our heritage from Greece that in retrospect the shift in centers of civilization from Asia to Europe constitutes for us one of the most important facts of history.

⚹ *Reasons for Greek primacy—a general statement*

That people in the Grecian peninsula should have amassed enough economic surplus to have permitted the establishment of a high level of civilization is astonishing in view of the small size of their territory and the innate poverty of their land. Greece comprises an area comparable to that of the

State of New York, but only 20 per cent of its surface can be cultivated because of mountains. Furthermore, it lacks mineral resources for industrial production or navigable rivers for internal transportation. What it does have are numerous ports, easy access to navigable seas, and a location which favors communication with neighboring lands.

In spite of the limited amount of arable land, the Greeks produced specialized agricultural goods which became articles of trade in commerce with other regions. They also began the manufacture of industrial products for export and performed certain services, like shipping and banking, for foreign peoples. In return for what they sold abroad, Greeks received foodstuffs and raw materials. To a large extent the skill of Greek workmen, the ability of Greek enterprisers to organize production, the acumen of Greek traders, and the aptitude of Greek protagonists to establish a foreign demand for Greek goods made possible the obtaining of what could not be produced in sufficient supply within Greece itself.

Between the sixth and third centuries B.C. Greece was in varying degrees the "workshop of the ancient world." As is usually the case when the products of skill and enterprise are exchanged for raw materials and foodstuffs, the producers of primary products devoted more hours of labor to each unit of value than did the producers of finished goods and services. Hence the Greeks reaped an important economic benefit from the exchange. In many respects, therefore, the economic position of Greece was analogous to that of Crete in the heyday of its primacy in Aegean culture or to that of Great Britain in the nineteenth century. As in these instances, the continued welfare of Greece depended upon the supplying of goods and services to foreign markets, upon mastery of the sea, and upon the fact that her customers did not produce those things which could be obtained from Greece. Ultimately Greece failed to maintain these conditions, for other peoples acquired Greek techniques of production and of trade and could match her in the arts of war. They became economically independent of Greece and with their superior resources and power were able by force to strike down a divided country. When this happened, Greek civilization relative to that of other countries declined.

As in the other cultures which we have treated, so also in Greece the accumulation of economic surplus was followed by a high level of civilization. In the case of the empire cultures of Hither Asia, however, surplus for cultural activity was located primarily in the hands of a small number of rulers, their officials, and priests. In Greece, on the other hand, surplus was acquired by landholders, manufacturers, traders, bankers, and shippers, who, although they effected a considerable concentration of wealth, constituted a fairly large number of persons. Because these people were free to use their wealth as they saw fit, a situation developed in which a comparatively larger segment of society than we have encountered hitherto was able to exercise its creative talents. Indeed many of the artistic and intellectual giants of Greece were the sons of well-to-do businessmen.

Wealth and its relatively wide distribution in ancient Greece were basic factors in the development of civilization in Greece, but there were also many more specific reasons for its accomplishment. One of these was the fact that Greeks came into contact with peoples of many different cultures and borrowed heavily from them. Thus they acquired a large body of knowledge, techniques, and standards of accomplishment from which they could proceed to greater heights. Furthermore, their economy was of a kind that led to the concentration of activity in urban centers, and, as we have seen, urban environment seems to have been best for civilizing accomplishment. In cities intellectuals and artists could congregate, exchange ideas, and find competitive inspiration. Then, too, the Greek lack of political centralization and the absence of a common religion, well-established priesthood, and a sacred book contributed to keeping the mind unfettered by dogma. The secular attitude of the Greeks helped to develop an inquisitiveness and a curiosity which led them to ask all manner of questions about nature, politics, philosophy, and art. They challenged traditional beliefs and permitted a considerable amount of individual leeway regarding them. In fact, Greek inquisitiveness seems to have gone hand-in-hand with a large degree of intellectual tolerance.

In the absence of an overweening religion, the Greeks were free to turn their attention to the most amazing thing which their sensory organs could encompass—to nature. Not only did they come to admire nature, but they attempted to idealize it in art, to understand it in their science, and to control it by their technology.

Apt as these generalizations about ancient Greece are, the reader should realize that they are based on the cumulative and culminating experience of the Greeks. They began to be apparent in the formative period of Greek culture from the twelfth to the sixth century B.C. They were clearly applicable to the period of greatest economic and cultural success from the sixth to the middle of the fourth century B.C. They become less sharply delineated in the period of the diffusion of Greek culture (the Hellenistic Period) from the end of the fourth to the first century B.C. They are introduced here merely to serve as guideposts for what is to follow.

⚓ The formative period of Greek culture, 1100 to 600 B.C.

The formative period of Greek culture was inaugurated by widespread changes, which resulted from mass migrations that in turn were probably caused by rapid increases in population, by drought, or by both. In any case, near the close of the second millennium B.C., as we have previously remarked, peoples in the food-gathering stage of economic development or people at least with only a partially settled agriculture pushed into districts that had a completely settled agriculture and a degree of urban life.

These migrations of the so-called "Peoples of the Sea" were amazingly far-flung, occurring all the way from the Danube Valley to the plains of China. In the Middle East the Medes and Persians, who had probably come from the Caucasus, moved into Iran and there developed, as we have seen, a power which was to be a threat to Greece. Somewhat earlier out of the Arabian desert had come the Aramaeans, the Habiru, or Hebrews, and the Phoenicians, but now the Aramaeans invaded Babylonia, the Hebrews moved northward into Palestine, and the Phoenicians began to establish trading posts in the Western Mediterranean and to gobble up sea trade. Into Greece came Dorians who pushed many of the occupying Achaeans out. This pressure led to the attack of the Achaeans on Egypt, 1198 to 1167 B.C., to the Achaean Wars with Troy,[1] which are the subject of the Homeric poems, and to the settlement of the Achaeans upon the coasts of Asia Minor. Thus at the end of the second millennium B.C. Greece was partly settled by newcomers, and many of the previous inhabitants were squeezed in among various peoples on the eastern shore of the Aegean Sea.[2] In these two groups Greek culture was to take form.

Civilization seems to have suffered as a result of the coming of these people of less advanced culture into regions occupied by people of more highly developed cultures. Archaeological evidence bears ample testimony to a decline in production in the last two centuries of the second millennium B.C., to a crudeness of craftsmanship, and to a reduction in trade. Yet this period was by no means a total cultural black-out. Assyria, which had suffered the least, was still collecting Sumerian, Akkadian, and Babylonian texts for its libraries. Astronomical studies continued to be made in both Assyrian and Babylonian temple observatories. Business practices were kept alive by peoples like the Phoenicians. Much craft lore was preserved, if not advanced, by workers throughout the civilized world. Nor among Greeks either on the mainland or along the coasts of Anatolia did the invasions result in a complete reversion to Neolithic standards. Pottery continued to be made on the wheel and to be decorated in the late Mycenaean manner. The Aegean arts of seamanship were never entirely wiped out. And cheap iron weapons and tools were being made to replace more costly bronze products.

So, although the beginnings of the Iron Age were accompanied by a dark age—darker perhaps and certainly more general than that which had ushered in the Bronze Age—

retrogression was not complete. Recovery from the effects of invasion required, however, the expenditure of an enormous amount of energy and nearly half a millennium of time. The Greeks who led the general revival were those who had settled on the coasts of Anatolia. As late as the sixth century B.C. the most active economic and cultural life of Hellas was to be found there, and not until the fifth century B.C. was the supremacy of mainland Greeks clearly established. Thus, although the greatest heights of Greek culture were attained on the European mainland, we must give consideration to Greeks on both sides of the Aegean.

One of Greece's most characteristic institutions, the city-state, appeared in the formative period among both the Ionian Greeks on the eastern shores of the Aegean and the Dorian Greeks of the mainland. The former found themselves crowded into small enclaves with the sea on one side, hostile natives on the other, and mountains between their settlements. The logic of their geographic situation dictated strong, local political regimes in which they would be the masters and natives (foreigners to them) would be without political rights. In mainland Greece, the topography of which is characterized by juxtaposed, cup-shaped valleys, the rims of the cups being lofty mountains, there was similarly an early tendency toward localism. There the aristocracy of the invaders took possession of the better lands in the valleys, created strongholds to protect their holdings, and established political regimes which limited rights of citizenship. So on both sides of the Aegean the city-state became a fixture—and such a firm one that political unification of the entire Greek people proved to be impossible. This fact is of utmost importance in understand-

ing Greek history. It meant that with many independent
rulers the assumption of divine rights by any one individual,
as in the Eastern monarchies, was made extremely difficult.
It meant, too, that the Greeks were seldom, if ever, able to
concentrate all their strength in aggressive foreign policy
or in defense and that they were badgered eternally by intercity wars which sapped their energies and ultimately invited foreign invasion.

As the institution of the Greek city-state began to take
shape, so, too, the economy of Greece started to assume its
characteristic form. At the beginning of the first millennium
B.C. Greek economic activity was largely agricultural with
concentration on the growing of grains and livestock for subsistence.[3] Gradually, however, commerce and industry began
to appear, particularly among Ionian Greeks. Situated as
they were on the coast of Asia Minor and hemmed in by hostile
natives, they found it impossible to expand their holdings of
land to take care of a growing population. In their predicament they began to specialize on the vine and the olive in
order to get goods which could be used in exchange for grain.
Subsequently they started the making of industrial articles in
order further to supplement their food supply through commerce. And as they did so they acquired from their neighbors much of the technological heritage of Egypt and of
Mesopotamia.

With the production of wine, olive oil, and industrial goods
beyond their own needs, the Greeks of Anatolia engaged in a
search for markets for their goods and for sources of supply
for the things which they wanted. Ultimately they obtained
fish and grain from Italy and Sicily, the Balkan peninsula, the

Straits, and the Black Sea; they imported silver, gold, copper, and iron from a score of places; and they got large quantities of leather and wool from their neighbors in the interior of Asia Minor. They also indulged in piracy, but in time they developed commercial fleets, regularized trade, and got control of most of Aegean carrying and commerce. Then, as the scope of their economic life became broader, many of them migrated to places with which they did business. In this way, they spread their techniques and also aspects of their culture over a wide area.

Among the regions influenced directly by this development was mainland Greece itself. There large landowners began to give up grain culture for the more profitable vine and olive, thus adding two important goods to livestock as articles of trade. The production of wine and oil encouraged slavery, for vines and olive trees could be tended successfully by slave labor, and slavery contributed to the commercial character of the economy. Furthermore, industry was expanded, particularly pottery making, textile manufacture, iron and other metalworking, and shipbuilding. Concomitantly the mainland Greeks started to do their own carrying and trading and to make themselves independent of the Phoenicians, the traditional "peddlers of the sea."

Then in the eighth and seventh centuries B.C. a mass migration from Greece took place, a movement that was analogous to, but on a larger scale than, the expansion overseas of Ionian Greeks. The reasons for this exodus are extremely complex. Apparently overcrowding in the cities led to the establishment of colonies to which the urban poor flocked. It is also probable that slave labor in agriculture so worsened the conditions of poor farmers that they sought relief by going abroad. And then there were some migrants, as is always the case, who left the homeland in search of great riches or to satisfy their curiosity and love of adventure. But whatever the impelling forces may have been, the expansion was remarkable in its extent. Settlements were made in Sicily, in southern Italy, which came to be known as Magna Graecia, or Great Greece, in Gaul, in Spain, in the Straits, in the Black Sea region, in Cyprus, in Egypt, and in Libya.

The expansion of Greece was also important for its effects on the homeland. Now the resources of a much larger area than hitherto could be drawn upon for the benefit of Greeks—and this fact made possible the success of Greek industrial and commercial activity in the sixth to the fourth centuries B.C. Also, Greece acquired new techniques from the cultures with which she came in contact. Finally, close association with people of other cultures stimulated that inquisitive attitude of the Greek mind which dominated Greek intellectual life in the succeeding four centuries.

The formative period of Greek culture was marked, then, in its economic aspects by a great development of industry and commerce, by the expansion of Greece overseas, and by the consequent amassing of considerable economic surplus. In time accumulated wealth permitted Greeks to devote part of their energies to the extension of knowledge and to the production of works of art. In the creation of these intellectual and aesthetic aspects of life, Greece was a heavy borrower. From Aegean culture it inherited techniques of bronze metallurgy, methods of making pottery, and artistic standards of a high order. Similarly, from late Mycenaean architecture

Greece borrowed the basic pattern for its most noteworthy architectural structures—the temples—a pattern which persisted without fundamental alteration throughout Greek history. And from Mycenaean and Eastern sources Ionian Greeks adopted their earliest styles of painting and sculpture—adoptions which were less stiff and formal than the originals.

Greeks were, however, originators as well as borrowers, and their creative talents were clearly marked during the first five centuries of their history. For example, Greek religion did not follow the arbitrary absolutes of Eastern creeds but tended to glorify man. Thus local Greek deities, like Athena at Athens, were worshiped as symbols of glory and power of the city-states, while national gods, like Apollo, were revered for the degree to which they possessed human virtues. In fact, the discovery of man, that is, interest in all phases of man's life, was perhaps the Greeks' greatest achievement.[4] This glorification of man resulted in an extension of the individual's intellectual freedom and to an increase in the range of opportunities for alternative decisions in intellectual and aesthetic matters.

The Greeks were also originators in their extension of control over human environment. With the creation of the city-state a step was taken toward weakening arbitrary, tribal monarchy, toward government by the corporate action of citizens, and toward rule according to man-made law. In establishing control over physical environment, Greeks displayed creative ability, particularly in securing foodstuffs and raw materials from abroad through the development of trade. And although they were not renowned for technological innovations, they did improve agricultural methods, advance mining techniques, and develop iron metallurgy.

Perhaps the greatest genius of the Greeks was, however, in the realm of the intellect and in aesthetics. Indeed some of the most remarkable achievements of all time in these two areas are of Greek origin. In the formative period of Greek culture this particular propensity may be illustrated by the *Iliad* and *Odyssey*, poems sung by court minstrels in Ionia and possibly brought together by an actual Homer in the ninth century B.C. These epics are a glorification of Greek history; they show pride in workmanship and a search for technical perfection; they establish as virtues honor, courage, and discipline, and patriotism; and they glorify man and his power in winning the favor of the gods. Here were standards and views which were set early in Greek history and which influenced Greek life for more than half a millennium.

Greek culture from the sixth to the fourth century B.C.

With the expansion of Greece overseas, the formative period of Greek culture and of Greek economy came to an end. It was succeeded from the sixth to the fourth century by the most illustrious period of Greek history. In these two centuries Greece attained her maximum wealth and achieved her highest degree of civilization.

As has already been intimated, economic well-being in the second period of Greek history was largely dependent upon commerce with overseas areas; and commerce, in its turn, relied heavily upon shipping. Here the Greeks were particularly successful. They adopted the trireme, a ship with three banks of rowers. Invented as early as the eighth century, this vessel was able to make headway when winds were contrary and to carry as much as 250 tons' burden. As the size of the ship was thus increased, the Greeks improved their ports with break-waters and docks. They also developed their seamanship, navigators greatly extending their knowledge of Mediterranean geography, weather, and winds.

Gradually the Greeks encroached on the carrying of the Phoenicians, whose sphere of dominance became restricted primarily to the North African coast west of Carthage. Greek ships henceforth found their way to Naucratis in Egypt, the Syrian ports, to the Black Sea, the Italian peninsula, to Marseille, to Spain, and probably to the Isle of Wight off England. To these places Greek merchantmen carried textiles, pottery, iron weapons and tools, works of art, jewelry, and leather goods and brought back grains, dried fruits, salt fish, metals, hides, amber, ivory, wool, and lumber.

Many Greek ports were involved in this trade, and as time went on sea routes tended to converge at centers which enjoyed particular locational advantages—at such places as Miletus on the Anatolian coast and as Delos in the Aegean Islands. By the fifth century B.C. Piraeus, the port of Athens, was probably the most prominent focal point for shipping in mainland Greece, and the ports of Corinth, located on either side of the isthmus that forms the extreme wasp waist of Greece, were close rivals in importance. Here ships frequently broke cargo to avoid the stormy voyage around the coasts of the Peloponnesus or to trade in nearby markets. That the ports of Athens and Corinth did rank so high among Greek shipping centers is important to remember, for the great cultural achievements of Greece were realized in places like

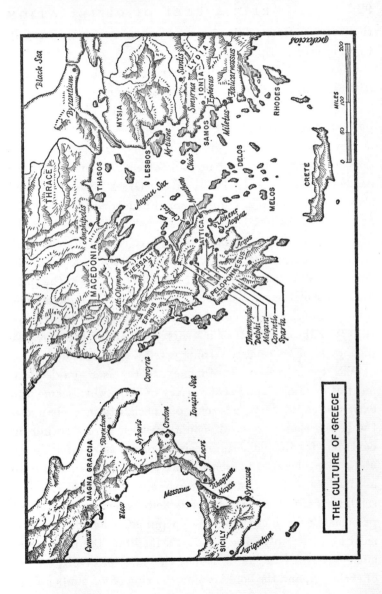

THE CULTURE OF GREECE

these rather than in the less commercial and more agricultural cities like Sparta.

Overseas trade had a direct bearing upon the development of Greek business institutions, especially upon the increased use of money, credit, and banking services. The first coins, pieces of metal whose quality and weight were guaranteed by a state, appeared in Lydia, a neighbor of the Anatolian Greeks, about 700 B.C. Not long afterward coins of large denomination began to be issued by Greek city-states, and shortly after 600 B.C. coins of small denomination put in an appearance. As has been previously stated, money oils the wheels of commerce. Thus it was in Greece. Small coins had, moreover, a new and almost revolutionary effect. With the extension of a money economy to the lowest classes, the small landowner or the small artisan could produce his goods for a wider market, and he could more readily store up surplus, either for the extension of his enterprise or for the purchase of objects of refinement—from jewelry to education.

With the increased use of money, there was also a development of banking. The great variety of coins which came into existence called forth a group of specialists in exchange— the money-changers—and soon these men were drawing up contracts for their clients, receiving moneys for safekeeping, and making loans. In essence what these bankers did was to accumulate wealth from many persons, even from those of small means, and to make it available to enterprising persons, presumably for increasing the output of goods or for augmenting the supply of services. To facilitate the transfer of moneys, Greek bankers developed the letter of credit, the bill of exchange, and the bookkeeping clearing of accounts among individuals. Bankers also made bottomry loans, whereby a skipper would borrow on his cargo but would be relieved of his responsibility to pay his debt if the cargo did not arrive at its destination. This was one of the first attempts at a dispersion of risks for a premium—an elementary form of insurance.

Finally, the use of money gave rise to a price system—to a relationship of market value between goods and whatever metals were used in coins, whether it was gold, silver, or copper. The price system in itself was extremely important, for as gold, silver, or copper coins commanded more or fewer goods, the burden of paying debts, of collecting from one's creditors, and of remunerating labor (for wages usually lag behind prices) fluctuated accordingly. In Athens, where most of these developments were carried farthest and where the famous Laureion silver mines produced a supply of silver, prices for grains, olive oil, and livestock rose by over 500 per cent from the end of the sixth century to the end of the fourth. To some of the effects of this price rise we shall refer later.

In spite of the fact that the state played a role in the development of the Greek monetary system, Greek economic progress was achieved under what was fundamentally a system of *laissez faire*. Especially in foreign commerce a policy of free trade was pursued, for Greek exports competed favorably for a long time with goods of foreign origin. Besides, customs barriers would have endangered the system of exchange by which Greece exported goods in order to get absolutely essential imports. What state intervention in economic affairs did exist was of a character to facilitate trade and not to impair it, like enforcement of the laws of contract, the establishment of standard weights and measures, and the building of roads and harbors. Moreover, the Greeks permitted anyone, including foreigners and slaves, to engage in commerce and often granted citizenship to the wealthiest of such groups.

Important as trade was to the economic life of Greece, industry was its ever-present companion. Without industrial production the Greeks would have had little to exchange for the foodstuffs and raw materials which they needed, and accordingly they could not have supported as large a population as they did. Indeed the production and exportation of goods are a substitute for emigration. When production and exportation were being conducted satisfactorily, Greece sustained a large population that contended successfully with powers like Persia, which embraced vastly more territory than she. When they were not being managed well, Greek migration took place.

The success of Greek industry has been the object of much speculation, for a simple explanation is not apparent. The peninsula had no particular advantages in natural resources, as we have seen, although it did have some outcroppings of iron ore, some copper, lead, and silver from the Laureion mines near Athens, and lumber for building and industry. Nor did its people introduce revolutionary techniques, although they displayed a certain inventiveness by the development of the ship with three banks of rowers and by the improvement of siege weapons. The Greeks did not employ any great amount of power from inanimate sources, although they knew about the expansive properties of steam, at least in the Hellenistic Period, the force of falling water, and the transmission of power by mechanical means. Furthermore, the demands of the market were never so great, sudden, or persistent that they could not be met by speeding up, or by adding to, the existing system of production.

Greek industrial success seems to have been based largely upon an ability to achieve low production and distribution costs. And this advantage was derived from a division of labor and from providing workers with superior equipment, particularly tools. The specialization of industrial tasks had reached a point by the end of the fifth century where, for example, in the woolen industry, there were workers who devoted themselves primarily to one of the following tasks: shearing, washing wool, carding, spinning, weaving, fulling, and dyeing. In most trades specialized tasks were performed in small shops, but in a few rare instances producing establishments were large, one entrepreneur at the Laureion mines employing a thousand slaves. As time went on, workers acquired better equipment, for tools made from iron were superior to those of bronze. Indeed a reasonable proportion of earnings went into investments for producers' goods—for goods with which to make more goods.[5]

Greek industrial success was also dependent upon the labor supply. In general, workers were divided into four main categories—freeborn citizens, freeborn foreigners or metics, freed slaves, and slaves. Slaves, obtained as a result of debt, war, or purchase, comprised an estimated one-third of the 300,000 persons living in Athens at the close of the fifth century. Whether or not they provided labor at low cost is disputed, but they certainly furnished a steady and fairly abundant source of labor. Free men were paid in fifth-century Athens enough to provide them with some margin beyond their basic expenditures, but their wages failed to rise as fast as prices with the result that between the sixth and the fourth centuries B.C. their real wages, that is, the amount of goods and services which they could actually obtain with their wages, declined, perhaps by one-half.

Such a situation usually acts in the short run as a stimulant to production, if there is an adequate market for goods, because employers find that labor costs are relatively low and profits are high. For a time this relationship of wages to profits—or profit inflation, as it has been called—reacted favorably upon Greek industry. But in time a lag of wages behind prices may have other consequences—it may lead to civil strife and to a limitation of the market because of the low purchasing power of workers. There is evidence to show that Greece experienced this result in the second half of the fourth century B.C.

In addition to those factors in Greek industrial development already mentioned, four others were certainly of importance. In many times and places throughout history societies have developed a contempt for participation in industrial and commercial enterprise. Although such an attitude prevailed in Greek cities like Sparta and Thebes, where the economy was largely agricultural and where government was in the hands of a rich agricultural aristocracy, it was by no means general. In industrial and commercial cities, like Athens and Corinth, business was not regarded as civilly degrading, and for the most part not degrading at all.

Second, the policies of city-states frequently had an

important bearing upon industrial development. At Athens, Themistocles's policy in the first part of the fifth century of devoting the state's profit from the mines of Laureion to the building of a fleet and a port gave industry and trade an enormous boost. Similarly Pericles's policy of rebuilding the city after the Persian Wars gave a stimulus to economic activity.

Third, Greek industry and commerce were conducted, at least in the fifth century B.C., largely for profit in a wide and impersonal market rather than for the satisfaction of known needs in a local market. Here was the germ, at least, of the capitalist system which on a much greater scale was to characterize the economy of Western culture. Fourth, the question arises of the extent of capital formation and investment during the sixth, fifth, and first part of the fourth centuries B.C. In the industrial and commercial cities capital accumulation was apparently extensive. For the most part the ancient Greeks do not seem to have indulged in extravagant personal expenditures but rather to have encouraged thrift and savings. At all events, we know of individual cases of large fortunes which were used in productive enterprise.

Before concluding this section on Greek economic development in its most prosperous period, something should be said about agriculture. In spite of the fact that industry and commerce were responsible for the larger portion of the economic surplus available in Greece, the majority of the Greeks found employment on the land. We have already mentioned the cultivation of the vine and the olive, to which might be added the fig, as commercial crops. Their development encroached upon the use of land for the growing of grain in which Greece became deficient. Yet total agricultural production probably increased during the period under review even though it did not meet the requirements of the expanding population. New land was brought into cultivation by irrigation, drainage, and forest clearance.[6] A greater abundance of iron tools served also to increase the productivity of Greek farmers. Then the Greeks learned by experience some of the uses of fertilizers. They realized the benefit of humus to the soil and spread the manure of their livestock on the land; and in addition they used chemical fertilizers—wood ashes (potash), nitrates, and lime. And they even practiced the rotation of crops to avoid leaving part of their land to lie fallow.

With the commercialization of agriculture there was a tendency toward the concentration of holdings in large estates. Perhaps the chief impelling element in this trend was the need for relatively large amounts of capital to finance the bringing into production of the vine, the olive, and the fig tree and for acquiring slaves to do the work. In any case, in states like Sparta where agriculture retained its position of primacy in economic activity, land became concentrated in the hands of a few—from 9,000 owners in the early sixth century to about a hundred in the middle of the second century B.C. Even in the more democratic states, large landholdings were not uncommon, although in some places, notably Athens, public policy favored the preservation of the "family" farm.

Thus it came about that through agricultural activity on a large scale, as well as from commerce and industry, economic surplus was amassed which allowed the Greeks to devote a larger part of their energies to intellectual and aesthetic matters. It was clearly in this middle period of Greek history from 600 to 400 B.C. that the greatest proportion of Greek wealth was so employed and that Greek civilization attained its acme. Then, for example, Greek cities had their greatest growth and built their most famous buildings.[7] Consequently, architects had great opportunities to put their talents to work,[8] and in doing so developed rules of proportion which account in large part for the beauty of Greek structures. They evolved excellent dimensions for their columns, worked out new styles of columns, the simpler Doric type giving way to the more ornate Ionic and Corinthian, and they found a happy solution to the place of sculpture in architecture. So far as sculpture and painting were concerned, Greek artists freed themselves from earlier rigidities and Oriental influences and made every effort to render human form in a more naturalistic manner, to glorify nature in every way. In both of these arts there appeared to be a fuller knowledge of anatomy,[9] a technical ability to portray likenesses,[10] and fine craftsmanship, which is an indicator of high aesthetic accomplishment.[11] In painting, important advances were made toward an understanding, although never a complete one, of perspective.

In letters this period saw the epic poet give way to the lyric poet,[12] the creation of the tragedy and the comedy,[13] and the appearance of historical prose. The tragedy was one of the most distinctive and most brilliant achievements of the Greeks. It portrayed the play of human emotions over human problems and through excellence of craftsmanship and concentration on a single theme attained a power seldom equaled in literary expression. Historical literature, the first literary prose of the Greeks, showed a desire to know other peoples,[14] to understand the forces at work in society,[15] and to present the past in an agreeable manner.[16]

In this same middle period of Greek history that inquisitiveness of mind to which we have already referred found expression in philosophy—in speculation concerning the origins of the universe, the facts and laws of nature, and the principles of human conduct. The first evidence of such speculation was found in Ionian cities in the sixth century B.C. and took such form as the atomic theory—that the universe was made by atoms floating in space which from inward necessity formed the universe as it is. From these Ionian beginnings two currents of thought are distinguishable, the one concerned with the collection and organization of observable data and the other with deductive reasoning about how we know things and what we know.

In the former current was Thales of Miletus who forecast correctly an eclipse of the sun in 585 B.C. and who introduced geometry from Egypt. Here also should be placed Pythagoras (582–507 B.C.) who advanced the system of theorems and proofs in geometry and proved the Pythagorean theorem which everybody studies in school.[17] Anaximander (611–547 B.C.) held that living organisms appeared first in water and then passed to dry land, that man was descended from the fish, and that the earth is suspended in space surrounded by heavenly bodies. Anaxagoras (500–428 B.C.) studied the structure of animals by dissection, and Hippocrates (460–377 B.C.) attributed the cause and cure of diseases to natural rather than supernatural phenomena and drafted the Hippocratic Oath which even today sets the ethical standards for medical practitioners.

In the deductive current of thought the first followers of the early Ionians were the Sophists. They were concerned with adult education and taught whoever could pay them. They concentrated their attention upon problems of government and upon methods of argument, having been the founders of dialectics, of disputation and the distinguishing of truth from error. They preached that knowledge is not absolute but is relative to man, for example, that good and evil are not absolutes but depend upon man's views. Socrates (469–399 B.C.) reacted strongly against this principle and argued that final knowledge was attainable by reason. His most famous pupil, Plato (427–347 B.C.), endeavored to deal with this problem of knowledge by stating (he could not prove it) that there are absolutes or superconcepts and that what we observe on earth are mere reflections of these absolutes or superconcepts. Before birth men are acquainted with superconcepts, and knowledge comes by remembering them— a task best performed by the philosopher. This doctrine of "dualism," so-called because what can be known is divided into two parts, has played an important role in Western thought where it has been used as an argument for the existence of God.

Plato's most famous pupil, Aristotle (384–322 B.C.), attempted to improve on his master's dictum by contending that things of this earth are more than reflections of superconcepts and that from material things we can form generalized concepts, that is, from knowing individual chairs we form a general concept of what constitutes chairs. He thus

found God in nature but at the same time brought speculation back to earth. His interests ranged so widely over all
manner of things that he has been called the first encyclopedist. He thought that the earth was spherical in shape and
that it was the center of the universe.[18] He was the founder
of the study of biology, of formal logic, of zoology, of botany,
and of comparative government. He believed that men were
social beings, that they naturally formed states, that the
state existed for the well-being of all, and that the moral
evolution of man was certain. Aristotle, coming as he did
at the extreme end of the middle period, provided a glorious
finale to Greece's most brilliant epoch.

Greek civilization was not, however, able to keep up such
high standards, nor were the Greeks capable of maintaining
conditions conducive to such great intellectual and artistic
achievements. Greece, like so many other famous cultures at
their peaks, became the victim of foreign aggression and civil
strife. The first important blow came at the beginning of the
fifth century B.C. when the country was attacked by Persians.
As we have seen previously, these people under Cyrus the
Great established a vast empire in the middle of the sixth
century and in their territorial expansion overran the Greek
cities of Asia Minor. Thus Persia came directly into conflict
with Greek interests in the Aegean and believed that it could
conquer Greece, divided as it was, at its will. When the
Ionian city of Miletus revolted from Persia and received aid
and comfort from mainland Greek cities and especially from
Athens, the Persians moved against mainland Greece. Of the
ensuing struggle little need be said except that Persia was

handicapped by having to wage war overseas, which made the use of its most effective arm—the cavalry—difficult, and by having to meet Greek naval power near its own bases. The first invasion of the Persians was stopped by the Battle of Marathon (490 B.C.) and the second, after the destruction of the city of Athens, by the Battles of Salamis and Plataea (480 B.C. and 479 B.C.).

Yet, saved from one fate, Greece became the victim of another. As so often happens after a victory of allies, the allies soon fell out. Athens, a commercial city, wanted to continue the war against Persia until the Eastern Mediterranean was completely free, but Sparta, an agricultural state, was willing to let well enough alone. To pursue their policy, the Athenians formed an alliance of Aegean and Ionian cities (the Delian League) to carry on the war and actually drove the Persians out of the Aegean. At this point Athens transformed the Delian League into the Athenian Empire and collected payments from her subjects, which were used in part to rebuild Athens (during the Age of Pericles, 461–431 B.C.). At the same time she directed her trade toward the West because of continued Persian and Phoenician hostility in the East. This brought the city immediately into conflict with Corinth, which had had the lion's share of Western business and which feared Athenian supremacy, and with Sparta, which depended on the West for grain. Gradually the opposing positions became solidified, and civil war in Greece ensued—the Peloponnesian War (431–404 B.C.)

This war weakened all the participants, but more important than that it resulted in the triumph of the forces of decentralization within Greece. City-state was embittered against city-state. When Persia demanded the Ionian cities as her reward for financial aid to Sparta, there was no unified Greek will to say her nay. Henceforth the unification of Greek cities of their own free will and accord was impossible. It is a sad commentary that Greece failed to reconcile the particularism of the city-state with the need for a powerful, unified state. Interests of particular persons, classes, and communities proved more powerful than the general interest. In this fact lay part of Greece's undoing. Intercity strife continued and was so extensive that of the eighty-five years between the Peloponnesian War and the conquest of Greece by Macedonia, fifty-five years were filled with conflict.

As though this were not bad enough, the general economic position of Greece began to deteriorate. The foreign demand for Greek goods was, relative to the population, declining, for many of those areas which had previously imported Greek goods had learned how to make them, and during the wars had made them. This was particularly noticeable in the Greek cities of Sicily and Magna Graecia and along the southern shores of the Black Sea. Here was a development that was a severe blow to Greece, for it ate away the very foundation upon which her economy was based. Nor were social conditions much better. Unemployment, the fall in real wages, the increase in the debtor class, the influx into the cities of war refugees and adventurers, and the concentration of landholdings provided fertile soil for popular movements of a revolutionary character. In fact, during the eighty-five years previously mentioned all Greek cities of any size had at least one war or social revolution every ten years. Thus it was that the most brilliant economic and cultural period in Greek history ended in economic depression, political anarchy, and social chaos.

➤ The economy of the Hellenistic Period, 338–30 B.C.

The crisis just described, which confronted Greece in the middle of the fourth century B.C., was of such dimensions as to invite other foreign intervention, this time from Macedonia. This country, which had acquired many Greek techniques and some Greek culture, was united by Philip in the middle of the fourth century B.C. and was soon given by its leader a potent military force composed of small landholders. This army was organized into solid phalanxes of great striking power, supported by heavy cavalry, bowmen, and corps of engineers equipped with the latest siege weapons devised by Greek scientists. With this force, Philip invaded Greece in 338 B.C. and was soon recognized by all but Sparta as leader in the land.

Immediately after Philip's conquest of Greece he effected an alliance of Greek cities to wage war under his leadership against Persia. This was his "great design." He hoped by conquering Persia to remove the most serious threat to his own position and at the same time to effect a unity of the civilized world. The execution of this ambitious plan fell to his son Alexander, later to be known as "the Great," for Philip was murdered as his first contingent of troops was leaving for Asia Minor. At the time Alexander was a mere boy of twenty years, and there were doubts concerning his ability to assume the responsibilities which had come to him. The army was, however, loyal to him and he soon dispelled

fears regarding his talents. He defeated the Persians in Asia
Minor (334 B.C.); then to protect his flanks and lines of
communication, he deprived the Persians of their naval bases
in Syria, Phoenicia, and Egypt (333 B.C.); and after a year
in Egypt strengthening his position and making certain of
food supplies for Greece, he attacked Persian forces in the
Mesopotamian plain. In the ensuing campaigns his success
was phenomenal—he never lost an engagement from the
Hellespont to the Indus River. In 331 B.C. he destroyed the
main Persian force near Nineveh and then overran all the
areas under Persian rule. In 324 B.C. he ascended the Persian
throne as "King of the World."

Alexander apparently had plans for developing his newly
acquired lands which were almost as ambitious as his dreams
of conquest. He did not regard his subjects as mere payers
of tribute but as people whose lot could be improved through
the exercise of his genius. He founded many cities along
Greek lines—including those established by his successors,
about three hundred—which were intended to be economic
and cultural centers. He placed Macedonians and Iranians
in the chief administrative posts and fostered a policy of
fusion between Orientals and Europeans, of which the mar-
riage of his soldiers to natives may be cited as an example.
He desired an increase in trade and gave an impetus to it and
to a money economy by putting in circulation the hoards of
precious metals which he captured.

What Alexander might ultimately have accomplished we
do not know, for he died at Babylon in 323 B.C. Immediately
his generals began quarreling over the succession until finally
a new series of states, constituted on the basis of a balance-of-

power concept, took form. Egypt fell to the Ptolemies, heirs
of Lagus, one of Alexander's favorite officers; the East went
to Seleucus, a Macedonian nobleman, whose successors were
known as the Seleucids; Macedonia and Greece fell to the
descendants of Antigonus, a favorite general; while a few
districts remained or soon became independent, notably the
compact kingdom of Pergamum in Asia Minor. In general
these arrangements were continued until upset by Rome in the
second century B.C.

The consequences of Alexander's conquests were so far-
reaching that his triumphs have been used to mark a dividing
line in the history of antiquity. The Hellenic Period, when
the focus was upon the Hellenes and their glorious achievements, now comes to an end, and the Eastern Mediterranean
moves into the Hellenistic Period, when attention was centered on the diffusion of Greek techniques, institutions, and
culture to the whole of the known world. Indeed the main
characteristics of the 200 years after the Macedonian conquest was the triumph of Greek culture over that of the East
and the effecting of a unity of culture which had never before
been equaled. In this entire process Greece profited less than
other areas, because she had the most to give.

Of particular importance to our analysis were foreign
borrowings of Greek techniques of production and distribution. In the field of agriculture, for example, Greek methods
of husbandry were spread through Greek treatises and were
widely practiced. The vine and the olive came to be cultivated
extensively on a kind of plantation basis. Fertilization and
crop rotation were spread. The growing of fodder crops became more general, which tended to make animal raising more

a part of settled agriculture and to reduce nomadism on the fringe of arable areas. The irrigation system of Mesopotamia was restored, and in Egypt nearly 500 square miles of good land were reclaimed by further drainage of the Faiyum.

In industry, too, changes following Alexander's conquests were tremendous. Perhaps the most notable effects were on the metal trades, especially iron, because of an increased demand for cheap weapons with which to carry on war and for cheap tools with which to attack nature. Thus an impetus was given to iron production on the southeastern shore of the Black Sea, in other places in Asia Minor, and in Italy where the ore of the island of Elba was exploited. Copper mining on Cyprus took a new lease on life, and silver became abundant enough to compete with the finest pottery for tableware. The manufacture of woolens increased in Asia Minor, where there was an abundance of sheep; cotton cloth, first known to the Greeks in the fourth century, was made in Egypt and Phoenicia; and linen and silk were produced on an expanding scale in Egypt and the East. Pergamum became famous for its parchments and Egypt for its papyrus. Shipbuilding developed, particularly at Tyre; the curing of fish, long practiced by the Greeks, was adopted by others; and the East generally produced a variety of goods for Mediterranean markets—dyes, bleaching materials, chemicals for paints, asphalt, petroleum, carpets, tapestries, and perfumes.

Trade also developed in the Hellenistic Period, but here again the centers of commerce tended away from mainland Greece. The heyday of Athens and the Piraeus was passed, and only Corinth, with its ports on both the Corinthian and Saronic Gulfs, was able to maintain for a time its prominence.[19] Of the Eastern Mediterranean ports Alexandria was one of the most important. As far back as the beginning of the sixth century B.C. Egyptian shipping to the East had grown to a point where an attempt was made to build a canal from the Red Sea to the Nile and where Phoenician sailors were sent on a successful mission to circumnavigate Africa with the intention of establishing an Asiatic empire. In time Alexandria became the clearing point for the products of Egypt and for goods obtained along the coasts of the Red Sea, central Africa, and even India. In the Aegean the island of Rhodes, off the southeastern tip of Asia Minor, was the chief shipping center for the greater part of the Hellenistic Period, although after the intervention of Rome in the second century B.C. the island of Delos surpassed her for a time. And in the West, commerce centered in Carthage, Syracuse, and ports of central Italy and southern Gaul.

Along the routes of commerce the Greeks gave an impetus to the use of money and banking services. Mints were widely established (the Ptolemies gave Egypt its first currency system), although the Greek drachma was universally recognized as the standard coin. Money-changers put in an appearance, and, as earlier in Greece, they accepted deposits for safekeeping, made loans, transferred funds on their books (Giro banking), and made bottomry loans. With an increase in the size of business enterprises, more funds were sometimes required than could be provided by the savings or borrowings of one man. To meet this situation, partnerships became more frequent, but the device of the joint stock company was probably not hit upon. The increased use of money had also an effect upon public finance. As taxes and tribute came to be paid in money rather than in goods, tax farming was resorted to, that is, the granting to individuals or groups of individuals the right to collect taxes in return for the prior payment to the state of specified sums. This procedure contributed to the development of banking; and so too did the practice of governmental borrowing.

Although the most obvious and most extensive economic consequences of Alexander's conquests were the spread of Greek techniques of production and of Greek business practices, the effects upon Greece itself were of major dimensions. Because of the increase in production abroad, a revival of trade, and a foreign demand for Greek goods at the beginning of the third century B.C., Greece was able for a time to return to her earlier position of being a "workshop" for others. Her export of goods and earnings from shipping, banking, and giving instruction in the ways of Greece provided a foreign balance which allowed her to import food supplies, raw materials, and luxury goods for the wealthy. Furthermore, economic development in the East and West provided Greeks with alluring opportunities for employment and for investment abroad and led to a new wave of Greek emigration which somewhat relieved the pressure of population at home.

The conquests gave the economy of Greece a shot in the arm—they relieved the economic and social crisis which had confronted Greece in the middle of the fourth century B.C. Unfortunately this return of prosperity was short-lived, for overseas areas, like Rome, began to produce many of those goods formerly obtained from Greece, and trade was hampered by numerous wars between and social disorders within the states of Alexander's empire. As the demand for her goods fell off, Greece was placed in her earlier predicament—a situation which was made still worse by Roman intervention. At the end of the third century B.C. in retaliation for an attempt to upset the balance-of-power system, Rome, now a great power, cut Macedonia off from Greece and the Aegean, isolated Syria, and made Rhodes and Pergamum, both her faithful allies, bases for the preservation of conditions favorable to her. Greece was thereby more effectively deprived of foreign markets and of opportunities for emigration.

Not only did Greece have to face a serious economic situation, but she had also to meet a social crisis of major proportions. Her population had been increasing steadily and created such a pressure upon the supply of goods that infanticide was widely practiced. The trend toward large landholdings had continued; real wages fell from the peak of post-Alexandrian prosperity; and private debt increased. In all the Greek city-states reform parties clamored for a redivision of the land, relief from debt, and higher wages. City-states, which were generally governed by the well-to-do, were reluctant to meet such demands, either because, as in Athens, such exceptional sources of revenue as in the mines of Laureion had petered out or because ordinary means of raising funds were limited by poor economic conditions. Although there were instances of city-states furnishing people free amusement and food by gifts or taxes on the rich, there were also many cases of strikes and civil uprisings. Troubles of this kind provided an opportunity for Roman intervention—an intervention which was actively sought by some of the wealthy classes.

In the Hellenized East a not dissimilar economic and social

situation developed by the beginning of the second century B.C. Here the age-old concept that the king owned all the land generally prevailed, and in addition the monarchs and their favorites, frequently Greek emigrants, came into possession of the more important industrial and commercial establishments. In Egypt this situation led to an attempt at a centralized control of the economy—to what has been called a "planned economy," but here as in other areas one thing was certain, the native lower classes benefited little from Hellenization. Wars between the succession states of Alexander's empire diverted energies from productive to destructive ends, impeded trade, and reduced economic activity.

Effective efforts at planned economy degenerated into ostentatious exploitation of the lower classes. Strikes and disorders, sometimes led by priests, were not rare in the second century. In desperation the Ptolemies resorted to inflation (about 170 B.C.) to meet the crisis. In Syria conditions were somewhat better because of trade over the caravan routes, but even here the economy was ailing. Only in the kingdom of Pergamum, extended now to include a large part of Asia Minor, and in Rhodes was there continued prosperity. Yet even in these places the future was not bright. Rome did not want its satellites to develop to a point where they might be politically dangerous, and the time came when both of these states were accused of treason and punished (about 168 B.C.).

All efforts to free Greece or other parts of the East from Roman interference only resulted in disastrous wars and eventual defeat. Indeed, the chief characteristic of economic history of the Eastern Mediterranean in the last two centuries

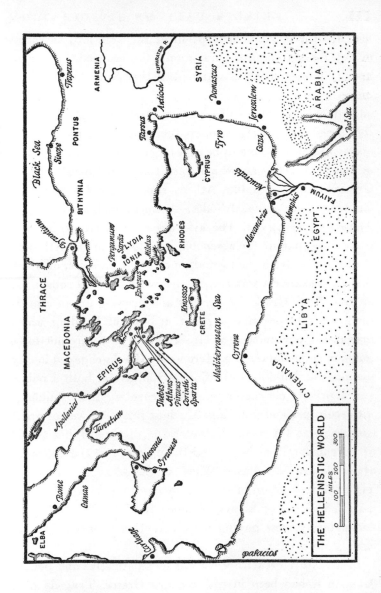

THE HELLENISTIC WORLD

MILES
0   100   200   300

ARMENIA

SYRIA

Damascus

EUPHRATES R.

Antioch

Tarsus

Jerusalem

Gaza

ARABIA

Red Sea

Trapezus

PONTUS

Sinope

Black Sea

BITHYNIA

CYPRUS

Tyre

Naucratis

Alexandria

Memphis

FAIYUM

EGYPT

THRACE

Byzantium

Pergamum

Sardis

LYDIA

IONIA

Miletus

RHODES

Ephesus

Knossos

CRETE

Mediterranean Sea

Cyrene

LIBYA

CYRENAICA

MACEDONIA

EPIRUS

Thebes

Athens

Piraeus

Corinth

Sparta

Apollonia

Tarentum

Messina

Syracuse

Rome

Cumae

ELBA

Carthage

palacios

B.C. was the gradual absorption of the area into the orbit of the Roman empire. Once again the economic center of gravity was moving westward—this time to the Italian peninsula.

As the economies of other areas came to surpass the material well-being of Greece, so too intellectual and artistic accomplishments came to be of a higher order in those places of the Mediterranean which were being Hellenized than in Greece itself. Cities like Athens surrendered leadership to Ephesus, Pergamum, Rhodes, Antioch, and above all to Alexandria in Egypt. The avid desire for learning in these places is apparent from schools established in the Athenian tradition and from the formation of great libraries, of which that at Alexandria was the greatest with 500,000 rolls, most of which were Greek. In the new and growing urban centers there was of necessity a great amount of building which provided opportunities for Greek artists that were not to be found in mainland Greece. Here we see Greeks engaged in city planning, in building Corinthian structures of both a public and a private character, and in decorating these buildings with sculpture and painting. Much of this work was of great excellence, as for example the statue of the "Dying Gaul" at Pergamum, "Laocoön" and the "Victory of Samothrace" at Rhodes, and "Venus of Milo" of the Alexandrian school. Yet the introduction of the arch began to make architecture more massive, statues were frequently of the colossal type, and painting, now paying more attention to landscape, became more ornate.

In literature creative work was less distinctive and powerful than it had been in fifth-century Greece. Tragedy died out, and the comedy of Aristophanes was supplanted by "the new comedy" of character and intrigue.[20] The mime, diatribe, and idyll were common, but most writing was in prose.[21] Here we find dull chronicles, histories, historical fiction, and travel tales, usually inspired by Alexander's exploits. In philosophy less attention was given to great cosmological, metaphysical, and epistemological questions than to emotional aspects of religion and to ethical escapism. This may have been because in the East the vast number of uprooted and poor people needed some rationalization of their misery.[22]

Already in fifth-century Greece, mystery cults had developed which preached that a knowledge of God could not be attained by the senses but rather by meditation, intuition, asceticism, or spells of ecstasy, that religion purifies the soul, and that a purified soul could get its rewards in heaven. These beliefs in the East were elaborated with a priesthood, initiation rights, sacramental meals, messiahs, powers of healing, virgin births, rituals, and symbols. Thus Greece and the East developed a whole body of religious practices which made progress in the West in the second century B.C. and some of which were adopted by the later Christians.

It was, however, in science that the greatest achievements of the Hellenistic Period were realized. In many of the Eastern cities, but especially in Alexandria with its observatory and medical school, the knowledge of Babylonia, Egypt, and Greece was fused for later transmission to the West. In general there was a trend away from speculation and system making to actual observation and investigation. Yet for the most part research was of a theoretical rather than of an immediately practical character. Perhaps the most notable

work of the period was in mathematics, for at that time
Euclid published his *Elements* (about 300 B.C.), Apollonius
of Perga (247–205 B.C.) prepared his geometry of cones,
Hipparchus (160–125 B.C.) invented plane and spherical
trigonometry, and only a little later Hero of Alexandria
(50 B.C.), who built the first "steam engine"—a toy pin-
wheel run by steam from a teakettle spout—worked out the
algebraic formula for a number of areas and volumes.

In this period, too, there were astronomers who held that
the earth rotates on its axis every twenty-four hours,[23] who
suggested that the sun rather than the earth is the center of
the Universe,[24] and who made elaborate categories of heavenly
bodies.[25] In medicine, Herophilus of Chalcedon (about 300
B.C.), the "father" of anatomy, named the upper portion of
the intestines "duodenum," the place where the most fashion-
able ulcers of our time are found, and Erasistratus (about
290 B.C.), the father of physiology, distinguished between
sensory and motor nerves and gave the valves of the heart the
names used today. In geography, the size of the earth was
computed almost exactly, the concept of latitude and longi-
tude was established, and the climatic zones were named. In
physics and engineering, Archimedes of Syracuse (287–212
B.C.) arrived at an almost exact value for π, developed hydro-
statics, and explained the principles of the lever, compound
pulley, endless screw, and burning mirror.

The Hellenistic Period was obviously not barren of important cultural advance toward a fuller understanding of
the world—an understanding that was essential for a more
extensive control of that world by humankind. Yet in this
period, it should be noted again, the best work was done in
the Eastern cities, which became the crossroads of trade and

thought, and in the century of this area's greatest economic prosperity. Here once more is evidence of a correlation between cultural and intellectual achievement and economic progress—a correlation that we found in the history of mainland Greece, of Egypt, of Babylonia, and of Sumer. We have witnessed again the role which the process of borrowing from others has played in the development of civilization, of the importance of conditions which allow individuals to realize freely and fully their potential talents, of the predicaments in which people are placed if their economies fail to keep pace in production with the growth of population, and of the curse of such disintegrating forces as social strife and war. When peoples become weak economically, socially, and politically, they fall prey to others. Such was the case of Greece and the East when the Romans came.

For Further Reading

Cambridge Ancient History (1923–1939), 12 vols.

Benjamin Farrington, *Science in Antiquity* (1936)

H. N. Fowler, *A History of Ancient Greek Literature* (1923)

F. M. Heichelheim, *Wirtschaftsgeschichte des Altertums* (1938), 2 vols.

A. H. M. Jones, *The Greek City from Alexander to Justinian* (1940)

M. L. Laistner, *Greek History* (1932)

R. W. Livingstone, *The Legacy of Greece* (1924)

M. Rostovtzeff, *The Social and Economic History of the Hellenistic World* (1941), 3 vols.

W. W. Tarn, *Hellenistic Civilization* (1927)

J. Toutain, *The Economic Life of the Ancient World* (1930)

H. B. Walker, *Art of the Greeks* (1922)

William L. Westermann, "Greek Culture and Thought," *Encyclopaedia of the Social Sciences* (1937), vol. I

V

ROME

ALTHOUGH Greece had lost its economic, political, and cultural vitality by the end of the Hellenistic Period, many of the seeds of its culture had taken root and were to flower in the Italian peninsula, whence, altered by local conditions, they were to spread again, especially in Western Europe. Indeed, the significance of Roman culture in world history is derived from the fact that it effected a new synthesis of Greek culture, added to it certain developments of its own, and through the process of diffusion provided fundamental elements in the culture of the people from the eastern Balkans to Poland and from Finland to the Mediterranean, as well as of the people in all parts of the globe colonized by Western Europeans.[1] It was from Rome that these people obtained to a large extent their religion, their state system, parts of their law, basic intellectual concepts, forms of art, and their emphasis on material welfare.

The place of Roman culture in world history

The diffusion of Roman culture is not, however, the object of our immediate interest. Rather we are concerned once again with those factors which allowed this culture to create an economic surplus and with the question whether or not there was correspondence between economic well-being, on the one hand, and satisfactory social arrangements and the production of works of a high intellectual and artistic quality, on the other. In the pages which follow, we shall see how Rome ultimately conquered all the known world, how she exchanged her industrial products for raw materials and foodstuffs, and how she succeeded in acquiring large deliveries of tribute from conquered peoples. Rome prospered most when her commerce and her collections of tribute were at their maximum, that is, in the first century B.C. and in the first two centuries of the Christian Era. Her period of highest civilization was also in these same centuries. Afterward, Rome failed to produce a vigorous and original intellectual life or to create forms of artistic expression able to stand the critical test of time.

Because Roman development was so dependent upon conquest, Roman history may be divided most usefully into segments for study according to the extent to which Rome exercised her dominion over others. Thus the first period reaches from the Bronze Age, through the establishment of an Iron Age economy and polity, to the time when Rome had conquered all the Italian peninsula, that is, from about 1000 to approximately 270 B.C. The second period extends from 270 B.C. to 14 A.D.—a period in which Rome triumphed over Carthage, Spain, Cisalpine Gaul, Greece, Macedonia, Transalpine Gaul, all North Africa, Egypt, Syria, Palestine, and parts of Asia Minor. The third period covers the first and second centuries A.D. in which Rome expanded over all the territory from the British Isles to the Caspian Sea and the Tigris River and from beyond the Rhine and Danube Rivers

to the African deserts. The fourth period, from the end of
the second century to the sixth century A.D., is characterized
by Rome's loss of authority throughout the Empire, by
foreign or barbarian invasions, and by the creation of small,
independent states even in the Italian peninsula itself.

As the rise of Rome is to be explained largely through
superior techniques of production and trade and through
conquests, so too, the fall of Rome is to be accounted for
largely by the loss of particular advantages of industry and
commerce and by the political disintegration of the Empire.
In the course of time Rome taught its techniques to people
in areas more richly endowed with natural resources than she;
her labor force, supported by "bread and circuses," came to
care little for work; and her capitalist class tended to spend
its surplus on luxuries rather than to invest it in productive
equipment. Conquered areas resented being eternally milked
by Rome, and a time finally came when they refused or were
unable longer to supply Rome with goods needed to sustain
life. Rome's wealthy ruling class fought over spoils to be had
from the peninsula or the Empire and became too divided and
too soft for imperial management. When the Roman Empire
fell apart in a general wave of economic, political, and social
disorganization, the Roman world, with the exception of
Byzantium, had little vitality left, and it entered the dark
ages of the early Medieval Period.

⚑ *Beginnings of Roman power, 1000 to 270 B.C.*

That Rome had to rely so heavily upon goods from its empire in order to attain its high economic position is to be

explained in part by geographical factors. The Italian peninsula, which has from time immemorial been aptly described as resembling the shape of a boot, is not richly endowed with natural resources. It is predominantly mountainous, with the Alps at the north being continued by the Apennines, which run from the top of the boot to the arch and under another name even to the very toe. Arable land is found primarily in the Po plain, along the coasts, and in the mountain valleys, and is limited to about one-third of the entire surface of the peninsula. In general, then, the land of Italy is analogous to that of Greece, although it is four times larger, or about the size of the state of Arizona.

In spite of limited arable land, Italy had in ancient times certain geographical assets which were conducive to economic development. At the beginning of the first millennium B.C. its mountains were well forested and ready to supply an abundance of timber for building and charcoal. Its foothills and sea marshes provided good pasturage. Its copper and iron deposits, particularly iron ore on the island of Elba, could be worked without great difficulty. In common with other ancient cultures it had excellent means of water transportation, and, being located near the center of the Mediterranean, it could easily get part of the trade between the highly developed Eastern Mediterranean and the unexploited West. Finally, although it was relatively safe from the concerted attacks of its neighbors, it could deploy forces in all directions.

Undoubtedly, also, the geographical position of the Italian peninsula had much to do with the fact that migratory peoples from several cultures went there as colonizers and that they were able to maintain contact with, and hence borrow steadily from, many sources. Toward the end of the second millennium B.C. there arrived in Italy a people from the Danube Valley, who brought with them knowledge of the use of iron. They were followed a little later by the Etruscans from Asia Minor, who imported their Eastern techniques and culture to the district north of the Tiber. Theirs was a policy of expansion, but in the sixth century B.C. they were stopped in the Po Valley by another invading people, the Celts. Then from the middle of the eighth century B.C. came Greeks who colonized the district southward from Naples, in the area which came to be known as Great Greece. Thus like Greece, Rome inherited from various cultures its productive techniques and much of its transmissible knowledge.

When the Etruscans arrived in the peninsula, the existing tribes were practicing settled agriculture, were making textile, bronze, and iron products, and must have been engaged in trade, for tin in the bronze had to come from afar. In spite of these evidences of economic progress, the Etruscans effected a real change in the economic life of Italy. They knew much about drainage and irrigation, and before long they had made such swampy areas as the Tuscan Maremma, the Roman Campagna, and the Pontine Marshes fertile and healthy areas. They built embankments along the Po to control that river, and they constructed the famous drainage canal, the Cloaca Maxima, near Rome's Palatine Hill, which furnished a model for the recovery of swampy land in the city. They increased agricultural production, which made possible the rise of cities, and the rise of cities led to the growth of industries. The Etruscans opened quarries for their buildings; they began mining in Tuscany and Elba, and they gave an impetus to the making of pottery on the wheel. Furthermore, they carried on trade with Greece, Phoenicia, Carthage, and Greek colonies in southern Italy. In brief, they developed an economy which made possible the accumulation of considerable economic surplus.

From the Etruscans the neighboring native tribes learned much, but the people of Latium were the ones to take fullest advantage of what they had learned. Their city of Rome grew into a commercial center and for a time was the seat of Etruscan power in Italy. Ultimately the people of Rome succeeded in freeing themselves from their erstwhile teachers, the last of the Etruscan kings, the Tarquins, being expelled from Rome at the end of the sixth century B.C. Then Rome, strengthened internally by the institution of a republican form of government which lasted until near the end of the first century B.C., began the conquest of the peninsula. By 270 B.C. Rome was in control of all the territory from the Straits of Messina, which separate Sicily and the mainland, to somewhat north of Pisa. The conquests had two important consequences in so far as Rome's status was concerned: first, they brought her into conflict with the powerful Greek state of Epirus, which sought to defend its interests in Great Greece; and second, they encroached upon the Carthaginian sphere of influence in Sicily. The defeat of Epirus made the world realize that Rome was a major power, while victory over Carthage in the ensuing period was to lead Rome on the path of foreign expansion, of overseas trade, and of collecting tribute from conquered peoples.

The economic surplus which Rome had to devote to war during the conquest of the peninsula resulted primarily from the use of Etruscan and Greek techniques. Then, as new lands were conquered, individual Romans got more land to which they could apply their technology, and the Roman world acquired the means of obtaining great revenue. A part of the conquered lands was given to favored individuals; a part, the "public land," was leased to patricians or to wealthy plebeians; and a part was left with former owners in return for a substantial payment. Moneys which went to the Roman treasury helped to finance further expansion, while lands which fell to Roman citizens allowed the building of great personal fortunes.

Some of the accumulated capital was invested in industry and commerce, and both of these branches of economic activity expanded during the wars for conquest of the peninsula. The need for military equipment stimulated metalworking and arms manufacture. The construction of great roads, for which Rome was to become famous, was begun—the Appian Way to the south and southeast, the Latin and Flaminian Ways to the east and northeast, and the Aurelian Way to the northwest—and because of them goods began to move overland in greater quantities. Trade with Greece and North Africa also grew, and Rome acquired at least a small merchant fleet of its own, although its range of action was limited by treaties with Carthage and other cities to the Tyrrhenian Sea. Finally, about the middle of the fourth century Rome established a coinage system—a sure sign of growing commercial activity.

In the first period of Roman history, from 1000 to 270 B.C., there was, then, considerable economic expansion, the amassing of surplus as is evidenced by the conduct of wars, and the concentration of a certain amount of wealth in the hands of craftsmen, patricians, and merchants. How much of this surplus was used for intellectual or artistic enterprise in the very early history of Rome we do not know. From what knowledge we do possess regarding the formative period of Rome's culture, it appears that more attention was given then to the extension of control over physical and human environment than to intellectual and artistic activity.

Of particular interest was the establishment of institutions which gave all citizens a sense of civic responsibility, ambition for individual material betterment, and rules for the conduct of basic human relations. An important factor in the development of the Roman pattern of behavior was the requirement, established under the stress of military exigency, that all landholding citizens between the ages of seventeen and sixty-five were liable for military service and for the cost of their own equipment. This drafting of the lower classes for military service was accompanied by reforms that gave the plebs a greater voice in government. Their members were given full citizenship,[2] were permitted to aspire to the highest offices in the state, could participate in the most important political decisions, could make laws by the early third century B.C. by means of plebiscita, and possessed in the Tribunes personal advocates of their own causes. Thus they came to feel that the welfare of Rome depended to a large extent on them; that they could improve their own lot; and that personal betterment was highly desirable.

Nevertheless, the plebs did not become political masters in Rome, nor was there any important leveling movement. In actual practice, the plebs elected patricians to the chief offices, and the Senate curbed the rights of the Tribunes by refusing advancement to those who proved obstreperous. Hence what was in form a representative republic was in actuality an oligarchy of the wealthy.[3] The Laws of the Twelve Tables, established about 450 B.C. and for long basic in Roman society, protected private property and facilitated capital formation by regularizing the transfer of property.[4] Under the existing system Romans could and did amass fortunes which helped finance the advancement of Roman civilization.

✎ *Rome becomes a world power, 270 B.C.–14 A.D.*

In the formative period of Roman culture a desire for economic progress, ambition for personal material improvement, and a tendency toward the orderly conduct of relations among citizens of the state were clearly established. In the next period of Roman history, from 270 B.C. to 14 A.D., the last of these goals was for a time not realized, but the former ideologies found expression in the great wars of Roman expansion—wars which effected important changes in Roman culture. In these two and a half centuries Rome fought Carthage, Macedonia, Syria, Greece, and Egypt and conducted campaigns against native tribes in northern Italy,

Spain, and southern Gaul. Although there were moments when Roman armies were sorely pressed, they emerged in the end victors over all their enemies. By the beginning of the Christian Era Rome had conquered most of the Mediterranean and was easily the greatest power in existence.

Of all the wars which Rome waged after 270 B.C. the three against Carthage were the most crucial to her development and to the future of the world.[5] They launched her on a policy of expansion outside the Italian peninsula; they inaugurated her policy of collecting tribute and otherwise of attempting to live off conquered people; and they determined whether Greco-Roman or Phoenician-Carthaginian culture was to be predominant in the West.

Rome became embroiled with Carthage by supporting the time-honored policies of the Greek cities of southern Italy to expand their commercial activities in areas dominated by Carthage and by threatening Carthaginian interests in Sicily. In the wars which followed, Carthage proved to be a formidable foe, and the outcome was not certain until the end. Carthage had a great potential for war, based upon the successful development of agriculture in North Africa, upon industries within her cities, and, most characteristic of all, upon an extensive commerce.[6] Furthermore, the Carthaginians had a powerful fleet, a cavalry superior to that of the Romans, a knowledge of all the Hellenistic machines of war, mercenaries schooled in Hellenistic methods of warfare, and a force of elephants, which like the modern heavy tank, was used to break through enemy lines. Yet, in spite of their strength, the Carthaginians were at last defeated. They were handicapped by having to do much of their fighting overseas and by having mercenary soldiers who, unlike the Roman citizen-soldier, lacked staying power in moments of crisis.

That Rome was able to overcome such a strong adversary speaks well for its economic strength. Even though the wars were costly in men and resources and though Italy was a battleground for twelve long years in the Second Punic War and was ravaged by Carthaginians under the brilliant leadership of Hannibal, the Italian peninsula experienced a movement toward further urbanization and industrialization. With the influx of capital from defeated countries, there was investment in new industrial enterprise. With labor moving from the country to towns, there was a greater working force available for domestic manufacture. And with the growth of cities, particularly of Rome, which became the largest city in the known world,[7] there was an increased demand for industrial products.

Urbanization stimulated the building trades and allied fields like quarrying, brickmaking, and lumbering. The construction of aqueducts to supply cities with water gave a fillip to stone masonry, engineering, and to plumbing, as in the manufacture and installation of lead pipes for the distribution of water to public fountains and individual buildings. Industry and war created a demand for iron and bronze tools and weapons. In Campania, the town of Puteoli, the modern Pozzuoli, became a center for the production of spears, javelins, swords, picks, scythes, and chisels, while Capua specialized in kitchen utensils, especially the ubiquitous pots and pans. During the first century B.C. the famous Arretine earthenware, which came to supply a large part of the West

with better grades, was established. Then there were also
jewelry making, goldsmithing, leather working, shipbuilding,
and a host of other important trades.

For the most part, industrial production was concentrated
in small workshops of the handicraft type, but some was
conducted upon large estates, where baskets, rough crocks,
and tables were made and where frequently olives and grapes
were put through presses. Some industry was established
by entrepreneurs on a relatively large scale, especially in the
case of mining, quarrying, brickmaking, and metalworking.
At all events, a division of labor took place which greatly increased output. We find specialized free workers organizing
collegia, a kind of benefit society, on the basis of their trades.
Thus by the end of the Republic, Rome had a thriving industry. Yet, Roman culture had produced little in the way of
new techniques, made almost no use of nonhuman energy, and
did not tap inorganic matter as a source of power. Its inability to advance technically placed definite limits upon its
industrial expansion.

The wars between 270 B.C. and 14 A.D. also effected changes
in Roman agriculture which were as significant for Rome's
welfare as changes in its industry. In the first place, there
was a trend toward the concentration of landholdings. The
well to do acquired large estates—the *latifundia*—by purchase or lease,[8] while the poorer farmers frequently sold their
holdings and moved to the cities or became agricultural laborers or small tenants. In the early period, as we have seen,
landholders were required to perform military service without
pay and were called to arms at any and all times of the year.
Under such circumstances, the small landholder was at a

tremendous disadvantage, for he was not in a position to get someone to carry on his work while he was away. If he were killed in battle, no one might be available to till his acres, and they would have to be sold. If he did return from the war with a whole skin, his fields would most likely be unkempt, his livestock driven away, and his buildings in disrepair. The poor returning veteran frequently felt that the sale of his land was the best way out of his predicament.

Henceforth in Roman history the question of large holdings was a serious one, but all attempts at reform, of which those sponsored by the brothers Tiberius and Gaius Gracchus were the most famous (133–123 B.C.), proved abortive. To provide labor for the great estates, owners employed not only landless freemen but, on a mounting scale, slaves from the Eastern slave markets and from captives taken in war. With the introduction of slave labor on the large farms, Italian agriculture underwent a profound change. Large landholders turned from the growing of grain to the cultivation of the olive and the vine and to the raising of livestock. So important did these branches of agriculture become that behind the plea of Cato, himself a landowner, for the destruction of Carthage was the hope that Carthaginian competition in oil and wine would be wiped out.[9]

With concentration on the growing of olives and grapes large landholders let drainage works fall into disuse and the land that had been reclaimed to return to malarial swamp, a condition which has only been rectified in recent times. Furthermore, Rome became dependent upon foreign imports for a significant part of its food supply, a situation that necessitated control of foreign areas to assure regular deliveries and that led to governmental sale of grain at a low price—to a policy of bread for the poor.[10]

Perhaps the outstanding economic consequence of Rome's foreign conquests was the enormous growth of trade. Now Rome dominated a large portion of the known world, and trade followed the flag. Improvements were made to such ports as Puteoli, the chief shipping center on the Tyrrhenian Sea, to Ostia, at the mouth of the Tiber and the port of Rome, to Brindisi, to Utica, and to many others. Roman roads were established throughout the peninsula, and new ones were constructed through southern Gaul from the Alps to the Pyrenees and across northern Greece from the Adriatic to the Aegean. Caravan routes in the Near East were improved, and at Rome itself the commercial district around the Aventine Hill was greatly expanded.

With the development of trade the Italian peninsula received more and more goods of all kinds from every point of the compass. From Sicily, North Africa, and the East came grain. From Gaul and Spain came copper, silver, lead, tin, wool, and hides. From the East came spices, perfumes, and carpets. From the Black Sea came salt fish. From Asia Minor came cloth and dyestuffs. And from Greece were brought in works of art and other luxuries. For these imports Italy could not export goods of equal value, although she did sell abroad metallurgical products, wool, lumber, naval supplies, and pottery. The balance of trade in goods was usually against her, but her deficit in goods was more than covered by tribute in kind or money and by the earnings on Roman property held abroad.[11]

Redressing the trade balance in this manner raises the

question of Rome's exploitation of her foreign territories. As we have already seen, Rome began with the First Punic War to demand tribute from conquered regions, initiating the policy to help meet the heavy debts incurred during the conflict and continuing it because of its lucrative nature. Tribute might be in the form of a portion of a grain harvest, royalties on mining, customs duties, income from lands either sold or rented, and seizures of precious metals from treasuries and temples. To the sums thus obtained should be added booty that was taken at random by the military and the income which favorite Roman citizens realized on property that they acquired overseas.

For the most part tribute was collected by tax farmers. Tax farming had come into being prior to the Roman expansion outside Italy when the collection of taxes and rents, the provisioning of the army, and the handling of all governmental expenditures had been auctioned off to the highest bidder. As Rome began its conquests of foreign areas, the system was simply extended. In many cases the tax farmers conducted their affairs without honesty, pity, or scruples of any kind, and they amassed large amounts of capital. Nor, as a rule, did the governors of the provinces attempt to rectify matters. On the contrary they often followed the examples set by the tax farmers and exploited their charges without mercy. Many of the officials, like Pompey, Crassus, Antonius, and Caesar, were able to create great personal fortunes which they could use for investment and for furthering their own political ambitions.

Through foreign trade, tribute collecting, and tax farming, Rome developed its major institutions for facilitating capital accumulation and investment. The tax farmers, known as "publicans," had money to lend. So too did money-changers who profited from laws requiring foreign traders to change their coins to Roman.[12] And so also did shippers who frequently financed the cargoes that they carried. So there developed such practices as the receipt of moneys for deposit, the transfer of funds, the acceptance of bills of exchange, trading in shares of companies, loans, and the sale of annuities. Probably bankers lent money principally for productive enterprise, but they also made consumption loans, helped finance ambitious persons like Caesar, and assisted Eastern cities in meeting their obligations. Although the maximum legal rate of interest was 12 per cent, established in the Twelve Tables of about 450 B.C. and thereafter frequently reiterated, rates for nonproductive and risky ventures ran as high as 48 per cent per annum. Furthermore, rates fluctuated widely, for the supply of money varied sharply as Rome brought in new quantities of precious metals from a campaign or failed for a long time to do so. And because Rome had to import so much grain, poor crops necessitated large purchases abroad that drained off money. Hence Rome had business fluctuations due to agricultural conditions and the amount of money obtained in conquest.[13]

Although Rome did a poor job of regulating the supply of money in order to minimize swings in business activity, she made one important contribution to business organization in creating the joint stock company. This institution came into being as a result of tax farming. When Rome made important conquests abroad and demanded very large sums in return for the privilege of collecting taxes for the state and

provisioning troops, it sometimes happened that no one individual had sufficient funds to make the advance payment.
Hence publicans banded together to take "shares" in the
enterprise and offered other shares to the investing public,
particularly to senators who were prohibited by law from
participating directly in tax farming. There thus came into
being an exceptionally important method for harnessing capital from many sources for speculative trading in shares of
business enterprise.

As a whole the two and a half centuries of Roman history
prior to the birth of Christ witnessed an enormous economic
development, not only in total goods available for consumption in Italy and growth of population but also in goods
available per capita of the population. Rome established important economic institutions for the accumulation and investment of capital, developed its own industrial capacity,
greatly extended its trade, and altered the character of its
agriculture. Rome was thus able to add to the resources of
the Italian peninsula by tapping supplies from other areas,
to increase production by a division of labor and by allowing
various localities in the Empire to specialize in those things
for which they were best suited, and to amass wealth which
could be used for development of the arts and sciences.

Perhaps the greatest weakness of Roman civilization at
this time was the lack of harmonious human relations. In fact,
so great was the hostility among social classes and aspirants
to political power that Rome's exalted position was threatened. Among the more important elements in the situation
were (1) there were the former small landowners who had
been pushed off their holdings and who felt resentment toward

large proprietors; (2) there were slaves who were treated poorly and who were antagonistic toward their masters; (3) there were the poor plebeians who felt that they were exploited by the wealthy and who were not content with "bread-and-circus" policies; (4) there were the equites, or knights, rich tradesmen, industrialists, and tax farmers, who were jealous of the patrician class; (5) there were the Senators, who held the real political power of the state and had the greatest prestige, but who were not necessarily the richest, as they were excluded from such profitable businesses as tax farming and shipping; and finally (6) there were the military men and high state officials who reaped large profit from their services and who frequently nourished extravagant ambitions.

During the early life of the Republic, political differences between plebeians and patricians were resolved by concessions and compromise. Now such a happy solution was not found, and from 125 to 27 B.C. Rome was afflicted by domestic strife initiated in the interest of some class or pressure group or combinations of classes or groups. The details of these internal struggles are exceedingly complicated and need not be repeated here. The chief facts to remember are that slaves and plebeians got little from their uprisings and that ultimately the struggle was among senators, knights, and commanders of various Roman armies who fought for control of the state in the hope of being able to reap the largest share of the spoils. It was this situation that gave rise to the meteoric career of men like Pompey, Antonius, and Caesar—and which led in many cases to their violent ends.

This state of affairs could not be allowed to continue, for

if Italy was devastated by civil war and if Roman armies fought one another to impotence, foreign provinces of the Empire might revolt and become independent of Rome's authority. Obviously no one in his right mind wanted to kill a goose while she was laying golden eggs. Finally, Octavius Caesar, better known by his title of Augustus, a grandnephew of Julius Caesar, restored order and was made *Princeps civitatis*—the first citizen of the state—a title which he held from 31 B.C. to 14 A.D. and which gave to his regime and that of his early successors the name of Principate. Augustus set about the reestablishment of Roman authority and governmental machinery so that Italy could continue to draw upon the production of the foreign provinces. This basic policy appealed to contesting Roman factions, and Augustus strove to carry it out with a strong but diplomatic hand. He curbed the power of the Senate but pacified its members by allowing them to retain their social privileges and rights to certain posts. He paid government officials so well that graft, now punished severely, was reduced to a minimum. He abolished tax farming, yet won the support of the knights by giving them certain privileges and by restoring peace and order for the conduct of business. He won the plebs by handouts of free grain and free entertainment on a scale more lavish than that of Caesar himself. He secured the support of foreign provinces by the restoration of order and by stopping predatory tax farmers and governors. And he minimized the possibility of military commanders rising against established authority by making them responsible to him, changing them frequently, and placing them and their armies on the frontiers where danger threatened.[14]

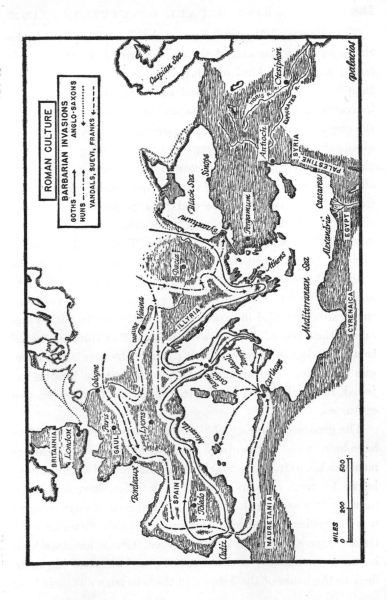

ROMAN CULTURE

BARBARIAN INVASIONS
GOTHS ———▶
HUNS —·—·—▶
ANGLO-SAXONS ·······▶
VANDALS, SUEVI, FRANKS —··—··—▶

MILES
0   200   500

The reforms of Augustus were of the utmost importance to the future of Rome and of the ancient world. First, they effected a different kind of control over human environment than had existed previously and one that greatly restricted individual action. The Republic was in essence abolished, for the new rulers assumed extensive personal powers which were given legal form when the Empire came into being toward the close of the first century A.D. Second, the policies of Augustus effected a compromise among social classes which perpetuated the existing class structure and thus permitted the upper classes, particularly the knights, to continue to enjoy a dominant position in society and in politics. Third, and most happily for Rome, Augustus prevented a disintegration of the conquered areas comparable to that of the Empire of Alexander the Great. He and his followers in the first two centuries A.D. succeeded in maintaining relative peace within and among the parts of Rome's vast holdings. Pax Romana, based to a large degree on force, made possible the turning of energies to peacetime production. As a result, all the Empire enjoyed relatively good economic conditions for nearly two centuries.

The greatest flowering of Roman culture is usually said to have been in the first century B.C., and in this period achievements in the arts, particularly in literature and architecture, began to rival accomplishments in politics, law, and engineering. It was then that Greek influence on Roman culture was everywhere apparent. Following Rome's victories over Great Greece, Carthage, and Greece, Greek ambassadors, hostages, craftsmen, traders, artists, and scholars began to flock to the banks of the Tiber and there to turn out their kind of products. Things Greek attained a style demand in Rome, Romans of "distinction" believing that they must know the Greek language and Greek culture. Attempts to satisfy the demand appeared on every hand. In literature we find that Livius Andronicus (about 284–204 B.C.) translated the *Odyssey;* Ennius (about 239–169 B.C.), the "father of Latin literature," wrote a historical epic of Rome in Greek hexameters and composed Latin comedies and tragedies using Greek models; Polybius (about 204–122 B.C.) wrote in Greek a history of the Punic Wars; and Terence (about 190–159 B.C.) produced works after the new Greek comedy. In architecture, the Romans adopted the Greek temple with its colonnaded porch and enclosed cella, simply placing the structure upon a raised foundation (podium). In painting and sculpture, Romans sought Greek works for decorating their buildings and immortalizing themselves. So highly, indeed, did the Romans hold Greek culture that they adopted Greek gods, giving them Roman names,[15] and endeavored through the creation of the Trojan War legend regarding the origin of Latium to trace their ancestry to Greek sources.

Although the first century B.C. witnessed social unrest, slave uprisings, and struggles among army commanders for control of the state, the fact remains that Roman economic production was relatively high, that Roman victories resulted in the seizure of hoards of wealth, and that large amounts of surplus were being devoted to the arts. One should remember that Augustus succeeded in establishing peace and order, that he spent vast sums embellishing his capital, and that Greek influence stimulated rather than choked off artistic activity.

The last decades of the Republic and the first of the Principate have been justly called the golden age of Latin literature. It was at this time that Cicero (106–43 B.C.) delivered his powerful orations and composed his delightful essays; that Caesar (100–44 B.C.) wrote his *Gallic Wars;* that Vergil (70–19 B.C.) produced Rome's greatest epic poem, *The Aeneid;* that the poets Horace (65–8 B.C.) and Ovid (43 B.C.–18 A.D.) flourished; and that the historian Livy (59 B.C.–17 A.D.) explained how Rome created her vast empire. It was also the period in which the encyclopedist Varro (116–27 B.C.) summed up Roman knowledge on many subjects; that Catullus (85–54 B.C.) wrote passionately of love in Alexandrian verse; and that Sallust (86–34 B.C.) produced his histories of Catiline's conspiracy and other contemporary episodes. For the most part these works were fresher and more vigorous than contemporary Greek literature. Latin materials were cast in Greek molds, but the final product had a made-in-Italy stamp on it.

In the other arts Roman achievements of this period were not up to the level of literature. Many Romans who could afford art had acquired their wealth suddenly and were unfamiliar with standards of quality. They did not hold artists in high esteem, nor did they distinguish between the major arts and the minor skills. In general, their tastes ran to works of ostentation and self-glorification, although their constant demand for more artistic surroundings increased and probably reached a high point of quality in the first century B.C. In architecture, where Rome showed the most originality, attention was given to better building materials, city planning, and new forms. Caesar is said to have found Rome a jumbled mass of buildings made from sun-dried brick, but Augustus boasted that he left it a city of marble. The use of concrete faced with brick and the arch resting on piers or columns allowed the construction of new types of structures. The ambitions of Caesar, Pompey, and Augustus led to the beautifying of the old Forum and the construction of new ones, majestic in their proportions. In sculpture there was evidence of greater technical competence and of a desire for a faithful rendering of real objects. The historical relief and portraits, particularly those of Augustus, attained a degree of expression and likeness that were never equaled unless perhaps in the reign of Vespasian (69–79 A.D.) and Trajan (98–117 A.D.). In painting, the selection of the peak of achievement is difficult to determine because of the paucity of remains, yet it would seem from the available evidence that this period marked a high point in technical performance and in freshness of treatment. The murals of the Villa of the Mysteries at Pompeii and of the villa at Boscoreale display concern for theme, composition, and detailed execution which is usually associated with the learned art of the Italian Renaissance.

Although Rome never produced a first-rate original philosopher or scientist, Romans gave attention in the first century B.C. to philosophies and scientific ideas acquired from the Greeks. Stoicism appealed to them because of its stern moral system and emphasis on self-discipline. It fitted in with the need for an orderly world, even if it hurt the lower classes. Epicureanism was even more popular because of its emphasis on material things. It was given brilliant statement by Lucretius (96–55 B.C.) in his *On the Nature of Things*. Neo-Pythagoreanism also enjoyed a vogue, for its contention that the human mind had fallen from a state of purity, that it lived in an evil world, and that it could be saved for an eternal life helped to explain obvious conditions and to hold out hope for the future. Yet Stoicism, Epicureanism, and Neo-Pythagoreanism instead of stimulating further exploration led to dogmatic positions.

In general the Romans were not interested in observing, classifying, and ordering data and were even less concerned with attempts to arrive at general statements regarding the physical world which would conform with observed information. If practical results could be obtained by trial-and-error methods or by rules of thumb, Romans did not fret about principles. Most of their science the Romans had derived from Greek or Hellenistic sources, and during the Roman Period most scientific advances were achieved not at Rome but in areas under the direct influence of Greece, particularly in Alexandria. It was from this center that a large portion of the science of the ancient world was passed on, notably to Arabic scholars and from them to Western Europe.

The chief contributions Rome made to science and Rome's greatest interest in science date from the end of the period under discussion. Instruction in medicine was given in Roman schools toward the close of the Republic, and the most famous medical treatise of Roman times, that by Celsus, appeared shortly after the birth of Christ. The leading Latin work on architecture and mechanics written by Vitruvius about 1 A.D. indicated that the writer had practical knowledge of windmills, cranes, water wheels, and, most important of all, a solution to the problem of the transmission of power by gears—an important element in the development of the machine. Furthermore, Romans picked up information about military machines—catapults, battering-rams, and slings—which was put to practical use. They introduced soap, probably from Gaul, and most significant of all, they abandoned the traditional calendar of 354 days, with its extra month every second year, and put into operation on January 1, 45 B.C., the solar calendar of 365 days.[16]

◁ The first two centuries A.D.

With the death of Augustus Rome's most brilliant period of civilization came to a close. There then followed two centuries in which Rome continued to enjoy imperial power and wealth and to realize important achievements in the arts. Yet in this period weaknesses in Rome's economic and political system became accentuated, and signs of decline in artistic and intellectual activity became apparent, particularly in the second hundred years. Population increases [17] seem to have outstripped growth in the supply of goods; control over human behavior deteriorated; no new important conquests were made; [18] natural resources were wastefully exploited; and Roman productive and distributive techniques were diffused to other areas. Rome's relative position was weakened as Roman economic and military techniques were adopted by others and as the exchange of goods between the Italian peninsula and the Empire declined.

The chief characteristics of Rome's economic system in this period may be illustrated from the field of agriculture. At this time Rome did not discover any new and important crops or domestic animals, nor did it develop any methods of field or animal husbandry by which to increase production. What it did was to practice known arts of agriculture and to diffuse them more widely. We have already noted that large land-holders in Italy began toward the end of the Republic to turn from grain to the olive and the grape. Now we can add that this policy upset the agriculture of places like Greece in which olive and vine cultivation was important, that other areas followed Italy in olive and vine growing, and that Italian wine became so overproduced that the Emperor Domitian (81–96 A.D.) ordered foreign farmers to "plow under" half their vineyards. Grain growing, on the other hand, having declined in Italy and Greece, was expanded in districts like Gaul, where land was being reclaimed from the forest, and in North Africa and on the fringes of Syria, where land had been used exclusively for grazing. In the latter cases, production at first was probably high, but in the long run these semiarid regions turned into "dust bowls" from which even to the present day they have not recovered, in spite of no appreciable change in rainfall or other climatic conditions. In the course of time, the grain supply of the older and larger cities of Italy became precarious. In the case of Rome, edicts were issued forbidding Egypt to export its grain to other than the capital of the Empire, but such measures depended on power for execution, and when ulti-mately Rome's power failed, her food supply was in jeopardy.

The diffusion of Roman agricultural methods and crops created in some cases competition between Rome and over-seas areas, in others difficulties in supplying the older re-gions, and in others problems of soil erosion. These issues were, however, no more important than questions of landholding. Because slave labor on large estates became very costly and agricultural management became inefficient, large landowners began to let out their land to tenants (*coloni*) in return for part of their crops and some labor service. The first evidence of such a development appeared on the estates of the Emperors, but it could soon be seen on other holdings, even in the East. The new estates tended to produce more for their own needs than for trade, which militated against the existing division of labor and exchange and thus helped ultimately in destroying the basis of Rome's economic system.

In many respects the history of industry in the first two centuries A.D. was similar to that of agriculture. No important industrial inventions were made and applied; no new techniques which might have revolutionized production were devised; and no radically new products were created. What the Romans did was to pick up industrial techniques and products in the Eastern Mediterranean and to diffuse them in the Western Mediterranean and to the lands north of Italy. This process of diffusion was apparent along trade routes, around camps of Roman soldiers where cities frequently grew up,[19] and at places where the presence of natural resources attracted Roman capital and Roman methods of production. As the process of diffusion went on, Italy turned away from the manufacture of staples and laid more and more emphasis upon luxury goods. Western Mediterranean lands profited most in this development, for they had resources of minerals and of wood superior to those of Italy and the Eastern lands. Hence the industrial phase of military potential increased in the West and declined relatively in the East and in Italy.

In many branches of industrial production the West went

through a minor industrial revolution in the first two centuries A.D. Not only was there a great development of mining, in many cases conducted by emperors or their agents, but also of metal industries. Swords, daggers, and armor of Spain became justly famous; iron tools and arms of Gaul attained high positions in quality and quantity; and the swords of Noricum found a large market in Italy. Pottery making was also greatly extended to the north and west of Italy. Until the end of the first century A.D. pottery from Arretium in Italy had a large export market beyond the Alps, but after that date production of pottery in Gaul and Spain, stimulated by business interests and workers who emigrated from Arretium itself, was so successful that Italian potters were unable to compete with it. In like fashion Roman methods of textile production were introduced into Gaul, which became one of the great cloth-producing provinces of the Empire, exporting its common wares to Italy. Italy and other of the old regions therefore produced relatively more for the luxury and style trades.

For the most part staple goods were produced in households or small shops, although in industries that required considerable investment or that had a large market production was frequently on a larger scale. The motive energy used in nearly all undertakings was provided by humans. Although Vitruvius, who wrote Rome's most important work on mechanics about 16 B.C., described an undershot water wheel with gears for the transmission of power, little evidence is available to indicate that this apparatus, which was certainly capable of general use, was employed at all widely. Labor was organized into collegia, whose primary purpose seems to have been to defend the interests of free workers against employers and to gain prestige for workers in a world whose upper crust looked askance at manual effort. The collegia indicated a division of labor, for among cobblers there were distinct societies for those who made different types of ᶠ gear, and

linen v

ᵛ the increased division of labor, with greater area ᵢalization in production, and with an increased population, Rome was more dependent than ever on imports. In fact, many parts of the Empire and particularly Italy could not have existed as they did without a system of multilateral trade. Hence Rome had to do everything possible to encourage trade. She extended her network of roads, both main highways and secondary routes, to the farthest corner of her dominions. Indeed not until the nineteenth century did countries outside the Empire, like Germany and Russia, have roads comparable to those in Roman provinces. Carrying by land was, however, expensive and slow. As was mentioned earlier, the horse harness was inefficient; horses were small; and horseshoes were simple sandals tied on. The ox, for its part, was not practical for long hauls because of its slow pace. Most bulky commerce moved by water, along rivers or upon the seas. Under the early Emperors traffic on the rivers of new provinces—on the Rhine, Danube, and rivers of Gaul and Britain—grew rapidly. River and land traffic tended, however, to converge upon the seas, and goods from the Baltic, North, Black, and Red Seas and from the Atlantic and Indian Oceans [20] gravitated toward the Mediterranean. Hence Ostia and Puteoli were undoubtedly the most active shipping centers, and large sums were spent in improving and enlarging their facilities and in making shipping safe from pirates and storms.

In many other ways Rome acted to aid commercial intercourse. Standard systems of weights and measures were established, and a unified coinage system was created for the whole Empire. Roman private initiative furnished the commercial world with banking services, maintained far-flung sales organizations, and endeavored to regularize trade through merchant guilds. Although the Empire was divided into eleven customs districts and a charge was placed on goods moving from one district to another and also on goods entering towns (the octrois),[21] tariff rates were low, ranging from 2 to 5 per cent ad valorem, and they took the place of other taxes. Some special measures were adopted to protect Italian agriculture and industry, but they were few in number and were seldom enforced. Rome's chief concern was in getting ample supplies for Italy from her Empire.

During the first two centuries after the birth of Christ, Rome continued to have considerable wealth and to produce important works of art and of the intellect. Yet, as in Roman economic affairs, so in its civilization, there were signs of weakness. Human relations deteriorated over struggles for the emperorship, with the appearance of such leaders as the insane Caligula and the pyromaniac Nero, and with the substitution of imperial dictatorship for republican institutions. The individual had less reason than earlier to have a sense of civic responsibility or to nourish great personal ambitions. There was less of a stimulus from the Greeks to attain perfection in the arts, and there was less freedom of thought, as is witnessed by the persecution of the Christians. In general, the range of opportunities for alternative decisions was becoming narrowed.

The field of literature illustrates well what was taking place. Upon the heels of writers of the golden age came a group of literary men who were more renowned for industry and form than for content. The rhetorical tragedies of Seneca (4 B.C.–65 A.D.), the tales of Petronius (died 66 A.D.), and the poetic satires of Persius (34–62 A.D.) were either trivial or lacked strength and form. The satirists Martial (40–104 A.D.) and Juvenal (60–140 A.D.), the educator and rhetorician Quintilian (died *circa* 96 A.D.), Rome's greatest historian, Tacitus (55–120 A.D.), and the brief renaissance of Greek letters led by the historian Plutarch (46–120 A.D.) raised the average of literary production of these centuries high enough so that they have attained the appellation of "silver age."

Yet the "silver age" of Roman literature was definitely inferior to the "golden age," and hardly more can be said for architecture. Although Roman builders made use of new forms like the rotunda, as in the case of Hadrian's Tomb, buildings of the first two centuries A.D. were distinguished primarily for their massiveness, size, and ornamentation. Even extensive construction throughout the Empire during the reigns of Trajan and Hadrian elicited little originality. Most of it, as seen in coliseums, aqueducts, and forums, was patterned on Roman styles. Painting also declined rapidly after the first century A.D., and sculpture, having attained excellence in portraiture and in reliefs picturing historical events, as in the scenes on Trajan's Column, deteriorated in

the second century to heroic equestrian statues, vulgar me-
morials, and meaningless ornamental pieces. In fact, it is
generally said that Rome produced no great work of literature or art after Trajan and Tacitus.

Legal scholarship was maintained on a somewhat higher
level after the birth of Christ than were the arts, perhaps because the field was newer and was freer from fixed forms. The
father of Roman jurisprudence is considered to have been
Scaevola (died 82 B.C.). He made an effort to state legal principles and had a great influence, for example, on Cicero, who
sought the "true law." With Augustus and the Empire a
group of the legal profession was employed as counselors,
teachers, and advisers to judges with the result that more
attention was given to legal scholarship. Gaius (died about
180 A.D.) wrote a model for many legal textbooks; Julianus,
a contemporary, prepared a digest of law; Paulus (died 222
A.D.) wrote on legal theory, identifying law with the equitable
and good; and Ulpian (died 228 A.D.) produced commen-
taries on laws and treatises on the functions of magistrates,
which provided nearly a third of the Justinian Code.

During the first two centuries A.D. Rome accomplished no
more in the field of science than in the century prior to the
birth of Christ. The most important scientific work by a
Roman in this period was Pliny the Elder's (23–79 A.D.)
Natural History. It was a compilation of what Rome knew
or thought of the physical world and was void of any serious
attempt at establishing the sequential relationship of observed data. On the other hand, scholars of the East, who
were in contact with Rome, strove to find out the ways in
which man and the world functioned. For example, Strabo

(63 B.C.–21 A.D.), a native of Pontus, traveled far and wide
to collect geographical information, described the formation
of alluvial deposits, and explained changes in the earth's
crust. Ptolemy (flourished 150–170 A.D.) of Alexandria provided the basis for trigonometry, gave π the value of 3.1416,
contributed substantially to geographical information, and
in the *Almagest* summarized ancient knowledge of mathematics and astronomy—a body of knowledge which went on
to Arabic and Western scholars and was adhered to until
modern times. Aretaeus (flourished 130–190 A.D.), a Cap-
padocian, was the first to describe diabetes and diphtheria.
And Galen of Pergamum (130–200 A.D.), the greatest of
ancient physicians, extended man's knowledge of anatomy and
physiology and presented the first coherent theory of the
interaction between external environment and the human organism—an achievement which ranks in the forefront of
Greek scientific thought.

In philosophy, the Romans were far from original, again
borrowing heavily from Greece. Of all Greek philosophies,
Stoicism, as taught by Seneca, Epictetus, and the Emperor
Marcus Aurelius, was in vogue primarily in the second century
A.D. Its main precepts were that an all-wise Providence controlled the universe by means of natural laws and that man
could live according to design of Providence by faithful ob-
servance of these laws. Of this philosophy Rome made good
use, justifying peace and justice by force. For a time Stoicism
was rivaled by Neo-Pythagoreanism, but both ultimately gave
way to Neoplatonism, whose chief exponent was Plotinus
(205–270 A.D.), which held that highest knowledge is not
attainable by human reason.

⚲ The economic decline of Roman Italy from the third to the fifth century A.D.

Within the economy of the Empire at the end of the second century A.D. there were many elements of softness, as we have seen. The greatest of these weaknesses was that Rome relied upon products obtained from the various provinces of the Empire for which goods of equivalent value were *not* given in return. The provinces were to a large extent sustaining Italy, and those north and west of the Alps, at least, were increasing their economic potential, while Italy's productive capacity was relatively withering away. How long Italy could continue to be supported by others in the manner to which it had become accustomed depended largely upon the degree to which she could satisfy the provinces by providing them with protection, peace, and economic progress and to the extent also to which she could bend them to her will by force.

The manifestations of this basic problem were many. Wealthy Romans failed to reinvest much of their capital in productive enterprise but, on the contrary, used a large portion of it for luxuries—for show places, circuses, and worse. Similarly they had little understanding of, or concern for, the lower classes, except to keep them quiet. The fact that they lived off others far removed from them made them callous to the effects of ruthless exploitation of resources or to the importance of technological advance. They "cropped" mines in wasteful fashion; they "mined" land, especially destroying grassland and forests in semiarid regions; they worked labor to exhaustion, discouragement, and apathy; they actually prohibited the introduction of new techniques;

and they failed to maintain drainage and irrigation works where they were absolutely essential.[22]

After the second century A.D. Rome did not conquer any new and wealthy areas which could be despoiled of accumulated riches. In the absence of booty and with increased governmental expenses the state had to rely more exclusively than formerly upon the proceeds of current taxation both at home and abroad. Tax rates were increased; tax rolls were periodically revised and tightened; and new taxes, even on land in Italy, were introduced under Diocletian (284–305). With higher taxes and no greater, and in some cases declining, production, the tax burden weighed heavily on people. A point was ultimately reached, as it usually is, where the consequences of tax evasion seemed less onerous than taxpaying. One of the groups particularly hard-hit by heavy taxes was that of the large landholders. Their assets were visible, which made tax evasion difficult; and furthermore, if they were members of municipal senates, as they often were, they were personally liable for the collection of taxes. Consequently many owners of large estates gave up producing for the market and became more self-sufficient. Sometimes the estate owners formed small armies, established their own judicial system, and had their own prisons. In fact, they declared their independence—a policy which struck at the very vitals of the Roman economic system of exchanging goods.

By heavy taxation Rome hoped to get the wherewithal to cover its expenses, but when this method proved insufficient, the state turned toward inflation (end of second century). This expedient, once begun, was resorted to time and time again in the third century, the value of some coins reaching

2 per cent of par. One of the results of this policy was that it destroyed the monetary unity of the Empire, cities and provinces issuing their own coins. Rising prices, which accompanied devaluation, meant that state officials, soldiers, and all others on more or less fixed salaries saw their real wages go down. Businessmen in the cities cried out against uncertainties created by debasement, lamenting the fact that loans made in coins of a given metallic content were repaid in coins worth a fraction as much. Even the *coloni* suffered, for most of their surplus went to lords, and goods in commerce which they needed were priced beyond their reach.

The trend toward declining production, heavy taxation, debasement of the coinage, and growing economic independence of the large estates was accentuated by a disorganization of the social and political structure by which relations among men were regulated. As the amount of goods available to the ruling classes fell off, these classes became engaged in a struggle to maintain their former economic status. Thus knights, bureaucrats, and commanders of armies lined up in contest against senatorial and privileged urban groups, while the lower classes threw their lot in with the side which would give them the most or merely resorted to grabbing what was nearest at hand. Clear evidence of this conflict was apparent when the Praetorian Guard put the emperorship up at auction. Shortly afterward, the Emperor Septimius Severus (193–211), a former commander of armies in the provinces, frankly adopted a policy of taxing or confiscating the property of senators and urban businessmen in order to pay his soldiers enough to secure their support.

Not long after his death an effort was made to effect

peace between the Senate and the armies, but emperors did not have sufficient authority to achieve their ends. From 235 to 285 only two of twenty-six emperors died a natural death, and at one time there were thirty claimants to the throne. Nor did the central government have enough power to maintain civil peace. Poverty-stricken landholders, bankrupt businessmen, urban workers, *coloni*, slaves, and riotous elements from the armies pillaged far and wide in Gaul, Sicily, Italy, North Africa, and Asia Minor. In 235, a band of brigands made a sweep through Italy. In 238, civil war broke out in North Africa. In 268, *coloni* in Gaul attacked several cities. And in 269, slaves in Sicily revolted.

Taking advantage of this turmoil, barbarians beyond the frontiers of the Empire broke through the cordons drawn against them. Goths went down the Danube and raided Asia Minor. Alamanni drove across the Danube. Franks broke through the Rhine barrier and pillaged as far as Spain (257 and 275). Saxon pirates preyed on the English Channel. Berbers plundered in North Africa. And Sassanians from Iran pushed into the Mesopotamian Valley and in 259 destroyed a Roman army and its emperor commander.

Wars, plagues, poor food supply, and a growing reluctance to raise children resulted in a decline in the population—a reduction estimated to have been for the whole Empire about one-third of the total population during the third century, and probably a larger proportion in Italy. The reduction in population led to enlisting barbarians in the armies, even for higher commands in the early fourth century, to new waves of invaders, and to the settlement of the newcomers within the Empire.

Efforts to bring order out of chaos were not wanting, but they were of little avail until Diocletian came to the throne (284–305). He had been commander in the East and had emerged the victor from a struggle between rival armies. He realized the absolute necessity of having a strong military force to support him and to this end united all Roman armies under his personal command, as Augustus had done much earlier. At the same time he increased the mobility and striking power of the military by using heavily armed cavalry and archers. [23] To enhance his prestige, he added to the practice of emperor worship and introduced a considerable amount of regal and courtly falderal from the East. He created a despotism in which his voice was supreme and that of the Senate a mere murmur. In order to minimize difficulties attending accession to the throne, he associated with himself a second and inferior emperor and two vice-emperors. For the purpose of strengthening his position, he abolished all semblance of local autonomy, and he persecuted dissident groups, particularly Christians. To increase production, he attempted to establish a "controlled economy" by regulating the collegia, supervising the large estates, and fixing prices.[24] Finally he revised the tax system in order to get more revenue, and he reformed the coinage in an effort to provide a stable currency.

With all these changes, however, Diocletian failed to reconcile rival groups of the population or notably to arrest the forces at work destroying Rome's economy. His system of two emperors and two vice-emperors merely created rivals to the throne. His economic controls proved unpopular and unworkable. In short, his entire program did little to stem

Rome's disintegration. All this became painfully apparent during the reign of Constantine (306–337) who succeeded him. This Emperor, although he was converted to Christianity and removed one sore spot in Rome's ailing anatomy by stopping persecution of his coreligionaries, was unsuccessful in realizing Diocletian's concepts for the restoration of Rome. The military returned again to the status of rival cliques; the frontiers were guarded by a few peasant soldiers; and the capital was removed from Rome to Byzantium, renamed Constantinople, which was a preliminary step to the division of the Empire into West and East.

In the last half of the fourth and during the fifth century the forces which were gnawing out the vitals of the Roman Empire took their most ferocious bites. The growing economic independence of local units, the breakdown of trade, and internal disorders continued their devastating work. Perhaps, however, the most voracious of all were the "barbarians." On all the frontiers pressures were increasing which the Romans could not control. Persians invaded Mesopotamia and Armenia; the Berbers ravaged North Africa; and the West Goths, pressed at their rear by Huns from Central Asia, were allowed to cross the Danube and settle in Roman territory. But almost immediately they turned on the Emperor of the East and by defeating him in the crucial Battle of Adrianople (378) gave the signal for a general rush of barbarians into the Empire. Some of these people infiltrated, finding employment as soldiers and even commanders of Roman armies. Leaders of this type not only learned military techniques from the Romans but had the advantage of numerous fifth columnists. Thus the so-called "barbarian invasions" were not just stampedes of wild human hordes who tore the Romans to bits. They were the coming of people who were familiar with Roman methods of warfare, were led by commanders often with Roman experience, and were equipped and supported by the products of industry established on Roman patterns. Thus Rome's diffusion of her techniques worked against her in the final showdown. To say that Rome fell is somewhat misleading. She wasted away.

The onslaught of the barbarians upon the Empire had a dramatic climax in spite of the fact that Rome was only a rotten shell. The West Goths invaded Italy, sacked Rome in 410, and finally wound up in Gaul and Spain. Vandals from eastern Germany plunged through Gaul and Spain to North Africa, whence they despoiled Rome in 455. The Huns preyed on the Eastern Empire, raided Gaul and Italy (451–452), and finally disappeared in Eastern Europe. The East Goths invaded Italy (489–493), established their chief as emperor, and were absorbed in the population. Jutes, Angles, and Saxons from northeastern Germany invaded Britain; Burgundians came into southeastern Gaul; and Franks moved into Gaul.

In the Eastern Empire there was also an infiltration of barbarians, but there the newcomers, when they attained the emperorship, were able to maintain their authority, the dignity of their office, and their frontiers. It was, indeed, the Eastern Empire, which lasted until 1453, that kept alive much of Eastern Greco-Roman culture during the Middle Ages. One of the early Eastern emperors, Justinian (527 to 565) succeeded in reuniting Italy to the Empire, but after his death invading Lombards from northern Germany undid his work, and Italy became a series of independent states. In this condition the Mediterranean world moved into the Middle Ages.

As the political and economic structure of the Roman Empire fell apart from the third to the sixth century A.D., Roman art and learning went to pieces with it. Literary men devoted their energies to copying, compiling, and commenting upon the works of classical authors. Philosophical writing was reduced to weaving aphorisms and moral maxims for schoolboys. Historical composition degenerated to the simplification of earlier writers and to anecdotal biographies. Scientific treatises were collections of fables and the reiteration of earlier second-rate writings. Legal scholarship declined sharply after the absolutism of Diocletian, and the law itself lost its vitality. The Justinian Code, prepared in the sixth century, summed up Roman legal developments and was a sort of last testament of Roman legal accomplishment. Painting, sculpture, and architecture became less original and more crude, artists failing to create new and vigorous forms of expression. Learning became more stereotyped, being restricted to the seven "liberal arts"—grammar, rhetoric, and dialectic (the trivium) and arithmetic, music, geometry, and astronomy (the quadrivium).

What creative activity there was in the three centuries being considered came more and more to be associated with Christianity. This religion, which had many of the characteristics of other Eastern cults, preached that one should treat one's neighbor as oneself and that salvation after death was the reward of a good life. Such doctrines had a particular appeal, especially to the lower classes, in a politically, economically, and socially disorganized society. Christianity, unlike any other cult, except its parent Judaism, taught that there was only one God and that its members should not bow down before Roman gods. Such an attitude was difficult for Romans to condone, because in spite of their traditional tolerance in religious matters, they believed that their welfare depended upon the extent to which they pleased their own deities. Yet Christianity spread in spite of persecution because of the relevance of its teaching to existing conditions, the ardor of its missionaries, and the diffusion of its doctrines through the dispersion (diaspora) of the Jews.

As Christianity grew in extent and popularity throughout the Empire, it developed a church organization for the government of its affairs. Before the weakening authority of the secular state, the Church assumed a position at least equal to that of the state, for it regarded all human institutions to have been created by God and hence limited by His will and His vicars on earth. Church and state were to share in governing men's lives, but the separation of jurisdiction was not made precise—a situation which gave rise to a lasting conflict in Western culture.

Christianity was, however, more than a political force— it came to dominate men's lives and to furnish inspiration for all their creative activity. In the realm of philosophy, the Church Fathers endeavored to find answers to the fundamental questions raised by classical philosophers. Their answers were as complete as they were simple. God was the ultimate reality; He revealed *truth;* and *His* will explained human history. These precepts necessitated proof of the existence of God, reconciliation of them with observed phenomena, and defense against heretics and persecutors. To these ends such early Church scholars as Origen (185–254), whose efforts centered on exegesis of the Scriptures and the preparation of a systematic Christian theology; Jerome (340–420), whose translation of the Bible, known as the Vulgate, became the accepted version; and Augustine (354–430), whose *The City of God* is generally considered as the most important Christian book after the Bible, worked with skill.

The most important product of Christian literature was the *New Testament*, written mostly in the second half of the first century A.D. But there were also sermons, Christian epics, romantic tales about saints, liturgy, and hymns which had merit. Christianity also provided inspiration for creative activity in the arts, for here was a medium of expression for the propagation of the faith and the glorification of God. Inasmuch as early Christians were from the lower classes, the first Christian art was restricted largely to craftsmanship. With the conversion of wealthier groups after Constantine's adherence to the faith, Christian influence was found in mosaics, which were replacing painting, in sculpture, and in architecture. Many of the products of this inspiration lacked technical perfection and beauty, for these were secondary considerations for Christians. Yet Saint Sophia in Constantinople, constructed by the Emperor Justinian in the sixth century, was one of the masterpieces of Greek Orthodox Christian architecture, and several churches built at Ravenna in the fifth and sixth centuries and adorned with mosaics had charm.

The most lasting and most important influence of Christianity was its ethical and social precepts. It preached the sacredness of human life to a world that had come to have scant regard for individuals of any class. It taught moral responsibility and personal freedom in an age which was woefully in need of both. It proclaimed the virtues of the family and the saintly ideal of abnegation. Although in practice many compromises had to be made between its doctrines and the real world, Christianity provided standards of human conduct which were of a high social order. Hence Roman civilization in its decline produced a religion which was to play an important role in Western civilization for centuries to come.

For Further Reading

Cyril Bailey (ed.), *The Legacy of Rome* (1924)

J. Carcopino, *Daily Life in Ancient Rome at the Height of the Empire* (1940)

Max Cary, *A History of Rome down to the Reign of Constantine* (1935)

J. Wight Duff, *A Literary History of Rome* (1936)

H. F. Jolowicz, *Historical Introduction to the Study of Roman Law* (1932)

M. Rostovtzeff, *A History of the Ancient World,* vol. II (1928)

————, *The Social and Economic History of the Roman Empire* (1926)

E. Strong, *Art in Ancient Rome* (1928), 2 vols.

R. Syme, *The Roman Revolution* (1939)

Tenney Frank, *An Economic History of Rome* (1927)

———— (ed.), *An Economic Survey of Ancient Rome* (1933–1940), 5 vols.

B. L. Ullman, *Ancient Writing and Its Influence* (1932)

E. L. White, *Why Rome Fell* (1927)

VI

WESTERN CULTURE

━━━━━━

WESTERN culture, so-called because it developed primarily in Western Europe, has been the most illustrious successor to Hellenic and Roman cultures and, according to the yardstick of civilization being employed here, has had one of the greatest civilizations of all time. It has achieved the most extensive and intensive control over physical environment that the world has ever seen. It has attained the greatest supply of goods and services in proportion to population for a longer period of time than any other culture. It has reduced to the lowest minimum the scourges of disease and other natural phenomena. It has produced greater social equality and absence of arbitrary government than any of the cultures treated heretofore in this study. It has created many of the world's most universally and generally recognized masterpieces in architecture, sculpture, painting, literature, and music. And it has practiced one of the highest systems of intellectual activity, that of science, or the ordering in a sequential fashion of knowledge about the physical and social world.

Its greatest weakness, and weaknesses it does have, has been its inability to curtail strife, particularly among its various political segments. It has not been consolidated politically in a way to prevent internal warfare, a fact which might conceivably be its undoing. What a commentary it will be upon its economic success, its aesthetic accomplishments, and its science, if its powers are dissipated in the destruction of what it has created!

◁ *A synopsis*

Of all the attainments of Western culture, none is more remarkable or more fundamental to an understanding of other achievements than the economic. In no other society has such a large population enjoyed equivalent well-being. The basic explanation of its material superiority is to be found in advances in technology, greater utilization of natural resources, a more extensive division of labor, greater investment in enterprise, a more efficient organization of economic activity, and an accepted view that material betterment is highly desirable. Among technological changes, the use of the machine driven by nonhuman energy stands out as the most significant. Indeed, man's control over physical things has been revolutionized and multiplied by the machine—a mechanical contrivance capable of producing goods many times more abundantly per unit of human input than any other method. In time the machine came to be made of, and to be driven by, inorganic materials. Thus the stored-up wealth of nature began to be tapped, and such stores provided more material and energy than were obtainable solely from organic sources. Yet, as this development took place, the supply of organic products, particularly foodstuffs, was increased, which permitted the maintenance of larger populations.

Western Europe and the Western Hemisphere, the second great center of Western culture, were richly endowed by nature with those resources, especially fertile land, coal, and iron, which were to play such an important role in economic activity. Improved methods of transportation, however, made it possible for Western culture to draw upon the natural resources of the entire inhabited world. These means of transportation account for Western culture's having removed India, China, and Russia-Siberia from their isolation and for having obtained from these and other areas both techniques and goods. Finally, improved methods of communication, particularly movable type and power-driven printing presses, allowed Western Europe easily to exchange new ideas.

Then, too, Western culture has had a remarkable organization of economic activity. A highly developed money economy has made possible an extreme division of labor. In fact, those districts within Western society which have the highest output of goods and services per capita are precisely those where the division of labor and material resources per operative are the highest. Furthermore, institutions, including the crucial one of interest taking, have grown up to effect a concentration of capital which has made possible the development of large enterprises and the exploitation of ever greater resources. There has also been a distribution of wealth such that, compared with other cultures, the masses have greatly improved their lot. Indeed they have had such a large share of purchasing power that investment in the pro-

duction of goods has been encouraged. The wage system—
the dependence of workers on a wage for obtaining the necessities of life—has made men work; and the profit system—
a desire of investors for gain—has stimulated efficiency in
the organization of the economy for the production of goods
and services. In general, Western culture has enjoyed great
economic expansion in spite of ups and downs; and economies
always function best where there is growth.

These economic characteristics of Western culture were
not always thus. They developed to the point described
through many centuries and many vicissitudes. Even a general account of this history is difficult to grasp, and consequently a brief statement of its major aspects will assist the
reader in understanding what is to follow. For this synopsis,
we must return to the end of the Roman Empire when Western
European economy was decidedly retrogressive.

At the beginning of the Middle Ages there was comparatively little trade between distant places; industrial production had fallen drastically; and agricultural output was primarily for local consumption. To make matters worse, the
Mohammedan conquest of a large part of the Mediterranean
at the beginning of the eighth century and the antagonism of
Christian Europe to this rival religion brought trade between
the Mediterranean and Western Europe almost to a stand-
still.

During these centuries, however, Venice and certain of the
southern Italian towns maintained commercial relations with
Constantinople and the Eastern Mediterranean and did not
suffer the same economic eclipse as their neighbors to the
north. Then, by the eleventh century, trade in the Mediter-

ranean revived, and this brought a new prosperity to the north Italian cities and to the French and Spanish towns of the Mediterranean coast. In the meantime, that is, from the ninth to the eleventh century, commerce between distant places developed in the North Sea and Baltic area. By the latter date the two main trading districts, that in the north and that in the Mediterranean, had made contact and were exchanging their various products. In both these centers and along the routes between them, the use of money became more extensive; towns grew and became more active; and industry for distant markets took a new lease on life. These changes brought with them greater economic surplus, which was used both for investment and for aesthetic and intellectual activity. There was a continued development of economic life, and in the twelfth century there was a revival of learning and the arts known as the "twelfth-century renaissance."

The course of economic progress appears to have been broken with the opening of the fourteenth century. Famines, the Black Death, failures among Italian businessmen, the Hundred Years' War from the middle of the fourteenth century to the middle of the fifteenth century, and social disturbances raised havoc with economic activity. Far-reaching as these disasters were, however, they failed to push Western Europe into another dark age. They retarded but did not destroy the new economy. Recovery was certain by the latter half of the fifteenth century with the Low Countries, France, and Italy leading the way.

At the beginning of modern times, *circa* 1500, the discovery of routes to India and the New World proved an enormous boon to Portugal and Spain, to England, and to the

Low Countries. But the countries in the Iberian peninsula soon lost their advantages, and their economies, based on a monopoly of trade with the New World and the importation of precious metals, foundered in the seventeenth and eighteenth centuries. Portugal and Spain were accompanied in their decline by Italy, which was situated off the new trade routes, and by the Germanies, which were devastated by civil strife during the Thirty Years' War (1618–1648). The Northern Netherlands, France, England, and Austria were the centers of greatest prosperity in the seventeenth century, and to them may be added the Germanies in the eighteenth century. Precisely in these areas the greatest artistic and scientific advances were made in these two centuries.

Toward the close of the eighteenth century, the accumulation of knowledge about mechanics and metallurgy had reached a point where it was possible to use machines driven by nonhuman energy to produce goods on a large scale. Similarly in the early nineteenth century knowledge was successfully applied to providing mechanical means for transporting these goods cheaply over vast distances. In a short space of time, it was as though each operative had acquired a thousand slaves to do his bidding, but these were mechanical slaves which could be renewed and augmented almost at will.

The chief advantages of the mechanization of industry and transportation fell initially to England, but Belgium and France soon benefited also. In the last quarter of the nineteenth century, Germany industrialized rapidly, and so too, but to a lesser extent, did Italy, Austria, and Scandinavia. Thus Western Europe became the "workshop of the world," obtaining raw materials and foodstuffs from overseas in exchange for manufactured products and services. From this activity, Western Europe acquired an economic prosperity that probably had never before been equaled. The large supply of goods and services available per capita in the countries involved provided a surplus that could sustain a larger population and a more numerous body of scientists, scholars, and artists. Whether or not their work will stand the test of time so that the nineteenth and twentieth centuries will appear to the man of the future as a period of great cultural achievement is difficult to foretell. Perhaps impressionistic painting, the music of Verdi and Wagner, and some architecture will be universally and consistently regarded as of a high order. Of one thing, a degree of certainty can be expressed, that scientific accomplishments have led to a knowledge never before possessed by man and to a control over physical environment never before equaled. Its accomplishments may mark this era as a high point in Western culture.

Whatever the decision of the future may be in this regard, it is clear that the position of Western Europe as the center of Western civilization has weakened. Toward the close of the nineteenth century and more definitely as the twentieth century progressed, certain trends began to indicate that Western Europe was losing its economic primacy. Technology became more mobile, spreading rapidly from its place of origin to new areas. Similarly capital for investment became more flexible and sought opportunities for profit on a growing scale outside Western Europe. Thus other areas, notably the Western Hemisphere and particularly the United States, the Far East, Africa, and Russia, began to mechanize their industry and to adopt improved methods of agriculture.

Some places with extraordinarily rich natural resources, proximity to large markets, or an efficient labor supply soon learned that they had advantages over Western Europe which they could exploit to their own benefit. Furthermore, antiquated productive equipment, social tensions, political rivalries, and devastating wars hindered progress in the older industrialized regions of Europe.

The fact that Western Europe was not keeping pace economically with the rest of the world was at first not disturbing, for the change was gradual, and Western Europe was receiving large "tribute" in the form of interest on investment from the rest of the world. By the end of World War II, however, after Western Europe had been deprived of so many of its overseas investments to fight the war and had seen such a large part of its property destroyed, its predicament was clear. A large part of the world was taking over the economic ways of the West and in the process was adopting much of Western culture. Economic preeminence was passing to regions outside of Europe, particularly to the English-speaking area of North America. As yet these areas have not had time to demonstrate convincingly that they will lead Western culture to new heights. Experience of the past, as recorded in this book, seems to suggest a strong probability that such will eventually come to pass.

The Middle Ages

An analogy has frequently been drawn between the plight of Rome in the fourth and fifth centuries A.D. and that of Western Europe in recent decades. In the two situations there are, indeed, elements of similarity. Both regions experienced a relative decline in productivity and a decline in the rate of population growth. Both suffered from a reduction in the receipt of goods or payments from foreign lands. Both were wracked by war, and both saw the center of economic primacy move to other places, Rome, to Constantinople, and Western Europe, to the Western Hemisphere.

Here the obvious similarities cease, for Rome's predicament was much worse than anything Western Europe has so far experienced. The tendency in Rome's case was for the new economy to become overwhelmingly agricultural, with production conducted to meet local needs. Labor became bound to the soil, and, in most parts of the old Western Roman Empire, it was also bound to overlords by pledges of fealty. Industry and trade on a large scale and for a large market also declined, the depth of commercial activity being reached in the ninth century.[1]

Whether the presence of Mohammedans of Arabic culture in the Western Mediterranean or the local self-sufficiency of Western Europe accounts primarily for this condition need not detain us. The fact is that the use of money, the division of labor, and the organization of business for profit, all declined. Land was almost the sole object of economic activity and condition of wealth, while the relatively small feudal estate was the basis for the only effective political organization of the time. From such an economy and from such political units came little economic surplus which could be devoted either to learning, to the arts, or to economic progress.

Western Europe's emergence from this economic morass of the Dark Ages, which may be considered to have extended from the fifth to the tenth century, was slow indeed. But from the ninth to the eleventh century signs of recovery were visible in two widely separated areas, in Italy and its neighboring waters and in the North Sea and Baltic region. In both districts economic revival was clearly initiated by trade in which piracy played an important role.

Certain Italian towns, notably Venice, Naples, and Bari, showed the first evidence of change. Arabs had never succeeded in dominating either the Aegean or Adriatic Seas and did not cut trade between Italy and Constantinople, nor did all Italian merchants allow their religious scruples to prevent them from making a profit from trade with Arabian infidels. By the end of the ninth century, commercial exchanges were being made between Italy and North Africa or Syria, while enterprising Venetians found profit and perhaps pleasure in capturing female Slavs, hence the word slave, along the Dalmatian coast for Arabian harems. By the end of the eleventh century, Venice had established trading posts with special privileges in all those commercial centers of Asia and Europe which were under the jurisdiction of the rulers of Constantinople and had almost a monopoly of the carrying trade among them. She also expanded her commercial activities to the mainland, even to the Alps, for the city was completely dependent upon imports for her food and raw-material supplies. What was of extreme importance was that she made money profits from her Eastern trade and paid for at least part of her mainland supplies in money. Thus Venice as carrier and trader gave new life to the institution of money, reestablished an economy organized for profit, and renewed the use in business of writing and numerical notation.

At approximately the same time that these changes were

taking place in northern Italy, commerce, as has already been stated, was beginning in the North Sea and Baltic region. The Northmen of the Scandinavian peninsula took to the sea in order to obtain a livelihood, for their land was mountainous and poor. They became excellent seamen and explored far and wide, even reaching the Western Hemisphere in the eleventh century. For the most part, they operated as pirates rather than as settlers, but from piracy they passed by the ninth century to legitimate trade. Danes and Norwegians operated chiefly in France, the Low Countries, and the British Isles, while Swedes turned their attention more particularly to Russia. The latter established trading posts (*gorods*) and opened up routes which reached Constantinople. Over these arteries of commerce, they sent women to Arab harems, along with amber, furs, and honey. In return they obtained spices, wines, silks, jewelry, and other luxuries. They also used money, if one may judge from Arab and Byzantine coins found in the Baltic area, and they brought back certain Byzantine cultural influences, Russia having obtained Christianity and architectural forms from the East.

Although the future of Baltic–Black Sea commerce was dark because of the arduous land journey and the difficulties of protecting shipments, other opportunities were open to the North Sea–Baltic trading area, especially commerce with the Mediterranean. Contacts between the two districts were certainly made by the eleventh century and grew thereafter. By that time, however, events were taking place which led to further development in the two main commercial regions. In the northern center, trade with nearby lands became more intense, and the people began to specialize in manufacturing goods for their neighbors. Flanders, for example, turned to the production of woolen cloths. In the tenth century, this region was supplying Northmen with textiles for their trade and soon began to import wool from England to supplement its own supply. From the beginning of the twelfth century, Flemish cloth was being sold in Novgorod, and Italian markets were obtaining it in exchange for silks and spices. Production of cloth became a *city* industry and was the foundation for the prosperity and growth in Flanders of Ghent, Ypres, Bruges, and in northern France of Lille, Douai, and Arras. Strangely enough, however, the Flemings themselves concentrated on production rather than on trade, leaving the latter activity to Scandinavians, Italians, and members of the hanse—merchants of a group of North Sea and Baltic cities.

In the Mediterranean perhaps even more important economic progress was being registered. Now in addition to Venice and southern Italian cities, certain ports, notably Genoa and Pisa, became interested in extending their commercial activity. These cities of the west coast were, however, hemmed in by Mohammedan Arabs and resorted to war to break the barriers against them. Inspired by a combination of commercial and religious considerations—both strong incentives—the Genoese and Pisans made headway, getting a foothold in Sicily and then in 1087 capturing an important Mohammedan town in North Africa which rendered them rich booty. Overjoyed at their good fortune, the Pisans devoted some of their plunder to the construction of their famous cathedral which stands as a symbol of their success against Islam.

Shortly after this event, Mediterranean trade was further stimulated by the Crusades—Christian wars against the Mohammedans for control of the Holy Places. Beginning in 1096 and continuing for at least two centuries, the Crusaders pushed Islam back. Furthermore, one of the Crusades was successfully directed against the Greek Orthodox capital at Constantinople, a move which weakened Constantinople's control of the Eastern Mediterranean. Nor did the Mohammedan Turks, when they extended their sway in this region, interfere seriously with Christian trade, for they reaped much benefit from it. Thus commerce in the Mediterranean was greatly extended, and it contributed to the economic growth of cities in Italy, southern France, and eastern Spain. To be sure, Venice, Genoa, and others were keen business rivals and at times fought for special advantages, but such strife did not invalidate the fact that in the Mediterranean Sea trade over a large area and an extended money economy were taking shape.

As in the North Sea and Baltic regions, so in the Western Mediterranean greater commercial exchange led to manufacturing for new markets, and both trade and industry stimulated the division of labor. The Italian cities, which had served as suppliers and carriers for all the Crusades except the first, saw a fillip given to ship construction, to the production of weapons and tools, and to the processing of foodstuffs. Furthermore, when Crusaders from beyond the Alps returned home, they demanded Eastern luxuries for which they had developed a taste while at the front. To supply the demand thus created, Italians began to produce some of these luxuries at home, as for example silk cloth at Lucca and jewelry at Venice. When these goods were shipped northward, they were paid for in metallurgical wares and in wool cloth, which was dyed and finished by special processes in Italy, especially at Florence. In such ways, commercial ties between the North Sea–Baltic regions and the Mediterranean were knit more closely, and industry in both places was developed anew.

As in antiquity so now, most commerce moved by water, for it was the cheapest means of transportation for bulky goods. Gradually larger ships were built, even up to 600 tons' burden, and sailing was improved in the thirteenth century by improvements in the rudder that permitted more successful tacking into the wind. By the beginning of the fourteenth century, fleets from Venice and Genoa made annual voyages through the Strait of Gibraltar to the markets of Bruges and London. Traffic on inland waterways also increased, and by the twelfth century canals were being constructed, particularly in the Low Countries, to connect rivers.[2] Land transportation was advanced by the opening of new routes, the construction of bridges, and in the tenth century by the construction of a horse collar that would allow the animal to pull without having his wind shut off.

In spite of all these changes, the volume or ease of medieval trade should not be exaggerated. Land transportation was slow, roads were poor, and tolls were numerous. Water transportation also had its limitations, and probably the entire fleets of either Venice or Genoa hardly exceeded the tonnage of the largest twentieth-century liner. Yet the growth of trade was sufficient to lead to economic change in the twelfth and thirteenth centuries—change which in the depth of its impact may be compared with the industrial revolution of the nineteenth century.

Of all the new economic developments none was more basic than a more widespread and intensive use of money. In this respect, Western European development was running true to historical form, for in all our previous investigations the appearance of a convenient medium of exchange always accompanied increased commercial activity. In this case, a coinage system was created by Charlemagne in the early ninth century which was widely used and which exists at the present time wherever the pound sterling is employed. He established the pound (*livre, libra*), which was divided into 20 shillings (*sous, soldi*), that in turn contained 12 pence (*deniers, denari*) each. For the most part, the coinage of money was as usual a prerogative of a political authority and at that time was in the hands of princes who had traditionally provided what money was needed on their estates. This was perhaps an unfortunate situation, for princes debased their money when they needed funds to pay off their debts irrespective of the consequences their acts would have on others. By the end of the twelfth century and the beginning of the thirteenth, there was such anarchy in money that trading cities like Venice initiated reforms to provide a more stable medium of exchange. Yet no scheme has ever been devised to prevent debasement by the issuer, and the fluctuations in the value of money were to be a perennial problem.

The great number of princes and autonomous cities meant that there were many different types of coins. Hence the ubiquitous money-changer came again upon the scene. He found his business profitable, although somewhat disreputable. Nevertheless, he was soon receiving sums for safekeeping, was effecting transfers of funds between debtors and creditors, and was making loans. Merchants, too, engaged in these banking practices, and in Italy great banking families, like the Bardi and Peruzzi, became established. As had happened in both Rome and Greece, the bill of exchange came into use to facilitate the settlement of business accounts; the letter of credit put in an appearance; and bottomry loans to reduce risks to shippers began to be issued. Most important of all perhaps, for it provided a material incentive for the accumulation and investment of money, was interest taking.

Now the Church was opposed to having money lent at interest, for, prior to the commercial revolution which we are describing, loans were usually sought when famine was at hand or by lords for extravagances beyond their means. This meant that lenders usually exacted high rates from borrowers. All interest taking was dubbed usury, that is, a charge for using a loan, but the term was coined when rates were actually usurious. As investments began to be made to render useful economic services, ways were found to circumvent the Church's antipathy toward interest taking. Money was loaned with the expectation of profit, and the long and slow process of overcoming the Church's compunctions got under way.

With the growth of commerce, new types of business organizations were created or old ones reincarnated. Among them were the fairs where merchants from distant points came together to buy and sell. They were opened to all—a freedom that was exceedingly important in facilitating exchange. These institutions began to appear in the eleventh century, and their number increased during the twelfth and thirteenth centuries. The most famous of the fairs and the only ones to attract merchants from all Europe in the twelfth and thirteenth centuries were those of Champagne in France, which lasted the year around. But other places had similar meetings which varied in size with the amount of trade carried on— Stourbridge, near Cambridge, England, Bruges, Ypres, and Lille in Flanders, and then Lyons and Geneva, which in the fourteenth century outstripped Champagne.

In addition to fairs, there were the merchant guilds, which came into being in the eleventh and twelfth centuries. They were, unlike the fairs, restrictive in character, and were designed by their members, the merchants of a given town, to create a local monopoly and to protect their general interests. As a rule, only guildsmen could buy, sell, or manufacture goods in a town, and foreign traders could do business only with them. Guilds sought privileges and rights from local princes, negotiated with guilds in other towns, encroached upon the powers of local government, and established rules aimed at preventing sharp practices. Above the guilds there were the hanse, associations of guilds or merchants of various towns, or leagues of towns, which began to be formed in the twelfth century for facilitating trade. The most famous of these groups was the Hanseatic League composed of cities along the Rhine, the North Sea, and the Baltic, but there were also the hanse of northern France and Flanders, the Merchant Staplers of England for the export of wool, and others. Below the guilds there were various combinations of interests for the conduct of business, especially partnerships, multiple partnerships, and *commenda*, in which there were sleeping partners, those who invested their funds but who took no active part in management.

At almost the same time that expansion of trade was leading to the creation of new forms of business organizations, the growth of industry was having a similar effect in the realm of manufacturing. Craft guilds began to be established, which, like merchant guilds, aimed to secure monopoly for the production of certain articles, to obtain a degree of political authority, and to lay down regulations for the conduct of their business. These bodies indicate a considerable division of labor, for there were guilds for weavers, dyers, fullers, butchers, and bakers—the number of specialists increasing as time went on. If the members of a guild produced for a relatively large market, especially an export market, the master was a businessman rather than a mere artisan. He often used borrowed capital, employed men as wage earners, sold his wares in an impersonal market, and sought to make a profit. The fact that in the middle of the fourteenth century the town of Ghent in present-day Belgium had 4,000 weavers out of a total population of 50,000 indicates that a considerable degree of specialization in production had been realized and that goods were being produced for more than local consumption.

Those who produced for large markets were interested in technological improvements. Thus we find them in the twelfth century using water power for the first time for other purposes than making flour. They employed it for driving fulling mills and grinding bark for tanning leather. In the thirteenth century, they used it for sawing lumber; and in the fourteenth, for turning grindstones in metalwork. Furthermore, the twelfth century saw the adoption of wind-mills for grinding grain, the making of paper in France and Spain, the installation of improved looms, and the reeling of silk by improved machines. In the thirteenth century the wheel rather than the distaff was used in spinning; and gun-powder began to be employed—an explosive which was to revolutionize the art of war, to increase the vulnerability of fortified castles, and thereby to weaken the position of lords. Also the clock escapement, a contrivance to control the action of wheelworks, was conceived and applied successfully in the following century, while general clockmaking provided the foundation for precision work which was so essential to the development of intricate industrial machines.

With the employment of borrowed capital, hiring of labor for wages, production for an unknown, uncertain, and impersonal market, and the desire for profit, the essentials of the capitalist system came into being. Furthermore, mechanical advances and more extensive use of inorganic power were hints, at least, of the direction in which Western European industry was to move. To a very large extent, however, agriculture was immune to changes in commerce and industry. Its technology remained largely unchanged, and its production was for the most part geared to local needs. To be sure, the increase in population which took place in the twelfth century gave rise to a search for new lands that could be brought into cultivation, as, for example, reclamation of swampy areas in Flanders, the Netherlands, and Germany by means of dikes; and it led also to the colonization by Germans of districts to their east and northeast. Furthermore, there was some specialization in the growing of crops, as in the case of silk in Italy, wool in England, and wine in the Bordeaux area.

The chief effects of the new economic developments in rural areas were the more extensive use of money and the beginnings of the breakdown of feudalism. As wealthy businessmen began to indulge in luxuries of one kind and another, nobles realized that they had to have even more extravagant things if they were to retain their social position. The deliveries of goods and services by serfs to lords did not suffice to provide luxuries, for they could not easily be turned into money. When in the twelfth century the prices of luxuries rose abruptly, lords were hard pressed for funds and in their extremity began to free serfs from certain obligations by means of cash settlements or by substituting money payments for dues in goods and services. Furthermore serfs might seek alternative employment in cities and could obtain their legal freedom by residence there for a year plus one day. Thus an inroad was made in the institution of serfdom in the twelfth and thirteenth centuries, although remnants of it were not completely abolished until the nineteenth century. As so many times in earlier instances, the growth of commerce and of a money economy led directly to greater legal liberty for the individual.

Another important concomitant of economic development from the eleventh to the thirteenth century was the growth of cities. By 1400, Milan, Florence, Venice, and Paris had populations of 100,000 or more, while London had 50,000 inhabitants and many places had 20,000. In general, the newly built parts of cities were close upon an older fortified place, or burg, and were hence called suburbs, or *Neubürge,* or *faubourgs.* The trading citizens of these new districts were known as bourgeois, as distinguished from the leading citizens, or nobles of the burgs. In most places the newly rich sought political power and often obtained it by force. They sponsored legal systems based on evidence, equity, and business considerations; they opposed arbitrary forms of feudal justice and noncommercial phases of Church (canon) law; and they founded schools so that their sons might acquire elements of education necessary for the conduct of business. In cases where they got political power, they levied taxes for city fortifications and other improvements, thus restoring the public character of taxation; they lent money to city governments; and they sometimes took over the collection of city revenues in return for loans, a practice that gave rise to at least one famous bank, the Bank of St. George in Genoa.

Such evidence as exists shows clearly a considerable economic development in Western Europe from at least the eleventh to the thirteenth century. It was, moreover, a kind of development which permitted the accumulation of surplus that could be used to support cultural and intellectual activity. It was, indeed, accompanied by a definite revival in the arts and learning, the so-called twelfth-century renaissance, as we shall see in more detail later.

Economic advance did not, however, go forward entirely unimpeded. Bad harvests and wars periodically resulted in arresting commerce and industry, and then in the fourteenth and early fifteenth centuries a concatenation of events led to a general economic setback. Perhaps Western Europe had reached the limits of its technology, resources, and business institutions, but in any case economic activity appeared to have reached a plateau by the first decades after 1300.

Then a series of misfortunes began. From 1315 to 1317 a great famine ravaged all Europe—the worst European famine to that date of which we have any record. This catastrophe was followed by the Black Death, which from 1347 to 1350 is believed to have carried off a third of the population of Western Europe, to say nothing of those who perished before and after those years. The calamitous character of these events was reflected in the bankruptcies of several of the leading bankers of Italy, the beginning of the decline of the Champagne fairs, civil strife in Italy, political anarchy in the Germanies, and the Hundred Years' War (1337–1453) which ruined France and exhausted England. A shortage of labor led serfs to demand and get the abolition or commutation into money payments of more of their obligations to lords; craftsmen to seek a voice in government; and journeymen—the wage earners—to agitate against guild masters.

Disastrous as these events were, the destruction should not be exaggerated. Europe was not laid waste—it was only retarded in its development. In Italy, for example, commercial relations with the Levant were maintained, and Eastern goods continued to be sent to the north through Alpine passes or via Gibraltar, even though in reduced quantities. In the Low Countries, too, production never ceased altogether, and conditions were such that individuals could continue to amass fortunes. Moreover, recovery in France under stronger royal authorities was relatively rapid, and England not only survived but seems to have developed some industry while cut off from European suppliers. Spain and Portugal also weathered the storm, the latter undertaking explorations for new lands along the African coast and for a new route to India. Only the Germanies continued to suffer seriously because of political unrest, but even here Rhenish cities conducted a profitable commerce, and a printer at Strasbourg is credited with having invented movable type. In the later half of the fifteenth century recovery from the depths of stagnation was fairly general. Indeed this period can be considered as a prosperous one, a time in which new wealth was being accumulated that was to support the artistic and intellectual accomplishments of the Renaissance.

At the very end of the fifteenth century an all-water route to India and the discovery of the New World were to give a new fillip to Europe's recovery. They were to lead to a new expansion of commerce, to a further development of the use of money, and to greater industrial production. These events at the very end of the Middle Ages were to usher in a new era of economic progress—an era that historians designate as the beginning of modern times. This progress was to follow along the lines of economic institutions which had become clear from the eleventh century onward.

Economic development, 1500–1800

From time to time in history a number of factors have combined in such a way as to inaugurate far-reaching changes in the course of human activity. This was the case regarding the discovery of an all-water route to India and the Far East and the finding of the New World. Prior to these momentous events, Europeans had made important improvements in navigation, shipbuilding, and seamanship. They had acquired the compass, which was probably a Chinese invention, by the twelfth century and were using it generally in the fifteenth. They had procured from the Arabs the astrolabe, a fore-runner of the quadrant and sextant,[3] which permitted voy-agers to determine their positions relative to the sun and stars. They had better charts, maps, and astronomical tables. And many of them believed, along with Eratosthenes of ancient Alexandria and contrary to popular belief, that the earth was round and not flat. They had developed more sea-worthy ships, like the caravel, and some of their seamen were being trained for long voyages, particularly in a school established by Prince Henry the Navigator of Portugal.

Among other factors which undoubtedly contributed to the making of the great discoveries were a religious zeal cre-ated by the "crusades" of the Portuguese and Spaniards against the Moors and current stories about "lost Christian kingdoms," somewhere out in the unknown, which ought to be saved. Then there was a generally awakened interest in discovery, which was combined with love of adventure and sheer curiosity about what was beyond the seas. There was a desire for gain, whetted by the knowledge that Venice had made great profit from Eastern trade and that a new trade route to the East would secure a part of this rich melon.[4] Fi-nally in Western Europe individual feudal princes had amal-gamated enough territory under their authority at the ex-pense of other princes to provide both men and equipment for expensive explorations.

Motivated by such considerations, Europeans began early in the fifteenth century to venture forth into the Atlantic. The Madeira Islands were settled about 1420; possession of the Canaries was contested by Spain and Portugal be-

ginning in 1425; and the Azores were discovered or redis-
covered about 1430. In 1434 an expedition inspired by
Prince Henry the Navigator of Portugal rounded Cape
Bojador, and about ten years later another group rounded
Cape Verde. In 1488 the Portuguese Diaz sailed around the
Cape of Good Hope; and four years later the Genoese, Colum-
bus, sailing under Spanish colors, reached the New World.
Between 1497 and 1499 Vasco Da Gama, a Portuguese, made
the first all-water round trip to India. Thenceforth discoveries and explorations came thick and fast. John and Sebas-
tian Cabot, Genoese sailing for the English in 1498, explored
the North American coast from Delaware to Labrador;
Amerigo Vespucci, a native of Florence, explored the Brazil-
ian coast for Portugal (1501); Magellan circumnavigated
the earth under the Spanish flag (1519–1522); and subsequently followers of these men penetrated into the interiors
of the New World and staked out claims for their sovereigns.

The area opened to Europeans by the early discoverers
was enormous. It comprised the Persian Gulf, India, China,
the Malay Peninsula, the Spice Islands, the coast of Africa,
almost all the coast of the Americas, the interior of Mexico,
the West Indies, Madagascar, the Philippines, and some of
the west coast of South America. Here was an empire which
exceeded in size the wildest dreams of the most enthusiastic
expansionist.

More important than territorial extent were, however, the
economic potentialities of the overseas territories for the
economic development of Western Europe. Some of these potentialities were soon visible, for early experience indicated
that great profits could be made in trade with the newly dis-

covered places. Da Gama returned from his first voyage with a cargo on which was realized a profit exceeding by sixty times the entire cost of his expedition. Magellan's voyage, although only one of his five ships returned, showed a handsome gain. And early Spanish conquerors brought back fabulous amounts of gold and silver which they obtained from the Indians of Peru and Mexico and which were to contribute mightily to the growth of Europe's money economy. Finally, the alleged wealth of the New World aroused the imagination of Europeans to such a pitch that they began one of the greatest colonizing movements in all history—a movement that can be compared with the migrations of the Peoples of the Sea at the end of the second millennium B.C. or with the barbarian migrations at the end of the Roman Period.

Of all these developments, the growth of trade with distant places had one of the most profound effects upon Europe's economy. In the absence of adequate statistics, it is impossible to give an accurate picture of the growth of European commerce with the rest of the world during the three centuries from 1500 to 1800, but some impression of the increase can be had from scattered data. For example, Portugal is reported to have received 1,300 tons of pepper from the East in 1503, and Portuguese royal profits from Eastern trade are said to have amounted to as much as $1,250,000 (1940 value) per year. Spain imported nearly a billion dollars (1932 value) of gold and silver bullion from 1503 to 1660; and English trade in the eighteenth century is thought to have quadrupled. Certainly, the new trade was relatively very great and brought economic advantage to those parts of Europe which were most directly engaged in it. These areas

THE DEVELOPMENT OF TRADE
IN EUROPE, LATER MIDDLE AGES

were located on Europe's Atlantic seaboard, and gradually their commerce came to exceed that of Mediterranean cities —a shift that is known as the commercial revolution of the sixteenth century. On the Atlantic coast were to be found, then, districts where economic progress was most pronounced and incidentally where the greatest intellectual and artistic achievements were to be realized. Nevertheless, north Italian and south German cities still had accumulated wealth, retained a considerable amount of trade, and contributed to the intellectual life of the sixteenth century.

Of course, the economic situation of the various European states on the Atlantic seaboard differed profoundly and changed fundamentally as time went on. At first Portugal had a complete monopoly of the all-water trade route to the East, and this commerce appears to have provided that country with enough profits to satisfy it or to have absorbed all its available energies. Inasmuch as the Portuguese made no serious attempt to distribute their Eastern wares to the North, that business fell to others and particularly to the Dutch. Before the end of the sixteenth century, these people began to go directly to the East for goods previously obtained from Portugal. This practice not only undermined the very foundations upon which Portuguese prosperity rested, but it also led to the establishment of Dutch colonies in the East from which Dutch merchants radiated in search of trade. From this commerce the Dutch reaped a great harvest and became in the seventeenth century one of the most important mercantile and shipping peoples of Western Europe.

Like Portugal and the northern Netherlands, Spain obtained large benefits from trade, but from trade with the

New World rather than with the East. Yet unlike these states, she discovered in her new possessions vast riches of precious metals either in hoards amassed by Indians or remaining in the ground awaiting industrious miners. At first she obtained large supplies of gold and silver through robbing the natives, but as this source was drained dry, she turned to mining such deposits as that at Potosí. Unfortunately for Spain this wealth did not lead to extensive economic development at home. Much of Spanish bullion went abroad in payment for goods needed for domestic consumption or for export to the colonies, but it did permit a relatively small class of benefici- aries to enjoy considerable luxury and to support an artistic movement in the sixteenth and early seventeenth centuries. In the seventeenth century, however, Spain lost its position of economic eminence as others gobbled up much of her co- lonial trade. The supply of bullion from the New World de- clined; her government fell into the hands of weaklings; and the fortunes of war turned against her. By the eighteenth century, she had become a second-rate economic and political power.

England and France, for their part, did not early acquire territories overseas with which they could have at once a profitable trade or from which they could obtain large quan- tities of gold and silver. For the most part their colonies were such that production in them had to be developed before they were of much value. By the end of the seventeenth and in the eighteenth century *la mise en valeur* of the colonies was being realized, and the motherlands were receiving raw materials and foodstuffs in return for exports of manufac- tured goods. In the same centuries, moreover, these two powers encroached upon Dutch commercial prerogatives in the East and the English, at least, made large inroads into Spain's trade with Latin America. But no sooner had the position of the Dutch been weakened, than the English and French became involved in a long struggle for world empire. In this "Second Hundred Years' War," which came to an end with the close of the Napoleonic Wars (1815), the English had extended their sway to include India, Canada, South Africa, Ceylon, islands in the West Indies, and many smaller possessions. And what was most important, England rose to the position of the leading mercantile nation of the world.

As first Portugal and Spain, then the Netherlands, and finally France and England vied for first place in the economy of Western culture, economic changes were bringing about the development of modern capitalism. In brief, this system may be described as one in which individuals own capital in the form of land, productive equipment, raw materials, money, or other valuables, who hire labor for wages in an impersonal market, and who with their capital and hired labor strive to produce goods and services on a large scale in the hope of reaping a profit. Although all the components of the definition are essential to the institution itself, all did not have equal roles in the historical process whereby modern capitalism came into being. Perhaps the most crucial component— the one that acted as a catalyst upon the others—was money, particularly money that could be lent out at interest. Inasmuch as money is a form of capital that has relatively great mobility and that through interest charges can be made to earn a profit without much supervision by the owner, it can be amassed from many small sources until a large amount is obtained and can be put to work with other factors of production to earn a profit. Its use facilitates the creation of large-scale new business ventures, and they lead, in turn, to the establishment of impersonal relationships between employer and employee and between producer and consumer.

Money had been the medium of exchange in many cultures prior to the one which we call Western, but never before had it succeeded in penetrating so deeply and permeating so extensively all economic life. This is a fundamental fact. The reasons for it are many, but two of the more important ones can be mentioned briefly. The first is that the supply of money in relation to the supply of goods and services became very great. The production of gold and silver in Europe increased dramatically in the late fifteenth century and then again in the sixteenth as bullion came in from the New World. One estimate has it that the stock of precious metals in Europe tripled between 1500 and 1650. The second reason is that Western Europe developed a system of credit instruments—bills of exchange, promissory notes, letters of credit, and bank notes—which increased the supply of exchange mediums far beyond anything possible with coins and far beyond anything that had existed previously. Just by way of example, let us cite the case of the Bank of Sweden which extended its issue of bank notes by over 1,300 per cent from 1737 to 1762.[5]

Such increases in the supply of money as have been mentioned led to a greater exchange of goods for money and to the establishment of a more extensive price system—two developments which tended to impersonalize relations between employers and employees and between producers and consumers. Furthermore the increased supply of money in the sixteenth and the first half of the seventeenth century led to a rise in price levels which for Spain and England are estimated to have been in the neighborhood of 300 per cent. Subsequently prices reacted downward until the beginning of the eighteenth century, when they went up once again to the period after the Napoleonic Wars. Thus with a money economy and its concomitant price system a new element of fluctuation was introduced into economic life on a larger scale than hitherto. This was significant, for as prices rose, wages failed to keep pace. The result of this situation was that with labor costs of production reduced, profits tended to increase; and this "profit inflation," as it is called, was a buoyant factor in economic development.

Incidentally another aspect of money should be mentioned here. Over long periods of time money has tended to lose value, or to put it in another way, prices in terms of a given medium of exchange have gone upward. The reasons for this phenomenon are complex, but perhaps the most important single explanation of it is that individuals and governments are inclined to spend beyond their ability to meet their obligations and, in order to free themselves from debt, demand and get cheaper money. Furthermore, it is clear that the capitalist system operates more satisfactorily when prices are reacting upward than when they are headed downward, and this fact makes for easy-money policies.

In any case, as modern capitalism developed, it brought into being new methods and organizations for performing its vital services: (1) the issuing of coins by state mints, (2) the issuing of bank notes by banks, (3) the transfer of funds,

(4) the accumulating of capital, (5) the investment of capital, (6) the minimizing of certain risks by diversification or insurance. Gradually, states, while retaining their power of debasement, standardized coins, established monopolies in the striking of them, and took steps to prevent wildcat operations. Banks, as we have seen, began to issue bank notes in the seventeenth century and by then were performing yeoman service in transferring funds by bills of exchange, by "*Giro* banking" (bookkeeping cancellations of debts and credits by order of their clients), and by other methods.

Capital was accumulated for investment by banks and then in the seventeenth century by joint stock companies. Here was a business institution, long out of use, which could draw upon the savings of many investors who took shares in an enterprise in the search for profits. Then states, which reestablished the public function of taxation as against the feudal practice of taxes for the benefit of the lord, accumulated wealth from their levies and sometimes turned it to investment ends. Finally bourses, or exchanges, came into being in which investors and entrepreneurs could meet more readily to borrow or loan or to buy and sell actual goods or shares of stocks or bonds. Attempts were made to minimize risks by the creation of a futures market, the sale of annuities that would assist men in their old age, and by the sharing of risks, particularly as regards sea and fire losses, through the process of insurance.[6]

The development of modern capitalism had a decided effect upon methods of production. In agriculture there was a tendency toward specialization of production to meet urban food demands and to provide industry with some of its raw materials, like wool. Furthermore, relationships between the actual owners of land and peasants became less personal and more and more regulated by money payments. In brief, the entire feudal system, which had been so drastically altered in the fourteenth century, slowly disintegrated under the impact of a money economy. In the nineteenth century, its last vestiges disappeared, and agriculture was carried on according to the tenets of capitalism.

Changes in industrial production were more dramatic than in agricultural production, and they were to have much more far-reaching consequences. Indeed the comparative advantages which Western Europe was ultimately able to achieve in industry account in large part for the supreme economic position which it attained in the world in the nineteenth century and for the relatively great economic surpluses which it created.

Although the history of European industrial development is a long one and its beginnings reach back to the Middle Ages and in technology even to ancient times, it is especially marked by rapid changes which began to take place in the sixteenth and seventeenth centuries. The demands of overseas areas for European industrial goods exerted pressure upon producers to turn out increasing quantities of such things as ships, ships' stores, hardware, and clothing. Simultaneously the growth of political states and intense rivalry among them increased demands for military supplies, while civilian demand went up, as it usually does when prices are rising. To meet the new demands, there was greater investment in industry, a more extensive division of labor, and an increasing tendency for consumers to obtain what they wanted in exchange for money in an impersonal market.

For the most part the existing guild system was not able to rise to the new challenge, for it was too rigidly bound to the idea of maintaining a monopoly over a small market. A different kind of enterprise system was needed, one which would put greater emphasis on volume of output and less on quality and price restrictions. Private business for profit grew rapidly, facilitated as it was by the existence of more money, instruments of credit, and banking services. Entrepreneurs found, moreover, that the guild system had effected a considerable division of labor which could easily be developed further and that guilds had trained labor which they could not absorb and which was ready to be hired for wages. Western Europe was, furthermore, richly endowed with natural resources necessary for industrial expansion. It had iron ore, charcoal and later coal for smelting the ore, lumber and resin for ships, and wool for textiles. Finally, Europe had been developing the field of mechanics and had machines, like the hand loom, that could be put to use to produce more goods with less labor per unit of output.

Perhaps it was not until the nineteenth century that Western Europe experienced substantial effects of the mechanization of industry—of what has been called the "industrial revolution"—and not until then did industrial output exceed in value agricultural production. Yet as early as the sixteenth century, if not before, the signs of industrial production organized on capitalist lines were unmistakable. One of the early indications of the trend was the growth of a system by which a business venturer bought materials, placed them in the hands of workers who processed them for him at a price, and then disposed of them in distant markets. Certain guild masters, instead of themselves making goods for local sale, engaged in such practices, but usually nonguild members, especially merchants, were the pioneers. In some instances, the laborers worked up the goods in their homes on their own machines, whence came the terms "domestic system" or "putting-out" system for this form of enterprise, but they might lease machines from their employers or they might congregate in a building owned by the merchant and operate their employers' machines, whence the more appropriate term of "merchant-employer system."

This type of industrial organization was frequent in the textile industries, particularly in knit goods. In this latter field a mechanical contrivance was developed in the late sixteenth century which was a harbinger of things to come. An Englishman, named Henry Lee, invented a machine which was capable of 1,500 stitches a minute as compared to 100 stitches of a hand knitter. Here was a device which could easily be operated at home and that could enormously increase output. It was just the type of thing that an entrepreneur could provide for homeworkers in order that they might more cheaply turn his raw materials into finished products.

In shipbuilding, coal mining, beermaking, salt refining, and certain other industries, where relatively large investment and continuous operation was required, production was concentrated in one place and was financed by the more wealthy, or by partners, or later by stock companies. In these industries, technological advances began to appear, such as the use of coal for heat, pumps for removing water from mines, hoists for lifting heavy loads, and tracks to reduce the friction of wheels. The steam engine itself evolved from a machine to pump water from mines, and the first steam engines used for traction were employed at the heads of coal mines.

The growth of mechanized, capitalist industry was so important in England from 1540 to 1640 that some scholars speak of an "early industrial revolution" there between these years. In the seventeenth century, France made several advances in mechanized industry, as did the Netherlands and southern Germany. Then in the eighteenth century, the invention of the spinning jenny, the fly-shuttle, and the power loom revolutionized the textile trade. The discovery of a method to smelt iron ore with coke and of machines for working iron more readily entirely changed the role of metallurgy in Europe's economy. The invention of the steam engine made possible the production of cheap and mobile power —power that was available for all kinds of manufacturing processes at almost any location. At last Europe had developed a means of using on a large scale inorganic materials —the stored-up resources of nature—for its economic needs.

In the first three centuries of modern times when these developments were taking place, the modern political state was coming into being. This new form of political organization was, of course, much concerned with its economic potential and its military power. Accordingly the youthful states of Europe adopted economic policies which they believed would operate to their best interests. What the specific policies were varied according to circumstances, but they comprised attempts by each state to retain for itself all or the lion's share of commerce with its colonies or spheres of influence, to prevent colonies from producing goods which the motherland could supply, and to encourage production in the colonies of goods, particularly raw products and bullion, for which there was a large demand in Europe. In addition, some states, notably Spain, tried to hoard bullion within their boundaries in the belief that such a policy would redound to their benefit. Some stressed the development of domestic commerce by breaking down internal customs barriers and by improving transportation; others emphasized the necessity of greater agricultural and industrial production; and all sought larger populations in the interest of military might.

These measures have been lumped together indiscriminately under the term "mercantilism," yet no one state pursued each or all concepts with equal vigor. Portugal and Spain stressed bullionism; the Netherlands, free trade and commerce; and England and France, the actual production of goods and services. England, particularly, placed emphasis upon the desirability of exchanging with other regions its manufactured goods for raw materials and foodstuffs. This concept of productivity as the basis of national wealth stood the test of time, for indeed Western Europe attained its position of economic eminence by becoming the "workshop of the world." In one sense this position of economic supremacy of Western Europe was an aim of mercantilism, for the aggregate of mercantilist policies envisaged the economic development of Europe at the expense of the rest of the world.

From what has preceded, the conclusion can be drawn that economic well-being in Western Europe in the period between

1500 and 1800 stemmed from the exploitation of overseas areas, increased commercial activity, greater industrial production, the extension of the institutions of a money economy, and the greater organization of economic life according to the tenets of modern capitalism. Thus the degree to which successful development along these lines was realized in various parts of Western Europe is a key to determining which places enjoyed the greatest supply of goods and services. In regions of greatest economic development fortunes could most easily be amassed which could be devoted to advances in civilization.

Civilization in Western Europe to 1800

The period from the fifth to the tenth century A.D. has been usually regarded as the "dark age" of Western Europe. In these years this area produced relatively little of an indigenous character in the fields of art, architecture, literature, philosophy, or science. In the early part of this period the chief center of civilization was at Constantinople where Byzantine culture preserved much of ancient Greek and Roman learning and added important contributions of its own. Byzantine influence made itself felt in Italy, particularly at Ravenna and Venice, with which Byzantium had political and commercial ties. Yet such influence was so limited that it led to no great period of creative activity in the whole of Western Europe.

Nor did Arabic civilization, which developed in the sixth century and spread rapidly in the seventh, provide the leaven necessary to start a new cultural movement in Western Europe, although certain of its currents had by the eighth century made their mark in Spain, southern France, and Italy. Troubadour literature of Languedoc, for example, reflects borrowings from Arabian literature; much of Greek philosophy, including parts of Aristotle's, was made known to the West through Latin translations from the Arabic; and Arabic scientists taught Western Europeans some of the natural science which passed current for centuries. The West learned of Galen and Hippocrates from Arabic sources and held Arabic medicine in high esteem. Similarly Western Europe acquired from Arabic civilization knowledge in the fields of chemistry, physics, astronomy, geography, and mathematics, including "Arabic" numerals.

The first really creative period indigenous to Western Europe began in earnest in the eleventh century and reached its height in the twelfth and thirteenth centuries, when commerce was expanding and towns were growing. Up to that time architecture had followed Roman models in the Romanesque style, but now it acquired originality with the Gothic form. Sculptors and painters began to throw off the rigidities and formalism of Byzantine art and to attain form, expression, and movement. Those trends were apparent in Gothic sculpture, particularly in scenes of the Last Judgment, which so frequently surmounted the main entrances of churches, and in paintings by masters like Cimabue (1240–1301) and Giotto (1266–1336).

French creative activity found literary expression in the love lyrics of the troubadours, which reached their peak in the twelfth century, and in the chansons, like the *Chanson de Roland* [7] in France, the tales of King Arthur and his Round

Table, *Tristan and Isolde* in France and Germany, and
Reynard the Fox in the Low Countries.[8] For the most part
these works were written in the language of the common people rather than in Latin, and they treated subjects closer
to the experience of Western Europeans than had the more
classical works of earlier times—facts which contributed to
their vitality and success. Upon their heels followed a literary
movement in Italy that owed much to the liberal patronage
and general encouragement of the Holy Roman Emperor,
Frederick II (1211–1250). To his Sicilian court he invited
many of the leading writers of his time and there gave impetus
to a movement whose most remarkable product was Dante
Alighieri (1265–1321). Dante's *Divine Comedy* is an amazing synthesis of the reflection of his part of the Middle Ages
and is generally considered to be one of the greatest epics of
all time.

The twelfth century also gave a strong impetus to learning. By its end the universities of Bologna, Paris, Montpel-
lier, and Oxford, at least, were well established. To the
"schools" came teachers, who had previously given instruction as best they could to groups of followers, to debate
theological questions. Peter Abelard (1079–1142) in his
Sic et Non presented arguments for and against most of
the established beliefs of the Church, and St. Thomas Aquinas
(1225–1274) in his *Summa Theologiae* presented logical
proofs for the existence of God, the immortality of the soul,
and all fundamental propositions of religion. Such intellectual activity, called scholasticism, did not result in an or-
thodoxy of belief but rather in a heterodoxy—a variety of

views and an attitude of questioning. It opened men's minds to the possibility of other than established doctrines, to the world of man, and to physical science.

Interest in the physical world proceeded at first along lines drawn by ancient authors like Aristotle, Galen, and Ptolemy. Soon, however, these "authorities" began to be questioned and corrected as a result of observation and experimentation. Roger Bacon (1214–1294) exemplified the new tendency in his pleas for the use of the "experimental method" and in his personal quest for knowledge. By the beginning of the fourteenth century a correct explanation had been made of the rainbow, revising Aristotle's view; attention was being given to the subject of dynamics through the study of projectiles; and efforts were being expended not only to organize existing knowledge of geography but to extend it by explorations. Furthermore the study of alchemy resulted in the accumulation of information about inanimate nature; the practice of medicine, to closer observation of anatomy and disease; and the development of pharmacy, to an alphabetical listening and explanation of the believed efficacy of drugs.

Following the flowering of cultural activity in the twelfth and thirteenth centuries, sometimes referred to as the Twelfth-century Renaissance, some retrogression seems to have set in. Architectural design, literary form and artistic styles seem to have been worn thin by repetition, and creativeness appears to have lost much of its earlier vigor. Yet the recession was not of long duration, nor was it general in all regions. In Italy, the then wealthiest part of Europe, a new cultural movement got under way in the late fourteenth century which was to dominate and give form to Western European artistic and intellectual life for the next three centuries. The movement was called humanism, and the period of its flowering has been called the *Renaissance*.[9]

Humanism sang the praise of Greek and Roman culture, lauded belles-lettres, and stressed the importance of amenities in literary discourse. To a degree it was a reaction against the emphasis which in the twelfth and thirteenth centuries had been placed upon theology, metaphysics, medicine, and physical science. Humanists wanted to capture the purity of classical Latin, and in doing so made Latin a dead language and consequently helped to establish vernaculars as literary languages. They strove to find, edit, and give currency to Greek and Latin texts, thus reviving much of the learning of these civilizations. And they scoffed at the culture of the entire period from the fall of Rome to their day for the lack of the very things which they admired, an attitude which laid the basis for the mistaken view that all the Medieval Period was a "dark age."

Gradually the influence of the humanists began to make itself felt in Italy and subsequently in the lands to the north and west. In architecture the graceful Gothic gave way before a revival of Roman forms with their horizontal lines and massiveness—to the creation of the Renaissance style. Sculpture lost the imaginative quality found in Gothic statues and became more lifelike or more naturalistic. Painting began to achieve the reproduction of what the eye actually saw rather than what the artist conjured up in his mind about what he saw and could only imperfectly represent because of crude techniques. Literature dealt to a greater extent with passions and frailties of the flesh and less with ethereal ideals and otherworldly virtues.

In order to have a vivid image of the changes which were taking place one has only to compare the massive architecture of St. Peter's in Rome with the ethereal Gothic cathedral at Chartres, the stylized sculptures of that same cathedral with the realistic work of Michelangelo (1475–1564), a stiff Byzantine mosaic with the almost photographic quality of a Raphael (1483–1520), the saintly *Summa Theologiae* of St. Thomas Aquinas or essentially religious *Divine Comedy* of Dante with the love sonnets of Petrarch (1304–1374), or the earthy tales of Boccaccio (1313–1375) and Chaucer (1340–1400) with the humanistic works of Erasmus (1466–1536).

Inasmuch as Renaissance humanism stressed values which were at odds with established religion, it contributed directly to the Protestant revolt from the Church. And its praise of Roman history and political concepts influenced the statecraft of rising national political states. Yet, its emphasis on the arts and belles-lettres detracted from scientific investigation. This detraction was, however, never complete. Questioning of accepted views went on and with it efforts to find new answers. Doubts about traditional views of geography were instrumental in encouraging explorations; and explorations resulted in startling discoveries. The *Notebooks* of Leonardo da Vinci (1452–1519) bear ample witness to the scientific interests of this extraordinary person and of the society around him. His work in mechanics, botany, zoology, and anatomy indicate a desire for knowledge of the physical world and for acquiring that knowledge by observation, experimentation, and the ordering of data in meaningful relationships. Nevertheless, science of the Renaissance was only a preliminary to the great flowering of this branch of intellectual activity in succeeding centuries.

Undoubtedly the finest achievements of the Renaissance were precisely in those fields which humanists held in highest esteem and which they cultivated most assiduously—in arts and letters. These achievements constituted for most of Western Europe real "golden ages." Italy had such masters in painting as Masaccio (1401–1428), Leonardo da Vinci, Raphael, Michelangelo, and Titian, as well as a host of others. In sculpture she had Donatello (1386–1466), Michelangelo, and Benvenuto Cellini (1500–1571). In architecture she produced such structures as the dome of the cathedral in Florence, St. Peter's in Rome, and the great colonnade around the court in front of it—the work of Bernini (1598–1680). In literature she had Ariosto (1474–1533) and Tasso (1544–1595). In music, she was honored by Palestrina (1524–1594) and the originators of the modern opera. And in political science, she could claim Machiavelli (1469–1527), author of *The Prince.*

During the last half of the sixteenth and the first half of the seventeenth century, Spain had its "golden age" with such masters of the pen as Cervantes, Lope de Vega, and Calderón. In painting it could boast such leaders as El Greco and Velazquez. The Netherlands, both north and south, had their golden age over a longer period, with the Van Eyck brothers and Memling in the fifteenth century, Brueghel and Erasmus in the sixteenth century, and Rubens, Van Dyck, Rembrandt, and Vondel in the seventeenth century. England attained a literary height during the reign of Elizabeth (1558–1603) with Shakespeare, Francis Bacon, Spenser, Marlowe, and Ben Jonson. Portugal produced its greatest literary work, the *Lusiads* by Camoëns, in the sixteenth century, and in the same period Germany had such great painters as Cranach, Dürer, and Holbein, and a number of classical scholars. France had its Rabelais, Montaigne, and Ronsard in the sixteenth century and its great dramatists, Corneille, Racine, and Molière in the seventeenth.

In this very partial list of great figures of the Renaissance appear the names of a large proportion of the most illustrious painters and writers of Western culture. Undoubtedly there was a spirit of the times which inspired these men and a competitive attitude which spurred them to action. Yet it is at least a curious coincidence that several Western European countries achieved such cultural heights at a time of rapid economic development and that the great vitality waned as economic conditions worsened. The last of the great Renaissance figures appeared in Italy in the late sixteenth century; Spain's economic decline was accompanied by a falling off in literary and artistic accomplishment; and the same can be said of Portugal, Germany, and the Netherlands. In England and France intellectual activity continued, although it took a new turn, and these were the very places where economic decline was the least. Finally, sight should not be lost of the fact that those areas of most intense intellectual life after the overseas discoveries were on the Atlantic seaboard.

In spite of the remarkable achievements of the Renaissance, considerable agreement among students of Western culture can be found for the proposition that the most distinctive and, relative to other cultures, the greatest accomplishment of the West has been not in the field of arts or letters but rather in the realm of the physical sciences. In no other culture has there ever been such an extensive sequential ordering of information about the real world, and in no other has there been a comparable understanding of, and control over, things physical.

Because one of the most distinctive and remarkable achievements of Western civilization has been the development of physical science, its history becomes as much a part of our concern as painting, sculpture, literature, or philosophy. At first glance profound differences might be thought to exist between the natures of aesthetic and scientific accomplishment. On closer observation, however, apparent distinctions appear to be minor. While established styles of art may hamper fresh expression and lead to sterile copying, so also may traditional beliefs restrict scientific investigation. In both fields, advances are dependent upon a large body of data and skills developed through time. Finally, progress in both realms depends in part upon a social attitude which demands vigorous and fresh treatment and upon economic conditions which permit or stimulate cultural activity.

In general, advances in the physical sciences in the sixteenth, seventeenth, and eighteenth centuries occurred in those parts of Europe where economic expansion was taking place and where ideas were freely exchanged. It was centered at first chiefly in Italy, in the Low Countries, and in scattered commercial centers but later tended to find most congenial conditions in England and France. To the career of Leonardo da Vinci, we should now add that of Galileo (1564–1642), who advanced the study of motion and of astronomy; that of Torricelli (1608–1647), who determined the weight of atmosphere and contributed to the development of the barometer; that of Vesalius of Brussels (1514–1564), who pushed forward the study of anatomy; and that of Mercator (1512–1594), the Flemish geographer, who developed a system for projecting the earth's form upon a flat surface.

To be sure, there was Copernicus (1473–1543), a native of Cracow, who advanced the theory that the sun is the center of the solar system and that planets revolve around the sun; but Cracow can be classed as a commercial center, and furthermore, Copernicus got much of his training in Nuremberg and in Italy. Then there were in the early period Tycho Brahe (1546–1601), a Danish astronomer, and Kepler (1571–1630), his assistant, who through observation and calculation proved Copernicus to be correct and who described mathematically the orbits of the planets.

Subsequently there were Descartes (1596–1650) of France, who preached the virtues of the scientific method; Harvey (1578–1657) of England, who discovered the circulation of blood; Robert Boyle (1627–1691), who raised chemistry from pharmacy and alchemy to the level of a natural science; and Sir Isaac Newton (1642–1727), who set forth the general theory of gravitation and who devised calculus, the mathematics necessary to describe motion. Leeuwenhoek (1632–1723) of the Netherlands was the first to discover bacteria and the first to describe human spermatozoa; Priestley (1733–1804) experimented with gases and discovered oxygen; Cavendish (1731–1810) isolated hydrogen; and the Frenchman Lavoisier (1743–1794) may be said to have developed quantitative analysis. Linnaeus (1707–

1778) of Sweden systematized botany; a score of workers established the science of geology; and Benoit de Maillet (1656–1738) suggested that existing species of living things had been produced by changes in preceding species, thus anticipating the theory of evolution.

These and many other discoveries not only extended man's knowledge of the physical world, but they also laid the foundation for a still vaster expansion of natural science in the nineteenth and twentieth centuries. Such knowledge had two important consequences: first, it permitted man greatly to increase his control over nature; and second, it led to new concepts regarding religion, the universe, and human behavior. Scientists seemed able to establish that the physical universe operated according to laws which could be described mathematically and which appeared to be eternal, immutable, and natural. In fact, Newton went so far in describing the world as a machine that he effected one of the most influential intellectual transformations of all time—the Newtonian revolution.

If natural, eternal, and immutable laws regulated the physical world, it was possible that the same was true of the world of man, or at least so many persons came to think. Accordingly a search for these laws characterized much of the activity of philosophers at the end of the seventeenth and in the eighteenth centuries. Among the pioneers was John Locke (1632–1704), who attempted to explain the laws by which stimuli, or sensations from material things, combined to form the mind, memory, thought, and character of the adult. Similarly, he endeavored by the use of reason to discover "laws" regulating the formation of social institutions and political organizations. In the same vein Diderot (1713–

1784), the editor of the *Encyclopédie*, and his fellow workers on this great compendium of eighteenth-century knowledge sought to describe human behavior.

As corollaries to these efforts to portray the universe as a mechanical contraption, God's existence was doubted or flatly denied. Voltaire (1694–1778) argued brilliantly that God should be regarded merely as a first cause—a "watchmaker" who had set the mechanism going and then had left the machine to run by itself according to eternal, immutable, and natural laws. Then so much faith came to be placed in nature that with Jean Jacques Rousseau (1712–1778) we encounter an advocate of a return to it and a condemner of the artificialities of man-made institutions. Thus much was done to question irrationally held beliefs and to prepare the way for wider acceptance of knowledge arrived at by empirical means.

Unquestionably the outstanding artistic accomplishments of the eighteeenth century were not to be found in architecture, sculpture, painting, or belles-lettres,[10] for each of these branches was bowed down by traditionalism. Few would deny that the most distinctive literature came from the pens of rationalists, who were primarily concerned with social science, history, or philosophy, and that the most illustrious artistic achievements were in the field of music. Probably it is correct to state that the chief centers of musical output were in Germany and Austria, where economic development had reached a point at which it could support the arts. Handel (1685–1759), although he was attracted to England, was born in Prussia and received part of his training in Hanover; Bach (1685–1750) was a Saxon; Beethoven (1770–1827),

a native of Bonn; and Mozart (1756–1791) and Haydn
(1732–1809), Austrians. The works of these men are generally recognized as being at the summit of all musical literature.

The great accomplishments in the arts and sciences from
the Dark Ages to 1800 might never have been realized had
there not been a remarkable extension of "control over human environment," that is, an expansion of the range of opportunities for developing the creative talents of individuals.
With the breakdown of feudalism from the Black Death onward, individuals began to acquire more freedom of action,
and human conduct came to be regulated by more generally
recognized laws. Many of the religious curbs on thought
were diminished by scholastic disputations, by the Protestant
revolt from the Church in the sixteenth century, and by
philosophies urging the use of the scientific method. Physical
scourges were reduced by better sanitation and medicine;
and even warfare probably was not worse than it had been
earlier in spite of new tensions among the growing national
states and the appearance of the forerunner of "total warfare" during the French Revolution.

Finally, progress toward political equality and toward the
participation of all in public affairs was remarkable. In
some places, for example, in Venice and Genoa, oligarchical
republics were established; and in others, like France, national monarchies replaced feudal rule. In most countries the
new national kings attempted arbitrary government, but
their efforts met stiff opposition. The belief developed that
the individual had certain "natural, eternal, and immutable
rights," like life, liberty, the pursuit of happiness, and the

possession of property, and that no authority could infringe upon these gifts of nature. If one did violate them, then revolt against established authority was permissible. Furthermore these doctrines were accompanied by demands for the extension of political privileges to a larger proportion of the population and for the formation of public policies by the duly elected representatives of this broadened electoral base. With the Glorious Revolution in England in 1688, with the American Revolution of 1776, and with the French Revolution of 1789, these demands were in part realized, and the principles upon which they were based were more firmly established. Thus the nineteenth century was ushered in by a political system which increased the individual's responsibilities and gave him an opportunity to improve his condition and his society.

From this survey of Western culture in the period between the fall of Rome and 1800, certain facts emerge which require especial emphasis. Economic progress appears to have been given an initial impetus by commerce between distant places. Commerce, in turn, increased when one area possessed or could obtain goods that were known to be in great demand in another area and when transportation of these goods was possible both politically and economically. Consequently, technological advances in transportation and in the production of goods, as well as political arrangements permitting commercial intercourse, were crucial matters. Furthermore, one of the striking aspects of this history was the effect which a relatively small volume of trade had in changing economic institutions. This phenomenon is to be explained primarily by the fact that trade greatly extended the range of economic opportunities and brought into being a class of businessmen—the merchants—whose task it was to make the most of business opportunities. For a long time merchants were pioneers in accumulating and investing capital in both trade and industry with the avowed purpose of reaping a profit.

With the growth of trade and of industry, towns increased in size, and in the urban areas there took place a division of economic and intellectual labor. Here we find again the chief centers for the achievement of those things which we consider to be marks of civilization. We also have found a considerable correspondence in time and place between general economic well-being and the creation of masterpieces in the arts and sciences. Again we introduce our frequently repeated caveat that economic surplus to support intellectual and artistic activity could be amassed by a few at the expense of the many and that the problem of such accumulation depended upon political and social conditions as well as upon economic factors. Finally, we have encountered once more the question of why and how a society develops goals, patterns of thought, and styles of expression. Our analysis is not aimed to deal with these questions in detail, but we have indicated shifts from the Dark Ages to a Christian-Gothic-scholastic culture, then to the humanistic-classical-Renaissance pattern, and finally to the mechanistic-scientific. From the last point, we move to a consideration of the history of Western culture in the last 150 years.

For Further Reading

J. N. L. Baker, *History of Geographical Discovery and Exploration* (1937)

Shepard B. Clough and Charles W. Cole, *Economic History of Europe* (1946)

A. Dopsch, *The Economic and Social Foundations of European Civilization* (1937)

E. J. Hamilton, *American Treasure and the Price Revolution in Spain,* 1501–1650 (1934)

C. H. Haskins, *Medieval Culture* (1929)

C. J. H. Hayes, *Historical Evolution of Modern Nationalism* (1931)

J. Huizinga, *The Waning of the Middle Ages* (1924)

Archibald Lewis, *Naval Power and Trade in the Mediterranean, 500–1100 A.D.* (1951)

E. Lipson, *Economic History of England* (1929), 2 vols.

John U. Nef, *War and Human Progress* (1950)

Henri Pirenne, *Economic and Social History of Medieval Europe* (1937)

————, *Medieval Cities: Their Origins and the Revival of Trade* (1925)

J. H. Randall, *The Making of the Modern Mind* (1940)

Henri Sée, *Modern Capitalism* (1928)

Percy Sykes, *History of Exploration* (1935)

John A. Symonds, *Short History of the Renaissance* (1894)

H. O. Taylor, *The Classical Heritage of the Middle Ages* (1911)

The Cambridge Economic History of Europe (1942 ff.)

The Cambridge Medieval History (1913–1936), 8 vols.

James Westfall Thompson, *An Economic and Social History of the Middle Ages* (1928)

Lynn Thorndike, *History of Magic and Experimental Science* (1923–1941), 6 vols.

————, *A Short History of Civilization* (1948)

A. A. Tilley, *Studies in the French Renaissance* (1922)

A. A. Wolf, *A History of Science, Technology, and Philosophy in the Sixteenth and Seventeenth Centuries* (1935)

VII

WESTERN CULTURE IN
THE NINETEENTH AND
TWENTIETH CENTURIES

———

THE historian of the mid-twentieth century who studies
the economic development of Western culture of the last 150
years is struck at once by the enormous quantitative increase
in the production of goods and services. Just how great the
increase was is difficult to state in precise terms, but for Western Europe it has been perhaps in the order of six to eight
times. At least what evidence we have indicates more than a
doubling of French national income from 1850 to World
War II, a quadrupling of that of Germany for the same
period,[1] a tripling of that of Italy from 1860 to 1938,[2] and
an eightfold increase in that of the United States from 1869–
1878 to 1829–1938.[3] Such increases made possible a rise in
the population of Europe from some 187,693,000 people in
1800 to over 530,000,000 in 1938 and in the United States
from 4,000,000 in 1790 to 140,000,000 in 1946. Over the entire period there was an increase in income per capita of
population of some two to four times.

The last 150 years

This expansion in physical production, in population, and in goods and services per capita does not have a parallel in all history. Such being the case, the curious man of the second half of the twentieth century can properly demand that the historian explain how this absolutely unique phenomenon came to pass.

The main lines of explanation are not difficult to perceive. First, much of the increase in production resulted from the mechanization of industry. By means of the machine, driven by mechanical power, the individual operator could produce a quantity of goods that by any other system would have required tens, hundreds, or even thousands of workers. The single workman came to have a multitude of mechanical slaves to do his bidding.

Second, the increase in goods was made possible by using on a scale never dreamed of before the stored-up riches of nature—the mineral resources of the earth. Hitherto most of man's economic wants had been supplied by plants or animals, supplies which were limited by available manpower, arable land, rainfall, and the growing qualities of strains and breeds. In the last century and a half, however, man has been able to dip into the "capital" of nature and thereby to have a new and additional, although not inexhaustible, supply of materials for his use.

Third, improved methods of transportation and communication made possible the tapping of resources over a larger portion of the earth's surface. Products from far-distant places, whether lumber from high mountain ranges or from jungles, grains from remote corners of the earth, or coal deep underground were made accessible to men. And as though this were not enough, new knowledge about the growing of crops and the raising of animals permitted a phenomenal increase in the production of agricultural products on land that had been tilled for centuries.

In brief, then, the great increase in goods and services between 1800 and 1950 may be attributed primarily to the mechanization of industry, the greater use of inorganic matter, improved agricultural technology and the tapping of resources over a greater portion of the globe. In this generalization, however, three aspects of economic development in the 150 years in question are not adequately recognized. The first of these is that the new economy required a greater division of labor or specialization of tasks than had ever before been realized; the second is that the division of labor and the bringing into the market of greater supplies of goods from a larger portion of the earth was made possible by improved means of transportation and a vast extension of trade; and the third is that trade could not imaginably have been conducted without money and elaborate banking services.

Great as was the increase in goods available for human use in Western culture after 1750, all parts of that culture did not experience economic expansion at the same time or in the same proportions. The movement was most marked at first in England, then in Belgium and France, and subsequently in Germany, Italy, the United States, and other nations. It was so located, however, that Western Europe achieved a predominant world position of economic and political-military power. In 1821 Great Britain and France alone took 92 per cent of American cotton exports, which was about one-half of the United States' total crop. In 1870, the United Kingdom, Germany, France, Belgium, Italy and Sweden had 61 per cent of the world's manufacturing production and the United States 23.3 per cent of it. In 1840 the United Kingdom alone had 32 per cent of world trade, France 10 per cent, and the United States 8 per cent. Furthermore, the states of Western culture acquired such economic-military power that they were able to exert their authority outside that culture almost at will. In fact, they extended their domination over a large portion of the globe in a second great wave of imperialism between 1870 and 1914.

As time went on, however, Western Europe failed to maintain its *relative* economic position in Western culture or, for that matter, in the world. The United States moved rapidly to the fore, with most of the world aping or trying to ape Western culture's productive technology. By 1913 the United Kingdom, Germany, France, Belgium, Italy, and Sweden accounted for 41.9 per cent of world manufacturing production, the United States for 35.8 per cent, and Russia for 5.5 per cent. In 1936–1938 these percentages were 29.7, 32.2, and 18.5, and for Japan 3.5 per cent.[4] Whereas the percentage of world trade accounted for by the United Kingdom, Germany, and France was 43 per cent in 1890, these same countries did only 36 per cent of that trade in 1913, and 26.9 per cent of it in 1938. The entire trend toward the development of industry outside of Western Europe is exemplified by these statistics.

Other elements in the European picture also looked black —and one of the blackest was nationalism. Within Europe a system of political states had developed in which each nation claimed its sovereign right to act in its own interests irrespective of the welfare of others. For the most part, the people of Europe had an emotional, if not a fanatical, feeling toward their nation-states. Nationalism made difficult and, indeed, almost precluded interstate cooperation: and it was a serious contributor to war. In this regard an interesting parallel can be drawn between Europe and ancient Greece. Like Greece, Western Europe has failed to achieve political unification; and internecine strife has weakened it to a point where invasion by another culture is a real threat. Such has become the economic, political, and military plight of Western Europe that it is almost certain to see the leadership in Western culture pass to the English-speaking people of the northern part of the Western Hemisphere. Up to the eve of World War II, however, the peoples of Western culture were much better off materially than were those in any other culture, and especially well off were the populations of the United States, Canada, New Zealand, the United Kingdom, and Switzerland.

The question now remains whether or not this economic well-being per capita has been accompanied by remarkable achievements in the arts and sciences. As to the latter, there can be no question. Never before has man been able so fully to order events sequentially, to explain their interrelationships, or to control his physical-material environment. As to accomplishments in the arts, there is great difference of opinion. Standards of aesthetic judgment are so highly subjective that a scientific appraisal of the art of the last 150 years is nigh impossible. Nor in this case, because it is so

recent, can we turn to historical literature to discover whether
or not works of art in the period have been universally and
consistently considered great in comparison with the creations of other cultures. We shall have to await a future historian with the advantage of more hindsight than we possess to have recourse to this method. Yet, we may hazard a
guess regarding his findings. It may well be his judgment that
relative to earlier periods within Western culture, artistic
accomplishments in the West during the last 150 years have
considerable merit. Impressionism in painting, the symphony
and opera in music, and functional building in architecture
will probably stand the test of time.

Our future historian's severest strictures of recent Western culture will probably relate to politics and philosophy.
So far Western nations have failed to settle their differences
peaceably, and unless they overcome this weakness they may
destroy themselves in combat or they may become so weak
that peoples of another culture will prey upon them. Nor at
present does Western culture have a philosophical system
to which its members all adhere and for whose ends its people strive. Material betterment seems to be the predominant
goal of life, but such a goal leads to differences and conflicts among both social and political groups and even between Western cultures and other cultures. How man is to
get along with man, and group with group, are the major
unsolved problems of our day. Indeed, they have been major
problems throughout all history.

✑ Mechanization of industry

As has already been said, the last 150 years have witnessed the most rapid and extensive technological changes

to be found in an equal period of time in all man's recorded past. This statement can be made without fear of contradiction in spite of the fact that, because technological innovations are dependent upon earlier achievements, no two changes are strictly comparable. The scope and effects of automatic textile machines, of the steam engine, of cheap methods of making steel, of the electrical dynamo, of the internal-combustion engine, and of nuclear fission have been enormous. In the author's opinion they surpass accomplishments in the thousand years prior to 3000 B.C.—a millennium which saw the development of copper and bronze metallurgy, the harnessing of animal power, the making of brick, and the use of the wheel in pottery and transport.

The explanation of the more recent changes in technology is of necessity extremely complex, for a myriad of factors and forces combined in the right proportions and with a favorable timing to produce the new ways of making things. The major factors involved here can be classified as follows: (1) capital or surplus which could be used to develop, construct, and put into operation new techniques; (2) the capitalist spirit or the desire for gain and material benefit from the use of new methods of production; (3) technical knowledge which made possible improvements in performing certain tasks; (4) markets or a demand for goods greater than could be met by older forms of production; (5) workers who were willing and able to operate the new machines; (6) a sufficient and steady supply of raw materials.

By the eighteenth century these factors were present to a considerable degree in various parts of Western Europe but particularly in England. There capital had been amassed through trade, handicraft production, seignorial payments, tax collecting, and banking, and methods had been developed for putting capital to work. The capitalist spirit had grown apace as land as a source of wealth, social prestige, and political power had given way to trade and industry. Technical knowledge had increased through experience with handicraft contrivances and through speculation and experimentation regarding the physical world. Markets had grown with the expansion of Europe overseas, with the development at home of better transportation, and with a tendency of individuals to specialize in the production of one or a few things and to satisfy their other needs by purchases. Workers were available at money wages from handicraft industry, from agriculture, and from the not fully employed child and female segments of population. Finally, a sufficient supply of raw materials was made possible by increasing production of them and by commerce.

Conditions seem, indeed, to have been ripe for technological change, and change did take place. This was the beginning of the "industrial revolution"—a revolution which did not occur so suddenly as is frequently thought and which even yet has not run its full course. Although no one factor acted as the sole catalytic agent to effect the mechanization of industry, the invention of machines and processes of production were so crucial to the "revolution" that they deserve our special attention.

What we call inventions come about from the accretion of many details, which, when finally put together, result in something new. They do not pop suddenly and fully developed from the head of some isolated genius but are related to some problem or process for which a solution is being sought. In turn, the actual adoption of inventions depends upon the need for them, their effectiveness in meeting this need at lower costs, the ability of workers to use them, and the willingness of those with capital to put them into operation.[5]

In the eighteenth and early nineteenth centuries those persons who were accumulating details about machines and industrial processes were mostly handicraftsmen rather than highly trained engineers and scientists. Inventions came in those crafts where rudimentary machines were already in use, where productive methods were not rigidly fixed by tradition or law, and where workers could easily adapt themselves to the new contrivances. And they appeared and were put to work in industries where the demand for goods was great in relation to supply, especially in staple industries with mass markets. Thus, among the first machines of the "industrial revolution" were the spinning jenny, the fly-shuttle, the power loom, the circular saw, the rotary planer, puddling in iron smelting, screw- and nail-making machines, the coking of coal, the harrow, iron plows, and the slide rest for metal-turning lathes.

Successful inventions tend to have an accumulative effect, for changes in one aspect of production create pressures for changes elsewhere. Thus, many of the early machines were designed to be activated by human energy, but they were so constructed that the application to them of energy from inorganic sources was an obvious step. Accordingly power from the existing water wheel was used on them at an early date, but because the location of water wheels was restricted to places with falling water and because the supply of water depended in many cases upon the whimseys of weather, there was need for a supplementary or substitute source of energy. This need was met by the invention of the Watt steam engine in the 1770s and the subsequent development of that machine.

The steam engine was itself an agent for rapid change. It came to produce power cheaply from an inorganic substance in abundant natural supply; it was mobile enough to be used where there was sufficient fuel and water; and it was adaptable to many uses, both as a stationary power unit or as a portable machine that could propel itself. Its adoption made possible the factory system on a large scale, cheaper and faster transportation, and an increase in the production of goods per unit of human input.

The application of more power and the attainment of higher speeds in industry necessitated, in turn, the construction of machines from more durable materials than wood or cast iron. Steel seemed to be the answer to this problem, but steel was too expensive until Bessemer (1856) and Siemens and Martin (1866) developed their respective processes of steelmaking. Their inventions led to a 50 per cent decline in the price of steel from 1856 to 1870.

In another respect, too, the new methods of making steel were an important landmark in industrial history. They illustrate a movement which had been taking shape and which was henceforth to be highly characteristic of technological change, that is, for inventions to be made by trained scientists rather than by handicraft mechanics. Industrial machinery and processes were becoming so complicated and highly specialized that the worker was seldom able to provide solutions to any except the more rudimentary problems which he encountered. In fact, nearly all basic technological changes after the middle of the nineteenth century must be attributed to men of science. They were responsible for the technical development of electricity and the internal-combustion engine. They created countless new machines and increased the efficiency of earlier ones. They effected the "chemical revolution" which permitted the production of useful goods from materials in abundant supply. They were able to effect an "agricultural revolution" that has increased productivity of land and animals beyond the dreams of our forefathers. And they are credited with the "physical revolution," which has extended greatly man's control over nature and which through nuclear fission bids fair to provide mankind with an important new source of energy and of materials. In brief, men of science have facilitated the output of more goods and services per unit of human input through better and fuller use of nature's resources.

➤ The agricultural revolution

Important as was the mechanization of industry in providing members of Western culture with a surplus which would permit them to devote time and energy to artistic and scientific activity, it was by no means the only source of increased wealth. It was accompanied, indeed, by agricultural changes which have been glorified by the term "revolution." This revolution began, like the industrial revolution in England, in the eighteenth century. At its origins it consisted primarily in the use of machines drawn by animals, the introduction of new crops, better care of animals, and selective breeding. As it progressed, it was characterized by the application of science to a wide range of agricultural problems and to extensive mechanization of agricultural processes.

Farm machinery was improved and made more useful, especially with the development of the tractor. Many new crops were introduced, such as the sugar beet, and older ones, like the potato, were more widely cultivated. Better grains were grown, and legume crops, like alfalfa, which are less exhaustive of the soil than others, were more generally grown, while hybrid seeds proved highly productive and resistant to drought and blights. Animal and plant diseases and pests were brought under greater control by the use of new medicines, insecticides, and fungicides. Breeding was greatly improved through more rigid controls, artificial insemination, and the discovery that individual stock had different abilities in passing desired qualities to its offspring.

Furthermore, science discovered the chemical requirements of plants, which led to extensive changes in the use of fertilizers, especially commercial fertilizers. New attention was given to land utilization, that is, to those particular types of land which were best suited to different kinds of agriculture, and subsequently to reforestation, irrigation, and drainage projects. Finally, greater care was exercised in the processing of foodstuffs, such as the pasteurization of milk, the storage of grains, and the making of sugar from beets.

These various changes had dramatic results. The production of grains per hectare just about doubled in Western Europe during the period under consideration, and the average would have been greater if so much marginal land had not been kept under the plow. Less food was necessary for draft animals after the use of the tractor became generalized.

More emphasis was placed upon animals that produced food, like cattle, poultry, and pigs, than upon those which were raised in part for an industrial raw material, like sheep. And these changes effected some alteration in human diets, with a trend toward protective foods, which are high in vitamin content, and away from foods eaten mainly for their calorific value.

The agricultural revolution had profound effects upon costs of production for different types of crops. Through the use of machines, grain, for example, could be grown most economically on large tracts by means of extensive agriculture, while poultry and truck farming could be conducted profitably on small holdings by intensive farming. The importance of these changes became clear in the 1870s and 1880s when American grain came into Europe on a large scale. In some European countries the reaction to overseas agricultural competition was to protect traditional crops by tariffs or subsidies; in others, it was to concentrate efforts on those products in which intensive agriculture had a comparative advantage. If the former policy were adopted, prices for basic commodities were higher than on the world market; if the latter course were taken, the country in question had to export its surpluses and import what it did not produce. High prices meant a disadvantage; while reliance on imports meant dependence on foreign suppliers and their willingness to take industrial goods.

In Western Europe farming was mostly on a small scale and of the intensive variety, for in this way a larger total output of farm products was possible and greater employment opportunities were provided. Much of the land was in small holdings and what was not was usually cultivated in relatively small units by tenants or sharecroppers. Thus European farmers had difficulty in mechanizing agricultural work, because the size of their enterprise was not great enough to carry machinery. Western Europe lost its early agricultural advantages based on technological advances to other areas of Western culture, but its agricultural production per worker remained high compared with that in non-Western cultures.

✑ Trade and the division of labor

With all its advances in technology Western Europe would not have been able to achieve a position of economic primacy in the world without trade. Only by means of an enormous expansion in the exchange of goods was it possible for Western Europe to get foodstuffs and raw materials for its use in return for manufactured goods. World manufacture is estimated to have increased 740 per cent from the period 1870–1880 to the years 1936–1938 and intranational trade to have expanded even more than this, but international trade grew by only 300 per cent.[6]

Important in the growth of commerce were improved methods of transportation, a greater use of money as a medium of exchange, bank credit for financing commerce, and a vast network of distributing services, but the points which require emphasis in the present context are (1) that commerce greatly furthered the division of labor and (2) that it led to multilateral trade among the various nations of the world.

How extensive the division of labor became can be judged by a comparison of the percentage distribution of the gainfully employed in a highly industrialized country like Germany with similar percentages in a nonindustrialized country like Mexico.

	Year	Agriculture, Fishing	Mining	Manufacturing	Commerce & Transport	Administrative, Domestic Service, Etc.
Germany	1933	29	4	36	19	12
Mexico	1930	68	1	13	7	11

The figures for Germany, if broken down still further, would show how very little of the production of the average worker was consumed directly by him and how dependent he was upon the market for his numerous needs. Consequently the individual operative in industrialized regions was better able than workers had ever been before to specialize in some aspect of production and by integrating his efforts with others to increase his productive capacity. Indeed, there has been a high correlation between those countries with an extensive division of labor and those with high incomes per capita.

Similarly multilateralism in international trade meant that various parts of the world could specialize in the production of various goods in which they had a comparative advantage and through exchanges get the things which they did not produce. Thus the United States could acquire raw rubber from the Malay Peninsula, the Malay Peninsula could obtain consumers' goods from the United Kingdom, and the United Kingdom could receive grain from the United States. Each area got materials that it needed, and if the system worked perfectly, there would be a multilateral balancing of debits and credits.

As the workshop of the world Western Europe enjoyed many advantages. One peculiar one was that the terms of trade favored industrial areas rather than agricultural, or, to put it another way, prices of industrial products were higher than foodstuffs or raw materials when measured in terms of human input. Yet this situation gave nonindustrialized countries an incentive to manufacture for their own needs and hence to weaken Western Europe's economic position. Furthermore, Western Europe's primacy was undermined by another factor. In its exchange of manufactured goods for foodstuffs and raw materials, Western Europe did not export goods equal in value to its imports, the difference being made up by various "invisible" items such as emigrant remittances, earnings on insurance, shipping services, tourist expenditures, and earnings on investments. In the final balance of payments Western Europe had for a long time a creditor position and kept investing overseas, a practice which hastened industrialization elsewhere.[7]

The industrialization of the rest of the world had an impact upon Europe's position and upon a pattern of trade which had come into existence. This pattern was also distorted by sudden technological changes or political events, like war. Thus the industrialization of the United States altered Europe's economic role in South America; the two world wars deprived Western Europe of many of its overseas markets; and these same conflicts led Europe to spend so much of its overseas investments that it no longer had large earnings from them to help pay for imports. Finally, depressions, especially that of the 1930s, caused a decline in international trade far greater than in domestic commerce as nation after nation sought in the emergency to protect its national economies.

Tariffs and, later, import quotas and currency controls were employed either to hasten industrialization or to meet the exigencies of sudden dislocations. While trade could usually become adjusted to mild protection, it could not cope with absolute limitations. Beginning with World War I the continuation of the system of multilateral trade and payments was threatened; and in the 1930s and 1940s multilateralism was almost destroyed by restrictions of various kinds. Western Europe suffered particularly from this turn of events because of its large dependence on foreign areas as markets and as sources of supply. Only with the existence of multilateralism can Western Europe retain some semblance of the economic position which it had previously enjoyed.

Desire for material betterment—savings and investment

Another important element in the economic expansion in Western culture during the last 150 years was the desire of people for material betterment. Ever since the appearance of humanism in the Middle Ages, a concept had taken form and assumed importance in Western culture that a perfectly legitimate interest of man was man, that man's full potentialities could not be realized by scorn for physical well-being, that man should seek a full, creative life here and now because of uncertainties regarding the hereafter, and that eternal salvation, if such there be, was not reserved for the poverty-stricken. Certainly by the end of the eighteenth century one of the fundamental ideologies of Western culture was to seek better material conditions on this earth.

In the presence of such an ideology, there were few indeed who did not strive consciously to improve their lot. The capitalist spirit was a vigorous agent of economic growth. It accounts, in part, for those captains of economic activity who specialized in making decisions which would increase their wealth. It accounts, in part, also for the willingness of large numbers to save from their current earnings in order to invest their surplus in the hopes of greater supplies of goods in the future. This process of savings and investments was about as crucial to economic expansion as were inventions, for without it a hand-to-mouth existence is inevitable. What the rate of capital formation was throughout the 150 years under consideration it is impossible to say, but it has been estimated to have been 12 to 15 per cent of national income in the United States for the years 1869 to 1938.[8]

Whatever the rate of capital formation was, however, it was enormous compared with that of any other culture, either contemporary or past. Savings were possible in large part because the supply of goods increased so much more rapidly than population that people could maintain a standard of living to which they had become accustomed and at the same time put something aside for the future. Furthermore, an increase in real wages and a growing population meant a greater demand for goods, which in turn encouraged the making of new investments. In brief, savings did not cause additional hardships, and a comparatively widely distributed, increased purchasing power stimulated the flow of savings into productive enterprise.

Obviously not all savings went directly into capital goods for the production of still more goods. A considerable amount was devoted to other than economic activity, which allowed achievements to be made that are marks of civilization. Large sums were devoted to education and research which permitted the talents of a larger proportion of the population to be developed than ever before, which provided scientific training for further control of environment by man, and which stimulated artistic creativeness. Yet the proportion of available earnings used for noneconomic purposes did not prevent economic expansion, as they did perhaps in Egypt in the third millennium B.C. when the great pyramids were constructed. Western culture has to a large extent in the last 150 years been able to eat its cake and have it too. It has had savings for capital investment and also for intellectual and artistic accomplishments. To an imposing degree it has combined economic aims with intellectual or artistic activity through functional, that is, economically useful, art and through science.

The greatest waste of savings has come from preparation for and waging of war—especially the two world wars of the twentieth century. Even though the countries involved were able to regain prewar levels of production within a decade after the conflicts, thus indicating the recuperative powers of Western culture's economies, the retarding effects of war on Western Europe have been tremendous. Furthermore, business fluctuations, which seem to be engendered by the operation of our capitalist system, have troughs during which our human, technological, and natural resources are not fully utilized and during which savings necessary for economic investment or achievements of an artistic or intellectual character are very low. Finally, the concentration of economic power in the hands of a few and the dependence of proletarians on wages for existence have resulted in conflicts of interest that have occasioned waste through strikes, lockouts, and inefficiency. The general historical picture of the last century and a half is a bright one, but it has its somber parts and some very black spots.

⤳ Natural resources and the location of industry

The bringing together at any one place of materials necessary for the production of goods is of great importance to the understanding of the development of industry in Western Europe and the subsequent growth of economic activity in the United States. Although steam shipping, steam railroads, motor trucks, and high-power electric-transmission lines have reduced the cost of transporting goods and energy and have made the location of industry more flexible, proximity to materials which are bulky and heavy in relation to their value or the value of what is made from them is a decided advantage. In fact, even today industrial activity tends to be concentrated relatively near them or in places to which they can be brought cheaply, usually by water.

Of all such raw materials coal was king and iron ore, queen. In 1938, four-fifths of inorganic energy used in industrialized European countries came from coal,[9] and almost all machines and tools were made from steel. Any industrial map of Western culture will show how industry has grown up around coal and steelmaking regions—in the Ruhr of Ger-

PERCENTAGE DISTRIBUTION OF THE WORLD'S MANUFACTURING PRODUCTION [a]

Period	United States	Germany	United Kingdom	France	Russia	Italy	Canada	Belgium	Sweden	Finland	Japan	India	Other countries	World
1870	23.3	13.2	31.8	10.3	3.7	2.4	1.0	2.9	0.4	...	—	11.0		100.00
1881–85	28.9	13.9	26.6	8.6	3.4	2.4	1.3	2.5	0.6	0.1	—	12.0		100.00
1896–1900	30.1	16.6	19.5	7.1	5.0	2.7	1.4	2.2	1.1	0.3	0.6	1.1	12.3	100.00
1906–10	35.3	15.9	14.7	6.4	5.0	3.1	2.0	2.0	1.1	0.3	1.0	1.2	12.0	100.00
1913	35.8	15.7	14.0	6.4	5.5	2.7	2.3	2.1	1.0	0.3	1.2	1.1	11.9	100.00
1913 b	35.8	14.3	14.1	7.0	4.4 c	2.7	2.3	2.1	1.0	0.3	1.2	1.1	13.7	100.00
1926–29	42.2	11.6	9.4	6.6	4.3 c	3.3	2.4	1.9	1.0	0.4	2.5	1.2	13.2	100.00
1936–38	32.2	10.7	9.2	4.5	18.5 c	2.7	2.0	1.3	1.3	0.5	3.5	1.4	12.2	100.00

League of Nations, Industrialization and World Trade, 1945, p. 13.

a Includes finished products, semimanufactures, like unworked metals, pulp, coke, fertilizers, and cement, as well as manufactured foodstuffs, like flour, canned goods, and sugar.

b The second line for 1913 represents distribution according to frontiers established after World War I.

c U.S.S.R.

many, in the north and east of France, in the Midlands in England, and in the eastern part of the United States. As it happens, Western culture has a great advantage over other areas in the matter of coal and steel, for over half the known coal resources of the world are located in it, and prior to World War II some two-thirds of world production came from Western Europe, the United States, and Canada. Also these same regions contain more than half the known iron ore reserves of the world and in 1947 produced some 80 per cent of the world's steel. These advantages contributed to the concentration in Western culture of the lion's share of the world's manufacturing production.

The accompanying table makes clear not only the predominant industrial position of Western culture but also shifts in the relative industrial importance of various regions. Of particular significance to our study are (1) the percentage decline of manufacturing production in Western Europe, especially in the United Kingdom and France; (2) the percentage increase of manufacturing in the United States; and (3) the percentage growth of industry in areas outside Western culture like Russia and Japan. These changes are altering the power potentials of states and the economic base of cultural achievement.

Why these shifts took place can be explained in broad terms by the extent to which factors of production came to be present in the more recently industrialized areas. More specifically they can be understood primarily from an investigation of the "process of diffusion" of mechanized industry —from a study of those elements of industrialization which were propelled outward by the earlier industrialized areas and of those forces in the so-called backward areas which led to borrowing from industrialized areas. Basic to such a study is the migration of machines and techniques.

The first mechanized processes to be adopted by nonindustrialized areas were those closely connected with handicraft production. Information about the new machines was easily obtained, was readily understood by handicraft workers and engineers, was applied without great capital investment, and pertained to industries for whose goods there was a large native demand. Thus textile machines were in the first group to migrate, along with such things as the circular saw, tools for installing and repairing simple machines, and more rudimentary forms of transportation.

From such beginnings the skills of workers and engineers gradually increased so that they could use more complicated machines and techniques. What was then introduced depended upon the natural resources, transportation facilities, market opportunities, and cultural desires of the area in question. By the middle of the twentieth century it was no exaggeration to say that technological knowledge could be quickly and widely diffused. No one area could long maintain an industrial advantage based solely on knowledge.

Why the earlier industrialized countries permitted this diffusion of industrialization is not difficult to discern. They were willing to sell machinery and knowledge to other areas to maximize profits, to get materials which they wanted, or to lift backward areas to their level of civilization. For similar reasons, these same areas were willing to invest capital in backward countries, thus providing these regions with a start that would have been greatly delayed if reliance had been placed solely on native capital formation.

With the diffusion of Western technology there has arisen

in backward regions a realization that industrial production is necessary for rapid economic growth and for the conduct of modern warfare. And most important of all, perhaps, there has spread throughout the world a cultural belief in the desirability for material improvement. The world is becoming mechanized and to a degree Westernized—and in this process Western culture has lent a helping hand. Whatever the virtues of such a policy may have been, that policy has tended to weaken the relative economic position of Western Europe and to shift the economic center of gravity in Western culture to the northern part of the Western Hemisphere.

⬧ *Population and labor*

Thus far in our discussion of the economic development of Western culture we have considered the role played by technological advances, by capital accumulation and investment, by the desire for material betterment, by the division of labor made possible through specialized production and trade, and by raw materials. We have also endeavored to show how Western Europe achieved first a position of world economic primacy by means of a fortunate combination of the factors of production and also how this position was weakened as other areas became industrialized. We come now to a consideration of the relationship of population to economic growth and to civilization.

As an historical proposition one may say that given equal technology, an equal desire for material improvement, and fixed natural resources, a population which is relatively sparse will increase its productivity per capita more rapidly and to a greater degree than a population which is relatively dense. In the former case, trade and a division of labor develop early which permit capital accumulation, while in the latter case a large portion of the population remains in self-sufficient agricultural pursuits, trade is slight, and capital accumulation difficult and slow. If it is a truism that a population which approaches the limit of its resources, including its technology, tends to be more static economically than a population far from such limits, it is a truism which deserves reiteration.

At the beginning of the nineteenth century Western Europe's population was not close to the limits of its resources, for technology and trade were making new resources available. This situation was conducive to economic growth and economic growth to increases in population. From 1850 to 1900 the rate of population increase was 1 per cent per annum, but by the beginning of the twentieth century the *rate of increase* began to decline, particularly in Western Europe, and by the 1930s the populations of some Western European states were nearly stable. Western Europe was approaching the limits of its economic resources, and many people, especially members of the middle and upper middle classes, wanted to protect their own standards of living by having fewer children.

In the United States and Canada limits of economic resources to population were much farther away than in Europe, and in these countries population increases were especially dramatic and have not reached their climax. The population of these countries rose from about 5,000,000 in 1800 to 160,000,000 in 1949, and the rate of increase is about

1.4 per cent per annum, compared with about 0.5 per cent in all of Europe. These countries benefited from a large immigration, the United States having received 38,000,000 Europeans from 1830 to 1930. In an economic sense, Europe was thus exporting a valuable, for emigrants were mostly of productive ages and the country of origin bore the cost of rearing that was not made up by emigrant remittances. Such movements of population undeniably assisted the United States and Canada in their economic growth *vis-a-vis* Western Europe. Incidentally immigrants brought with them heterogeneous skills and ideas, and heterogeneity is believed to have been a stimulating factor economically and culturally in the New World.

If the United States and Canada had an advantage over Western Europe in that their populations were relatively sparse compared to their economic resources and in that they received immigrants of productive ages, they profited perhaps less from the fact that the mean ages of their populations were lower. From Roman times to 1800 life expectancy at birth did not vary much from twenty-five years. Just before World War II, however, life expectancy at birth in Western Europe was about 62.2 years and in the United States 63.8 years. Undoubtedly a higher mean age of the population has meant greater productivity per capita, for older persons have greater skills and make fewer mistakes than younger ones and a smaller proportion of the population is in unproductive early ages. In time the mean age of a population may become so high that speed and strength are markedly diminished or that a large proportion of the population will be unproductive because of old age. This danger has only recently seemed very real. Of greater concern has been a fear that, whereas an increased mean age may permit a larger use by artists and scientists of their skills, it may lead to less innovation of new styles and ideas and to stereotypes.

Whatever the effect of longer life of populations may be culturally or economically, it is still a fact that those things which Western culture considers to be marks of civilization are products of urban societies. Even with improvements in transportation and communication and with many rural areas having urban tastes and patterns of life, this generalization holds true. Thus it is significant that with the various population changes already discussed, there was a steady trend toward urbanization. In 1800 only about 20 per cent of the population of Western Europe lived in towns of more than 2,500, while in 1948 some 60 per cent lived in such communities. In 1790 only 5 per cent of the population of the United States lived in towns of 2,500 or more, while in 1940, 57 per cent lived in agglomerations of this size or greater and 29 per cent lived in cities of 1,000,000 or more.

Thus general population changes tended to favor economic development, particularly in the United States and Canada, and to enlarge the environment in which cultural activity flourishes. At the same time the laboring elements of the population were undergoing changes which made them more productive. Not only was labor distributed occupationally so that it could use its energies more effectively, but it was given an opportunity to master economic arts to an ever-increasing degree. Western culture embarked during the nineteenth century upon a program of mass education that had never been equaled. More people were taught to read and write, illiteracy dropping from more than 60 per cent of the population at the beginning of the nineteenth century to less than 6 per cent of the population at the mid-point of the twentieth century. Literacy permitted workers to extend their knowledge of economic activity—to learn about their trades from the printed page.

In formal educational institutions attention was given to vocational training so that workers could acquire the fundamentals of their crafts quickly and at a low cost to society. Apprentice training, in-training, "growing-up" in a trade, and similar practices meant the development of skills within a larger portion of the population. The simplification of tasks performed by a larger percentage of labor through the mass production of goods on assembly lines required less specialized training for the rank and file but higher skills for the few.

Accompanying these changes, there was also, for a long time, at least, an increasing willingness or necessity on the part of labor to turn out goods and services. As ownership of the means of production shifted from workers to capitalists, workers became largely dependent for existence upon their wages. If they did not work steadily, they did not eat, while formerly they had been able to loaf for considerable periods of time if they had amassed enough surplus for their needs. In short, the wage system drove men to work. Furthermore entrance to most trades was free, and the worker who wanted to better his lot had the possibility of advancement to more remunerative callings.

Even after the development of highly organized labor unions with their hostility to "speed-ups," with their pro-

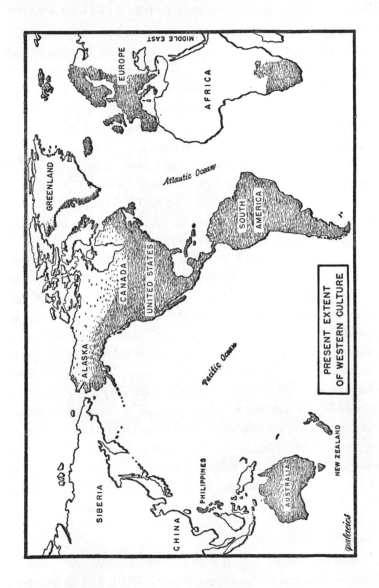

PRESENT EXTENT
OF WESTERN CULTURE

tection of certain trades and positions for their members, with such uneconomic practices as "feather bedding," and with such economically wasteful weapons as the strike, the above generalizations remained essentially true. The power of labor and its unproductive policies will have to become much stronger than they are before a radical decline in productivity can be attributed to them.

Nor under the system which has prevailed has labor failed to better its lot. The number of hours worked per annum has fallen from 4,000 in 1800 to 2,000 in 1937, and the real wage of labor has about doubled since the beginning of the nineteenth century.

The most important impediments to production, so far as labor is concerned, are to be found in its social attitudes, such as hostility to the capitalist system and to capitalist employers, opposition to the introduction of new methods of production, reluctance to leave a depressed economic area for a more active one, and awareness of the lack of adequate incentives for greater output. In general, however, the most important hindrances to the productivity of labor are to be found in the absence of opportunities to work accompanying depressed periods of economic activity. Indeed one of the major problems in Western culture has been so to order economic life that the fullest use of economic resources may be possible and commensurate with the desires of its population for goods.

⌇ Concentration of wealth

Remarkable as was the expansion of economic production in Western culture from 1800 to 1950, first in Western Europe and subsequently in the United States, Canada,

Australia, and New Zealand, it was not so astonishing as the remarkable increase in production per capita with its augmentation in goods available for individual use. Contrary to what had so frequently been the case in previous cultures, populations did not grow so rapidly that economic gains were absorbed by new consumers. But wealth was not so evenly distributed that increases per capita resulted only in a marked improvement in the standard of living for all. In fact, wealth was concentrated enough so that surplus could be used for providing leisure to a comparatively large number of persons for the pursuit of artistic and intellectual activity.

Just how concentrated wealth was can be illustrated in the cases of Great Britain and Paris, France.

GREAT BRITAIN, 1924–1930 (PERSONS OVER 25 YEARS OF AGE)

Capital Group	Percentage of Total Number of Persons	Percentage of Total Capital for Group
£100 or less	76.3–78.6	3.6–6.1
£101–£1,000	15.5–17.3	10.4–11.1
£1,001–£5,000	4.2–4.6	17.0–17.7
£5,001–£10,000	0.8–0.9	10.0–10.3
£10,001–£25,000	0.5–0.6	14.4–14.8
£25,001–£100,000	0.2–0.3	18.6–19.1
Over £100,000	0.04–0.05	23.2–23.8

DISTRIBUTION OF PERSONAL WEALTH AMONG SOCIAL CLASSES IN PARIS, 1930, 1931

Size of Fortune (fr.)	Number of Persons	Amount of Wealth (fr.)	Per Cent of Total Number of Persons	Per Cent of Total Amount of Wealth
0–50,000	2,663,431	8,136,958,826	94.1	7.65
50,001–500,000	128,929	21,306,896,762	4.7	20.02
500,001 and over	32,416	76,968,910,305	1.18	72.31

Clearly, concentrated wealth permitted its owners to purchase the works of artists and intellectuals, to subsidize their activities through educational institutions, and to provide leisure to family members for work that was not immediately economically rewarding. Furthermore, the state through the great extension of its scope of action provided its members with many cultural opportunities—through its educational structure, its theaters and musical activities, its buildings, and its adornment of public property. In America, public-school buildings have been referred to as the "cathedrals" of the modern age, and a similar statement could be made regarding public building generally in Western culture. What is more, tax-supported cultural undertakings aimed to develop talent wherever it could be found, irrespective of the social and economic status of persons involved. Probably never in history had a parallel attempt been made to discover and train the best-qualified persons in society and to provide them with opportunities for creative activity.

Finally, business has probably supported the arts and sciences to a larger degree than in any of the other cultures which we have studied. Inasmuch as economic enterprise has been organized on such a vast scale, it has built great office structures, imposing factories, and enormous railroad stations, stores, and warehouses. To the erection of these buildings some of Western culture's leading architects and masters of the plastic arts have lent their talents. So, too, in the design of products and in advertising, gifted persons have found opportunities for expressing themselves or for making a living which allowed them to engage in creative work of their own.

Between business and physical science, relationships have been even closer. As has already been indicated, most of the great inventions and discoveries of the past hundred years have been the work of scientists. Men of affairs wait breathlessly for the findings of scientists doing basic research to extend man's knowledge of the world around him; and scientists have collaborated with businessmen in efforts to find solutions to practical operational problems. It is not an exaggeration to say that recent science and art in Western culture are highly functional, that is, that they are useful and are integrated with economic activity.

⚐ *Achievement in the arts*

Economic surplus which permits leisure for artistic and intellectual pursuits is not a guarantor of great achievements. It is fundamentally a permissive factor, albeit an essential one. The quality of artistic and scientific production depends on a myriad of other factors, such as the abilities of the artist or scientist, opportunities for the exchange of ideas, and the presence of a competitive spirit which drives artists and scientists to improve their work. Much depends also on the standards of workmanship and the patterns of style which are inherited from the past, borrowed from other cultures, or newly generated. The effectiveness of actual accomplishments will also hinge upon the degree to which creative work is in harmony with the basic ideologies of a culture and upon the extent to which it appeals to such fundamental psychological drives as hunger, sex, and fear, or to human experience in society relevant to these forces. The quality of scientific activity depends, on the other hand, upon the degree to which scientists make possible an extension of human control over physical and social environment for the welfare of all.

In any attempt to evaluate aesthetic and intellectual achievements of Western culture in the last 150 years, one has inevitably to face the knotty problem of determining the boundaries of that culture. Although the chief areas of the Western world are easily recognized, there are borderline districts, like Russia, where Western influence has been very great and where many of the leading artists and scientists have behaved much like Westerners but in which the general pattern of culture is not Western. Similarly individual workers in cultures even more different than Russia from the West may have received training in Western culture and, as borrowers, may have created great works in the Western tradition. Whatever may be the merits of including such persons in any general survey of artistic and scientific accomplishment, we have excluded them here as not being relevant to our basic concern.

Artists in Western culture of whom we shall treat were all subjected to classical art forms as passed on by humanists of the late medieval and early modern period and as transformed during the seventeenth and eighteenth centuries. Yet many of them in the period now under consideration broke away from slavish attachments to classical traditions. As persons in a dynamic and changing environment they altered the old styles and generated drastically new ones. In literature, for example, men of letters liberated themselves from classical rigidities and developed a new form of expression— the novel—in which they found boundless freedom. Novels might be short or long, descriptive or narrative, and limited or unlimited in time and place. They might deal with any subject—history, everyday life of people, social problems, or the inner workings and emotions of the human mind. Brilliant use of this literary device was made by men like Victor Hugo (1802–1885), Alessandro Manzoni (1785–1873), Gustave Flaubert (1821–1880), Thomas Hardy (1840–1928), and Thomas Mann (born 1875).

Trends, similar to those discernible in the novel, were also to be found in poetry and drama. Among many distinguished masters, we may mention Johann Wolfgang Goethe (1749–1832), a great and sensitive poet, who in *Faust* could analyze the ruminations of the mind over a universal problem, and George Bernard Shaw (1856–1950), a genius in the dramatization of social and political issues and in the satirical questioning of many of Western man's irrationally accepted concepts.

In music, as in literature, there were breaks with tradition and a search for new forms obtained by volume, discordance, and rhythm. In opera there were Richard Wagner (1813–1883) and Giuseppe Verdi (1813–1901); in symphonic and chamber music, Johannes Brahms (1833–1897), Frédéric Chopin (1810–1849), Hector Berlioz (1803–1869), and Claude Debussy (1862–1918); and in jazz, George Gershwin (1898–1937).

In the plastic arts a great transformation took place. By the early nineteenth century, pioneer painters were seeking to convey an impression of what they saw (impressionism) rather than a mere photographic likeness, and successors endeavored to express what they felt or thought about what they saw (expressionism, futurism, and surrealism). New strength and originality were thus added to art by such men in painting as Eugène Delacroix (1798–1863), Édouard Manet (1832–1883), Claude Monet (1840–1926), Pierre A. Renoir (1841–1919), Henri Matisse (born 1869), and Pablo Picasso (born 1881); and in sculpture by Auguste Rodin (1840–1917) and Aristide Maillol (1861–1944).

Architecture suffered from ornate and gaudy buildings constructed along Renaissance lines, like the Opera House in Paris (1861–1875), from stereotypes of the Gothic style, like the Riverside Church in New York, or from copies of classical and neoclassical buildings, like government offices in Washington, D.C. Yet architecture was able partially to free itself from earlier models and by becoming highly functional developed styles of grace or of imposing mass. Urban dwelling units in the Netherlands planned by Berlage (1856–1934), the skyscrapers of Rockefeller Center in New York built in the 1930s, the work of Auguste Perret, the public buildings of Charles Le Corbusier (born 1887), and factories, like the Industrial Tape Corporation's plant at New Brunswick, New Jersey, are all cases in point.

How well these structures, or for that matter, how well literature, painting, sculpture, and music, will stand the test of time we cannot say. Judgment at present is warped by a great conglomeration of inferior work which all too frequently blinds us to the good. Mass education has led to a plethora of art for consumption by those who can read and write but who are untutored in regard to fine, artistic craftsmanship or who lack appreciation for other than the simple and obvious. It may be that this fact has led to a deterioration of literature through the popular novel, of the theater through the moving picture, of music through jazz, of painting through advertising art, and of architecture through hasty construction to meet rapidly expanding needs. The fact remains, however, that art for the masses has not prevented great accomplishment, nor is this art entirely void of appeal to basic psychological forces.

❧ Control over physical and human environment

Whatever the final verdict may be regarding art in the years since 1800, one may be certain that future judgments concerning the extension of human control over physical environment will be extremely favorable. Much of this extension was of an economic character and has already been dealt with in our discussion of economic progress, but part of it was noneconomic and of a more purely scientific nature.

As we have already seen, the nineteenth century inherited a desire to learn more about the physical world and a tradition of collecting observed data, of arranging those data in sequence, and of investigating meaningful relationships among them. This tradition was strengthened by the high place given to science in the dominant philosophies of the time, from the positivism of Auguste Comte (1798–1857) to the pragmatism of William James (1842–1910) and the instrumentalism of John Dewey (born 1859). Even those who criticized complete reliance on rationalism, like Immanuel Kant (1724–1804), Friedrich Hegel (1770–1831), and Henri Bergson (1859–1941), recognized the fabulous accomplishments which could be achieved in its name.

One cannot contemplate the names of the great scientists

of the last 150 years without being amazed at the contributions made to the extension of man's control over his physical environment. One stands in awe before such men as Michael Faraday (1791–1867), who got mechanical motion from electrical current; Justus von Liebig (1803–1873), who developed soil chemistry and discovered the basic chemical ingredients of plant food; Thomas Edison (1847–1931), who invented the incandescent electric light; Louis Pasteur (1822–1895), who worked out the chemistry of fermentation; Charles Darwin (1809–1882), whose theory regarding the origin of the species stimulated work in genetics; Pierre Curie (1859–1906) and his wife Marie (1867–1934), who discovered radium; Albert Einstein (born 1879), who developed the theory of relativity; and many more too numerous to mention here.

Some notion of the progress made in science can be gleaned from the fact that, while Aristotle was able to enumerate about 500 species of animals and Linnaeus in 1758 could list 4,236 species, Pratt in 1911 was able to name 522,400. Nor should we forget that naturalists have succeeded in making plants and animals produce more things useful to man, astronomers in measuring with a high degree of accuracy stars millions of miles from the earth, physicians in freeing man from many maladies, chemists in deriving useful substances from hitherto useless materials, and physicists in splitting the atom and thus in forming new material. Nor should we fail to recognize the enormous achievements of engineers in transporting electricity over long distances, in bridging rivers, in mining nature's capital, in irrigating deserts, and in building a great variety of machines. So great have been advances in science and in the application of scientific knowledge that modern man controls nature more completely than any of his predecessors.

Understanding of human behavior that is essential for an increase of man's control over human environment has also been remarkably extended since 1800. Psychologists have been able to describe the physiology of human action and of learning, to understand more thoroughly the processes of action, and with Sigmund Freud (1856–1939) to penetrate even into the subconscious mind. Economists from Adam Smith (1723–1790) to John Maynard Keynes (1883–1946), and not excluding Karl Marx (1818–1883), have sought comprehension of the operation of the ever-changing economic system. Political scientists from at least John Stuart Mill (1806–1873) have acquired a more complete understanding of politics. Sociologists have greatly extended our knowledge of how men live together in society. And historians have vastly augmented our information of how people have behaved in the past.

Fortunately Western culture has adopted several principles of conduct which have been conducive to the attainment of higher levels of civilization. For example, life, liberty, and the pursuit of happiness have come to be regarded in Western culture as rights that cannot be taken away without due process of law. Participation of the individual in the political affairs of the state has been enormously expanded, and this movement has increased the individual's sense of civic responsibility and has led to the adoption of public policies for the development of the individual's talents and for the improvement of his welfare. The state has taken on not only the task of educating its members but also of looking out for their physical, mental, and moral well-being. Indeed the concept of the welfare state has to a marked degree become generally established throughout large parts of Western culture.

Against these positive factors in human organization must be set some very important negative ones. Tensions have developed among groups which consider themselves rivals for material betterment, as, for example, between labor unions and employers or between the "have-not" and the "have" nations. Changes in living habits, particularly those which provide anonymity and freedom from family discipline, have resulted in antisocial conduct. In some extremely congested urban areas and in periods of great social upheavals, crime and other manifestations of social disorganization are tremendous. Then, too, among negative factors in human organization must be included extreme forms of nationalism—a blind, irrational, and emotional attachment to one's nationality and a belief that the advancement of that nationality's place in the world is the goal of human endeavor. To this kind of chauvinism must be attributed a large portion of the blame for international friction and for the disastrous world wars of the first half of the twentieth century.

Unfortunately social science has been less successful in coping with these problems than physical science has been in dealing with physical questions. To be sure, the social scientist has learned much about the behavior of individual humans and of human groups, has adopted the scientific principle of arranging data sequentially, and has shown relationships among these data. Accordingly he has become able to explain how a society is constructed, how its ideologies, institutions, and leaders influence behavior; how its economic apparatus functions; how its political life is managed; and what the links are among all these aspects of social existence. Indeed the social sciences have become more empirical and less speculative than they once were—and more analytical and less exhortatory. Yet social scientists, unlike natural scientists, are unable except in a limited sense to control their subject matter, which is humankind. For the most part public policies are fashioned by leaders rather than by social scientists, and the influence of the latter upon the former is usually indirect.

When all is said and done the nineteenth and twentieth centuries will probably not be considered in the future to have been so bleak as many contemporaries have thought, and they may, indeed, seem very great compared with the seventeenth and eighteenth centuries. In any case the center of artistic and intellectual activity will be found to have been in Western Europe where, during the major part of the period, most of the wealth was and the strongest traditions for creative activity were.

In retrospect and in prospect

In the preceding pages we have seen pass in rapid review a large part of the recorded economic past of those cultures which had the greatest influence on Western culture.[10] Our fundamental purpose has been to investigate what correspondence, if any, there has been between periods of economic well-being when there was leisure for other things than just eking out a living and the attainment of the highest stages of civilization. The pursuit of this purpose involved us at once in a statement regarding the meaning of civilization. We found that in Western culture, at least, degrees of civilization are measured by the extent of control over physical and human environment and by the quality and quantity of intellectual and aesthetic accomplishments. Our task also involved us in a study of factors of economic growth, of factors of economic decline, and of those specific economic conditions which influence directly stages of civilization.

We have now reached a point where general statements regarding our findings are in order. As a rule, the historian is more reluctant than other social scientists to make general statements concerning human behavior, for he knows what a small portion of human experience is available to him and how crucial may be a bit of missing data. Furthermore he recognizes that no situation is ever exactly duplicated in time, for people change and relations among techniques, resources, ideologies, and leaders are so complex that the chances of exact repetition are infinitesimal. Yet in spite of the fact that history never repeats itself exactly, enough similarities occur in analogous situations to permit generalization. Finally, it seems appropriate that general statements based on historical data be formulated by historians rather than by other social scientists who are less familiar than they with the records of the past.

In both economic development and in the attainment of relatively high degrees of civilization, advance depends to a large extent upon the range of opportunities for alternative decisions. In intellectual and aesthetic accomplishments the extent of this range is determined by the amount of economic surplus which permits of freedom from working exclusively to obtain the necessities of life. Economic surplus should be sufficient or so concentrated as to allow the training of a group of scientists and artists numerous enough so that they may stimulate one another to creative activity. Furthermore economic surplus should be adequate to permit the educating of a large part of society so that it can appreciate or make use of the intellectual and aesthetic creations of scientists and artists. One reason why so many intellectual and artistic achievements have been realized in cities is because these conditions have been most fully met in them.

The range of opportunities for alternative decisions depends also upon the degree of freedom of thought and action which exists in society. There is considerable evidence to indicate that freedom fluctuates along with economic surplus, but economic well-being is in this case not necessarily the chief determining factor. In the formative period of any culture, patterns of behavior become more or less firmly established, and if these patterns restrict opportunities for alternative decisions, they invariably detract from the attainment of a high state of civilization. If, on the other hand, patterns of behavior favor individual freedom, encourage civilizing creativeness rather than, for example, the satisfying of animal appetites, and if they promote the use of the best techniques and the adoption of the highest artistic standards of the past, they add immeasurably to the extent to which human energies are devoted to the building of a higher civilization.

The idea of individual freedom as a factor in civilization is related to our concept of "control over human environment" as a mark of civilization. By this is meant the organization of society in such a way that differences among individuals and groups may be at a minimum and that what conflicts do exist may be settled in an equitable manner according to previously established rules. Finally, we mean by "control over human environment" a situation in which society is so organized that the civilizing talents of all are developed to their fullest extent and in which policies which might prevent further advance are not pursued.

Our concept of "control over physical environment," although it is meant to be broad enough to include the prevention of disease and the forecasting of the weather, is closely related to economic progress—to increasing the supply of goods and services available per capita in society. Economic progress we found to be dependent on a variety of factors in a variety of combinations. In two periods, in the millennium prior to 3000 B.C. and in the four and a half centuries since 1500, the development of new techniques were crucial in economic progress. In other periods existing techniques were adopted by people who had not had them previously, or they were used more intensively and extensively by those who already possessed them. We found also that savings and investment were important in economic progress, especially investments in commerce and in capital goods. We ascertained, too, that economic progress was greatly advanced by a division of labor that permitted individual workers to specialize in those tasks for which they had the greatest facility. The division of labor had trade as a concomitant, and trade permitted drawing upon the natural resources of a large area and the specialization of different areas in the production of those things for which they had a comparative advantage. Commerce was, in turn, facilitated by means of transportation, communication, money, and banking services for transferring credit.

In general, economic decline can be accounted for by the reverse of those factors of economic progress which we have just mentioned. Retrogression may, however, come about by a rapid increase in population without a corresponding increase in the supply of goods, or by devastating war, or by social disorder. It may result from savings being diverted from investment in capital goods to expenditures on art and knowledge, which we consider to be marks of civilization. In fact, this process helps to explain the long-term fluctuating character of civilization. Finally, economic decline may follow the diffusion of one culture's techniques to another culture. In this case, the second culture may be able by means of its acquisitions and other resources to surpass the first culture and possibly to prey upon it.

Our study seems to show clearly that economic well-being is one of the necessary conditions for a high stage of civilization. At least, we have found a remarkable correspondence between periods of high productivity of goods and services in a culture and its periods of greatest achievement in controlling human environment and in creating intellectual and aesthetic works of a high order. No matter how important the economic factor is, however, other forces are obviously involved. A culture inherits part of its art forms, social goals, and ideologies from preceding cultures; it borrows part of them; and it creates part of them. No matter how it gets them, it behooves man to select his standards well and to foster those forces which lead to greater accomplishment, if he would raise civilization to a higher plane.

Now that the end of our study has been reached, one may legitimately ask whether or not our findings can be used to answer those questions regarding the position and future of civilization in Western culture which most perplex modern man. Can they be employed to evaluate the place of Western culture in the present world? Can they throw light upon shifts in leadership within Western culture? And can they serve as a gauge that will indicate probable developments in the near future?

If we answer these questions in the affirmative and then apply our principles to the present situation, what do we find? First, we may decide that by our definition of civilization Western culture has now the highest civilization of any culture in the world. It has the greatest control over both physical and human environment and is accomplishing more in the arts and sciences than any other culture. It has the highest incomes per capita and in view of rapid population increases elsewhere will probably continue to have them. So outstanding are its economic accomplishments and its civilizing achievements that its technology, many of its institutions, and some of its ideologies and art forms are being adopted by peoples in other cultures. In fact, the diffusion of Western culture throughout the world is one of the important phenomena of our time.

Second, it is apparent that major changes are taking place in the location of the greatest amount of economic well-being and military power within Western culture. Western Europe has clearly lost the position of overwhelming economic superiority which it enjoyed a hundred years ago. Other areas have forged ahead economically because of their superior resources, superior organization, or greater desire for material betterment. In the meantime Western Europe has not been able to draw upon the natural resources of the entire world in the proportions that it had earlier, partly because of war, partly because of the growing industrialization of backward areas, and partly because of its inability to pay for imports. Economic leadership has definitely shifted to the English-speaking people of North America. Consequently our findings suggest that the center of civilization in Western culture will be transferred to the United States and Canada, but this change will take place at best gradually and then only if ideologies and styles leading to high civilization become firmly established in the New World. If earlier patterns are repeated, however, Western Europe should still have a period of great civilizing accomplishment. Only if that area is further wracked and ruined by war is it probable that such a development will not take place.

Lastly, it is possible that as peoples outside Western culture acquire and perhaps improve upon Western culture's methods of producing goods and of waging war, these peoples will become embroiled with the West in war. At least in all the cases which we have passed in review in these pages, cultures with inferior civilization but with growing economic power have always attacked the most civilized cultures during the latters' economic decline. The chance that this may happen again is so great that enormous amounts of human resources will undoubtedly be dissipated during the coming years in preparation for war. Indeed, the threat of war, as well as actual war, is the greatest present impediment to advances in civilization both in Western culture and in all the leading cultures of the world.

For Further Reading

G. A. Baitsell, *Science in Progress* (1945)

Charles A. Beard, *Idea of National Interest* (1934)

J. H. Clapham, *An Economic History of Modern Britain* (1938), 3 vols.

T. Craven, *Modern Art* (1934)

Merle E. Curti, *The Growth of American Thought* (1943)

B. F. Fletcher, *A History of Architecture on the Comparative Method* (1948)

Helen Gardner, *Art through the Ages* (1948)

S. Giedion, *Mechanization Takes Command* (1948)

Industrialization and World Trade (League of Nations, 1945)

Simon Kuznets, *National Income* (1946)

P. H. Lang, *Music in Western Civilization* (1941)

Paul Mantoux, *The Industrial Revolution in the Eighteenth Century* (1927)

National Income Statistics of Various Countries (United Nations, 1949)

Karl Nef, *An Outline History of Music* (1935)

The Network of World Trade (League of Nations, 1942)

V. L. Parrington, *Main Currents in American Thought* (1930), 3 vols.

W. A. Robson, *Civilization and the Growth of Law* (1935)

Bertrand Russell, *A History of Western Philosophy and Its Connections with Political and Social Circumstances from the Earliest Times to the Present Day* (1945)

Henri Sée, *Histoire économique de la France* (1942), 2 vols.

The Economic History of the United States (1945 ff.)

A. Sartorius von Waltershausen, *Deutsche Wirtschaftsgeschichte, 1815–1914* (1923)

Chester W. Wright, *Economic History of the United States* (1941)

Erich W. Zimmerman, *World Resources and Industries* (1951)

NOTES

___

Chapter I, Pages 1–19

1   Effective distance means actual distance augmented by impediments to transportation and communication, like mountains, swamps, or seas.

2   John Q. Stewart, "Empirical Mathematical Rules Concerning the Distribution and Equilibrium of Population," *The Geographical Review,* vol. xxxvii, no. 3, 1947, pp. 461–485.

3   Ellen C. Semple, *Geography of the Mediterranean Region,* Henry Holt and Company, Inc., New York, 1931, Chap. XI. Also Graham Clark, "Forest Clearance and Prehistoric Farming," *The Economic History Review,* vol. xvii, no. 1, 1947, pp. 45–52.

4   In the United States in 1937, 52 per cent of energy used for heat and power was from coal, 32 per cent from petroleum, 10 per cent from natural gas, 4 per cent from water, and the remainder from wood, wind, etc. It has been estimated that 97 per cent of the energy used was provided by machines and animals and 3 per cent by humans. The United States used 1.6 times more energy per capita than the United Kingdom, 2.5 times more than Germany, 11 times more than Japan and 250 times more than China. *Energy Resources and National Policy. National Resources Committee,* Government Printing Office, Washington, 1939, p. 41.

5   The word "exploitable" is used to imply that the popula-

tion has the knowledge and skill necessary for utilizing the resources.

6   S. C. Gilfillan, *The Sociology of Invention,* Follett Publishing Co., Chicago, 1935, and *Technological Trends and National Policy. National Resources Committee,* Government Printing Office, Washington, 1937.

7   Colin Clark, *Conditions of Economic Progress,* Macmillan & Co., Ltd., London, 1940, pp. 29 and 179.

8   *Industrialization and Foreign Trade,* League of Nations, New York, 1945, p. 116.

9   John Maurice Clark, "The Relation of Wages to Progress," *The Conditions of Industrial Progress,* Industrial Research Department, University of Pennsylvania, Philadelphia, 1947.

CHAPTER II, Pages 20–44

1   Arnold J. Toynbee in *The Study of History* presents a list of twenty-one cultures which presumably were the major cultures of all time. Once he introduces a culture like the Eskimo to his list, he invalidates its selectivity, for many cultures can be found which can equal the Eskimo in every respect.

2   Knowledge of Paleolithic civilization comes mainly from archaeology and anthropology. The former endeavors to reconstruct the story from remains of man's activity. The latter relies more upon comparative studies of groups now living in primitive conditions. Examples of people at the Paleolithic stage have been found in central Africa and in the jungles of Malaya. See V. Gordon Childe, *Man Makes Himself,* C. A. Watts & Co., Ltd., London, 1941 and *What Happened in History,* Penguin Books, Inc., New York, 1946; G. Renard, *Life and Work in Prehistoric Times,* Alfred A. Knopf, Inc., New York, 1929, and G. P. Murdock, *Our Primitive Contemporaries,* The Macmillan Company, New York, 1934.

3   The Neolithic Age cannot be given terminal dates for all parts of the world. Neolithic civilization has persisted in certain areas down to the present. The North American Indian had a

Neolithic culture when the first white settlers arrived on these shores.

4   Bellows to produce a forced draft did not come into use until 1500 B.C.

5   Richard Lefebvre des Noëttes, *L'Attelage et le cheval de selle à travers les âges,* Picard, Paris, 1931, 2 vols., and *La force motrice animale à travers les âges,* Berger-Levrault, Paris, 1924. Modern experiments with the ancient harness indicate that the horse could pull only about half a ton.

6   Romans had a horseshoe which was tied onto the hoof. Probably it was not very efficient.

7   Gravitation, an inorganic source of power, had been used previously to assist in rolling or sliding materials downhill and in carrying a floating vessel downstream.

The water wheel was not invented until near the end of the first millennium B.C.

8   A city may be defined as "a function of the concentration of population in such a manner that the greater the density and the greater the population and the continuous area exhibiting this density, the more of a city it is."

9   Documents of about 2500 B.C. attest to an extraordinarily high return of barley.

10   Similar conditions were found in the early civilized parts of China on the Yellow River. Two major differences were the open country to the north through which invaders might come and outlet of the rivers into the Pacific Ocean, a body of water so vast that with existing techniques of navigation the Chinese were not able to use it as an artery of commerce to distant lands.

11   See below, p. 63.

12   Sumerians wrote on clay tablets, pressing out their characters with reeds. The characters were usually of wedge shape or cuneiform. Egyptians wrote on papyrus reeds pressed together to form a smooth paperlike mat.

CHAPTER III, Pages 45–77

1   This was effected by heating sand and natron (sodium carbonate), which was found in its natural state. The process of making clear glass was essentially similar to that of making a glaze for pottery (faïence). Glass was at first cast or molded. It was not blown until the Roman Period.

2   Dates prior to 2000 B.C. are difficult to determine with accuracy. Consequently chronologies employed by various scholars differ for these early periods. The author has endeavored to use those dates most generally accepted and those which indicate the correct relations of time in the civilizations treated. The date for the Code of Hammurabi, crucial to early chronology, has recently been changed from about 1950 B.C. to about 1750 B.C.

3   Some scholars are of the opinion that the cultural developments of the Sixth Dynasty (2420–2270 B.C.) were greater than those of the Third.

4   There had been improvement in the political situation under the Eleventh Dynasty (2100–2000 B.C.), but the reign ended in war and famine.

5   Crete may also have paid tribute for a short period.

6   This was the religious reform of Amenhotep IV, who took the name of Ikhnaton (1375–1358 B.C.).

7   The Bible story of the building of the Tower of Babel stemmed without doubt from the construction of a ziggurat.

8   Semitic refers to a linguistic and not a racial group.

9   An example of their ruthlessness was their forceful displacement of entire populations in an effort to consolidate their power.

10   This civilization is sometimes referred to as Minoan after Minos, legendary king of Knossus in Crete. It is also called Cretan. Both of these names seem too restrictive. The term "Aegean" accords best with the facts in the case.

11   The islands extending southeastward from Attica.

CHAPTER IV, Pages 78–117

1   These wars are believed to have occurred between 1194 and 1184 B.C. Archaeologists have identified nine cities on the site of Troy for the period between Neolithic and Greco-Roman times. The dates assigned to the seventh city, of which Homer sang, are 1350 B.C. and 1184 B.C.

2   Thracians moved into Asia Minor, forcing the Phrygians into the Anatolian highlands, Lydians into the southwestern corner of Asia Minor, the Philistines to the Syrian coastal plain, and peoples who were to be known as Etruscans to the Italian peninsula north of Rome.

3   Much of our knowledge concerning the agriculture of this period comes from Hesiod, *Works and Days,* written about 735 B.C.

4   In this connection we should mention that the Sophists in the fifth century B.C. preached the doctrine of the "supremacy of the individual," that is, that the welfare of the individual was to have precedence over all other considerations.

5   Incidentally, reduction in the cost of iron compared with bronze allowed members of the lower classes to possess weapons with which they could advance their interests with force.

6   In time the Greeks realized that the depletion of the forests was making them short of lumber and was resulting in soil erosion. They then attempted reforestation but had only a moderate success.

7   William L. Westermann, "Greek Culture and Thought," *Encyclopaedia of the Social Sciences,* The Macmillan Company, New York, 1937, vol. I, p. 20.

8   The most dramatic example was the reconstruction of Athens under the direction of Pericles following the Persian Wars, that is, about 461 to 431 B.C. This enterprise, which involved the construction of the Parthenon and other buildings on the Acropolis, led to the employment of one of Greece's greatest architects, Ictinus, and one of her great sculptors, Phidias.

9   An early landmark was Myron's "Discus Thrower" of about 450 B.C.

10   See the works of Phidias in the fifth century B.C.

11   These qualities are visible in the work of Praxiteles, fourth century B.C., and of Scopas, a contemporary.

12   Among the better known are Sappho of Lesbos (about 600 B.C.) and Pindar (522–448 B.C.).

13   Among the tragedies should be mentioned those of Aeschylus (525–456 B.C.), Sophocles (496–406 B.C.), and Euripides (480–406 B.C.). Among the comedies the most distinguished were those of Aristophanes (448–388 B.C.).

14   This was true of Herodotus (484–425 B.C.). He attempted a universal account of the past of known Asiatic peoples and of the deeds of the Greeks and Persians in the Persian Wars.

15   This was the case of Thucydides (465–396 B.C.) who wrote of the Peloponnesian War. He was the most careful workman and the most penetrating analyst among the Greek historians.

16   Xenophon (430–354 B.C.) illustrates this tendency. He was a student of rhetoric, the art of writing and speaking with grace and ease. The Athenian orator Demosthenes (384–322 B.C.) was an eminent rhetorician.

17   For those who may have forgotten, it can be restated thus: In a right-angle triangle the square of the hypotenuse is equal to the sum of the squares of the other two sides.

18   This is the geocentric theory as opposed to the heliocentric theory, that the sun is the center of the universe.

19   Corinth was burned by the Romans in 146 B.C., but it was revived by Julius Caesar and Augustus.

20   Menander (342–291 B.C.) was perhaps the leading writer of this school.

21   The last Greek epic was by Apollonius of Rhodes (about 194 B.C.), who wrote on the well-worn theme of Jason and the Golden Fleece.

22   The five important ethical schools were as follows: (1) The Cynic, whose most important thinker was Diogenes of Sinope

(412–323 B.C.) and whose main program included attacks upon social institutions, particularly on private property.

(2) Cyrenaic, which preached the pursuit of pleasure.

(3) Skeptic, led by Pyrrho (360–270 B.C.), which taught that unhappiness results from an inability to attain knowledge and that one should therefore renounce all efforts to attain it.

(4) The Stoic, whose master was Zeno (336–264 B.C.), which held that God is good, that man should live according to the law of God, and that if one were unfortunate, one should bear up like a man.

(5) Epicurean, whose leader was Epicurus (341–270 B.C.), which contended that the end of life is pleasure but that this pursuit should be avoided if it caused pain.

23   Heraclides of Pontus (388–315 B.C.).

24   Aristarchus (310–230 B.C.). He also computed accurately the distance from the earth to the moon.

25   Hipparchus (160–125 B.C.). He elaborated the geocentric theory and the theory of epicycles to explain the movement of the planets, moon and sun—a theory which was not upset until modern times. He also used the Babylonian system of dividing the circle into 360 degrees for purposes of observation, and he is said to have invented an astrolabe.

CHAPTER V, Pages 118–162

1   Rome's impact on the aspirations, institutions, and technology of Eastern Europe was not great. Here certain Hellenistic traditions were kept alive during the Middle Ages at Byzantium, whence they were diffused to the eastern Balkans and Russia.

2   The question of Roman citizenship was a complicated one. At first citizenship was a hereditary right of the freeborn residents of the city. As Rome expanded she attached some cities to herself and gave its freeborn full participation in the Roman Popular Assembly, but, of course, these citizens had to be present in Rome to exercise their privilege. In some cases conquered people were given half-citizenship or probationary citizenship. Full

citizenship to all freeborn in Italy was not granted until after the Civil War of 90–88 B.C. Citizenship was later extended to freeborn outside the peninsula after a given number of years of military service. Only in 212 A.D. were all freeborn citizens of cities given Roman citizenship.

3   The population was divided into five classes for voting purposes, with each class having one vote. This system limited further the power of the plebs.

4   Early Roman law also established the idea that equity is superior to written law, that arbitrary decisions of judges must be minimized, and that the accepted customs of states, *jus gentium,* must be respected.

5   These were the First Punic War, 264–241 B.C., the Second Punic War, 220–201 B.C., and the Third Punic War, 149–146 B.C.

6   Carthage had amassed a considerable economic surplus. The city itself is estimated to have had a population of about 750,000 at the time of the First Punic War and to have been one of the largest cities, if not the largest, in the Mediterranean world. In spite of considerable economic success, Carthage did not display much creative ability of an intellectual or artistic character. In its culture, as in that of the Phoenician, the "making of money" seems to have been the goal for which men strove. Archaeological remains seem to indicate that the most creative period in Carthaginian art was in the fourth and third centuries B.C., which was also the period, or just after the period, of greatest economic success. See S. Gsell, *Histoire ancienne de l'Afrique du Nord,* Paris, 1913 *ff.*

7   Estimates of Roman population are not accurate. For the Augustan Age at the end of the first century B.C. they vary from 650,000 to 1,200,000. The maximum population during the first century A.D. is placed at 1,600,000. Before the Second Punic War there were 270,000 male Roman citizens of military age. In 209 B.C. there were about 137,000, but in 163 B.C. there were 337,000 and in 130 B.C., 317,000.

8   Laws limiting the amount of public land which any one person might hold were not enforced.

9   The great Roman works on agriculture, those by Cato, Varro, Vergil, and Columella, were concerned mainly with farming on medium and large estates. A medium estate comprised roughly 150 acres in olives and 60 acres in vines, with other land given over to pasture and some grain. Incidentally, Greek influence on Roman agricultural methods was very great.

10   This practice was begun with the grain laws of Gaius Gracchus in 123 B.C.

11   The payment for imports by coins or by gold and silver bullion was extensive enough to be prohibited by law. For example, Cicero, during his consulship, instructed customs officials at Puteoli to seize gold and silver marked for shipment from Italy (about 63 B.C.).

12   Laws of 217, 104, and 89 B.C. stipulated the weight and purity of Roman coins.

13   After the conquests and reorganization of parts of Asia Minor, Syria, and Palestine by Pompey, 66–62 B.C., the income of the Roman treasury more than doubled, and the publicans were particularly active. In 62 B.C. the interest rate for legitimate transactions in Rome was 6 per cent. Six years later during Caesar's conquest of Gaul, money in Rome was in short supply, and interest rates were at the legal maximum or above.

14   The army was reorganized so that the main force was composed of citizens of Italy. Foreign provincials were usually attached to these forces. Men were shifted around periodically to prevent the creation of an organization hostile to Rome. Augustus established a Praetorian Guard for his protection. He also created a small standing navy.

15   Unfortunately they also introduced the orgies of the cult of Dionysius, or Bacchus, which had such enthusiastic devotees that the Senate took restraining measures (186 B.C.).

16   Every four years a day was added. At about this time the seven-day week was adopted by Caesar, and its use spread rapidly in the Mediterranean world. Earlier the Romans had worked seven days and had gone to market on the eighth. The Greeks divided the month into three ten-day periods.

17   During the second century A.D. the population of Rome is thought to have been between 1,200,000 and 1,600,000 at its highest point and the population of Italy to have been about 14,000,000.

18   In these two centuries the Empire was increased by the acquisition of the greater part of Britain, by land between the Rhine and the Danube, by Dacia, which was north of the lower Danube, by part of Armenia, Assyria, Mesopotamia, and a portion of Arabia.

19   Such cities as Strasbourg, Cologne, Mainz, Vienna, Belgrade, and Budapest were built on the sites of Roman camps.

20   Discovery of the action of the monsoons of the Indian Ocean made possible by the first century A.D. a round trip to India in one year. Trajan (emperor 98–117 A.D.) connected the Nile and Red Sea, but the extent to which this canal was used is disputed.

21   The customs districts were Italy, Sicily, Spain, Britain, Illyricum, Syria, Egypt, North Africa, Bithynia-Pontus-Paphlagonia, and the provinces of Asia.

22   The theory has been advanced that agricultural production declined because of a fall in the fertility of the soil. This theory has some foundation in fact in deforested and dust-bowl areas. Yet in Egypt where the Nile floods bring alluvial soil to the land and where fertility is thereby maintained, agricultural production declined in the later Empire.

23   This practice was learned from Sassanian armies. It had been tried by the Romans as early as 260 but was greatly extended by Diocletian.

24   The famous "law of the maximum" was issued in 301. This attempt at price control was unsuccessful.

CHAPTER VI, Pages 163–216

1   Some trade over long distances was always maintained, especially trade in salt, pepper, incense for church services, and grain in times of famine.

2   At this time there were no locks. Dams were constructed at intervals to maintain a water level on canals. These dams were

passed by means of an inclined plane and winches to pull up or let down the barges.

3   They, in turn, probably obtained it from the Far East.

4   Contrary to views which were held for a long time, the fact that Turks conquered Constantinople in 1453 seems to have had little to do with the case. Turks had been active along many of the Near Eastern trade routes to India for a long time and did not relish the thought of destroying a trade which was profitable to them.

5   A bank note, like any other credit instrument, is in essence a promise by someone to pay a stipulated sum. A bank may issue paper notes to a greater amount than the gold and silver in its coffers, even if it promises to redeem the bank note in bullion, because *all* holders of its bank notes would not usually demand the bank to meet all its promises to pay at one time. Probably the first true bank notes were issued by the Bank of Stockholm in 1661.

6   For a bona fide insurance contract, five elements must be present: (1) the insured possesses an interest of some kind that can be estimated in money terms; (2) the insured is subject to risk of loss or impairment of this interest; (3) the insurer assumes risk of loss; (4) such assumption is part of an arrangement to distribute actual losses among a large group bearing similar risks; (5) as consideration for the insurer's promise, the insured makes a ratable contribution, called a premium, to a general insurance fund from which claims of loss are paid.

7   It was written prior to the close of the eleventh century.

8   *Reynard the Fox* was probably not written until the thirteenth century.

9   The use of the word "renaissance" for the period in question has been severely criticized because it implies the absence of a revival in the twelfth and thirteenth centuries. The term has become too well established to be abruptly discarded, but clearly the *Renaissance* was a new *Renaissance*.

10   Even the works of Reynolds, Romney, and Gainsborough

in England can hardly be considered an exception, although they gave England what is sometimes called the golden age of English painting.

CHAPTER VII, Pages 217–264

1   Colin Clark, *Conditions of Economic Progress,* Macmillan & Co., Ltd., London, 1940, pp. 107 and 95.

2   F. Coppola D'Anna, *Popolazione, Reddito, e Finanze Pubbliche dal 1860 al Oggi,* Partenia, Rome, 1946, pp. 57–58.

3   Simon Kuznets, *National Income,* National Bureau of Economic Research, Inc., New York, 1946, p. 32.

4   From League of Nations, *Industrialization and World Trade* (1945), p. 13.

5   S. C. Gilfillan, *The Sociology of Invention,* Follett Publishing Co., Chicago, 1935.

6   *Industrialization and Foreign Trade,* League of Nations, New York, 1945, p. 14.

7   For example, Great Britain's foreign investments in 1913 were said to have been £3,763,000,000, or nearly 25 per cent of national wealth.

8   Simon Kuznets, *op. cit.,* p. 53.

9   In the United States in 1945 only about 50 per cent of energy used came from coal. Forty-five per cent came from petroleum and natural gas, and 5 per cent from water power. *Historical Statistics of the United States, 1789–1815,* Government Printing Office, Washington, 1949, p. 155.

10   We have omitted from our consideration the culture of China, India, Byzantium, and Arabia and of the Mayans and Incas because data regarding their economic development are few and because the inclusion of them would have extended this book beyond desirable limits.

INDEX

285

Lead, 28, 31
Leaders and environment in developments in history, 17–18
Le Corbusier, Charles, 252
Lee, Henry, 198
Leeuwenhoek, 210
Lefebvre des Noëttes, Richard, 267
Levant, 184
Libya, 87
Liebig, Justus von, 254
Lille, 174
  fairs of, 179
Linnaeus, 210, 254
Literature, beginnings of, 55
  early Christian, 161
  in Greece, 89
  in Rome, 140, 148–150
  of Western Europe, 202–203
  (*See also* Arts; Renaissance)
Livius Andronicus, 139
Livy, 140
Locke, John, 211
Lombards, 158
London, 176
Low Countries, 167, 168, 176, 184
Lucca, 175
Lucretius, 141
Lusiads, 208
Luxor, Great Temple of, 57
Lydia, 92
Lyons, fairs of, 179

M

Macedonia, 111
  conquest of Greece by, 104
Machiavelli, Niccolò, 207
Madeira Islands, 186
Magellan, Ferdinand, 187, 188
Magna Graecia, 87, 104, 122, 123, 138
Maillet, Benoit de, 211

Maillol, Aristide, 252
Malthus, Thomas, 51
Man and Civilization, 2
Manet, Édouard, 252
Mann, Thomas, 251
Manufacturing (*see* Industry)
Manufacturing production of world, percentage distribution of, 237
Manzoni, Alessandro, 251
Marathon, Battle of, 103
Marcus Aurelius, 151
Marlowe, Christopher, 208
Marseille, 90
Martial, 149
Marx, Karl, 255
Masaccio, 207
Material betterment, desire for, in nineteenth and twentieth centuries, 233–236
Mathematics (*see* Science)
Matisse, Henri, 252
Medes, 82
Mediterranean Sea, 26, 67, 173, 175, 176
  (*See also specific entries like* Aegean Sea; Hellenistic World)
Memling, Hans, 207
Memphis, Egypt, 55
Menander, 270
Mercantilism, 199
Mercator, 210
Merchant guilds, 179
Merchant Staplers of England, 179
Mesopotamia, 28, 65, 67, 69, 85, 108, 157
  cultures of, civilization in, 58–64
Mesopotamian Valley, 155
Messina, Straits of, 123
Metallurgy, technological advances in, 28